ADVANCED SYSTEM DEVELOPMENT/FEASIBILITY TECHNIQUES

ADVANCED SYSTEM DEVELOPMENT/FEASIBILITY TECHNIQUES

J. Daniel Couger
Professor of Computer and Management Science
University of Colorado, Colorado Springs

Mel A. Colter
Assistant Professor of Management Science
University of Colorado, Colorado Springs

Robert W. Knapp
Professor of Business Administration
University of Colorado, Colorado Springs

John Wiley & Sons

New York Chichester Brisbane Toronto Singapore

Library of Congress Cataloging in Publication Data:

Couger, J. Daniel.
Advanced system development/feasibility techniques.

Includes index.
1. System analysis. I. Colter, Mel A. II. Knapp,
Robert W., 1935- III. Title.
T57.6.C677 658.4'032 82-2818
ISBN 0-471-03141-0 AACR2

Printed in the United States of America

10 9 8 7 6 5 4 3 2 1

To my son-in-law and daughters-in-law, Mark, Lita, and Sherry

J. D. COUGER

To my parents, Archie and Jessie

M. A. COLTER

To my parents, Frank and Helen

R. W. KNAPP

ABOUT THE AUTHORS

J. Daniel Couger is Professor of Computer and Management Science at the University of Colorado at Colorado Springs. He is the author of 13 books and more than 60 journal articles. He spent 13 years in industry as a system designer and as a manager of MIS. He has served as consultant on system design for more than 20 organizations, including IBM, Dow Chemical, and Hewlett Packard.

Dr. Couger has lectured throughout the United States and on six continents on the subject of system design. In 1977 he was selected as U.S. Computer Science Man of the Year. He is a Fellow of the American Institute for Decision Sciences. He is listed in *Who's Who in America* and *Who's Who in the World.*

Mel A. Colter is an Assistant Professor of Information Systems and Management Science at the University of Colorado at Colorado Springs. After receiving his Ph. D. in management science from the University of Iowa, he directed systems development in industry before returning to academia in 1976. As a faculty member, he has taught many courses in the systems area, including systems analysis and design, data base, and a variety of introduction and language courses. His major research interests involve the life cycle processes necessary to allow effective systems development and management. He has published a number of articles that have appeared in various conferences and journals.

Dr. Colter is an active consultant to a number of organizations, including the System Development Corporation and the United States Air Force, primarily in the areas of structured systems analysis and design. He has presented numerous lectures and training sessions on structured techniques in the United States, Europe, and South America.

Robert W. Knapp is Professor of Business Administration at the University of Colorado at Colorado Springs. After obtaining his Ph. D. from the University of Michigan, Dr. Knapp worked for three years as a staff economist with the General Motors Corporation. He taught for two years at the Graduate School of Business Administration at the University of California at Los Angeles, and has been at the University of Colorado for 14 years, where he teaches courses in business economics, finance, business and society, and small business.

Coeditor with J. Daniel Couger of the book *System Analysis Techniques*, Dr. Knapp's publications have also appeared in books and in professional and trade journals in economics, marketing, management, and small business. Dr. Knapp has served as consultant to several Fortune 500 firms and to a score of medium and small firms.

PREFACE

This book was prepared as a textbook for a course in system analysis. Specifically, it was designed for the course IS5 of ACM Curriculum '81: *Recommendations for Programs in Information Systems at the Undergraduate and Graduate Levels.* J. Daniel Couger has been a member of the ACM Curriculum Committee since its inception in 1967, and Mel A. Colter served on one of the subcommittees during the update of Curriculum '73 to the 1981 version. Copies of that curriculum are available from ACM, 1133 Avenue of the Americas, New York, NY 10036.

Although designed primarily for course IS5, this book is also referenced in the ACM revised curriculum recommendations for two other courses, IS3 (Systems and Information Concepts in Organizations) and IS8 (Systems Design).

Some schools combine their system analysis and design course, so this book would be useful as one of the two textbooks for such a course.

This is not a second edition of our prior book, *System Analysis Techniques* (Wiley, 1974), but a companion volume. The earlier book was widely used both in the United States and overseas as a textbook for the systems curriculum. The papers in that book were the landmark publications on the key concepts and techniques and their application for analysis of business/managerial uses of the computer. Those papers represented first, second, and third generation system development techniques. They have not been repeated in this book. Each has been summarized, however, in the survey paper that precedes each section of the book.

The key papers on fourth and fifth generation techniques are included in this book. The book is divided into six sections. Sections 1 to 3 provide *survey* papers on the respective techniques of generations 1 to 3. Sections 4 and 5 provide the landmark papers covering the fourth and fifth generations of techniques. The sixth section is devoted to feasibility techniques, both theory and practice. Landmark papers are provided for Section 6. Most systems books are weak on the subject of feasibility analysis, and these papers provide solid background and workable application approaches.

To facilitate its use as a textbook, questions/problems are provided at the end of Sections 1 to 3 and at the end of each paper in subsequent sections. An instructor's manual is available, providing solutions to the questions/problems and suggestions on teaching approaches.

This book will also be useful to practitioners. *Systems Analysis Techniques* was adopted by two major computer book-of-the-month clubs whose primary market is practitioners. We believe this book to be equally useful for practitioners who want to ensure that they stay current in the field.

We give special recognition to the other members of the ACM curriculum committee for the many hours they spent in designing the revised curriculum: Jay F. Nunamaker (chairman), William Cotterman, Gordon B. Davis, Benjamin Diamant, Andrew Whinston, and Marshall Yovits. Benn Konsynski also made valuable contributions, both to the ACM Committee and to this book. We appreciate the efforts of our secretarial staff, Anne Reints and Lynn Munroe, and the editorial assistance of Toni Knapp.

Colorado Springs, Colorado

J. Daniel Couger
Mel A. Colter
Robert W. Knapp

CONTENTS

INTRODUCTION 1
EVOLUTION OF SYSTEM DEVELOPMENT TECHNIQUES 6

SECTION 1 First Generation Development Techniques for Computer-Based Systems 15

Section Introduction: Survey Paper 17

SECTION 2 Second Generation Development Techniques for Computer-Based Systems 35

Section Introduction: Survey Paper 37

SECTION 3 Third Generation Development Techniques for Computer-Based Systems 53

Section Introduction: Survey Paper 55

SECTION 4 Fourth Generation Development Techniques for Computer-Based Systems 69

Section Introduction 71

Evolution of The Structured Methodologies — 73
Mel A. Colter

Software Engineering — 97
Barry W. Boehm

Structured Methodology: What Have We Learned? — 122
Chris Gane and Trish Sarson

Structured Analysis (SA): A Language for Communicating Ideas — 135
Douglas T. Ross

Structured Design — 164
W. P. Stevens, G. J. Myers, and L. L. Constantine

Structured Systems Design — 186
Kenneth T. Orr

Business Information Analysis and Integration Technique (BIAIT)—the New Horizon — 213
Walter M. Carlson

Business Information Characterization Study — 223
David V. Kerner

Business Systems Planning — 236
IBM

PSL/PSA: A Computer-Aided Technique for Structured Documentation and Analysis of Information Processing Systems — 315
Daniel Teichroew and Ernest A. Hershey, III

The PSL/PSA Approach to Computer-Aided Analysis and Documentation — 330
Daniel Teichroew, Ernest A. Hershey, III, and Y. Yamamoto

A Requirements Engineering Methodology for Real-Time Processing Requirements — 347
Mack W. Alford

An Extendable Approach to Computer-Aided Software Requirements Engineering — 365
Thomas E. Bell, David C. Bixler, and Margaret E. Dyer

Software Requirements Engineering Methodology (SREM) at the Age of Two 385
Mack W. Alford

Plexsys: A System Development System 399
Benn R. Konsynski and Jay F. Nunamaker

SECTION 5 Fifth Generation Development Techniques for Computer-Based Systems 425

Section Introduction: Survey Paper 427

Automation of System Building 437
Daniel Teichroew and Hasan Sayani

SECTION 6 Cost/Effectiveness Analysis Techniques 447

Section Introduction: Survey Paper 449

Cost/Benefit Analysis of Information Systems 459
James Emery

Techniques for Estimating System Benefits 489
J. Daniel Couger

Index 501

ADVANCED SYSTEM DEVELOPMENT/FEASIBILITY TECHNIQUES

Introduction

J. Daniel Couger

The impetus to prepare these two volumes resulted from the most embarrassing event in my professional career. In September 1972, I was asked to be a speaker at a conference of European computer science academicians. The conference was held in a beautiful setting at the historic University of Newcastle Upon Tyne, Great Britain. The lovely and peaceful campus gave me a sense of tranquility immediately upon arrival. I was completely unprepared for the traumatic experience that I would suffer during the next three days of the conference.

Each of the principal speakers gave 1½ hour presentations three days in succession. The audience response during my first presentation belied their true feelings. I was surprised, therefore, as their feelings quickly surfaced when I asked for questions and comments.

E. W. Dijkstra of the Netherlands raised his hand for the first question. Professor Dijkstra was one of the principals in the development of structured programming and of ALGOL. I was pleased by his interest but dismayed by his comment. It went something like this,[1] "Professor Couger, in your country you have developed academic degree programs on the subject of business data processing. Most of us here do not consider BDP worthy of being called an academic discipline." Several participants rose to my defense but it was obvious that the large majority was in agreement with Professor Dijkstra. Neither side conceded the argument and the discussion ended in an impasse.

I considered altering my second day presentation to respond to the Dijkstra criticism but because all papers were prepublished in the proceedings, I decided against it. The response to my second presentation was similar to that of the preceding day. As soon as I finished and asked for comments, Professor Dijkstra was the first to raise his hand. "Professor Couger," he said, "I have the same response as yesterday. Your talk today has done nothing to convince me that BDP

[1] I'm repeating the essence of the comments because I didn't record the exact wording of each.

is a true academic discipline. You people have no rigorous mathematical or scientific foundation for your field. In contrast, computer science has numerical analysis as its foundational basis. We believe a computer science graduate can easily learn BDP in a two-week training program after he is employed." The next comment, by one of his colleagues, went even further. "You Americans are so influenced by U.S. industry that you will introduce any kind of a degree program industry wants, whether or not it is worthy of a university level curriculum. We understand you even have doctorate level programs in business data processing. Here in Europe, we believe academia should be separated and not tainted by industry. We don't let industry have any say in what we teach." Again, the majority of the audience was unresponsive to my defense of academic programs in design of management information systems. They continued to refer to the field as BDP with considerable disdain when they used the term.

That evening one of the Europeans visited me in my room. "I wanted to apologize for my colleagues," he said, "and to tell you the principal reason for the criticism you are receiving." I asked him to have a seat and he continued his explanation. "This is the fifth year of this conference. It is funded by IBM and in the four previous years the topics were exclusively computer science oriented. This year, IBM representatives told the organizing committee that they would continue to sponsor the program but that it should concentrate on business rather than scientific uses of the computer. The organizing committee was angry and originally determined to cancel the conference. They then decided to humor IBM and invite speakers on business topics for the formal sessions."

I began to feel a little better at my colleague's agitation with what had happened. He continued, "They decided the real conference would be held late in the afternoons and evenings—where computer science subjects would be explored in private discussions and away from the formal conference setting. So, they are dutifully attending the daytime sessions in deference to the financial sponsor but consider the *real* conference to be taking place after hours."

He shrugged his shoulders in frustration as he concluded, "You see, a small number—maybe 10 percent of us here—are as interested in MIS as you are. There are no European university programs in MIS and we are eager to learn about your programs in order to develop our own. We resent Dijkstra and his colleagues feelings of superiority about scientific uses of the computer."

I felt somewhat relieved by his comments but decided to discard my prepared paper for the third day session and write a new one in direct refutation of the criticism.

In the next day's speech I was successful on one major point, but unsuccessful on another. My first point was to demonstrate the advantages of cooperation with industry. "By having industry representatives on our national curriculum committee we ensured our curriculum was relevant," I told the group. "Nevertheless, we academicians have the final say on curriculum. For example, industry would like us to have two COBOL courses so graduates are prepared to enter the field as programmers, on the way to becoming analysts. Instead, our objective is to prepare analysts who have a good understanding of programming. We introduce students to business application programming using a business oriented language such as COBOL or PL/1. If we offered a second COBOL course, we would have to give up one of our system courses. We are more concerned with the long- rather than short-range benefits—to both the student and the company." I couldn't resist being a little vitriolic in some examples of the benefits of mutual cooperation between academia and industry. "The computer science department here at the well-known University Upon Tyne is considered one of the best in Europe. Yet, my university has much superior computers and teaching facilities—and we are just a second level university, without the resources of a Harvard or a Stanford. Industry contributions have enabled us to have better facilities than the best university in Great Britain. You sacrifice a great deal to remain aloof from industry!"

I continued, "However, the principal benefit in the industry/academia cooperative effort is the synergistic effort of producing together more than the two alone could produce." I gave several examples including the University of Michigan ISDOS project where teams were comprised of both industrial and academic researchers. I sought to prove my point by a European example of what they were missing by not interacting with industry. "Do any of your schools include lectures on the procedures for automatic code generation?" I asked. No one responded. "That is surprising," I said, "because here in England an industrial firm has produced such a system, the Hoskyns system. Your students are shortchanged by your refusal to interact with industry and keep your curriculum up to date with important industrial research contributions."

I felt I won this point but was frustrated at my inability to support my other primary argument—the position that MIS curriculum was worthy of being classified as a university level discipline. "Over 70 percent of all computer use is for business information processing," I told them. "By neglecting this area in your university curriculum, your countries are not benefiting from an academic approach to the field." Also, you have a very simplistic view of the difficulty of designing business information systems." I described some of the mathematical techniques used in designing computer based management information systems. Nevertheless, upon conclusion of my talk, one of Professor Dijkstra's colleagues summarized what was obviously the opinion of the majority of participants. "We accept that your field has produced an array of techniques for coping with the problems of business applications of the computer. Nevertheless, none are as rigorous as numerical analysis. We are not convinced that BDP is justified as an academic discipline."

On the plane trip home, I had to admit that our field had not produced a cohesive classification of techniques nor an analysis that demonstrated the theoretical foundation for MIS. I began work on the classification scheme immediately upon my return and gave it top priority for my research that year. My first paper, "Evolution of System Analysis Techniques," was published in *Computing Surveys*, the prestigious and rigorous ACM publication. The paper's acceptance (and the accompanying check for $1000) assured me that the research was considered valuable by our American computer science community. I then undertook to gather all the landmark papers for the MIS discipline and combine them into one volume easily accessible to academia and industry. The resulting book, *System Analysis Techniques*, was published in 1974. Royalty reports demonstrate that the book has been widely used as text for graduate level MIS curriculum both in the United States and abroad. I've received more congratulatory correspondence on this book than any other book I've written.

So, with some pride, I mailed a copy of the book to Professor Dijkstra with an accompanying note. I concluded, "I hope you will now agree that these materials substantiate that our discipline has a theoretical foundation just as sound and rigorous as the field of computer science."

For several weeks I eagerly awaited his response. It didn't arrive. In fact, he never replied. I assumed that he, and probably a number of his colleagues on both sides of the ocean, were unconvinced. On the other hand, degree programs in MIS are now implemented all over the world, particularly in the countries that are leaders in the computing industry, such as the United States, West Germany, and Japan.

I'm pleased that the first book has been considered valuable by both academia and industry. This book includes an expanded classification of the techniques in the field. It also includes landmark papers published since the publication of the first volume.

The tree chart below provides a perspective on the book and shows our broad classification of systems development and feasibility techniques. The book is organized to correspond to the delineation of techniques on the tree chart. Except for precomputer techniques, a section is allocated to each generation of techniques. (Precomputer techniques are covered in Section 1.) For the sections covering fourth and fifth generation techniques, the landmark articles are also included. Papers for earlier generation techniques are not reprinted from our earlier volume, *System Analysis Techniques*.

For those generations of system development techniques where landmark papers are not repeated from volume one, a survey paper is provided. Sections 1 through 3, therefore, consist of survey papers on the first three generations. For the fourth and fifth generation (Sections 4 and 5)

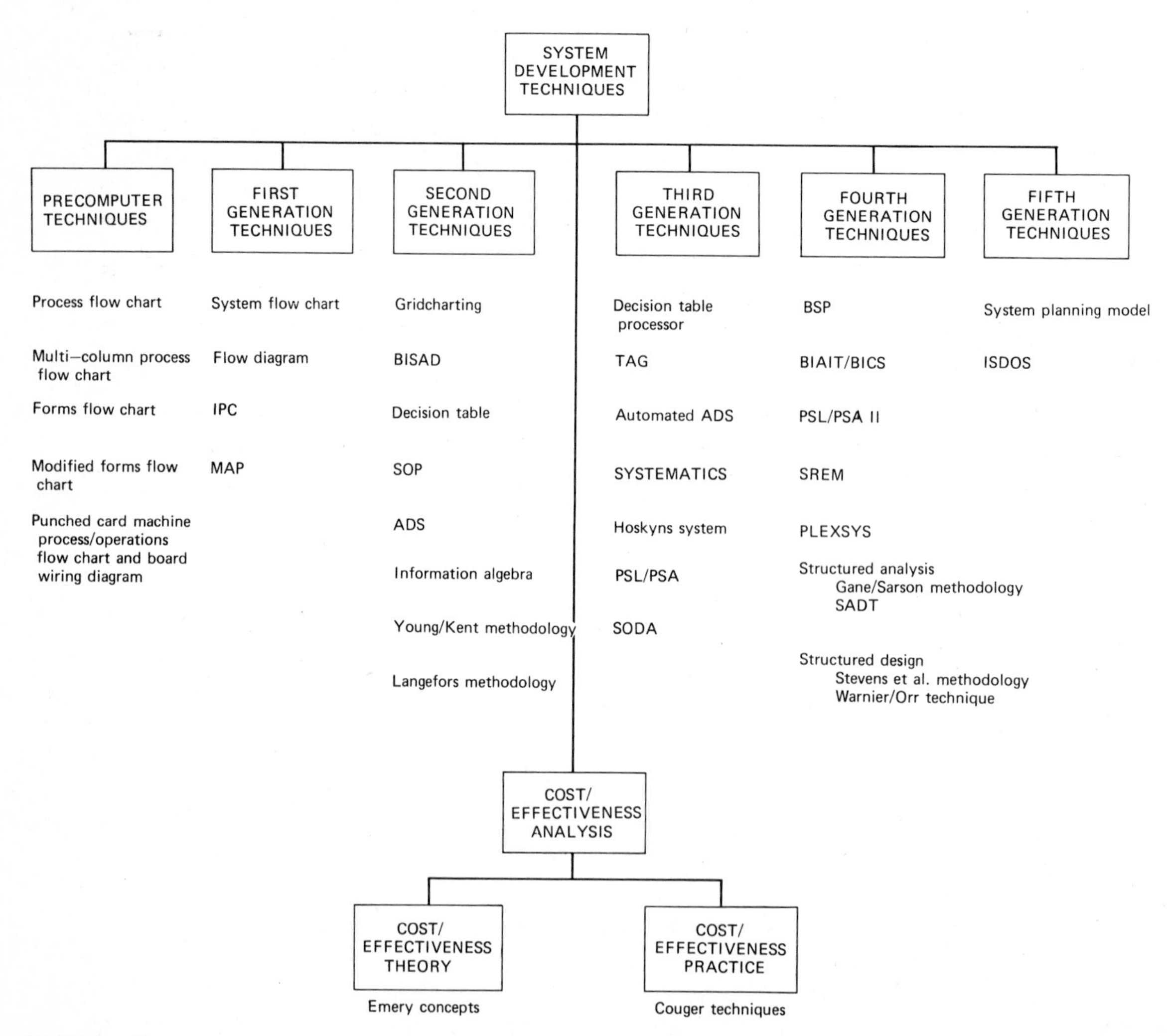

FIGURE 1 Five generations of evolution of system development techniques.

and for cost/effectiveness analysis (Section 6), the landmark papers are included as well as the survey paper. Only two papers are repeated from volume one. James Emery's paper on cost/effectiveness analysis is included because of its comprehensive coverage of theory. The other repeat is Daniel Teichroew's paper on ISDOS. As in the prior volume, the papers in this book demonstrate the rich theoretical basis of the system development techniques for business applications of the computer. One would hope that Professor Dijkstra and his colleagues have changed their views about our discipline. Certainly there is ample proof of the value of our field. The corroborative evidence pervades business and government. Ours is a sound, dynamic, and exciting discipline that is making a profound impact on business and management around the world.

Evolution of System Development Techniques

J. Daniel Couger

Systems analysis consists of collecting, organizing, and evaluating facts about a system and the environment in which it operates.

The objective of systems analysis is to examine all aspects of the system—equipment, personnel, operating conditions, and its internal and external demands—to establish a basis for designing and implementing a better system.

Understanding the role of the systems analyst is facilitated by reference to the steps in the system life cycle. For the purposes of this discussion, seven phases in the system life cycle are distinguished:

Phase I	Documentation of the existing system
Phase II	Analysis of the system to establish requirements for an improved system (the logical design)
Phase III	Design of a computerized system (the physical design)
Phase IV	Programming and procedure development
Phase V	System test and implementation
Phase VI	Operation
Phase VII	Maintenance and modification

Systems analysis, then, is concerned with Phases I and II of the system development cycle. The product of systems analysis is the logical design of the new system: the specifications for input and output of the system and the decision logic and processing rules. Phase III, the physical design phase, determines the organization of data, processing procedures, and the devices to be used. *System development* covers the span of Phases I through III. Some authors use the term *software engineering* to refer to Phases I through V. However, their descriptions of activities typically ignore Phase I of the life cycle. That is why we prefer the term *system development* and concentrate in this book on Phases I through III of the system life cycle.

Today's systems are complex in development. Previously, only subsystems such as the payroll system were computerized. Today, in the era of integrated systems, the scope of the system is enlarged many times. Payroll is a module in the accounting subsystem, which is only one of several subsystems in the finance system.

The early-day system approach was to design independent subsystems for interdependent activities. The payroll application was designed as an entity when it, in reality, was a part of both the finance system and the personnel system of the firm. Today the payroll module is redesigned to feed both of these major systems.

Until recently, most computer applications were operational-level systems. They provided the information needed by first-level supervisors and their subordinates. Today's systems include the tactical (control) and strategic (planning) levels, as well. The thrust of system development effort today is to expand systems horizontally and vertically.

The expansion in scope and sophistication of systems increases the complexity of system analysis and design. There are more "front-end" costs in designing for integration.

Figure 1 shows the change in system development emphasis and the associated costs for systems designed in the 1970s versus those of the 1980s. Both the amount of cost and the allocation between phases have changed. Phases I through III (the system development phases) now account for approximately 55 percent of the system development cost. In the 1970s, only 35 percent of development cost occurred in these phases.

In addition, the total cost of system development has increased. As shown by the cost curves in Figure 1, the primary increase in cost occurred in the early part of the system development cycle. The expanded scope and sophistication of present-day systems have produced this result. To properly prepare for a data base environment, more "front-end" analysis is required. On the other hand, better analysis and planning of systems results in lower costs in the subsequent phases. Figure 1 shows a reduction in costs in Phase IV (programming) and Phase V (testing and implementation) for the 1980s compared with the 1970s.

The increased scope and sophistication of today's systems also demand improved system development techniques. Unfortunately, improvement in system development techniques has not kept pace with the improvement in com-

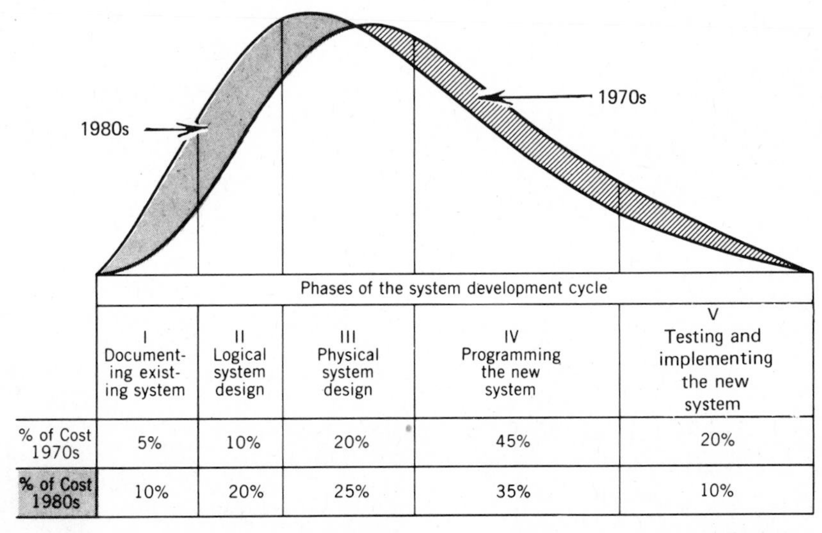

	Phases of the system development cycle				
	I Document-ing exist-ing system	II Logical system design	III Physical system design	IV Programming the new system	V Testing and implementing the new system
% of Cost 1970s	5%	10%	20%	45%	20%
% of Cost 1980s	10%	20%	25%	35%	10%

FIGURE 1 Comparison of costs of different generations of system development.

puting equipment. The comparison in evolution of hardware versus evolution of system development techniques is discussed next.

Lag in System Development Techniques

Concomitant with the increase in complexity of systems is the need for increased capability of system development techniques. Nevertheless, evolution of system development techniques significantly lagged hardware evolution. Systems analysts continued to use techniques developed for unit-record systems during the era of first generation computers. Computer-oriented techniques were originated in the 1950s, lagging hardware evolution by one generation, as shown in Figure 2. The lag diminished only slightly from 1960 to 1970.

In the latter 1970s and early 1980s, the gap began to close. Techniques especially suited for analysis and design of complex systems were developed. The fifth generation of system techniques has been developed almost in parallel with the fifth generation of hardware. System professionals are finally utilizing the power of the computer to aid in analysis and design of computer applications. As a result, systems are more cost effective. Such a change is vital because the cost of software is occupying a much higher proportion of the budget. Figure 3 shows the disparity between hardware and software costs.

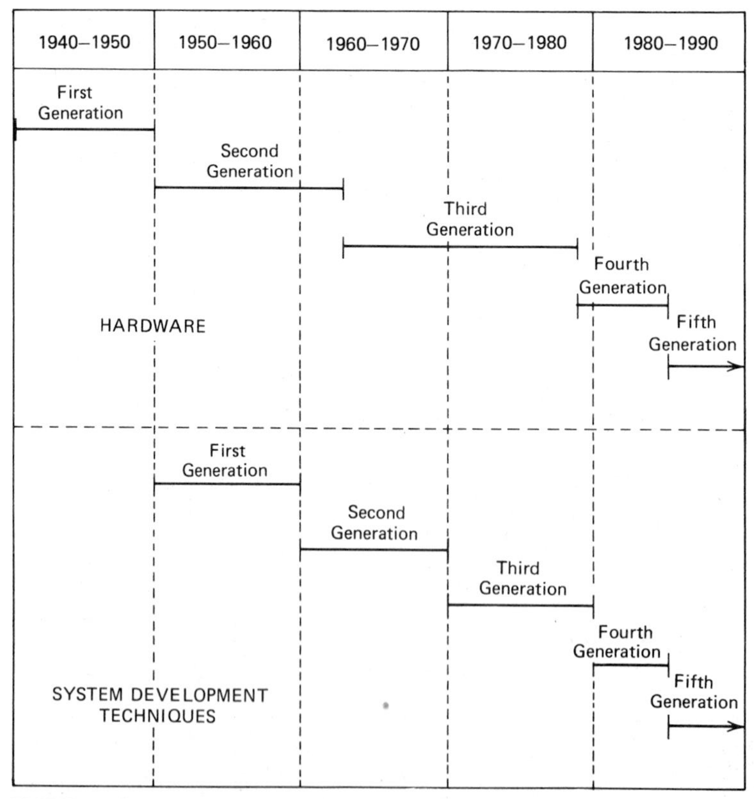

FIGURE 2 Lag in improvement of system development techniques compared with hardware improvement.

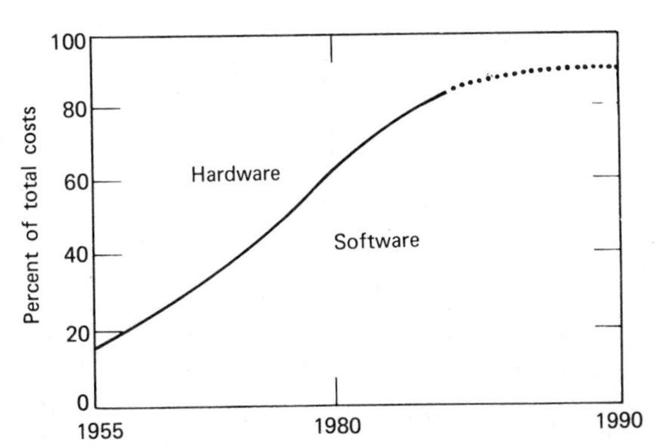

FIGURE 3 Hard/software cost trends. (*Source*: Extrapolation of a chart by Barry Boehm, "Software Engineering" in this book.)

The disparity has three principal causes. The continual decline in cost of hardware is the major impact on the cost curve. The continual increase in software labor cost is another significant factor. The lag in improvement of analysis/design/programming techniques is the third contributing factor. The projected flattening of the curve is based on the assumption that software development efficiency will increase. Such a result should be produced by the parallel development of fifth generation hardware and system development techniques.

Only the accelerating improvement in hardware has enabled the field to keep pace with demand for computing services. Without the accompanying decrease in cost, many companies would not have been able to utilize advanced computer techniques. Data base is a good example. The overhead cost of data base software is quite high. Faster, cheaper hardware offset the high software cost, permitting implementation of a data base for many companies that would otherwise have been unable to utilize the data base approach.

The improved fourth and fifth generation development techniques are bringing software costs in line with hardware costs, enabling computerization of many complex systems previously too expensive to computerize.

Figure 4 provides a schematic of the evolution of system development techniques. Three major categories are distinguished in Figure 4. The top portion of the schematic depicts the evolution of techniques for portraying and analyzing the *flow of information* through an organization. The central portion depicts the evolution of *mathematical and statistical* techniques. The lower portion depicts the evolution of techniques for *recording and analyzing resources*.

Although this book will concentrate on the top path, techniques for analysis of information flow and design of system logic, it will also emphasize the convergence of the three paths into the set of fourth and fifth generation systems development techniques.

To clearly identify the evolution of computer-oriented systems development techniques, Figure 4 also depicts precomputer techniques.

The convergence toward uniform techniques in the last decade should not imply that the field has moved toward acceptance of a single technique. For example, in the area of automated problem statement languages, two excellent versions are available with a half dozen new ones under development. Also, a variety of approaches to structured analysis, design, and programming are in existence. The differences in the principal methods will be delineated in later sections of the book. On the other hand, the convergence *does* reflect a major improvement in system development. A synergy of techniques has occurred. Rather than the hodgepodge of techniques representative of earlier periods, today there is a merger of techniques resulting in more powerful system development tools. Such an approach was necessary to handle the complexity of the wide-scope/in-depth systems being developed today.

Readers need not take my word for this conclusion. They can examine the literature themselves to draw their own conclusions. Figure 4 facilitates such an examination. The small circle to the left of each technique contains a number that refers to the bibliographic reference for that tech-

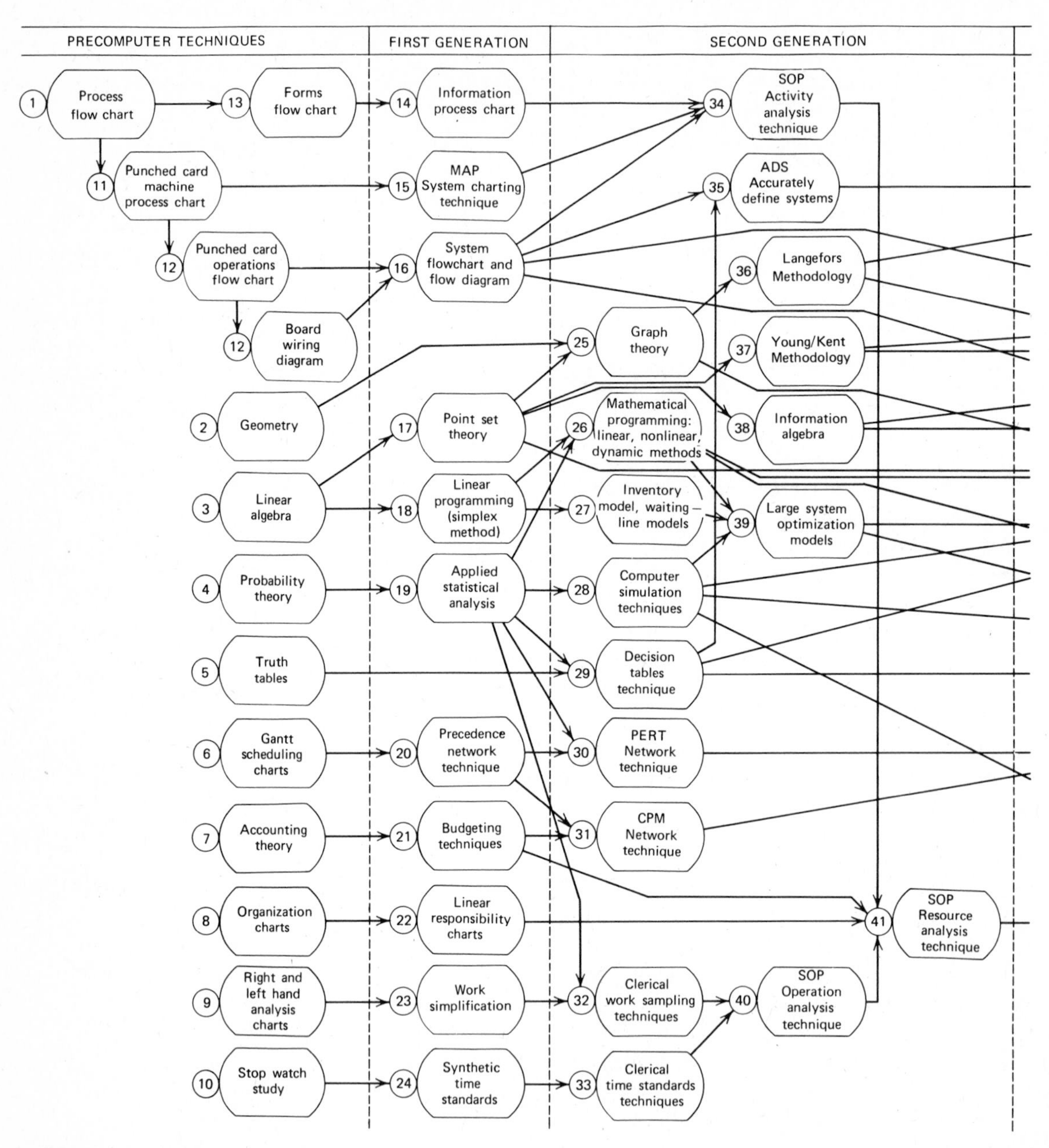

FIGURE 4 Evolution of system development techniques.

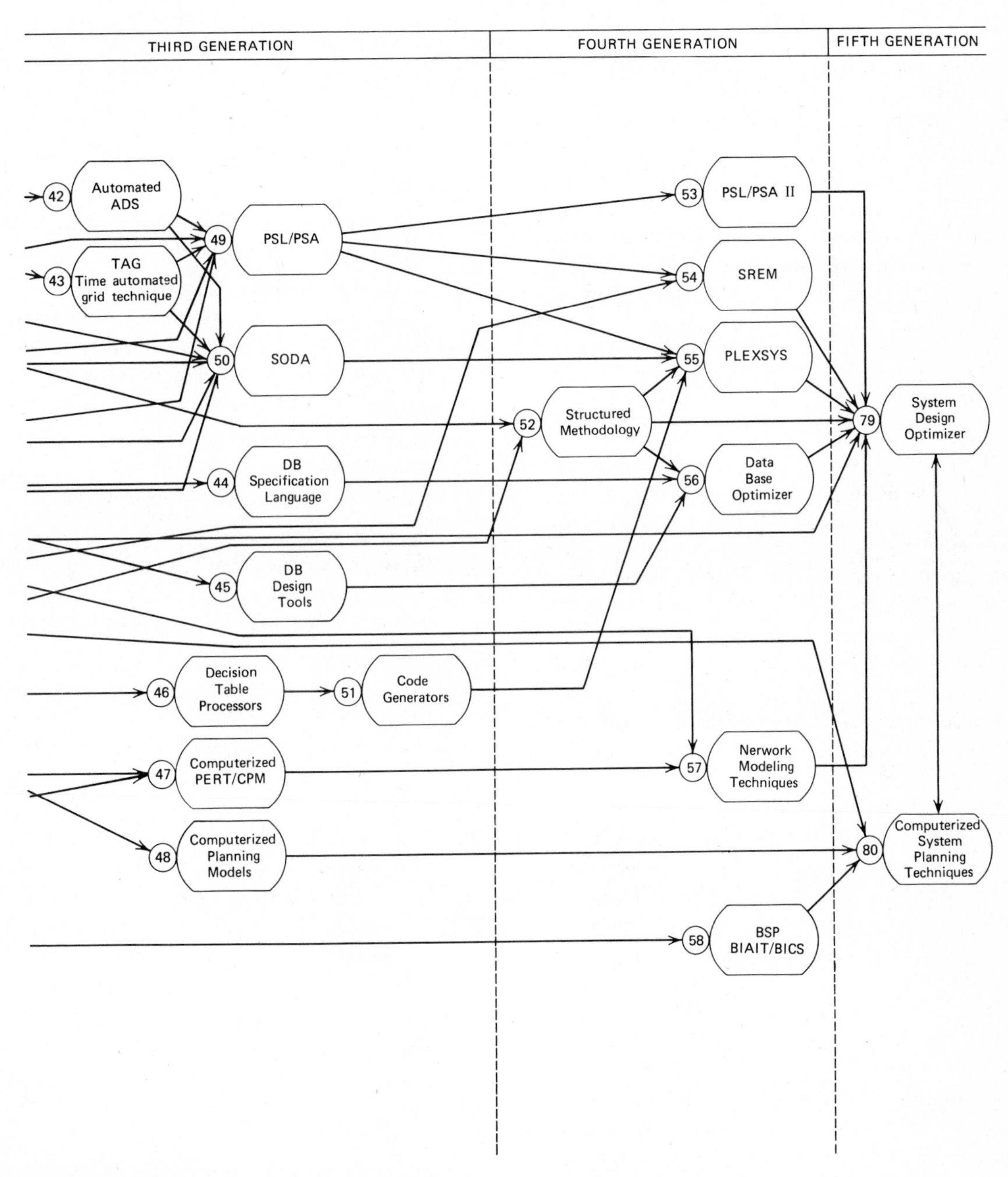

FIGURE 4 Continued.

nique. Landmark papers are cited whenever possible, for example, the CODASYL paper on Information Algebra. In other cases, a survey paper is cited, for example, the Industrial Engineering Handbook article on process flowcharts. More than one reference is provided in several cases. The reader may want only an overview rather than in-depth treatment of a subject. Several handbooks are referenced for that purpose. They provide an overview of more than one technique.

SUMMARY

Despite the lag in development of techniques for analysis and design of computer-based systems, some powerful tools are now available. Projections of future developments indicate that the gap between hardware improvements and software improvements will narrow significantly. Unfortunately, even when these new techniques are proven effective, some companies are slow to adopt them. It is important to broaden the dissemination of information about the progress in technique development to encourage adoption.

The process of expediting adoption of new techniques is facilitated by better understanding of the origin and application of each technique. The primary purpose of this book is to clarify the evolution of system development techniques, those used in Phases I through III of the system life cycle.

This book is divided into sections, one for each generation in the evolution of system development techniques. Such classification and explanation should enhance the understanding of each of the techniques used today and, equally important, how they will evolve into the next generation of techniques.

QUESTIONS

1. Distinguish the differences between logical and physical design.
2. What is the distinction between system analysis, system development, and software engineering?
3. Why do today's systems need more sophisticated system development techniques?
4. Hardware costs are continually diminishing. Why are system development costs higher in the 1980s than the 1970s?
5. Complete the following sentence: *Information algebra* is to business system analysis as _______ _______ is to computer science. (Refer back to the Introduction for your answer.)
6. Why are academic programs in Information Systems justifiable at the university level? (The Introduction provides two reasons.)
7. What advantages result from interaction of industry and academia concerning curriculum and research? (The answers are available in the Introduction.)
8. Why are data base systems expensive? Why have companies been able to afford them? What must happen for a larger number of firms to be able to utilize the data base approach?
9. Even with the availability of fourth generation system development techniques, system design may not be improved in some companies. What is necessary to change this situation?
10. From reading the article and analyzing Figure 4, why are the fourth and fifth generation system development techniques more powerful than those of the second generation?
11. If you wanted to conduct further research in any of the techniques shown in Figure 4, how would you go about it?
12. Analyze Figure 4. PSL/PSA is one of the computer-aided system development techniques. What quantitative techniques (math, stat, etc.) are utilized in the PSL/PSA algorithms?

13. Is PSL/PSA the only computer-aided system development technique? Explain your answer.

14. What is implied by the convergence of techniques in the fourth and fifth generation?

SECTION 1

FIRST GENERATION DEVELOPMENT TECHNIQUES FOR COMPUTER-BASED SYSTEMS

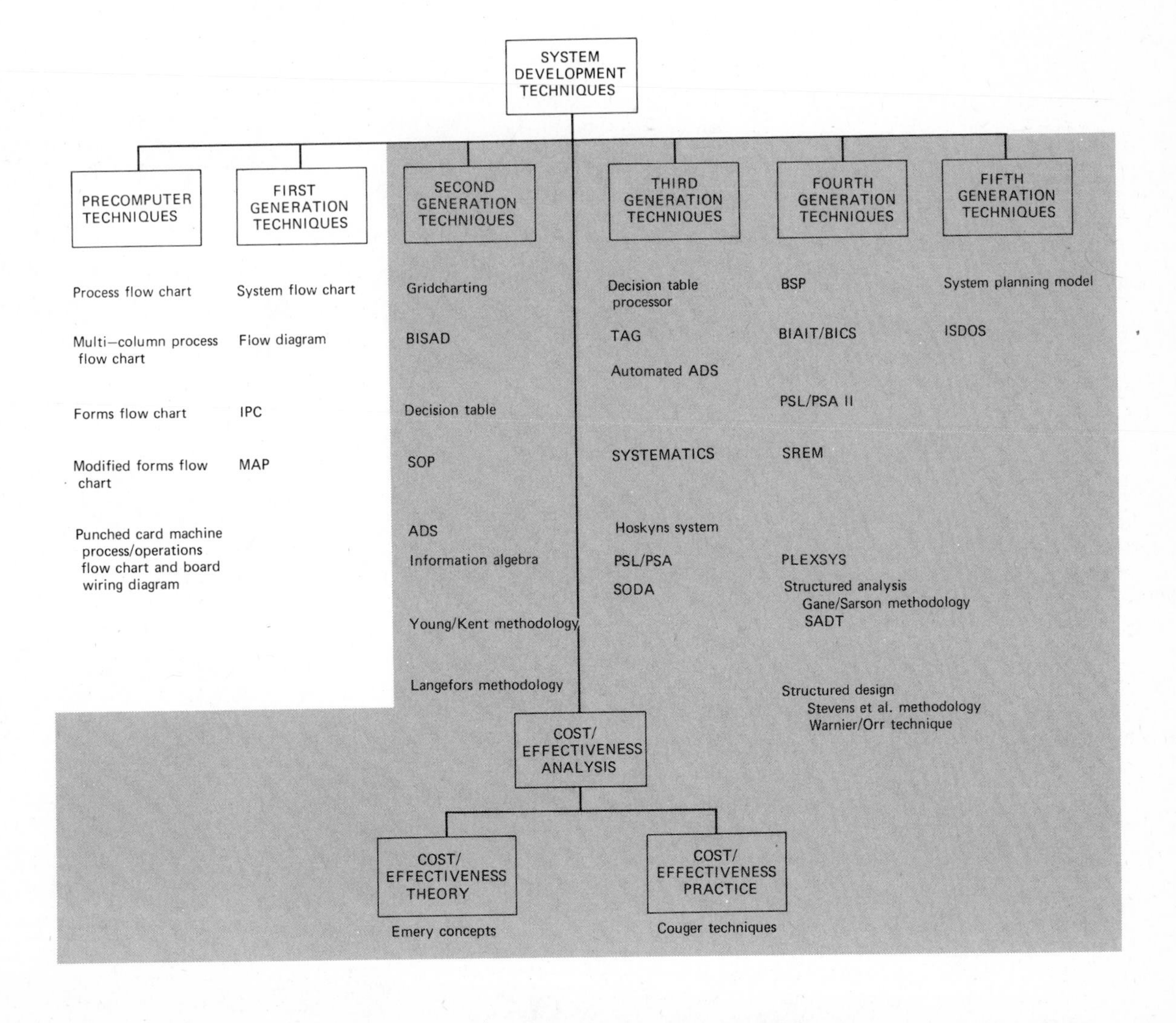

SECTION 1

FIRST GENERATION DEVELOPMENT TECHNIQUES FOR COMPUTER-BASED SYSTEMS

J. Daniel Couger

OVERVIEW

Techniques for portraying information flow evolved from process flow charts used by industrial engineers in the early 1900s. Process flow charts were used for a very different purpose, however. They depicted the flow of physical products through a production process, including the related material flows. Frederick W. Taylor was a leader in the development of the process flow chart.

With the introduction of computers, analysts did not develop new tools specifically for computer-related system analysis. Their approach was to modify techniques used for analysis and design of manual systems. It is important to understand this background in the evolution of system development techniques to explain why computer-oriented techniques lagged the development of hardware. Because each of our present-day systems contains some manual activities, it is important also to evaluate these techniques for manual system analysis in light of their applicability to current systems.

PRECOMPUTER TECHNIQUES

Techniques used for improvement of manual systems will be analyzed first. They include: process flow charts, multicolumn process flow charts, forms flow charts, punched card machine flow charts, and the modified forms flow chart.

Process Flow Charts

Figure 1 illustrates a typical process flow chart. Special symbols were utilized and soon became standards for each activity that occurred in the product flow. This portrayal permitted analysis of all activity at each stage of the product

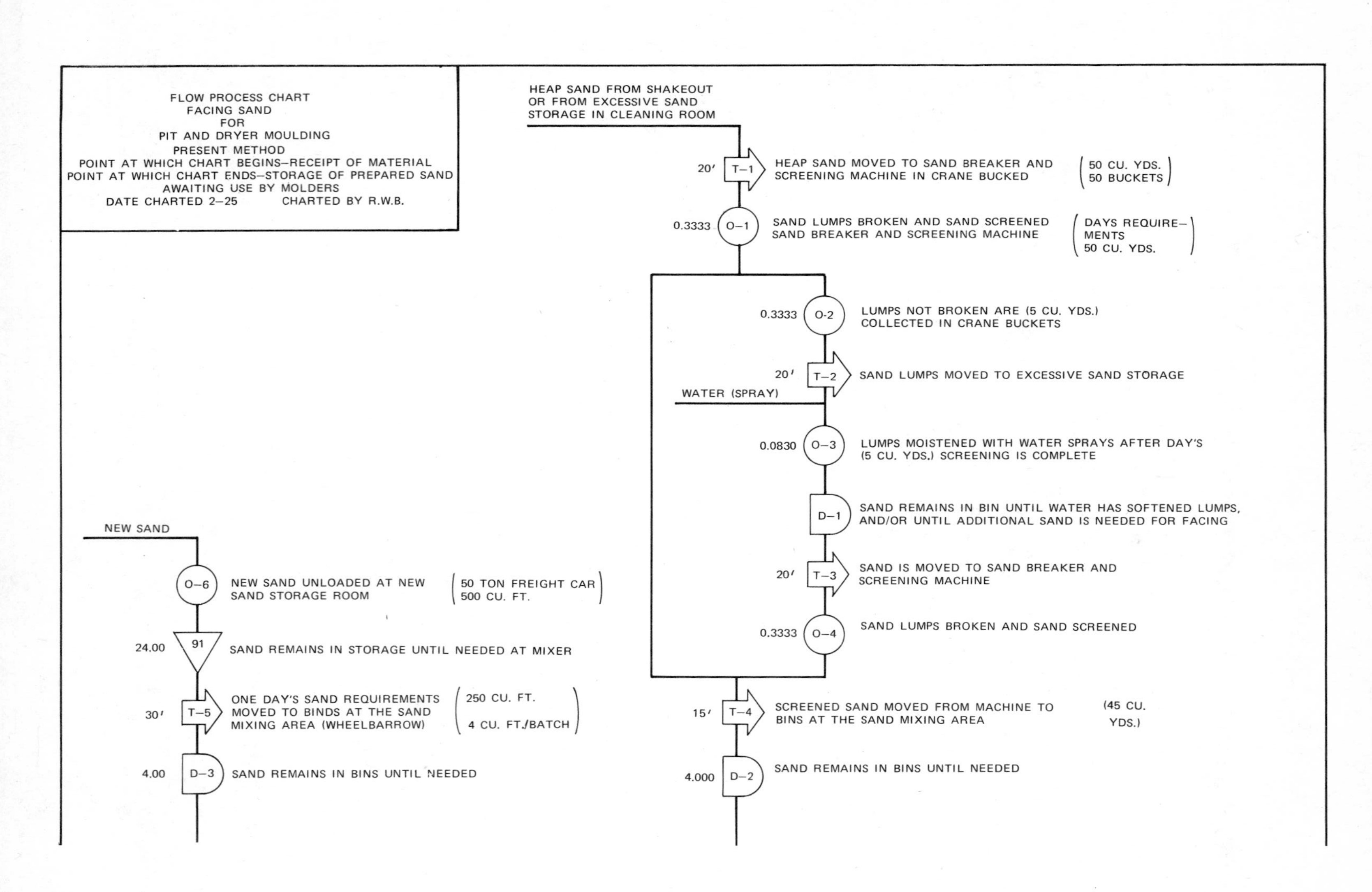
FLOW PROCESS CHART
FACING SAND
FOR
PIT AND DRYER MOULDING
PRESENT METHOD
POINT AT WHICH CHART BEGINS–RECEIPT OF MATERIAL
POINT AT WHICH CHART ENDS–STORAGE OF PREPARED SAND
AWAITING USE BY MOLDERS
DATE CHARTED 2–25
CHARTED BY R.W.B.
HEAP SAND FROM SHAKEOUT
OR FROM EXCESSIVE SAND
STORAGE IN CLEANING ROOM
20′
T–1
HEAP SAND MOVED TO SAND BREAKER AND
SCREENING MACHINE IN CRANE BUCKED
50 CU. YDS.
50 BUCKETS
0.3333
O–1
SAND LUMPS BROKEN AND SAND SCREENED
SAND BREAKER AND SCREENING MACHINE
DAYS REQUIRE–
MENTS
50 CU. YDS.
0.3333
O-2
LUMPS NOT BROKEN ARE (5 CU. YDS.)
COLLECTED IN CRANE BUCKETS
20′
T–2
SAND LUMPS MOVED TO EXCESSIVE SAND STORAGE
WATER (SPRAY)
0.0830
O–3
LUMPS MOISTENED WITH WATER SPRAYS AFTER DAY'S
(5 CU. YDS.) SCREENING IS COMPLETE
D–1
SAND REMAINS IN BIN UNTIL WATER HAS SOFTENED LUMPS,
AND/OR UNTIL ADDITIONAL SAND IS NEEDED FOR FACING
20′
T–3
SAND IS MOVED TO SAND BREAKER AND
SCREENING MACHINE
0.3333
O–4
SAND LUMPS BROKEN AND SAND SCREENED
15′
T–4
SCREENED SAND MOVED FROM MACHINE TO
BINS AT THE SAND MIXING AREA
(45 CU.
YDS.)
4.000
D–2
SAND REMAINS IN BINS UNTIL NEEDED
NEW SAND
O–6
NEW SAND UNLOADED AT NEW
SAND STORAGE ROOM
50 TON FREIGHT CAR
500 CU. FT.
24.00
91
SAND REMAINS IN STORAGE UNTIL NEEDED AT MIXER
30′
T–5
ONE DAY'S SAND REQUIREMENTS
MOVED TO BINDS AT THE SAND
MIXING AREA (WHEELBARROW)
250 CU. FT.
4 CU. FT./BATCH
4.00
D–3
SAND REMAINS IN BINS UNTIL NEEDED

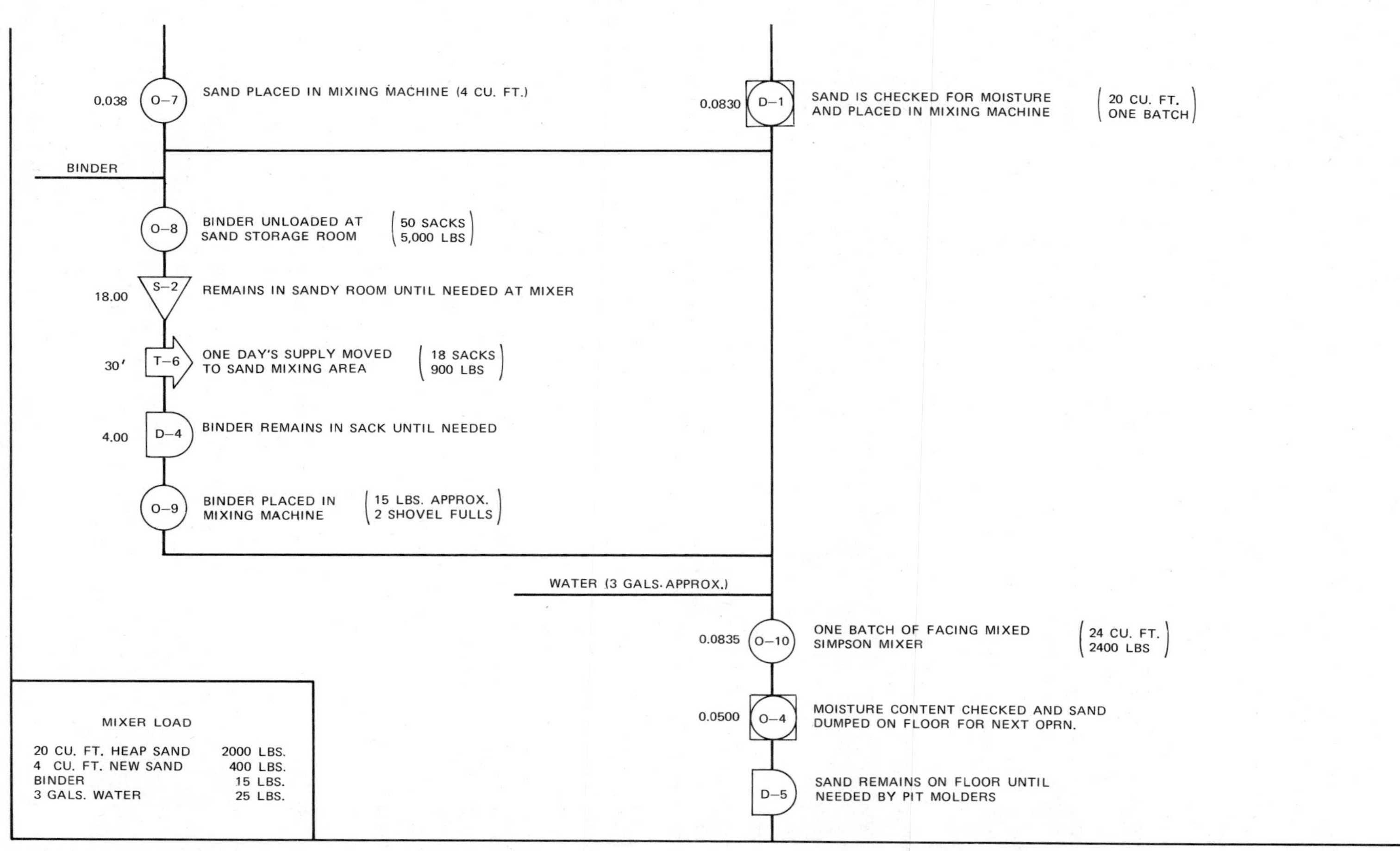

FIGURE 1 Process flow chart. (Courtesy McGraw-Hill Book Co.)

flow and highlighted areas for improvement. The symbols were defined as follows:

To change

Operation. An operation occurs when an object is intentionally changed in any of its physical or chemical characteristics, is assembled or disassembled from another object, or is arranged for another operation, transportation, inspection, or storage. An operation also occurs when information is given or received or when planning or calculating takes place.

To move

Transportation. A transportation occurs when an object is moved from one place to another, except when such movements are a part of the operation or are caused by the operator at the work station during an operation or an inspection.

To verify

Inspection. An inspection occurs when an object is examined for identification or is verified for quantity or quality in any of its characteristics.

To wait

Delay. A delay occurs to an object when conditions, except those which intentionally change the physical or chemical characteristics of the object, do not permit or require immediate performance of the next planned action.

To protect

Storage. A storage occurs when an object is kept and protected against unauthorized removal.

Combined Activity. When two activities are performed concurrently, or at the same work station, the symbols may be combined. The example shown represents a combined operation and inspection.

The role of the industrial engineer was to prepare the process chart for the existing process, then examine it for possible improvement. That role is little different than the analyst's role in examining a modern system. With only four major considerations—materials, operations, inspections, and time—the subject of material was analyzed first. All alternative materials, finishes, and tolerances were evaluated as to function, reliability, service, and cost. Next, the operations were reviewed for possible alternative processing methods and equipment. Inspections were reviewed for quality level, for replacement with in-process sampling techniques, or for combination on related operations. Time values were reviewed in terms of alternative methods, tooling, use of outside services, or special-purpose equipment. The objective was to eliminate, combine, or simplify operations to speed up the process flow.

Multicolumn Process Flow Chart

While the process flow chart enabled improvements in the overall flow, it had shortcomings. The chart depicted the flow of the product rather than the activities performed by individuals in the production process. The multicolumn process chart was developed to show what each worker did at each stage in the production process. Figure 2 illustrates a multicolumn process flow chart. This enhancement of the process flow chart provided another important source of information for systems analysis. In addition to recording the work of each person, the cycles of production and related activities were also identified.

It is not difficult to visualize the transformation of this industrial engineer's tool to a technique for displaying the flow of information. The multicolumn process flow chart's ability to handle (1) multiple activities and (2) cycles of process flow, were the principal reasons for its adaptation for portraying information flow.

SUMMARY

	PRESENT NO	PRESENT TIME	PROPOSED NO	PROPOSED TIME	DIFFERENCE NO	DIFFERENCE TIME
● ORIGIN						
ADD TO						
○ OTHER OPERATIONS	16					
⇨ TRANSPORTATIONS	2					
□ INSPECTIONS						
▽ STORAGES						
D DELAYS	12					
DISTANCE TRAVELED	60	FT		FT		FT

MULTI–COLUMN FLOW PROCESS CHART NO.____ PAGE __ OF __

JOB MAKING LADLE STOPPER RODS (ACTIVITIES OF GANG OF 3 MEN)
MAN OR MATERIAL
CHART BEGINS WITH BARE ROD (ABOUT 2¼" & ABOUT 8'0" TO 9'0" LG) STRAIGHTENED.
CHART ENDS COMPLETED ROD IN DRYING OVEN
CHARTED BY INDUSTRIAL CREW(4), DATE 6–15–

DESCRIPTION		EAGLE PENCIL NUMBER	1st CYCLE	2nd CYCLE
STOPPER ROD MAKER	1	744	1–3–3–3–3–3–9–10–12–D	1–3–3–3–3–3–9–10–12–D
FIRST HELPER	2	740½	1–4–6–8–D–D–D–11–D–D	1–4–6–8–D–D–D–11–D–D
SECOND HELPER	3	738½	2–5–7–D–D–D–D–D–D–13	2–5–7–D–D–D–D–D–D–13
	4	740½		
	5	737		
	6	740½		
	7	742½		
	8	740½		
	9	736		
	10	740½		
	11	744		
	12	740½		
	13	730½		
	14	740½		
	15	737		
	16	740½		

LEGEND

1. MOVE STOPPER ROD TO WORK PLACE
2. TRANSPORT COMPLETED ROD TO DRYING OVEN–A OUT 30'
3. APPLY MUD TO SLEEVE JOINTS
4. PLACE SLEEVES ON STOPPER ROD, ONE AT A TIME
5. PLACE COMPLETED ROD IN DRYING OVEN
6. PUT BULL NOSE ON NEXT ROD
7. RETURN TO WORK PLACE
8. SCREW RETAINING NUTS ONTO ROD
9. PRELIMINARY TIGHTENING OF NUTS
10. COMPLETE TIGHTENING OF NUTS
11. VISE–HOLD ROD, BY MEANS OF CHAIN, AGAINST TURNING WHILE NUTS ARE BEING FINISHED TIGHTENED
12. WIPE EXCESS MUD FROM ALL SLEEVE JOINTS
13. PICK UP COMPLETED ROD BY MEANS OF MONO–FAIL CHAIN HOIST

LAYOUT OF WORK AREA

ROD BEING WORKED ON BY MAKER AND FIRST HELPER
(A)
(B)
(C)

SECOND HELPER MOVES FINISHED ROD FROM POSITION (A) TO POSITION TO DRYING OVEN (B), A THREE–TIER OVEN, WHERE THEY ARE SLOWLY DRIED FOR ABOUT 24 HOURS, PROGRESSING AS INDICATED BY ARROW.
SUPPLY OF SLEEVES IS SHOWN AT (C).

FIGURE 2 Multi-column process flow chart. (Courtesy McGraw-Hill Book Co.)

Forms Flow Chart

As organizations grew and paperwork began to be a problem, the process flow chart was modified to depict forms flow. Figure 3 illustrates a forms flow chart. It was similar to the process flow chart in that special symbols were used to identify different activities. However, an additional feature was added—vertical lines to identify organizational responsibility for various activities. The objectives in improving the system were similar to those stated previously. It is obvious from the chart, however, that the quantity and distribution of copies of forms were highlighted as a target for simplification.

Punched Card Machine Flow Chart

Examination of Figure 3 reveals some shortcomings of this flowcharting approach: lack of identification of data elements and volumes. With the advent of mechanical processing of information (with punched card equipment), process flow charts were modified to portray the devices involved in data processing. Figure 4 is an example of such a chart. Pictures of punched card machines (later referred to as unit record equipment because a card stored a unit's records) were inserted on the process flow diagram. For example, a sorting function was depicted by a picture of a card sorter. A collation function was depicted by a picture of a collator and a listing function by a picture of a tabulator (which tabulated and printed).

One could quickly visualize the flow of records and the activities performed at each step in the process. By supplementing this flow chart with a tabulating operations documentation package (Figure 5), a good representation of data processing was obtained. The documentation included processing procedures, process diagrams, and board wiring diagrams. The latter was the forerunner of the computer program. The board wiring of a punched card machine could be changed to perform a variety of functions. A wiring diagram showed the connections required to produce the desired results.

Modified Forms Flow Chart

At the same time punched card equipment was being introduced to government and industry, special data capture equipment was introduced. For example, while typing a document such as a purchase order, a by-product punched paper tape would capture key data elements. By inserting this tape into the tape reader, portions or all of the original document could automatically be typed to produce the receiving report. Figure 6 shows how the forms flow chart was revised to include pictures of the devices and to show the flow of machine media: punched paper tape and punched cards. Note that this system analysis technique combined the punched card machine flow chart with the forms flow chart. The punched card machine department (typically referred to as the machine accounting department) became an important addition to the forms flow chart.

Although the modified forms flow chart adequately portrayed media flow (punched cards, punched paper tape, and forms) and showed the machine functions performed, it neglected three areas important for computer-oriented systems:

1. Detailed logic/procedure.
2. Data elements: fields, records.
3. Volume information.

Nevertheless, these systems techniques were the ones used with the first generation of computing equipment.

COMPUTER-BASED TECHNIQUES

The first system analysis techniques used to design computer-based systems were modified manual system techniques.

In the previous section on evolution of system development techniques, the left side of Fig-

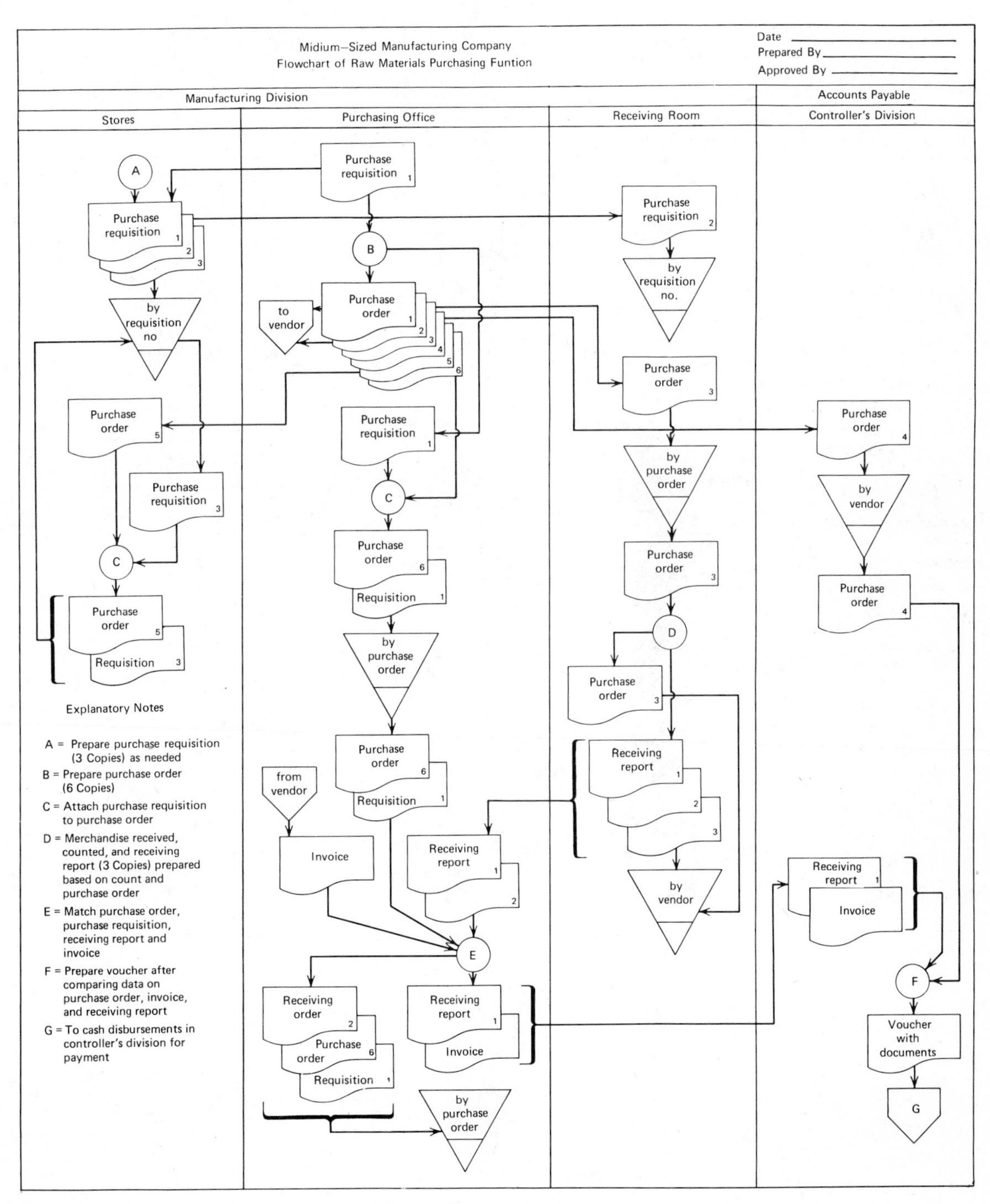

FIGURE 3 Forms flow chart. (Courtesy Allyn and Bacon, Inc.)

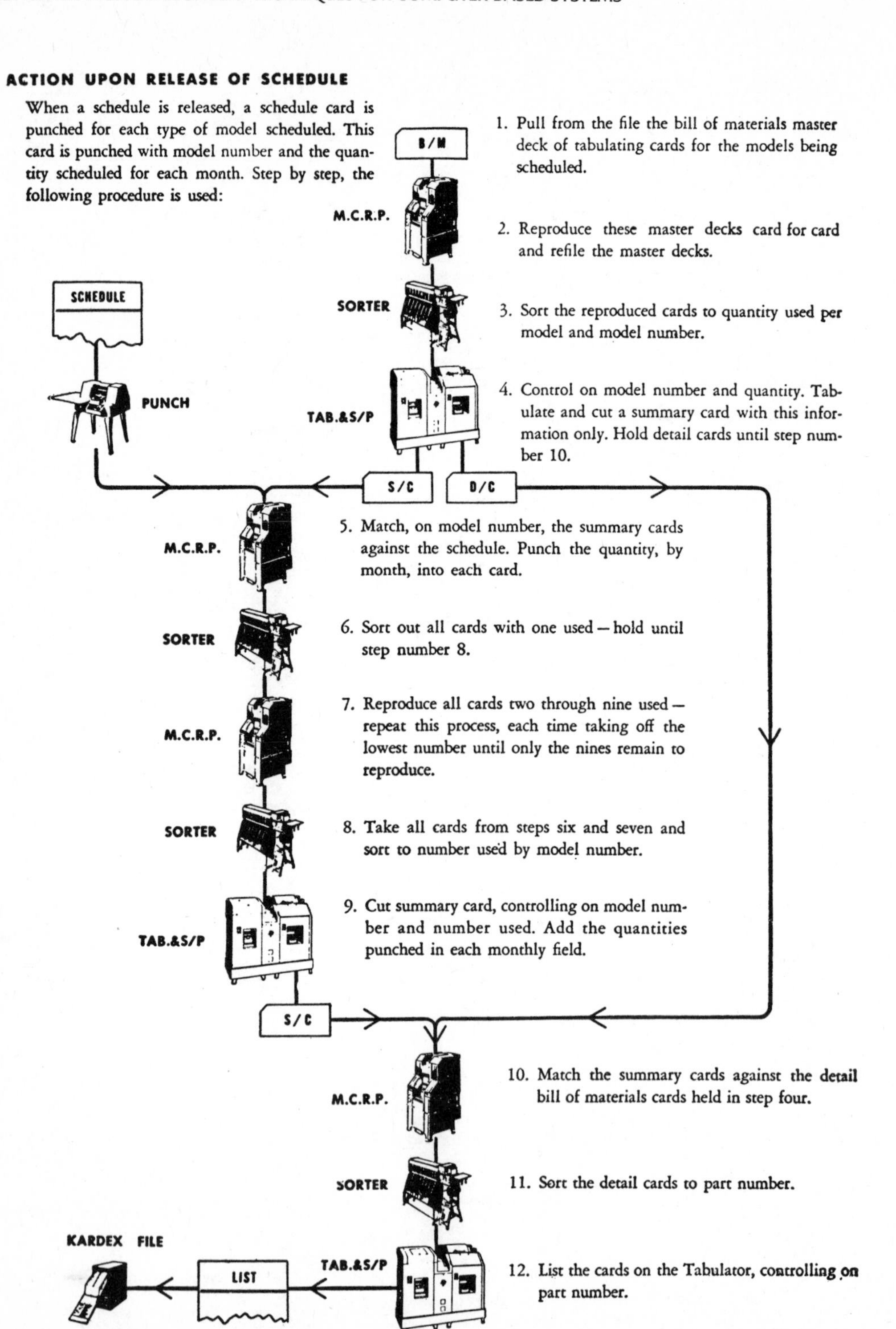

ACTION UPON RELEASE OF SCHEDULE
When a schedule is released, a schedule card is punched for each type of model scheduled. This card is punched with model number and the quantity scheduled for each month. Step by step, the following procedure is used:
B/M
M.C.R.P.
1. Pull from the file the bill of materials master deck of tabulating cards for the models being scheduled.
2. Reproduce these master decks card for card and refile the master decks.
SCHEDULE
SORTER
3. Sort the reproduced cards to quantity used per model and model number.
PUNCH
TAB.&S/P
4. Control on model number and quantity. Tabulate and cut a summary card with this information only. Hold detail cards until step number 10.
S/C
D/C
M.C.R.P.
5. Match, on model number, the summary cards against the schedule. Punch the quantity, by month, into each card.
SORTER
6. Sort out all cards with one used — hold until step number 8.
M.C.R.P.
7. Reproduce all cards two through nine used — repeat this process, each time taking off the lowest number until only the nines remain to reproduce.
SORTER
8. Take all cards from steps six and seven and sort to number used by model number.
TAB.&S/P
9. Cut summary card, controlling on model number and number used. Add the quantities punched in each monthly field.
S/C
M.C.R.P.
10. Match the summary cards against the detail bill of materials cards held in step four.
SORTER
11. Sort the detail cards to part number.
KARDEX FILE
LIST
TAB.&S/P
12. List the cards on the Tabulator, controlling on part number.

FIGURE 4 Punched card machine process chart. (Courtesy Univac.)

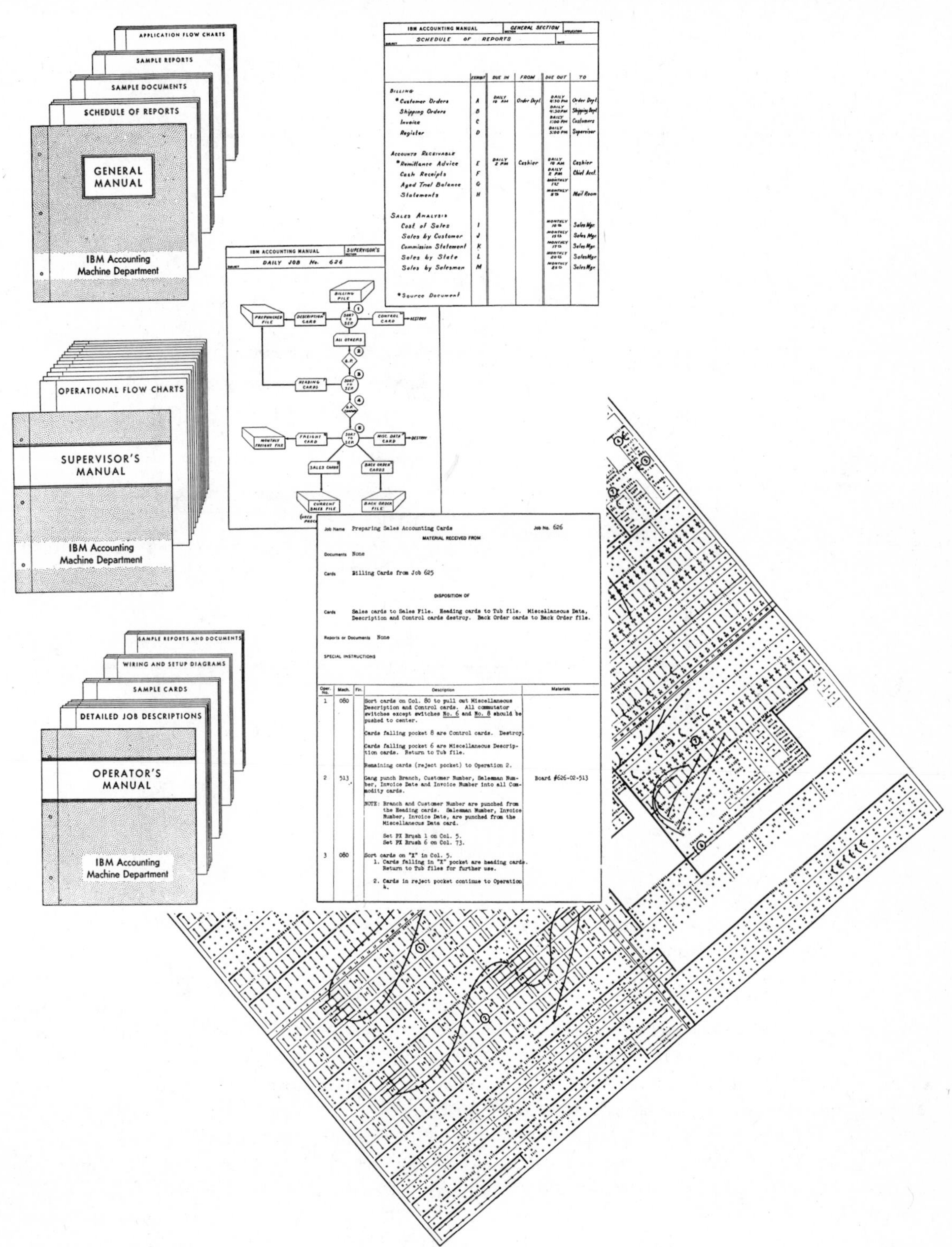

FIGURE 5 Punched card operations documentation package. (Courtesy IBM.)

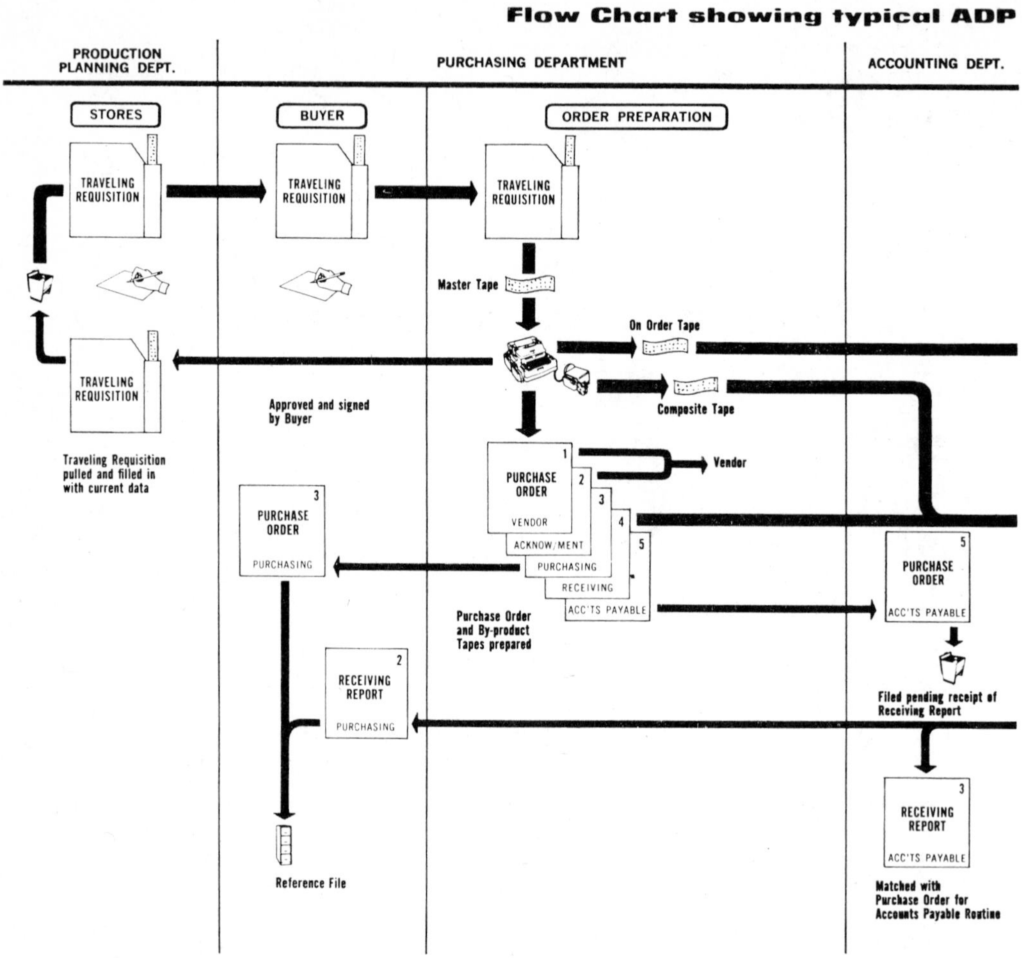
Flow Chart showing typical ADP
PRODUCTION PLANNING DEPT.
PURCHASING DEPARTMENT
ACCOUNTING DEPT.
STORES
BUYER
ORDER PREPARATION
TRAVELING REQUISITION
TRAVELING REQUISITION
TRAVELING REQUISITION
Master Tape
On Order Tape
TRAVELING REQUISITION
Composite Tape
Approved and signed by Buyer
Traveling Requisition pulled and filled in with current data
PURCHASE ORDER
VENDOR
ACKNOW/MENT
PURCHASING
RECEIVING
ACC'TS PAYABLE
Vendor
PURCHASE ORDER
PURCHASING
Purchase Order and By-product Tapes prepared
PURCHASE ORDER
ACC'TS PAYABLE
Filed pending receipt of Receiving Report
RECEIVING REPORT
PURCHASING
RECEIVING REPORT
ACC'TS PAYABLE
Reference File
Matched with Purchase Order for Accounts Payable Routine

FIGURE 6 Modified forms flowchart.

ure 2 showed how the process flow chart was the source of each of these techniques: multicolumn process flow chart, forms flow chart, modified forms flow chart, punched card machine process chart and documentation package.

There was sufficient rationale for such an approach, because most systems at that time were being converted from manual systems. In contrast, most system development projects today are revisions of earlier computerized systems. First generation systems techniques were not designed to capitalize on the advantages of the computer, but to flag unnecessary steps and delays in the manual system to justify computerization. Rather than depict the logic of a computer-based system, they showed the flow of information. They focused the analyst's attention on the things important for improving the efficiency of manual systems, but neglected to emphasize the things important to computer-based systems.

For example, there was no specification sheet for verbal information. Analysis of manual systems concentrated on paperwork flow. Verbal information flow was excluded or ignored. Computers had been in use for almost 10 years before

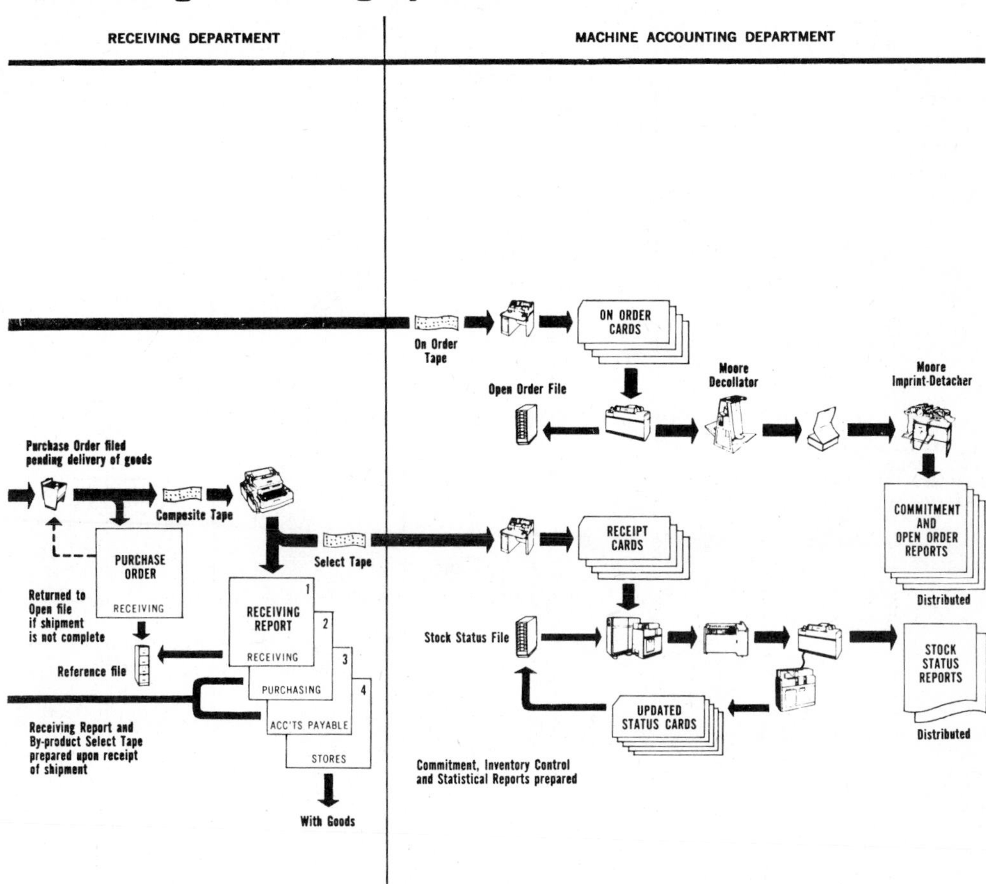

FIGURE 6 Continued

analysts developed a procedure for identifying verbal information so it could be input to the system.

Three principal techniques were used in the first generation of computer-based system analysis techniques: general flow chart/flow diagram, information process chart (IPC), and MAP.

Information Process Chart

Although the system development techniques of the 1950s had clear orientation for the computer, there is little doubt of their evolution from prior manual systems analysis techniques. For example, the first impression in viewing Figure 7 (information process chart) is the similarity of symbols to the process flow chart. On the other hand, several new symbols clearly distinguish this chart from its predecessors. The diamond symbols (search) and elliptical symbols (compare) show that the analysis was geared to computerization. Another feature characterized this technique as one that was computer-oriented. Entries and exits were highlighted.

In the information process chart (IPC), one line was used for each operation, with columns provided for indicating the fields of information on which the operations were performed. Cer-

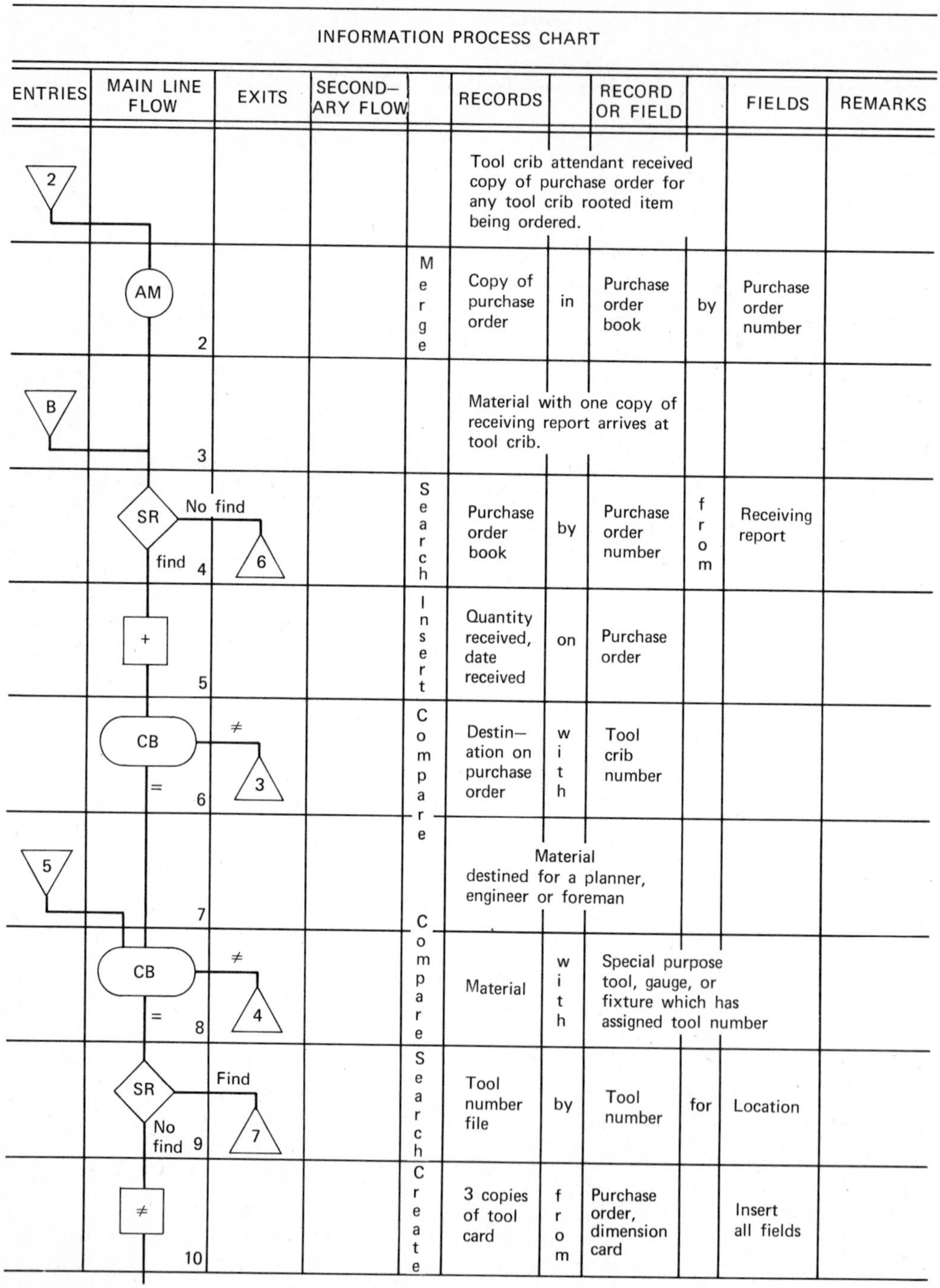

INFORMATION PROCESS CHART

ENTRIES	MAIN LINE FLOW	EXITS	SECOND–ARY FLOW		RECORDS		RECORD OR FIELD		FIELDS	REMARKS
2					Tool crib attendant received copy of purchase order for any tool crib rooted item being ordered.					
	AM 2			Merge	Copy of purchase order	in	Purchase order book	by	Purchase order number	
B	3				Material with one copy of receiving report arrives at tool crib.					
	SR find 4	No find 6		Search	Purchase order book	by	Purchase order number	from	Receiving report	
	+ 5			Insert	Quantity received, date received	on	Purchase order			
	CB = 6	≠ 3		Compare	Destin–ation on purchase order	with	Tool crib number			
5	7				Material destined for a planner, engineer or foreman					
	CB = 8	≠ 4		Compare	Material	with	Special purpose tool, gauge, or fixture which has assigned tool number			
	SR No find 9	Find 7		Search	Tool number file	by	Tool number	for	Location	
	≠ 10			Create	3 copies of tool card	from	Purchase order, dimension card		Insert all fields	

FIGURE 7 Information process chart (IPC).

tain verbs were specified and carefully defined to insure consistent understanding among all users of the charts. Although not widely used, the technique recognized the need for formal annotation which is necessary for computerized analysis.

MAP

As seen in Figure 8, MAP permitted a better overview of the flow of information, at the sacrifice of some detail provided in the annotation of IPC. In MAP, each horizontal level identified a type of document or file. A "transcription break" was used to show interrelationships between files. (The direction of the arrow shows the flow from one file to another.) Verbs were not as well defined as those used in IPC, and therefore MAP was not as useful as IPC.

Note also the lack of compatibility of symbols, characteristic of a new field. The industrial engineering profession had finally been able to standardize flow symbols after a number of years. In the early years of computing, standardization of symbols was nonexistent. In MAP, the decision symbol was a rectangle. The search symbol was not distinguished from an operation symbol; a circle was used to represent both activities. Nevertheless, both IPC and MAP were clearly superior to manual system techniques.

System Flowchart

Although IPC and MAP bore closer resemblance to earlier techniques, system flow charts and programming flow charts were quite different. Yet, there should be little argument that they evolved from punched card operation flow diagrams.

It may be of interest to note that the evolution of the term *flow chart* itself changed with the redirected emphasis on computer-based systems. The words *flow* and *chart* were separated when preceded by the word *process*. In other words, the emphasis was on the *process* flow. In the 1950s, the words were combined into the term

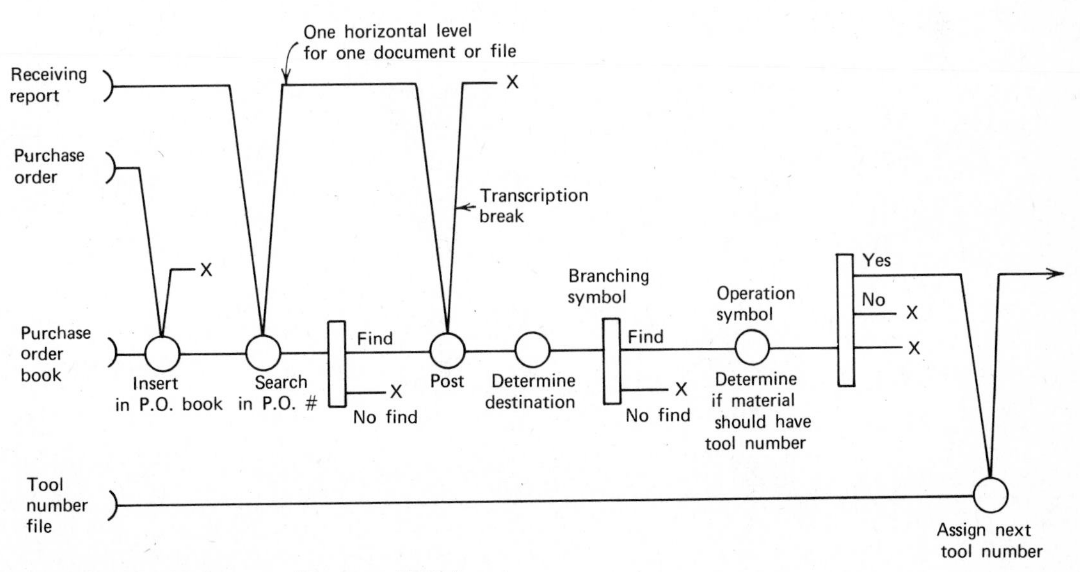

FIGURE 8 MAP diagram. (Courtesy NCR.)

flowchart. At the time, emphasis was redirected to information *flow* rather than physical product or paperwork flow.

It was not until the 1960s, however, that standardization of symbols for flowcharting finally reached fruition. Standardization of symbols within a department or company occurred much earlier, but documentation for packaging of software for sale was hampered for many years because of lack of industry standards.

Figure 9 shows the characteristics of the system flowchart which distinguished it from precomputer techniques. Most of the symbols were unlike previous flow chart symbols. The system chart identified (1) the input data for each step in the process, (2) the data processing activities, and (3) the output from each step. An annotation column permitted clarification of functions in each step.

The system flowchart provided an overview of the processing procedure, yet with more depth than first generation techniques. It conveyed "what is done" by the system. Nevertheless, a vehicle for depicting even greater detail was needed. In response to this need, the flow diagram was developed.

Flow Diagram

The flow diagram was designed to describe detailed data processing activities performed in-

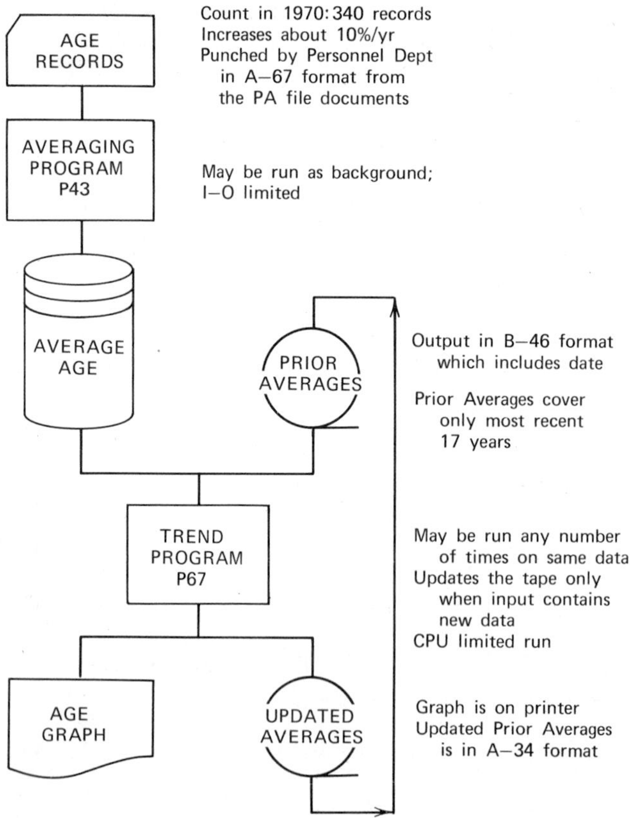

FIGURE 9 System flow chart with annotation. (Courtesy Ned Chapin.)

ternally within the computer. In addition, it described the process used to transform input data to output. Instead of explaining what was done (specified on the system flowchart), the flow diagram identified "how" the processing was done.

Because of the amount of information conveyed, many pages were produced. It was necessary to identify connections between pages and activities. Even with the increased amount of detail, it was often necessary to include annotation. Figure 10 illustrates a flow diagram with annotation.

Message Specification Sheet

During the later part of the 1950s a message specification sheet was designed. Figure 11 shows that it identified frequency, processing action, and relationship to files—important information for computer systems analysis. It also provided

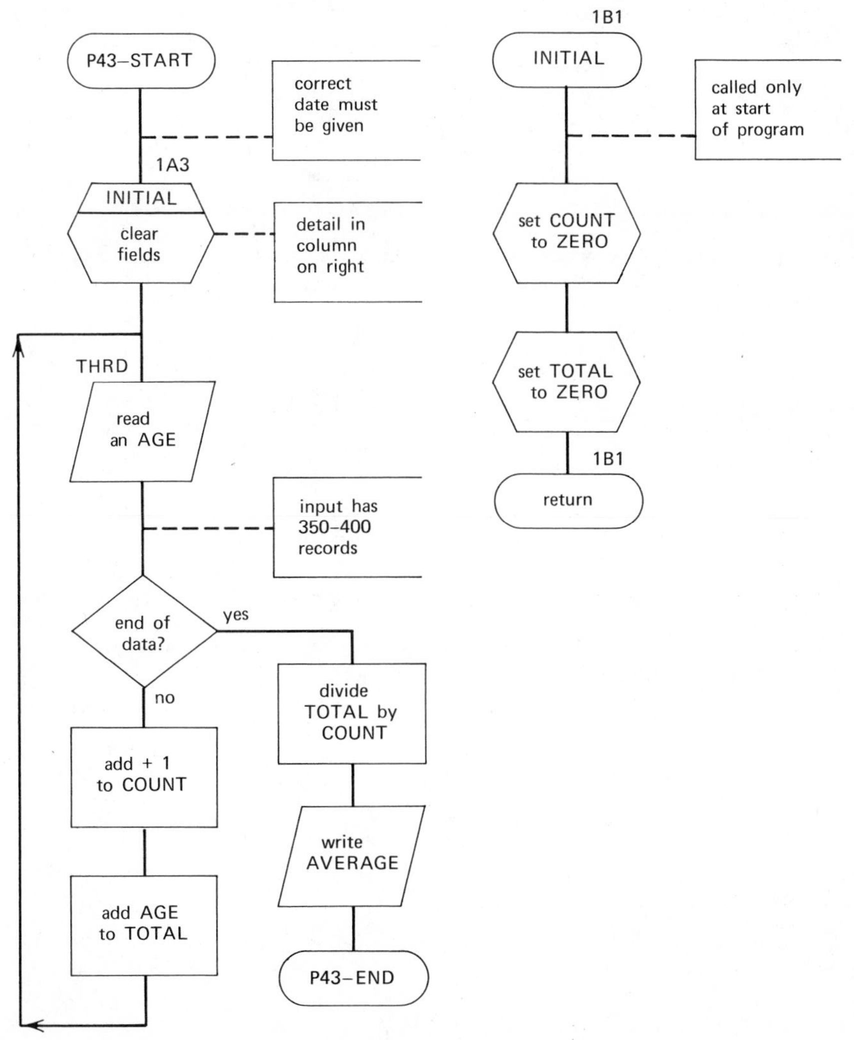

FIGURE 10 Flow diagram with annotation. (Courtesy Ned Chapin.)

IBM **Message Sheet**

MESSAGE NAME: CUSTOMER RECORD
MESSAGE NO: R0030
OTHER NAMES USED: CUSTOMER MASTER; BILLING MASTER
LAYOUT NO: 12
FORM NO: GRC 172
NO. OF COPIES: 1
MEDIA: 8½ x 5½ MANILA CARD
HOW PREPARED: TYPED - BOOKKEEPING MACHINE
OPERATIONS INVOLVED IN: 005-001, 005, 010 ; 010-005, 010 ; 012-010, 015, 075 ; 014-010, 020.

REMARKS: THIS MESSAGE IS THE MASTER FOR XEROXED BILL. AMOUNT REMITTED IS ENTERED ON THE BILL BY THE CUSTOMER, BUT THE FIELD IS NOT USED ON THIS RECORD.

CONTENTS

NO.	DATA NAME	FREQUENCY	CHARACTERS	A/N	ORIGIN
01	COMPANY NAME	1	25	AN	ON FORM
02	COMPANY ADDRESS	1	33	AN	ON FORM
03	TELEPHONE NUMBER	1	8	AN	ON FORM
04	CUSTOMER NAME	1	50	AN	012-075
05	CUSTOMER ADDRESS	1	150	AN	012-075
06	ACCOUNT NUMBER	1	7	N	012-075
07	AMOUNT REMITTED	1	6	N	SEE REMARKS
08	FIELD LABELS	1	35	AN	ON FORM
09	DATE	1-22 AVG 8	6	N	010-005
10	TRANSACTION NUMBER	1-22 AVG 8	2	N	010-005
11	CHARGE AMOUNT	1-22 AVG 7	6	N	010-005
12	PAYMENT / RETURN AMOUNT	1-22 AVG 1	6	N	010-005
13	BALANCE AMOUNT	1-22 AVG 8	6	N	010-005
14	PAYABLE INFORMATION	1	79	AN	ON FORM

DATE: 29 FEB 61 ANALYST: L H BAKER JR. SOURCE: COLLECTIONS PAGE: 1 OF 1
STUDY: ASSOCIATED RETAILERS INC.

M08-4404-0

FIGURE 11 Message specification sheet. (Courtesy IBM)

specific information about data elements, to facilitate conversion of messages to something processable by computers.

SUMMARY

Techniques developed prior to the computer era were used to design the first computer-based system—techniques that could be used to analyze forms flow and to develop punched card systems.

It was not until the 1950s that techniques were developed to capitalize on the unique abilities of a computer. System flowcharts, flow diagrams, IPC and MAP were the first computer-oriented tools. On the other hand, the symbology and the methodology of use of those tools clearly evolved from the forerunner manual system and punched card system techniques.

The second generation of computing hardware was introduced before these computer-oriented system analysis techniques became available. Analysis techniques lagged hardware evolution by one full generation at this time in computing history.

SECTION QUESTIONS

1. Why study precomputer techniques? Isn't this just "ancient history"?
2. Techniques used to develop the first generation of computer applications were suboptimal. Why?
3. List the differences and similarities of each of the precomputer system development techniques.
4. What information was not provided by precomputer techniques that is necessary for proper design of computer-based systems?
5. Although first generation flowchart techniques themselves are not useful for modern-day system development, in one respect the role of the analyst is unchanged. Explain.
6. What similarity exists between the precomputer techniques and IPC? What are the differences?
7. Was the message specification sheet a part of the first generation system development package? Explain.
8. From your earlier experience and coursework, which first generation techniques are still in use? Why?
9. Compare MAP to IPC.
10. Compare MAP to the system flowchart.
11. Why are the system flowchart and flow diagram used in tandem?
12. The symbology and methodology of precomputer techniques demonstrate that these techniques were clearly forerunners of the first generation computer-oriented system development techniques. Explain.
13. Assume you are transported through a "time warp" back to the early 1950s and are assigned the development of a computer application. Design a new system development technique that would utilize the best features of all the techniques available at that time. Explain your reasoning.

SECTION 2

SECOND GENERATION DEVELOPMENT TECHNIQUES FOR COMPUTER-BASED SYSTEMS

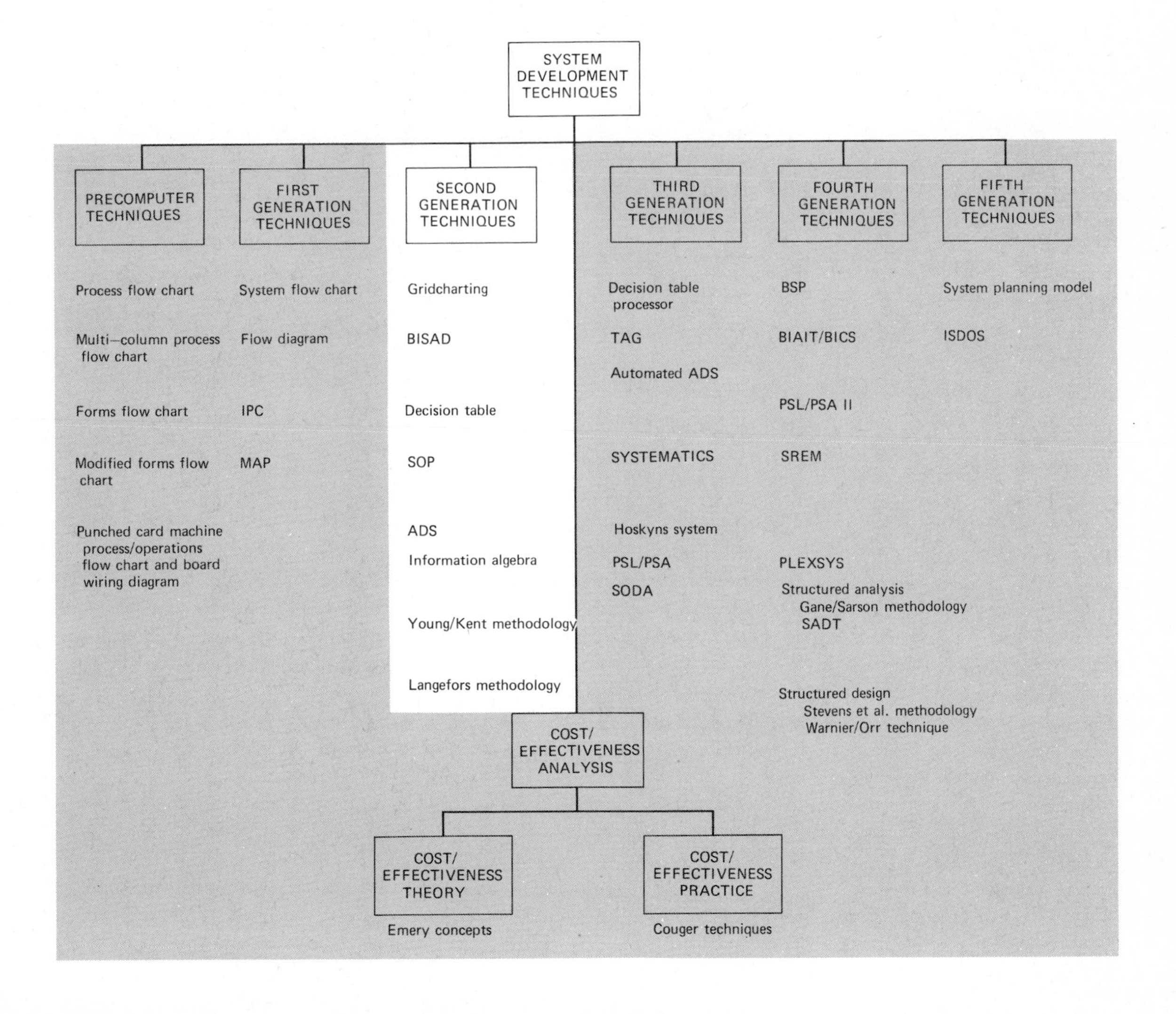

SECTION 2

SECOND GENERATION DEVELOPMENT TECHNIQUES FOR COMPUTER-BASED SYSTEMS

J. Daniel Couger

OVERVIEW

The second generation of system development techniques for computer-based systems was far superior to the previous generation. Nonetheless, development of system techniques still lagged hardware development by a decade.

Moreover, the lag was further accentuated. Despite the availability of new tools, many organizations continued to use outdated tools and techniques. For example, decision tables were proven advantageous by the General Electric Corporation early in the 1950s. Ten years transpired before they were widely used.

Another example is IBM's Study Organization Plan (SOP), first published in 1961. Few organizations utilized the approach until a decade later. It is enigmatic that companies quickly assimilated new generations of hardware but demurred when it came to upgrading their system development techniques.

If we modified Figure 2 (in the section on evolution of system development techniques) to show *availability* compared with *use* of techniques, an interesting but distressing picture emerges, as shown below. Although the first generation techniques were adopted rather quickly after development, widespread use of the more sophisticated techniques of the second generation lagged availability by an average of 5 to 10 years.

The second generation techniques described in this section include decision tables, ADS, gridcharting, BISAD, and SOP.

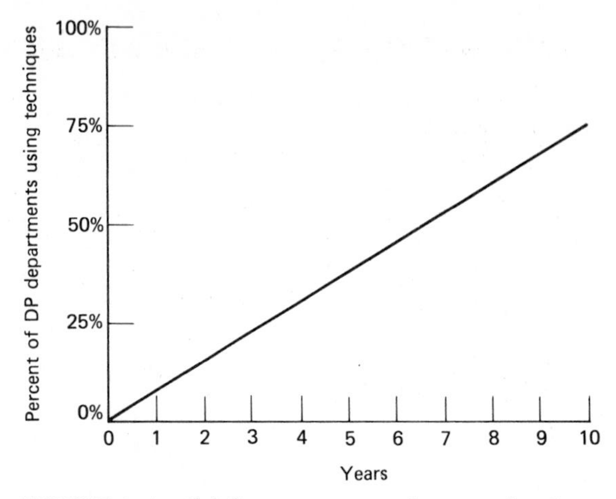

FIGURE 1 Availability versus use of system development techniques.

DECISION TABLES

Although system flowcharts and flow diagrams were adequate for the rather simplistic systems computerized in the early days of computers, they were inadequate for the more complex systems of the 1960s. An extension to the flow diagram (shown below) attempted to cope with a system where many decision alternatives existed. It was cumbersome to use, however.

The decision table technique was developed to supplement flowcharting procedures with a more straightforward approach. They reflect an "if-then" comparison. *If* a certain condition existed, *then* a specific action was taken.

In limited-entry form a decision table had four major sections, as shown in Figure 3.

The upper-left section was the condition stub. This area contained (in question form) all those conditions being examined for a particular problem segment. The lower-left section was the action stub. This area contained in concise narrative form all possible actions resulting from the conditions listed above.

The upper-right section was the condition entry section. It is here that the questions asked in the condition stub area were answered and all possible combinations of these responses were developed. Responses were restricted to "Y" to indicate yes, "N" to indicate no. If no response

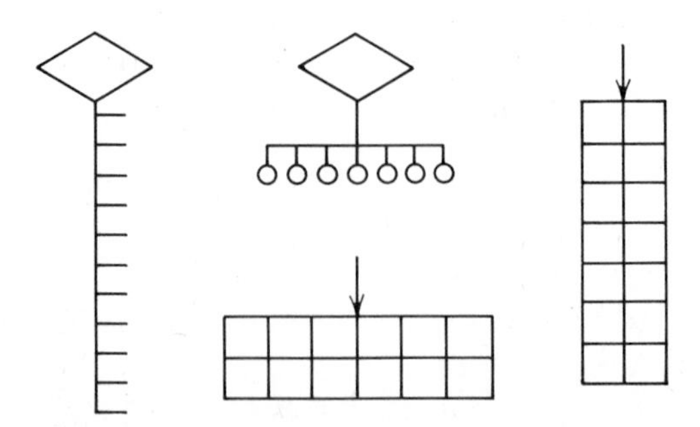

FIGURE 2 Flow diagram procedure for displaying logic where many decision alternatives existed. (Courtesy Ned Chapin.)

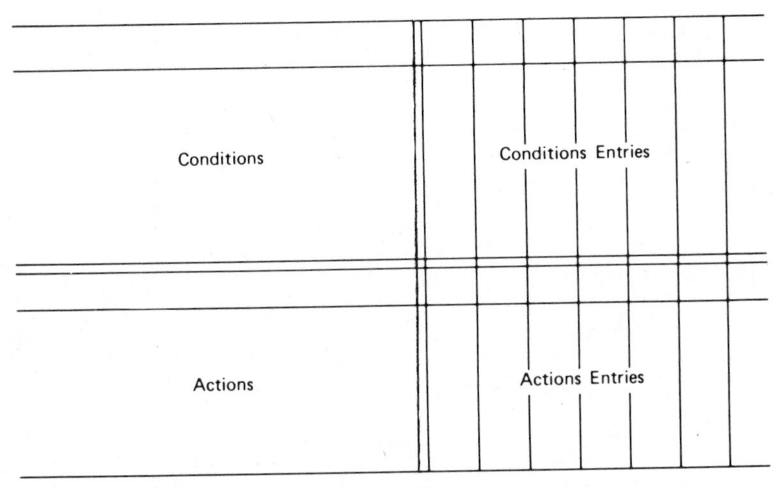

FIGURE 3 Components of a decision table.

was indicated, it was assumed that the condition was not tested in that particular combination. The remaining section was the action entry portion of the table. The appropriate actions resulting from the various combinations of responses to conditions above were indicated here.

For a limited entry table, the only acceptable symbol stub was an "X". A blank opposite an action in a given column was interpreted as "Do not take the action shown." One or more actions could be indicated for each combination of condition responses.

The various combinations of responses to conditions shown in the condition entry portion of the table and their resulting actions were called rules. Each was assigned a number for identification purposes in the rule portion of the table.

The condition half of the table was separated from the action half by a horizontal double line. A double vertical line separated the stubs from the entry portions of the table.

Three types of tables were designed. Limited-entry tables (Figure 4) were the most widely used type. In this form the rules regarding the placement of information in each of the four sections were fixed and inflexible. The condition and its state or value were restricted to the condi-

CREDIT ORDER APPROVAL PROCEDURE

		R1	R2	R3
C1	CREDIT LIMIT OKAY	Y	N	N
C2	PAY EXPERIENCE FAVORABLE	–	Y	N
A1	APPROVE CREDIT	X	X	–
A2	RETURN ORDER TO SALES	–	–	X

NOTE: BE CERTAIN THAT THE LATEST CREDIT INFORMATION IS USED.

FIGURE 4 Decision table example using limited entry format.

CREDIT ORDER APPROVAL PROCEDURE				
		R1	R2	R3
C1	CREDIT LIMIT	OK	NOT OK	NOT OK
C2	PAY EXPERIENCE	–	FAVORABLE	UNFAVORABLE
A1	CREDIT ACTION	APPROVE	APPROVE	DON'T APPROVE
A2	ORDER ACTION	–	–	RETURN TO SALES

FIGURE 5 Extended entry table.

tion stub; the condition entry could show only the response "Y," "N," or a blank. Similarly, the specific actions had to be fully identified within the action stub; permissible notations in the action entry section were limited to "X" or a blank.

In extended-entry tables the condition stub served only to identify the variables to be tested, whereas the condition entry defined the value or state of the variable. Similarly, in such a table the action stub only named an action; the action entry gave the specifics for the action named. Figure 5 shows in extended-entry form exactly the same information shown in Figure 4.

When limited-entry form and extended-entry form were combined into a single table, the resulting table was mixed-entry format. Figure 6 depicts the previous tables in mixed-entry format.

It is difficult to see why it took so long for industry to adopt the decision table approach. One reason was that its advocates were too insistent that it was superior to the flow diagram. Its most appropriate use was not as a substitute, but as a supplement for flow diagrams for portraying logic when many alternatives existed.

ADS

One of the important developments of the second generation was ADS. Unfortunately the acronym did not convey what the technique accomplished. ADS was an abbreviation for accurately defined system. It was developed by NCR and was used internally for several years before a manual was published for use by customers in 1968. ADS was an improvement over prior techniques because it provided a well-organized and correlated approach to system definition and specification.

ADS used five interrelated forms to provide the system (application) definition, shown in Figure 7. The process began with the definition of

CREDIT ORDER APPROVAL PROCEDURE				
		R1	R2	R3
C1	CREDIT LIMIT OKAY	Y	N	N
C2	PAY EXPERIENCE	–	FAVORABLE	UNFAVORABLE
A1	CREDIT ACTION	APPROVE	APPROVE	DON'T APPROVE
A2	RETURN TO SALES	–	–	X

FIGURE 6 Mixed entry table.

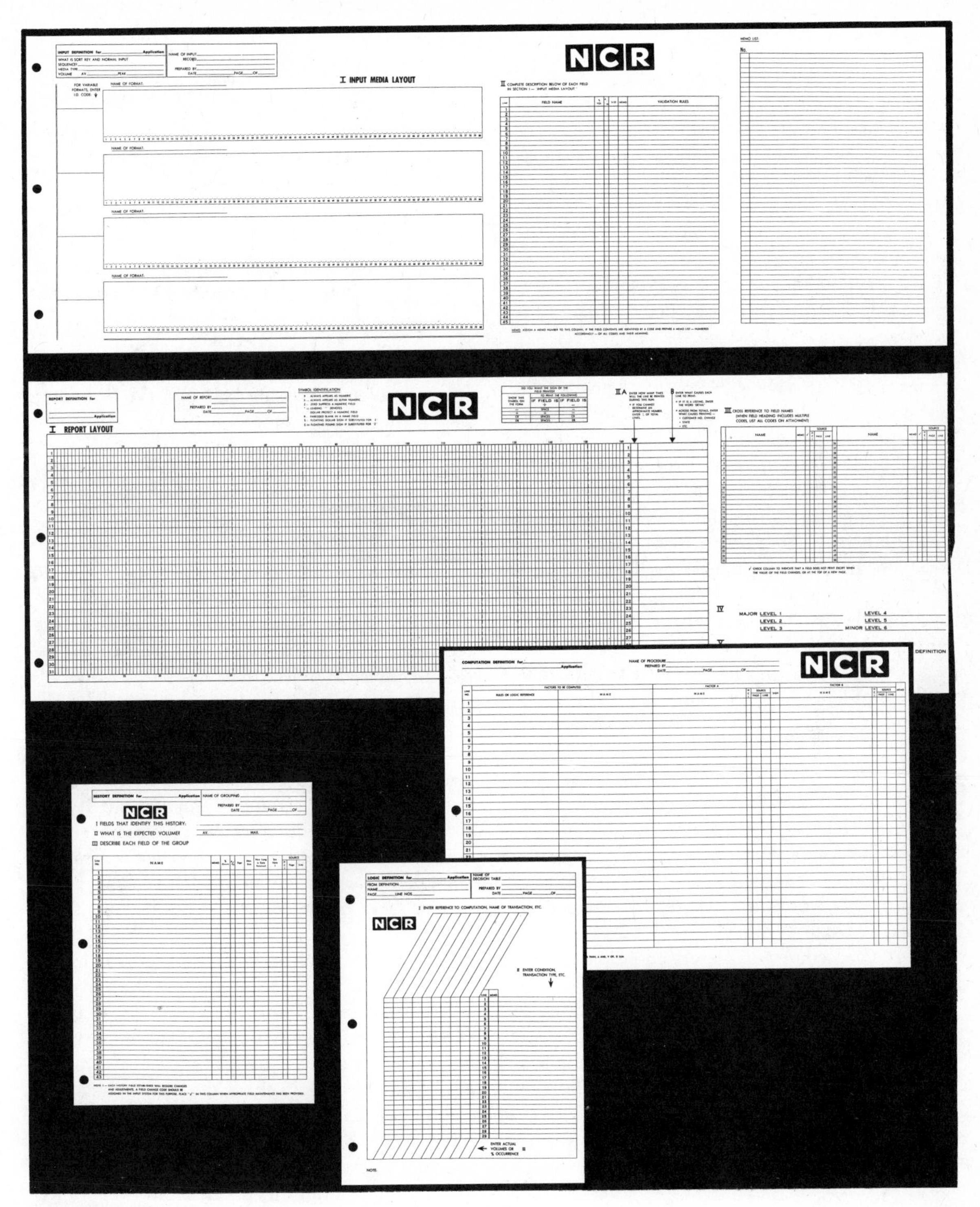

FIGURE 7 ADS interrelated forms. (Courtesy NCR.)

output. Next, inputs were defined—on the second form. The third form provided the definition of computations to be performed and the rules of logic governing the computation. The interrelationships of computations were also defined on this form, as were the sources of information used in the computation. The fourth form, the history definition, specified information to be retained beyond the processing cycle for subsequent use. The fifth form provided the logic definitions.

Within ADS, information linkage was accomplished in two ways. First, each data element was assigned to specific tag or reference. Next, each time the tag was used in the system, it was linked back to the previous link in the chain. All elements of data were chained from input to output, accomplished through the use of page and line numbers. Figure 8 shows these linkages.

The process of chaining facilitated identification of omissions and contradictions in the system. The final form used for ADS system definition dealt with logic operations.

Before reviewing the form itself, it is valuable to understand the relationship of logic operations to other segments of an ADS-defined system.

As shown in Figure 9, logic was not an integral part of the ADS report generation cycle. Rather, it was a supplemental definition that supported the computation function or any of the other forms where logical rules governed selection. In other words, a decision table was used to portray the logic of the system.

The procedure was closely coordinated with the next phase of development, physical system design. Once the information requirements were established, the system design phase determined the appropriate hardware mix to effect the system. It was also integrated with the following phase, programming. The two outputs were labeled A-SPEC and B-SPEC, described below.

A-SPEC. System description relating to run flow, hardware approach, timing estimates, uniform descriptions, and specifications of planned systems tests. All transactions and narrative of responsibilities and an overall function flowchart were prepared for each program specified. Also included were a detailed estimate of implementation in man-months and programming assignments.

B-SPEC. Macro flowcharts of each program, file labels and layouts, and individual test cases used in testing.

Although NCR did not provide the means for computer processing of ADS, the technique paved the way for such use through its systematic approach to system definition.

GRIDCHARTING

Gridcharting was developed to facilitate representation of the relationships between two groups of system elements. In this context, the term *system element* meant any defined operation or physical object that was part of the system.

Even when the number of system elements involved was high, representation by means of a gridchart remained clear. In addition, it was amenable to analysis and manipulation. It was thus a valuable tool for use in circumstances where it was necessary to consider rearrangement, or reduction in the number of system elements. Gridcharting served as a useful design check.

The manipulation of gridcharts was useful for obtaining an explicit set of relationships which was implicit but not obvious. It was applicable only when the two sets of system elements related by the second chart, say Y and Z, were in a continuing relationship of the "hierarchical" or "continuous inclusion" type. This was true of the relationship between components (i.e., subassemblies versus assemblies) and also for such groupings as data elements, files and functions, because functions access data elements via files.

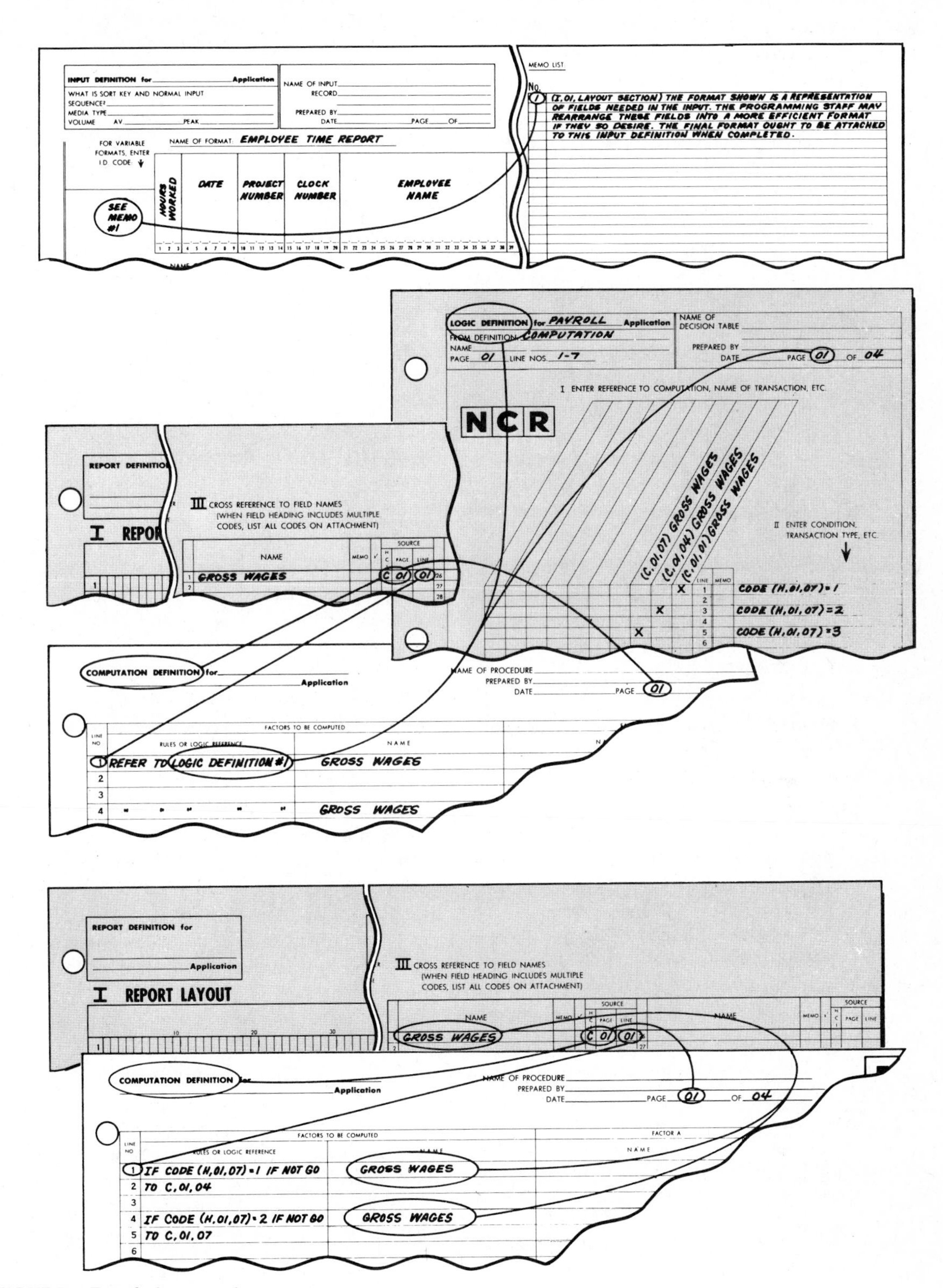

FIGURE 8 Data linkages with ADS.

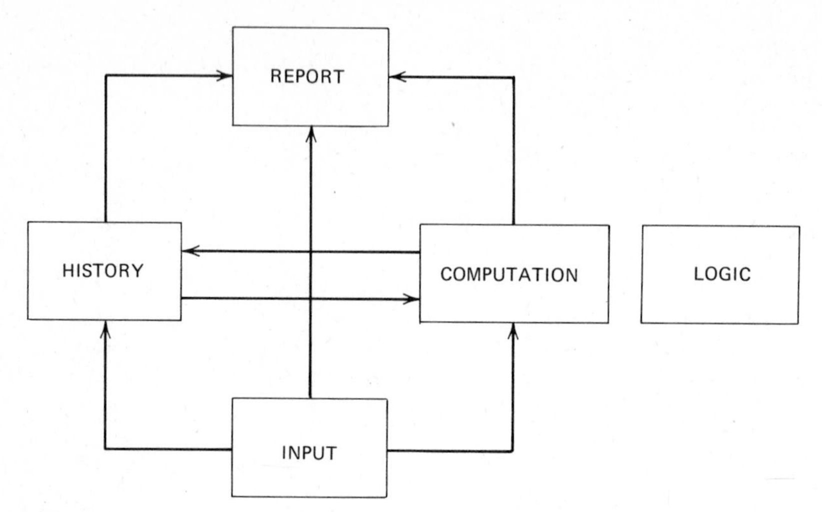

FIGURE 9 Report generation cycle.

In groupings with this type of relationship, a gridchart relating the first two sets of system elements could be combined with one relating the last two, to eliminate the middle set of system elements, thus providing a gridchart relating the first and last sets.

An example is shown in Figure 10. A chart of data elements against files was combined with a chart of files against functions to produce a chart of data elements against functions.

It would be physically possible to combine by multiplication the chart of data elements against functions with that of functions against files; but because in this case the ordering of the sets of system elements would not be correct, the resulting chart would not represent the relationship between files and data elements. Indeed, the result would, in the context of systems analysis, be meaningless.

The hypothetical system dealt with in this set of charts comprised the material requirements calculation, the production planning, the external order planning, and the internal inspection with respect to the manufacture of assembly "A", which is constructed from components "P" and "Q". Component "P" is purchased, and component "Q" is manufactured in plant from material "R".

It will be noted that the set of files dealt with was oriented more to a "paper" system than to an automated one. This was done in order to keep the example clear and simple.

When a one appeared in the chart of functions against data elements, it was implied that the data element in the row concerned was accessible to the function in the associated column via the proposed file design. If a zero appeared where access to a data element was required, the file design must be modified accordingly. Inspection, particularly in a case where more than one element is missing, often revealed the most economical modification.

If a figure greater than one appeared, redundancy might be indicated. In the example given, the figures greater than one did not indicate redundancy so long as the number of files and their named purposes remained as shown; they appeared only in the rows of those data elements. A group of three ordered sets of system elements meant that the technique was in general more applicable to design checking than to design itself. To "divide" one gridchart by another was not possible. However, first approximations to design requirements were often possible by means of a critical examination of the appropriate gridcharts. An examination of the functions against

		files					functions			
		stock	article	capacity	customer order	external order	mat. reg. calculation	production planning	extonal order planning	goods inward inspection
data elements	code number of assembly "A"	1	1	1	1	0	3	4	1	0
	number of "A's" in stock	1	0	0	0	0	1	1	1	0
	number of "A's" on order	0	0	0	1	0	1	1	0	0
	code number of component "P"	1	1	0	0	0	2	2	1	0
	number of "P's" in stock	1	0	0	0	0	1	1	1	0
	number of "P's" per "A"	0	1	0	0	0	1	1	0	0
	code number of component "Q"	1	1	0	0	1	3	3	2	1
	number of "Q's" in stock	1	0	0	0	0	1	1	1	0
	number of "Q's" on order	0	0	0	0	1	1	1	1	1
	number of "Q's" per "A"	0	1	0	0	0	1	1	0	0
	code number of material "R"	1	0	0	0	1	2	2	2	1
	quantity of "R" in stock	1	0	0	0	0	1	1	1	0
	quantity of "R" on order	0	0	0	0	1	1	1	1	1
	quantity of "R" per "P"	0	1	0	0	0	1	1	0	0
	reorder level for "Q"	0	0	0	0	1	1	1	1	1
	reorder level for "R"	0	0	0	0	1	1	1	1	1
	machine group (MG) code	0	0	1	0	0	0	1	0	0
	total capacity per MG	0	0	1	0	0	0	1	0	0
	committed capacity per MG	0	0	1	0	0	0	1	0	0
	capacity required per "A"	0	0	1	0	0	0	1	0	0
functions	material requirements calculation	1	1	0	1	1				
	production planning	1	1	1	1	1				
	external order planning	1	0	0	0	1				
	goods inward inspection	0	0	0	0	1				

FIGURE 10 Gridchart, relating data to files to functions for a manufacturing application.

data elements and functions against files charts of the above example, entry by entry, would not permit reconstruction of the files against data elements chart; but it would give a valid, although not necessarily the most economical, version of it. In complex cases, however, the version thus found could be of considerable value in that it offered a workable starting point.

Gridcharting is an example of specialized techniques developed to handle problems unique to computerization of systems.

BISAD

Honeywell developed BISAD to facilitate analysis of subsystems that were to be integrated into a computer-based system. BISAD represented Business Information Systems Analysis and Design. In BISAD tasks were organized into six logical steps which could be applied to obtain the solution to most information problems.

1. The first step, background analysis, was the foundation of accurate problem definition. The analyst gathered information, classified it into logical groups, and developed knowledge of the study area.
2. Functional analysis divided the total business into logical groups of task centers, allowing the analyst to study each segment of the business as a unit, describing the information requirements of each function, and developing the relationship of each segment to the other areas of the total business.
3. System design prototype developed a model of the business system by progressing from *what* must be done to *how* it must be done. The model was developed for each function and then the individual models were merged

to produce the total system. The development of the prototype system was to be accomplished in minimum time. The result was a picture of the system hierarchy described in business terms.

4. After the business system model was approved, the selection of a priority area could be made as the particular situation dictated. The analyst was then concerned with the design detail necessary to convert the prototype to a working system, involving both media and sequence of activities.
5. Once the working system was developed, BISAD stressed the importance of planning to make the system operational. Utilizing such aids as PERT (Program Evaluation and Review Technique), the analyst developed a plan and the controls necessary for the successful implementation of the system. Representing the system specifications, documentation gathered during the study was then presented for the final approval of management.
6. The analyst examined each of the activities; and when all the activities were completed, the functional model was formed as shown in Figure 11.

Having developed the functional model as described above, the analyst checked the model to see that it was complete.

The functional model represented graphically the activities of a data processing system. From the model, one could see the input to an activity, the files used, and the output generated. What was not shown was the way the input and the files were used within the activity to produce the output. Thus, the procedure required that a complete descriptive narrative accompany each flow diagram, including the functional model. However, there was another document that helped the analyst graphically represent the prototype. This was the information matrix. Figure 12 represents an information matrix for a sales order processing function.

The information matrix identifies five connections between the input, the files, and the output of the functional model. The matrix identified the internal or external source of the input to the model (generator functions). By following the natural flow of information, the input within the model could be shown as updating the master files used in the model or generating output from the model, or it could be used in conjunction with a master file to generate output. The output could then be shown as being used by a particular function within the company or by a function outside the company. These relations were identified by a check mark in the appropriate grid location on the matrix. When complete, the matrix was used to trace the flow of information through the activities.

Honeywell published the BISAD manual the same year that NCR published its ADS manual. The techniques were similar, but different enough for each to gain its own set of advocates.

STUDY ORGANIZATION PLAN

As computer use became more pervasive, new tools were required for Phase I of the system development cycle. Documenting only one system as a basis for improvement analysis took little time. Rather informal documentation procedures were used. But when several systems were to be integrated, a more thorough documentation approach was required.

IBM developed SOP (Study Organization Plan) for this purpose. ADS and BISAD concentrated on specifying system requirements, that is, the final part of Phase II of the system development cycle. They presumed completion of Phase I, the study of the organization and its information needs.

Several approaches were developed for Phase I. Philips, the Netherlands-based company, produced ARDI (Analysis, Requirements Determination, Design and Development, Implementation and Evaluation) through the work of

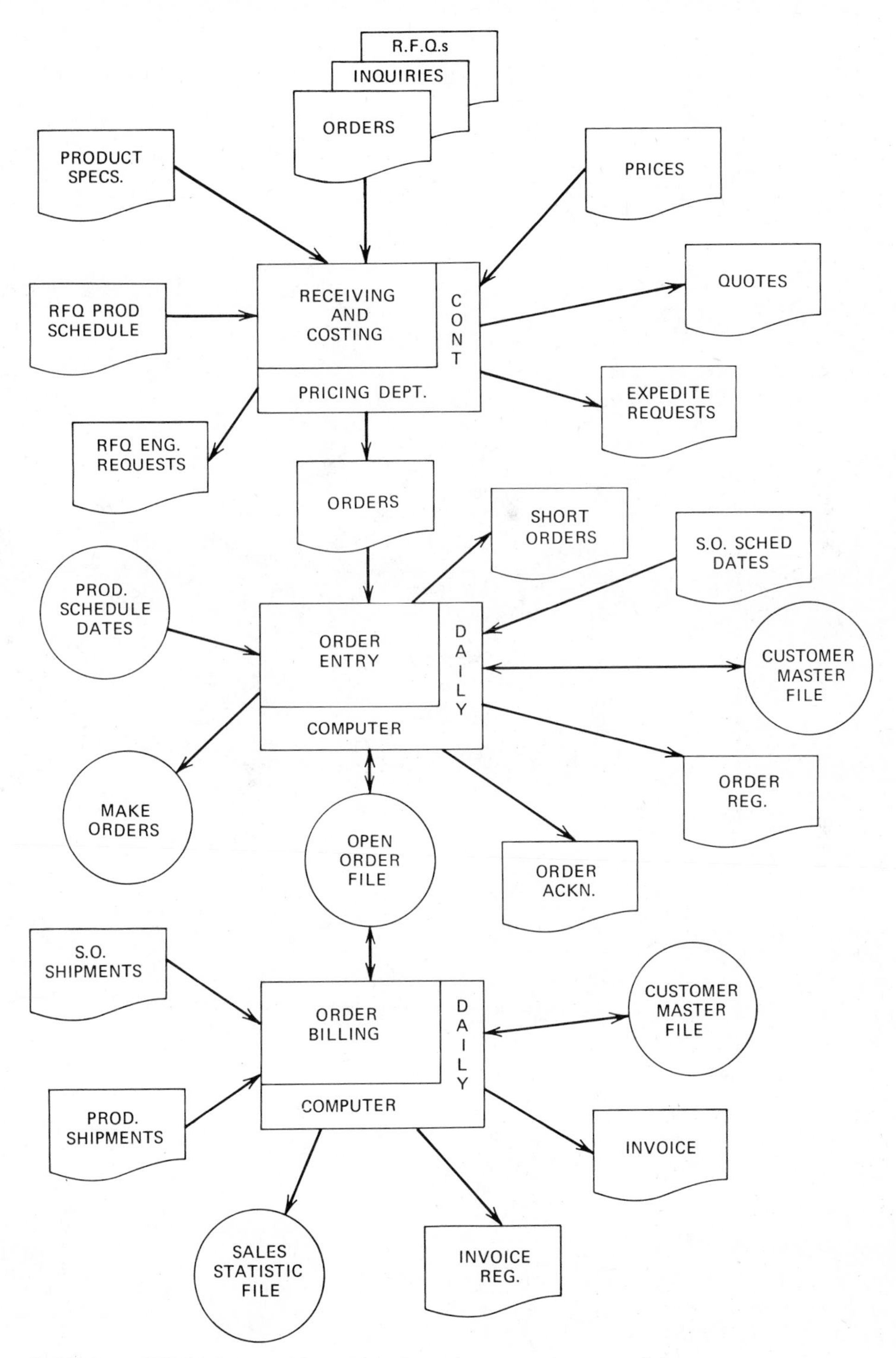

FIGURE 11 BISAD functional model (sales order processing example).

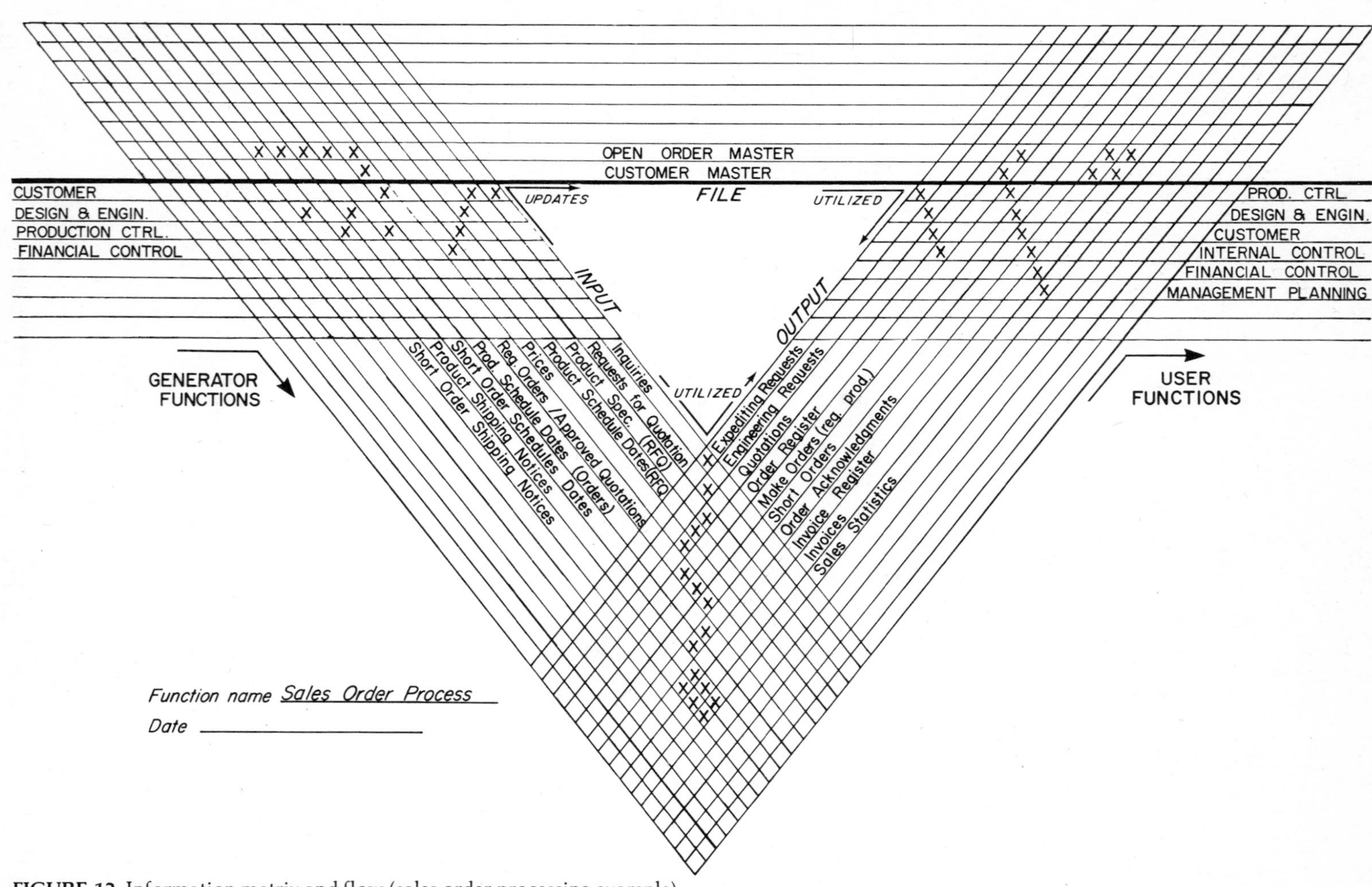

FIGURE 12 Information matrix and flow (sales order processing example).

Hartman, Matthes, and Proeme. IBM produced SOP (Study Organization Plan), through the work of Burton, Grad, Holstein, Meyers, and Schmidt.

ARDI is not included in our descriptions because it was a handbook of techniques. SOP was a more significant contribution to the field because it pulled various techniques together into an integrated approach. The fact of its integrative quality is evidenced by the three SOP components shown in Figure 13.

SOP was designed to gather data with which to analyze the information needs of the entire organization. Information was gathered and organized into a report containing three sections, shown on the left-hand side of Figure 13.

The *General* section included a history of the enterprise, industry backgrounds, goals and objectives, major policies and practices, and government regulations.

The *Structural* section contained a schematic model of the business, describing it in terms of products and markets, materials and suppliers, finances, personnel, facilities, inventories, and information.

The *Operational* section included flow diagrams and a distribution of total resources to represent the operating activities of the business. These charts showed how the resources of a business respond to inputs, perform operations, and produce outputs.

The appendix included the detailed working documents needed to explain operations, identify documents, and define the files in which the organization's information was stored. These documents were organized into four levels of hierarchy, depicted on the right-hand side of Figure 13.

The organizational structure comprising an activity, and a cost analysis of the activity, were recorded on the Resource Usage Sheet; and the flow of the activity itself was displayed on the Activity Sheet. Operations within the activity are further described on Operation Sheets, with detailed information inputs and outputs and information resources provided on Message and File Sheets, respectively. These forms permitted the analyst to describe the flow of incoming materials or information through the internal workings of the system to output products, services, or information.

The Resource Usage Sheet fit each system under study into its larger context. It showed the organization and structure of the business environment. It also provided a rapid analysis of costs for the organizational components being surveyed and showed the cost impact for each activity or system.

The Activity Sheet traced the flow of a single activity, breaking it down into its major operations. Each Activity Sheet presented as large a group of related operations as could be handled conveniently. It included a flow diagram of the activity with individual blocks representing various operations (each of which could be more minutely described on Operation Sheets). Key characteristics such as volumes and times were recorded in tabular form.

The Resource Usage Sheet and Activity Sheet worked together to provide a quick look at a business system.

For a closer look at the operation of a system, the analyst used forms that permit a more detailed documentation: The Operation Sheet and the Message and File Sheets.

Usually an Operation Sheet existed for each block on the Activity Sheet. It was used for recording the related processing steps that form a logical operation. It described what was done, with what resources, under what conditions, how often, to produce what specific results. Its primary purpose was to show relationships among inputs, processes, resources, and outputs.

The Message Sheet was one of two forms that supported the Operation Sheet. It described all inputs and outputs.

The File Sheet described a collection of messages, an information file. It identified the stored information the operation utilized. When describing what was done (on the Operation Sheet), the analyst could choose from several

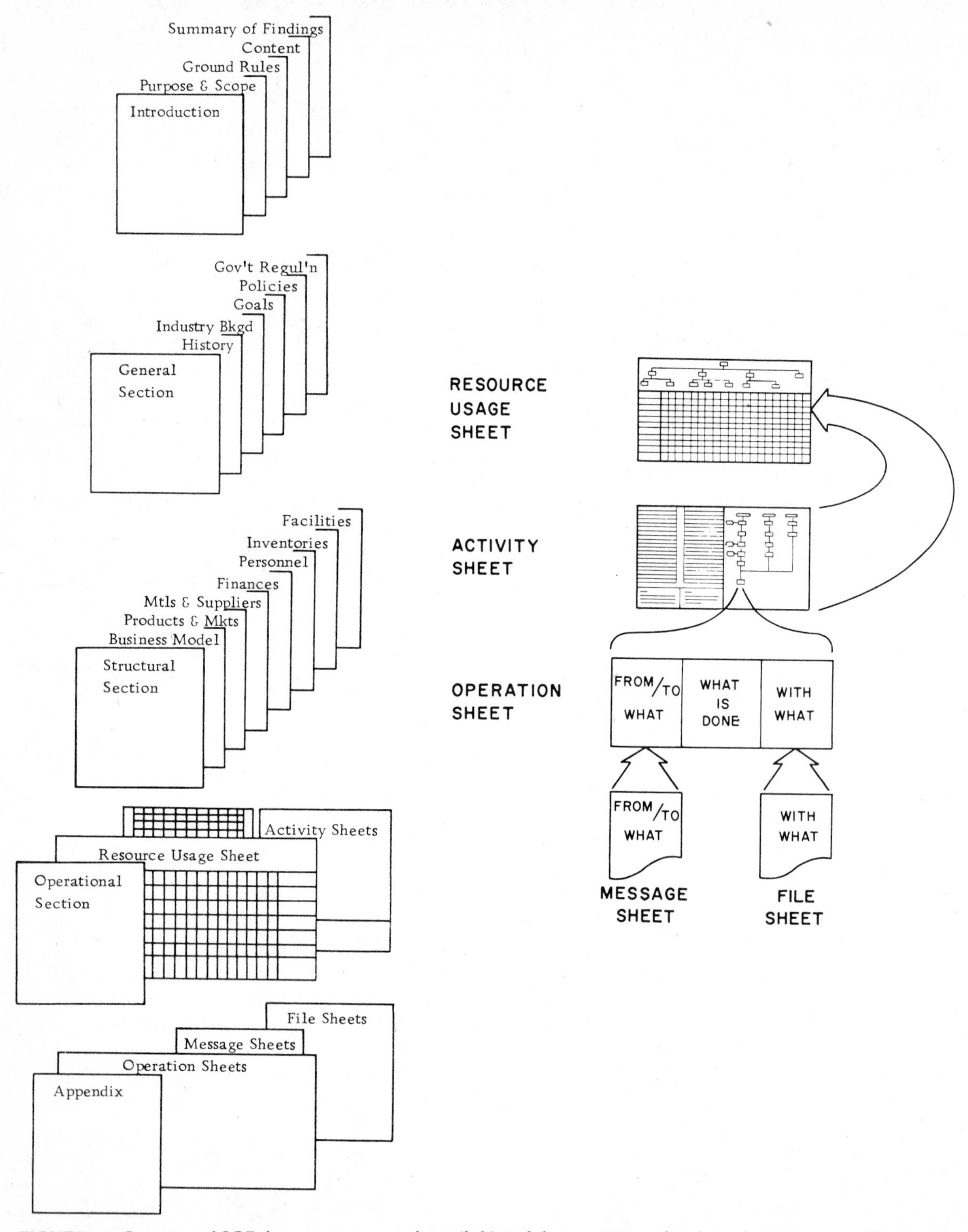

FIGURE 13 Contents of SOP documentation package (left) and documentation levels (right). (Courtesy IBM.)

levels of detail by exercising a choice of descriptive verbs.

It is readily seen that the comprehensiveness of SOP, and the interrelated documentation techniques, enhanced the possibilities for integration of systems.

SUMMARY

Second generation techniques were an order of magnitude better than the previous generation. The significant difference was their ability to handle large systems and to facilitate analysis for integrating systems.

Unfortunately, not many firms recognized the need for, and the advantages of, integrating systems. A relatively small number of computer departments utilized these techniques. System analysts continued to treat systems as though they were *in*dependent of other systems, when it was a rare situation where *inter*dependence was not the case. System analysts could be forgiven for this shortsightedness if there were no tools available for designing integrated systems. The description of tools provided in this section clearly demonstrates their capability for analysis and design for integration. Regrettably, another 5 to 10 years would pass before these tools were widely adopted.

SECTION QUESTIONS

1. What are the major differences between decision tables and the system flowchart? What are the similarities? Where might they be used in conjunction?
2. Examine the figures illustrating system flowcharts and decision tables. Why would DT be used for user documentation as well as program documentation?
3. ADS utilizes which of the prior development techniques? Explain.
4. Why is ADS superior to previous techniques? Give a comprehensive answer considering all of the ADS functions.
5. Gridcharting is a special-purpose technique. What is that purpose?
6. Where is gridcharting useful?
7. Examine Figure 10 and explain what the numbers refer to.
8. BISAD utilizes which of the prior development techniques? Explain.
9. What are the similarities and differences in ADS and BISAD?
10. Refer back to Figure 1 of the paper "Evolution of System Development Techniques." Identify which phase (or phases) were performed by each of the system development techniques discussed in Section 2. (Figure 1 did not have space for gridcharts; include them.)
11. What information is provided with the SOP technique that was not available with use of first generation techniques?
12. Develop a table comparing the information provided by SOP, BISAD, and ADS.
13. How would the total SOP package be of value in a presentation to management on a proposed approach to the M.I.S.? The answer is implied in this section, but needs some careful thought in order to provide a comprehensive answer.
14. Assume you work for a third computer manufacturer and are assigned the task of developing BEST, an approach that utilizes the best techniques from BISAD and SOP. Describe your version of BEST.

SECTION 3

THIRD GENERATION DEVELOPMENT TECHNIQUES FOR COMPUTER-BASED SYSTEMS

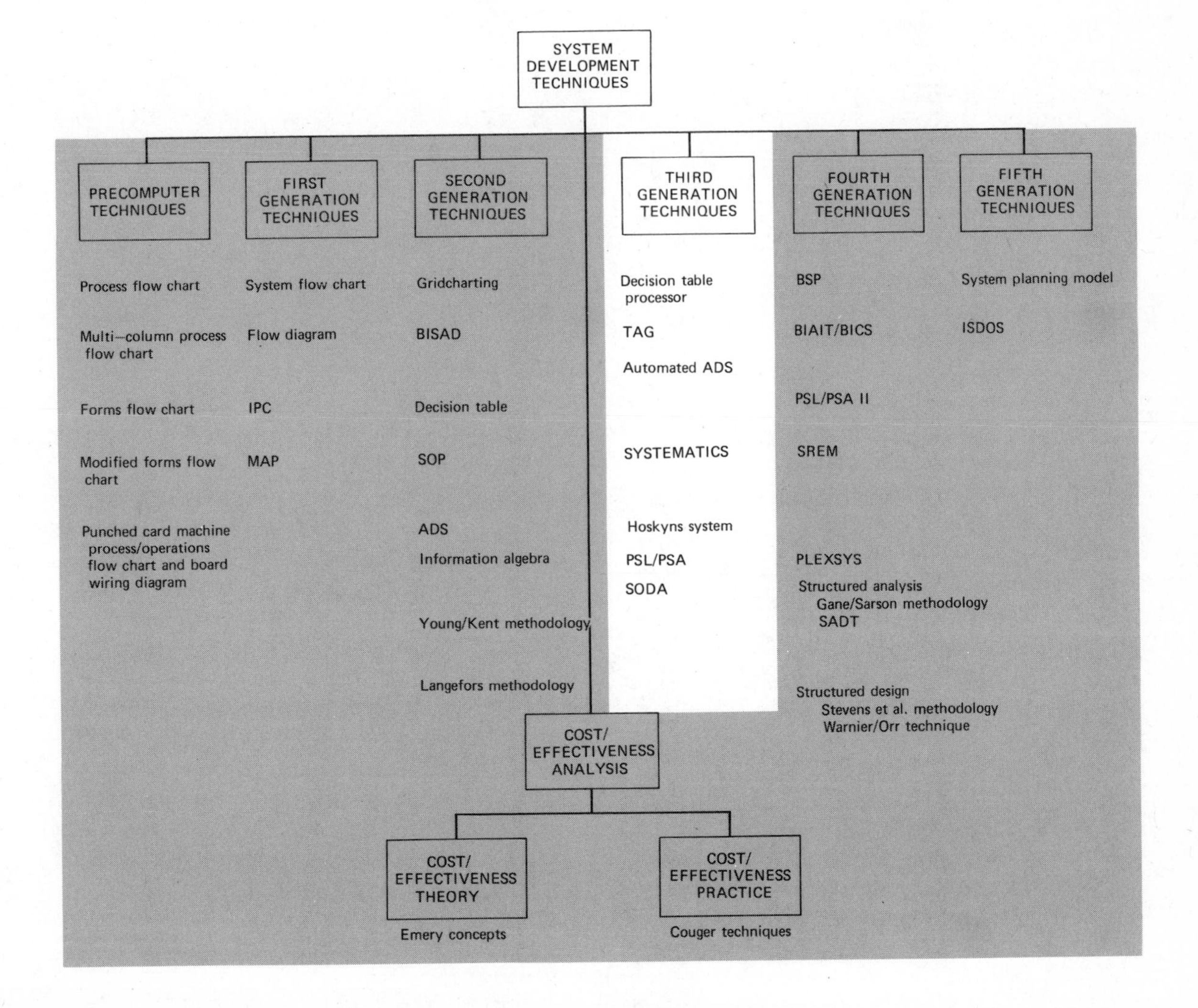

SECTION 3

THIRD GENERATION DEVELOPMENT TECHNIQUES FOR COMPUTER-BASED SYSTEMS

J. Daniel Couger

OVERVIEW

The major distinction of the third generation of system development techniques was the emphasis on computer-aided techniques. The scope and complexity of systems to be computerized had grown so much that analysts needed the power of the computer to assist in the development process.

Two parallel research efforts occurred. One concentrated on automating manual system analysis/design techniques. The second concentrated on building a theoretical basis for automation of systems development methodology. The results of these two activities were utilized by a third group during the latter part of this era. This group concentrated on merging the earlier developments into a optimal set of computer-aided techniques. These three activities will be described in this section.

Papers on data base specifications and data base design techniques are excluded from this book because they are covered, in most universities, in the data base/data communication course rather than the system analysis course. The bibliography provides references on these techniques for persons who wish to pursue this subject.

DECISION TABLE PROCESSORS

The logical approach inherent in a decision table made this technique readily adaptable for computer processing. In May 1959, CODASYL began a project to enable computer processing of decision tables. In September 1962, the product of this effort, DETAB-X, was made public. It consisted of a language supplement to COBOL-61, for use within the framework of decision tables. Little use was made of the product, however. In 1965 the Special Interest Group for Programming Languages (SIGPLAN) of the Los

Angeles Chapter of the Association for Computing Machinery appointed a working group to develop a decision table processor. Written in a subset of COBOL, the processor (DETAB/65) accepted decision tables coded in COBOL and converted them to COBOL source code. It was implemented on a variety of CDC and IBM computers.

However, the conversion algorithm was inefficient and the technique was not widely used until the next generation of processors evolved. The next approach consisted of a preprocessor, written in COBOL, which converted decision tables containing COBOL components to code acceptable to the COBOL compiler. A few processors were written in FORTRAN.

This use of the computer did not directly facilitate the work of the system analyst, but it speeded up the total development process and facilitated modification of the system.

AUTOMATED ADS

Although ADS provided a self-organized approach to cross-referencing system definition documents, modifications to the system were laborious to incorporate. Also, it was difficult to assess the consequences of a modification because ADS was based on backward referencing (i.e., from output back through intermediate processing to input). The logical sequence in designing a system is forward referencing.

In automating ADS, the documentation medium was punched cards instead of paper forms. Each line of the original ADS forms was represented by one to three punched cards. The card file was analyzed by programs which (1) checked for consistency and completeness, (2) checked back references, (3) generated back references that had been omitted, (4) produced a dictionary of names, (5) generated incidence and precedence matrices, (6) flagged errors, and (7) produced diagnostics.

The inevitable system modifications were readily incorporated. Cards for individual lines in the ADS definition were revised and replaced and the effects assessed by computer analysis. The automated dictionary gave forward as well as backward referencing to aid the systems analyst in redesigning the system instead of serving primarily as a documentation tool. Incidence and precedence matrices permitted analysis of the structure of the system and facilitated planning of implementation.

TAG (TIME AUTOMATED GRID)

Developed in 1962 by D. H. Meyers of IBM's System Research Institute, TAG was automated in 1966. To use the Time Automated Grid system, the analyst first recorded the system output requirements. Inputs were examined during later iterations of the program.

Once the output data requirements were fed into the system, TAG worked backward to determine what inputs were necessary and at what point in time. The result was the definition of the minimum data base for the system. With the aid of the reports generated by TAG, the analyst systematically resolved the question of how the required inputs were to be entered into the data flow. This approach enabled the analyst to concentrate on pertinent input elements and to bring them into the system at the proper place. Superfluous or repetitious data were identified and eliminated from the system. Discrepancies in the use of any data element were corrected.

When both inputs and outputs were defined to TAG, the next iteration of the program produced file format and systems flow descriptions. File contents and data flow were both based upon time—the time at which data elements entered the system and the time at which they were required to produce output.

To TAG, the elapsed time between these two moments created the need for files. The files defined by TAG indicated what data must be available in each time period to enable the system to function. The job definition depicted the flow of these files, as well as of the inputs and outputs, within and between time cycles. This approach provided an overview of the system, showing the interrelationship of all data in the system.

Knowing these interrelationships made it possible for the system analyst to determine whether the outputs desired were quickly and easily obtained, and thus economically justified. With knowledge of the availability of data elements in given time periods, the analyst determined where additional useful outputs might be obtained.

The upper portion of Figure 1 shows the principal form used for TAG. The form was divided into two horizontal sections, one for requirement titles, the other for data names. The characteristics of the input, output, or file being described were recorded in the requirements title section. Comments on the data requirements of the input, output, or file were detailed in the data name section.

The output of the TAG system was a series of 10 reports that documented input, analyzed data requirements, and provided file and dataflow definition. The key report was the timegrid analysis, which traced the appearance of each data element, by time, through all the requirements in the system (shown in the lower portion of Figure 2). The grid indicated those data elements that must be carried in files, enabling the analyst to identify the minimum data base requirements.

The other nine reports were: time/key analysis, user data, glossary of data names, document analysis, sorted list of data names, summary of unresolved conditions, serial file records, direct access records, and job definition.

The development of this semiautomated technique was significant in the evolution of system analysis techniques.

THEORETICAL BASIS FOR COMPUTER-AIDED SYSTEM ANALYSIS

The automated manual system analysis techniques proved useful (and are still being used). As could be expected, however, automating existing techniques proved to be a workable, but *suboptimal* approach. Several important theoretical concepts were developed during the same era, providing the foundation for the optimized techniques in later generations of system development methodology.

The key theory papers were included in our previous book, *System Analysis Techniques*, and it is not feasible to reprint that material here.

The landmark publications in our earlier book were:

1. "Information Algebra," CODASYL Development Committee
2. "Abstract Formulation of Data Processing Problems," John W. Young, Jr. and Henry K. Kent
3. "SYSTEMATICS—A non-Programming Language for Designing and Specifying Commercial Systems for Computers" and "The Use of Decision Tables Within SYSTEMATICS," C. B. B. Grindley
4. "Some Approaches to the Theory of Information Systems," Borje Langefors

Another paper in our previous book was "Problem Statement Languages in MIS," by Daniel Techroew. Again, space limitations prevent reprinting that paper. We will use portions of it as a means of showing the synthesis of concepts and techniques to provide the basis for the theory underlying computer-aided system development methodology. Figure 3 identifies the objectives of the techniques. In addition to the abbreviations for ADS and TAG, the other abbreviations represent IA (Information Algebra), YK (Young-Kent methodology), and LA (Langefors methodology).

IBM TIME AUTOMATED GRID TECHNIQUE (TAG) INPUT/OUTPUT ANALYSIS FORM PAGE 1 OF 2

DATE FEB 1

PREPARED BY

DATA TYPE	CODE	FREQUENCY	PERIOD	PRIORITY	PROGRAM SEQUENCE	REQUIREMENT TITLE	PEAK (P) VOLUME	PEAK (P) FREQUENCY	PEAK (P) SURVEY PERIOD	AVERAGE (A) VOLUME	AVERAGE (A) FREQUENCY	AVERAGE (A) SURVEY PERIOD	MINIMUM (M) VOLUME	MINIMUM (M) FREQUENCY	MINIMUM (M) SURVEY PERIOD	DESIGNED FOR
O		1	D	2		INVOICE	10300	1	W	9700	1	W	9200	1	W	P

COMMENT	DATA NAME	SIZE	A/N	CLASS	USE	RATIO	SEQUENCE	FORMAT	REMARKS	SIGNAL
	CUSTOMER-ORDER-NO	6	N	F	I	1		2		
	CUSTOMER-NAME	25	A	F	I	1		4		
	CUSTR-SHIP-TO-ADDRESS	75	A	F	I	1		6		
	OUR-ORDER-NO	6	N	F	F	1	1	1		
	PART-CODE-NO	6	N	F	F	5	2	10		
	PART-SIZE-AND-BRAND	3	N	F	F	5	3	11		
	PART-NAME	10	A	F	I	5		12		
	QUANTITY-ORDERED	3	N	V	F	5		7		
	QTY-SHIPPED	3	N	V	F	5		8		
	QTY-OUT-OF-STOCK	3	N	V	R	1		9		
C	SUBTRACT QTY-SHIPPED FROM QUANTITY-ORDERED GIVING QTY-OUT-OF-STOCK									
	INVOICE-NO	8	N	F	I	1		3		
	SOLD-TO-ADDRESS	75	A	F	I	1		5		
	PRICE	5	N	V	F	5		13	3, ZZ.999	
	LINE-EXT	6	N	V	R	5		14	2, ZZZZ.99	
C	COMPUTE LINE-EXT ROUNDED = QTY-SHIPPED * PRICE									

DATA TYPE (Column 1) = INPUT-I, OUTPUT-O, FILE-F

X20-1779-0 U/M 25

IBM TIME AUTOMATED GRID TECHNIQUE (TAG) INPUT/OUTPUT ANALYSIS FORM PAGE 2 OF 2

DATE FEB 1

PREPARED BY

DATA TYPE	CODE	FREQUENCY	PERIOD	PRIORITY	PROGRAM SEQUENCE	REQUIREMENT TITLE	PEAK (P) VOLUME	PEAK (P) FREQUENCY	PEAK (P) SURVEY PERIOD	AVERAGE (A) VOLUME	AVERAGE (A) FREQUENCY	AVERAGE (A) SURVEY PERIOD	MINIMUM (M) VOLUME	MINIMUM (M) FREQUENCY	MINIMUM (M) SURVEY PERIOD	DESIGNED FOR

COMMENT	DATA NAME	SIZE	A/N	CLASS	USE	RATIO	SEQUENCE	FORMAT	REMARKS	SIGNAL
	DISCOUNT-AMT	5	N	V	R	1		16	2, ZZZ.99	
C	SUM OF LINE-EXT * DISC-RATE = DISCOUNT-AMT									
C	IF SUM OF LINE-EXT GREATER THAN DISC-QUALIFICATION-AMT MOVE									
C	DISC-RATE-1 TO DISC-RATE ELSE MOVE DISC-RATE-2 TO DISC-RATE									
	DISC-RATE-1	2	N	V	F	1				
	DISC-RATE-2	2	N	V	F	1				
	DISC-QUALIFICATION-AMT	5	N	F	F	1				
	DISC-RATE	2	N	V	R	1		15	2, V99	
	COD-OR-CREDIT-CODE	1	N	F	F	1		18		
C	BLANK, 7, 8, OR 9 ARE INVALID									
	TOTAL-INVOICE-AMOUNT	8	N	V	R	1		17	2, $$$$, $$9.99	
C	SUM OF LINE-EXT MINUS DISCOUNT-AMT									

DATA TYPE (Column 1) = INPUT-I, OUTPUT-O, FILE-F

X20-1779-0 U/M 25

FIGURE 1 TAG input form. (Courtesy IBM.)

```
RESULTS OF ANALYSIS BY TIME-GRID TECHNIQUE



DATA                                                      DATA
NUMBER DATA NAME                   SIZE A/N USE           NUMBER DATA NAME                    SIZE A/N USE
    21 QUANTITY-ORDERED               3 N    FI               22 SHIPPING-INSTRUCTIONS         100 A    FI
    23 SOLD-TO-ADDRESS               75 A    FI               24 TOTAL-INVOICE-AMOUNT            8 N    VR


DATA NUMBER                          21   22   23   24

CYCLE

  1 CUSTOMER-ORDER                    5    1    1    0
    (      1) 8300...    1 X D
  1 WAREHOUSE-ORDER                   5    1    0    0
    (      2) 8300...    1 X D
  2 INVOICE                           5    0    1    1
    (      3)10300...    1 X D

SUMMARY CODES                         1    0    1    2


MEANING OF SUMMARY CODES

0 - RATIO OF INPUT = RATIO OF OUTPUT, INPUT AVAILABLE AT TIME OF OUTPUT
1 - PLURAL CYCLES - FILES
2 - SYSTEM GENERATED (VARIABLE RESULT)
3 - NO INPUT BUT OUTPUT, NOT VARIABLE RESULT
4 - NO OUTPUT BUT INPUT
5 - RATIOS NOT EQUAL
6 - OUTPUT REQUIRED BEFORE INPUT IS AVAILABLE
```

DATA NO.	DATA NAME	CODE	PAGE NO.
2	CUSTOMER-NAME	15	3
5	DATE-OF-ORDER	4	3
6	DISC-QUALIFICATION-AMT	3	3
8	DISC-RATE-1	3	3
9	DISC-RATE-2	3	3
12	INVOICE-NO	3	3
13	LINE-EXT	152	3
16	PART-NAME	13	3
18	PRICE	3	3
20	QTY-SHIPPED	153	3

```
1      1 I CUSTOMER-ORDER                8300
                   14    0    0    0     0    0    0    0
1      2 O WAREHOUSE-ORDER               8300
                   14   15   17    0     0    0    0    0
2      3 O INVOICE                      10300
                   14   15   17    0     0    0    0    0
3      4 O WEEKLY-SHIPMENT-REPORT        5000
                   15   17    0    0     0    0    0    0
```

FIGURE 2 Three of 10 outputs from TAG. (Courtesy IBM.)

YK, ADS, and TAG were problem statement techniques that used a practical, straightforward approach with very little attempt to develop a theory of data processing. They consisted of a systematic way of recording the information that an analyst would gather in any case. IA was more concerned with developing a theory. It used a terminology and developed a notation that was new to most analysts. LA started with a precedence relationship among information sets (files), but did not indicate how these were obtained. This technique, therefore, was more relevant to the analysis of a problem statement and to the design of a system. However, it suggested some desirable features of a problem statement technique. These approaches, on the surface, appear to be very different; but upon detailed examination, have some major similarities.

Objectives	IA	LA	YK	ADS	TAG
Nonprocedural	x			x	x
Abstract formulation		x	x		
Reduce work of systems analyst	x				x
Machine independent	x	x	x	x	x
Toward overall systems optimizing	x		x		
Manipulation of PS	x		x		
Machine readable PS			x		x
Maximize use of creative personnel					x
Standardization and documentation					x

FIGURE 3 Objectives of techniques.

All five approaches assume that the problem statement starts at output. What is required output? Therefore, a necessary part of the problem statement was the description of the desired output. As will be seen later, this requirement was implied by LA in the definition of TERMINAL SETS* and in YK by the definition of output DOCUMENTS. IA did not mention required output as such and, in fact, in the example given in the paper says, "The problem is to create a new pay file from . . ."

Data Relationships. A problem statement required some description about the data that would be processed. The most extensive data description facility was the one used by IA. This started with the concept of an entity that has a connotation of a physical entity in the real world such as an employee, a paycheck, or an order. Each ENTITY had PROPERTIES which described that entity; for example, an employee had an employee number, hourly rate, and so on. For any given ENTITY there was a value for each property. The PROPERTY VALUE SET was the set of all possible values that a property could have in the problem. The COORDINATE SET was the list of all PROPERTIES that appeared in the problem. A DATUM POINT was a set of values, one for each PROPERTY in the COORDINATE SET, for a particular ENTITY. The PROPERTY SPACE was the set of all DATUM POINTS, that is, all possible points obtained by taking the Cartesian product of all possible PROPERTIES. Once this PROPERTY SPACE was defined, further definitions dealt with subsets of this space. A LINE was a subset that was roughly equivalent to a record, and an AREA was a subset roughly equivalent to a file. Other subsets of the PROPERTY SPACE were BUNDLES and GLUMPS. The basic reason for this choice of data description was to use the concepts of a set theory as the formulation for a theory of data processing. (The authors of IA rejected data description by arrays as being too limited.)

In YK, the basic units of data were called ITEMS and corresponded to a PROPERTY in IA. The term INFORMATION SET was used for the set of all possible values of a particular item and was, therefore, equivalent to the PROPERTY VALUE SET in IA. The information that could

*Words that have particular meanings in any of the languages are typed in capital letters.

be provided for each INFORMATION SET was: (1) the number of possible values, (2) the number of characters or digits, and (3) relationships. The following relationships were defined:

Relationship	Description	Symbol	Graphic Symbol
Isomorphism	One-to-one correspondence		⟷
Homomorphic	Many-to-one correspondence	–	⟶
Cartesian product	P_jXP_k means a pair of P_i and P_k	X	
	Contained in	C	
Equal to		=	

The relationships were used to make statements such as: There is one employee number for each employee name and address. YK did not want to make any statements about the file structure; hence, there were no terms that corresponded to records or files. YK also provided a graphical notation for showing relationships.

In LA, there was no definition of data corresponding to data items. The problem definition started with collections of data that were called INFORMATION SETS. This corresponded roughly to the notion of a file in common terminology. LA introduced the concept of an elementary file in which each record contains a data value and enough "keys" to identify it uniquely.

ADS provided three forms on which data was described: reports, inputs, and history. Each of these forms provided space for some information describing the particular report or input: name, media, volume, sequence, and space for each variable. For each variable the forms provided space for name, how the value of the variable was obtained (INPUT, COMPUTATION, HISTORY) and a cross-reference, how often the variable appeared, and size (number of characters).

TAG provided one form that contained space for data describing the document (or file) and space for each variable. Figure 4 shows a comparison of the data required by YK, TAG, and ADS.

Processing Requirements. In standard programming languages, each program or subprogram includes statements that produce output, statements that test the conditions under which the output is produced, and statements which compute the values of the variables that appear in the output. A nonprocedural language should separate the statement of what output is to be produced when, from the statement of the procedure for producing the value of the variables that appear in the output. All five approaches followed this concept.

In IA the basic operation that the problem definer used to state processing requirements was a mapping of one subset of the PROPERTY SPACE into another subset. Two kinds of mappings were defined. One corresponded to operations within a given file. For example, suppose a tape contained time cards, sorted in order by employee number, one for each day of the week. A mapping could be defined which would take the set of (five) POINTS for each employee into one new POINT which would contain the total for the week.

The second type of mapping corresponded to the usual file maintenance operations in which POINTS from a number of input files were processed to produce new output files. These two types of mappings were called GLUMPING and BUNDLING, respectively. The actual computation of the PROPERTY VALUES of the new POINTS produced by a mapping was specified by a COORDINATE DEFINITION which must contain a computational formula for each PROPERTY in the COORDINATE SET.

FOR WHOLE DOCUMENT	YOUNG & KENT	TAG	ADS
1. NAME	✓	✓	✓
2. TYPES OF DOCUMENTS	INPUT, OUTPUT	INPUT OUTPUT FILE	INPUT REPORT HISTORY
3. WHEN PRODUCED	PRODUCING RELATIONSHIP	PERIOD (S, MI, H, D, W, MO, Q, Y) PRIORITY (NUMERIC) FREQUENCY	SELECTION RULES
4. MEDIA	NOT MENTIONED	NOT MENTIONED	FOR INPUT
5. SEQUENCING CONTROL MAJOR INTERMEDIATE MINOR	NOT MENTIONED	NOT MENTIONED	SEQUENCES MAJOR INTERMEDIATE MINOR
6. VOLUME AVERAGE (A) MINIMUM (M) PEAK (P)	 ✓ ✓ ✓	 ✓ ✓ ✓	EXPECTED VOLUME FOR HISTORY, INPUT AND REPORT
7. DESIGNED FOR P, A, M	NOT MENTIONED	DESIGNER'S CHOICE	NOT MENTIONED
8. OTHER DATA		REFERENCE AUDIT	
FOR EACH VARIABLE			
1. NAME	✓	✓	✓
2. HOW USED		FI–FIXED; INFORMATIONAL FF–FIXED; FUNCTIONAL VF–VARIABLE; FACTOR VR–VARIABLE; RESULT	MODIFIED BY –FORMULA –PARTICULAR VARIABLE HOW OFTEN? –NEVER –PARTICULAR CYCLE –FIXED TIME –LOGICAL CONDITION
3. COMPUTATION FORMULAS	DEFINING RELATIONSHIPS FOR VARIABLES ON OUTPUT REPORTS	NOT INCLUDED (MAY BE ADDED AS COMMENTS)	COMPUTATION FORM LOGIC FORM
4. SEQUENCING ROLE	NOT MENTIONED	✓	✓
5. VALIDATION RULE	NOT MENTIONED	NOT MENTIONED	✓
6. FORMAT A or N SIZE (NO. of CHAR.) FOR OUTPUT	 IN INFORMATION SET TABLE	 ✓ ✓ Ordering number of P for presence.	 ✓ ✓
7. NO. OF TIMES PER DOC. MINIMUM AVERAGE MAXIMUM	✓	RATIO	✓

FIGURE 4 Description of documents. (A checkmark denotes provision for including the information listed in the left-hand column.)

In YK the major unit of processing was a PRODUCING RELATIONSHIP; there must be one PRODUCING RELATIONSHIP for each type of output document. This PRODUCING RELATIONSHIP gave the conditions under which a document would be produced. This statement could contain conditions (Boolean expressions) that depended on values of data ITEMS or on time. For example, a PRODUCING RELATIONSHIP might be "a monthly statement is produced for a customer each month for all customers with a nonzero balance." A PRODUCING RELATIONSHIP could also state that Document D_2 was produced for each input D_1. The values of the data ITEMS that appear in the output documents were calculated using a DEFINING RELATIONSHIP. There must be one defining relationship for each data ITEM that appeared on an output document.

In LA the relationships were given for production of information sets and, hence, corresponded to PRODUCING RELATIONSHIPS. However, they were stated only as precedence relationships; for example, information sets *a, b,* and *c* were necessary to produce *d*. No functional relations were given. A problem statement could be represented by a graph as shown in Figure 5.

In ADS some basic information was specified about when reports were to be produced. In many instances, however, this was supplied by written notes. This information was analogous to the PRODUCING RELATIONSHIPS in YK. ADS required that each variable be identified as coming from input, computation, or history. A form was provided for specifying the computations; this specification was somewhat limited. Another form was used to state logical conditions and these could modify computations or input.

TAG provided for stating how often input would be produced by specifying a period. The available codes were second, minute, hour, daily, weekly, monthly, quarterly, and yearly. A priority could be assigned to distinguish a sequence ordering between two documents with the same period. TAG also provided a means for stating which data elements were to be computed, but it did not provide for stating the formula for the computation. (The formula could be included in the "Comments" section of the form, but it was not analyzed by the program.)

IA did not specify any additional information; LA assumed that the relative size of files was available. Both YK and ADS provided for specifying time and volume requirements.

YK defined two kinds of time: extrinsic (when an event occurs) and intrinsic (the time written on a document). Volumes of documents could be expressed in terms of averages over some time period. The operational requirements consisted of a volume for each document (input and output) and a time statement for each output document.

ADS permitted specification average and maximum volume in input, report, and history forms. In addition, each variable in HISTORY was characterized by how long it was to be retained; this could be a fixed number or may depend on a computation.

TAG provided for volume information for documents, size information on data elements, and repetition information on data elements within documents.

Data Description. IA was the only approach that associated data with the real world through the use of the ENTITY concept. It should be noted, however, that the IA language in itself did not depend on how the PROPERTY SPACE was obtained, that is, whether it was derived from

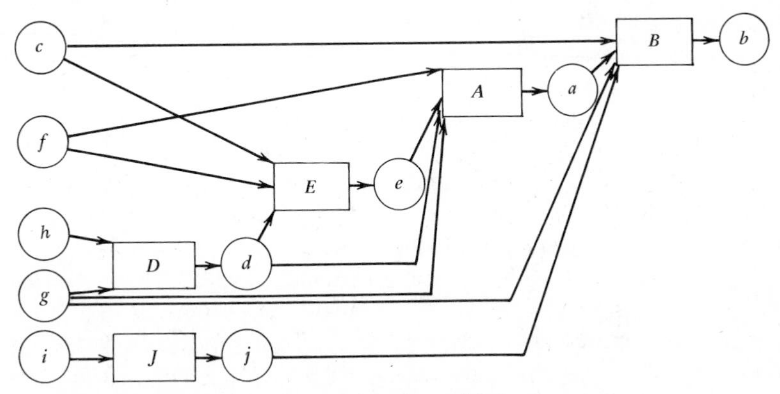

FIGURE 5 Graphical representation of a problem statement.

real ENTITIES or from a set of abstract variables. It was desirable to give the problem definer as much help as possible in defining data, and the analogy to the real world through entities was the best method available. Hence, it might as well be part of a problem statement language as long as it did not restrict the language in defining data abstractly.

It is important to distinguish between two possible uses of VALUE SETS. (A VALUE SET consists of all possible values of a variable.) The first use was for variables in which only one value would be in the machine at any one time, and the second use was for variables that could have many values at any one time. For example, the variable "warehouse number" might have many values in memory at one time, whereas the "quantity" of a particular part number at a particular warehouse would have only a single value at any particular time.

In the first case, the VALUE SETS could be used for validation of input data. ADS, for example, permitted validation rules to be given for each variable. In practice, validation is a complex process depending on combination of variables rather than on single variables; such rules were difficult to state on the ADS forms. It is more desirable to specify validation reports as outputs of the system; these then can include any processing specification permitted by the language for specifying variables on output reports.

The second use of VALUE SETS was in providing information about how much memory space was required. The basic question was how the problem definer stated the role the variables played. In COBOL the definition is through the structure definition in the DATA DIVISION and the use of OF and IN; in PL/1 nested qualifiers separated by periods are used. In YK the relationships among INFORMATION SETS were used to present this information.

The information in a problem statement must be sufficient to infer what variables have to be stored in the auxiliary memory and in the main memory. A variable must be stored in the memory if:

1. It appears in an update statement, for example of the form X (, ,) = X (, ,) + Y, where X might be "gross pay to date" and Y "the pay this week."
2. It is used in a statement without its value being processed, for example, "number of exemptions" in a payroll problem. This variable would appear as input on a new-hire transaction and in a change transaction, and would be used in pay computations.

ADS permitted the problem definer to specify variables to be available in HISTORY. These could be either intermediate variables that were used in a number of places or variables whose values the problem definer believed would have to be stored.

It is immediately clear from the preceding paragraph that one could not determine which variables fell into these categories unless the problem statement contained information about the time at which processing requirements occurred. In the first case (1), there should be some way of stating that payroll was computed weekly and the "gross pay to date" was cleared (set to zero) at the end of each year. Similarly, in the second use (2) it should be clear that a new-hire transaction occurred only once, whereas the pay computations occurred regularly.

Other Information. None of the problem statement languages had well-developed procedures for describing the time aspects of requirements, although YK, ADS, and TAG provided some capability. Some help in developing an acceptable "time" language could be obtained by studying the master time routines in simulation languages such as SIMSCRIPT or the executive systems for real-time systems.

Time specifications are required not only for determining which variables will be stored, but also for determining feasible and optimal storage organizations. The criteria used to determine optimality include both memory space and processing time. One important factor to be considered is organization of data to reduce memory space, by such techniques as header-trailer organization as used in hierarchical files and IDS. In order to do this, one must be able to infer the header-trailer relationships from such information as qualifiers. Another important factor is the question of what data should be stored semipermanently and which need to be held only temporarily. Again, the analogy to simulation may be useful: SIMSCRIPT, for example, distinguishes between PERMANENT and TEMPORARY ENTITIES.

The second part of the criterion is to reduce processing time. This can be done by reducing the number of accesses to external memory. Because a number of different types of processing requirements must be accomplished, the problem statement must contain both the values of each type and the time periods over which they occur so that accesses to auxiliary memories can be grouped whenever possible.

These deficiencies were corrected in PSL/PSA, to be explained next.

PSL/PSA AND SODA

In the early 1970s, several organizations began research on problem statement languages designed to make optimal use of the computer's capabilities. Two principal efforts were occurring concurrently: at Xerox and at the University of Michigan. The Michigan research produced a problem statement language (PSL) which was, in the words of Project Director Dr. Daniel Teichroew, "a generalization of Information Algebra, TAG and ADS."

After developing an automated version of ADS, the Xerox group decided to design its own problem statement language, SSP. SSP was not made operational, however, so this book will confine its analysis to PSL, which is operational in many organizations affiliated with the Michigan project.

Using the Michigan approach, the system analyst concentrated on what was wanted without saying how these needs should be met. The Problem Statement Language (PSL) was designed to express desired system outputs, what data elements these outputs comprised, and formulas to compute their values. The user specified the parameters that determined the volume of inputs and outputs and the conditions (particularly those related to time) that governed the production of outputs and acceptance of inputs.

The Problem Statement Analyzer (PSA) accepted inputs in PSL, analyzed them for correct syntax, and then:

1. Produced comprehensive data and function dictionaries.
2. Performed static network analysis to insure completeness of derived relationships.
3. Performed dynamic analysis to indicate time-dependent relationships of data.
4. Analyzed volume specifications.

The result of these analyses was an error-free problem statement in machine-readable format. The second output was a coded statement for use in the physical system design process.

Detailed description of PSL/PSA is omitted from this section because the more recent version of the system is described in the next section.

SODA (Systems Optimization and Design Algorithm) was a subject of PSL, designed as a vehicle for supplying a problem statement to a systems generator. PSL/SODA specified processes at a greater level of detail than did PSL, to develop precedence relationships in the process. Jay Nunamaker, a Purdue faculty member and Teichroew team member, was the project leader on the development of SODA.

SODA consisted of a number of submodels that were solved by using mathematical programming, graph theory, and heuristic procedures. The algorithm acted as a multilevel decision model with the decision variable of one level becoming a constraint at the next level and so on. The problem was partitioned into a multilevel structure, using a different set of decision variables for each level of the algorithm.

SODA selected a set of hardware and generated a set of programs and files to satisfy timing requirements, central memory, and storage constraints, such that the hardware cost of the system was minimized.

SODA was comprised of a set of computer programs that began with the initial statement of requirements (i.e., what the system is to do) and proceeded through the design and specification of the system. SODA was not concerned with the determination of which requirements were to be stated. The assumption was that the problem definer (PD) accurately identified requirements. The major components of SODA are shown in Figure 6 and are explained below.

SODA/PSL stated the requirements of the IPS (Information Processing System) independent of processing procedures. It also provided the capability for easily handling changes in requirements.

SODA/PSA analyzed the statement of the problem and organized the information required for SODA/ALT and SODA/OPT. This program also provided feedback information to the problem definer to assist in achieving a better problem statement.

SODA/ALT utilized a procedure for the selection of a CPU and core size and the specification of alternative designs of program structure and file structure.

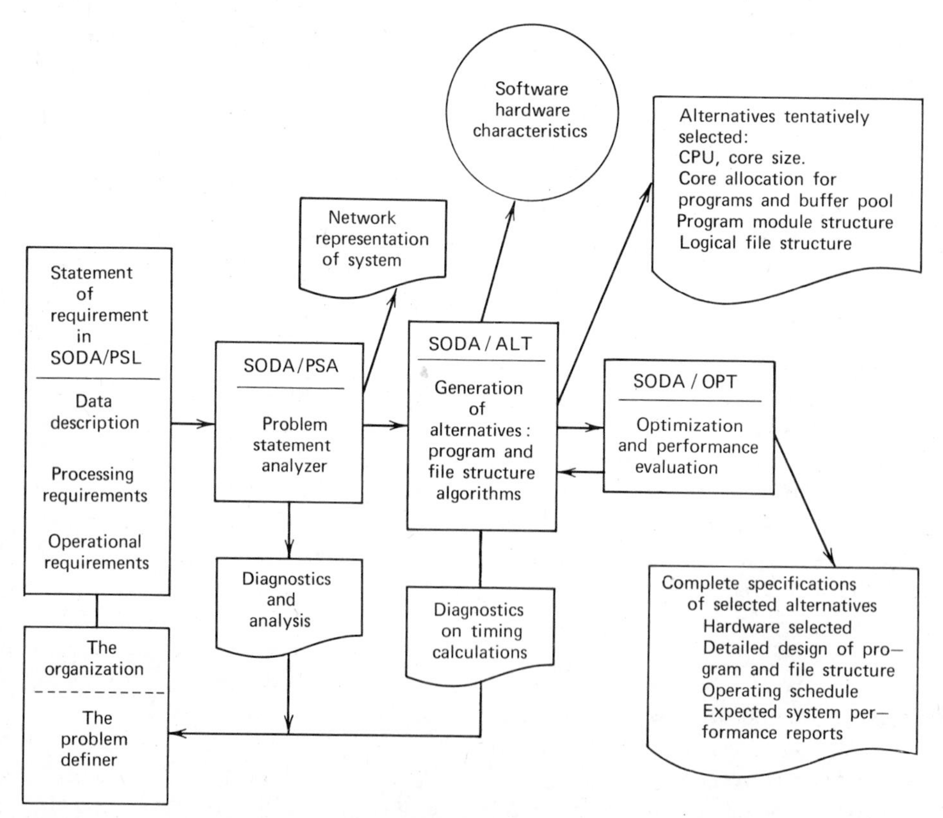

FIGURE 6 SODA: Systems optimization and design algorithm.

SODA/OPT contained a procedure for the selection of auxiliary memory devices and the optimization and performance evaluation of alternative designs.

The output of SODA was (1) a list specifying which of the available computing resources would be used, (2) specifications of the file structure and the devices on which they would be stored, and (3) a schedule of the sequence in which the programs must be run to accomplish all the requirements.

HOSKYNS SYSTEM

Parallel to the development of PSL/PSA and SODA was the development of the Hoskyns System, an automatic code generator. Designed by Hoskyns Systems, Incorporated, it was implemented in three British Corporations and then was introduced in the United States in 1972.

Figure 7 illustrates the Hoskyns System approach. A preprocessor automatically translated system specification matrices into COBOL programming statements, permitting program elements to be built and then consolidated into programs. Using the Hoskyns approach, the system was described in terms of programs and files, and the programs in terms of records and data elements. These sets of relationships were recorded in the form of matrices, as shown in the figure.

The first matrix provided the program file relationships in the system, completely defining the system flow. The second matrix stated the keys by which the files could be accessed and defined which record types existed in which file. These matrices also contained such information as file organization approach (i.e., index-sequential).

Taken together, these matrices provided the information necessary to generate the identification and environment divisions for all COBOL programs in the system, as well as the File Descriptions with record layout COPY statements for the data division. These divisions represented the envelope within which the programmer must write procedure coding. This envelope was automatically generated from the matrices by the COBOL generating processors.

The second matrix also provided the record specification. These record descriptions were held in a library, to be called into the File Descriptions of the programs by the previously described processor.

The remaining COBOL element was the procedure division. Decision tables, prepared by the system designer, listed the conditions and actions of the procedure. The decision tables

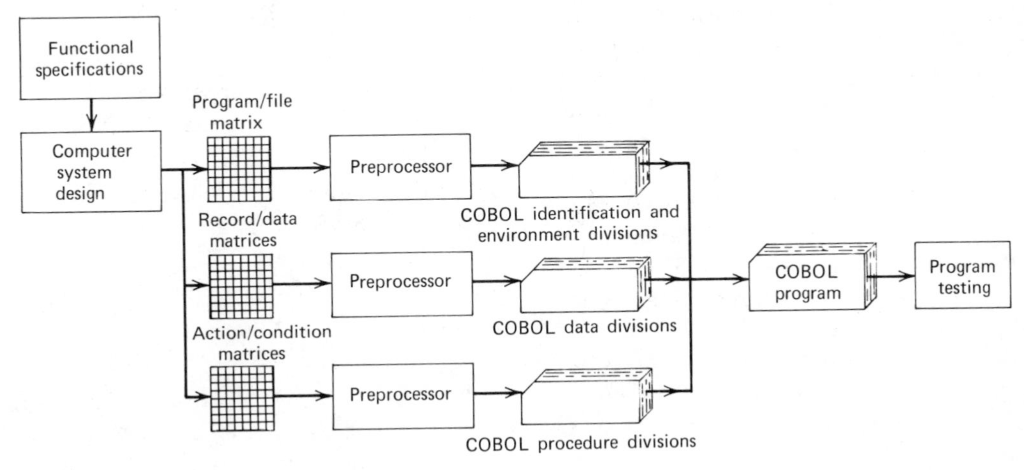

FIGURE 7 Hoskyns code generator system.

were input to the COBOL procedure division processor and were incorporated into the source program.

In summary, the Hoskyns system accepted system specifications and converted them to COBOL programs without manual intervention.

SUMMARY

After two decades of slow progress in designing special analysis techniques for developing computer-based systems, the progress rapidly accelerated. The first approach was automating Decision Tables and ADS. Next, TAG was developed. These techniques proved useful; nevertheless, few organizations adopted them. So, although the development of new techniques was accelerated, the delay in adoption showed little change from the trend depicted in our chart at the end of the introduction to Section 2.

Fortunately, some enterprising researchers were not discouraged by this situation. Recognizing the need to optimize computer-aided system analysis/design techniques, a number of researchers undertook the task of developing a foundational theory for such tools. Langefors, Grindley, Young, and Kent were at the forefront of this research. Teichroew, Nunamaker, and others on the Michigan research team synthesized the best features of prior theoretical and practical work to produce PSL/PSA and SODA. The Hoskyns code generator was also representative of the advanced techniques of the third generation of system development methodology.

Progressive companies and a few insightful government agencies recognized the advantages of these new tools and implemented them. Although the large majority of companies were slow to recognize the advantages of these new tools, their availability spawned another generation of improved techniques to be discussed in the next section.

SECTION QUESTIONS

1. Explain the parallel activities that represent the third generation of system development techniques.
2. What do automated ADS and DT processors have in common? What are the differences in output of the two techniques?
3. What are the features of automated ADS that distinguish it from the prior version of ADS?
4. What does TAG provide that was not properly considered in earlier system analysis techniques?
5. Prepare a diagram that shows the relationship of TAG input and output.
6. How does the system analyst use each of the 10 reports produced by TAG?
7. Prepare a table that evaluates ADS versus TAG.
8. One might see the trees and not the forest in reading about the comparison of problem statement languages. What is the importance of this material?
9. Explain Figure 5.
10. Prepare a table that compares the problem statement approaches of each of the five references to express data relationships.
11. Prepare a table that compares the problem statement approaches of each of the five references to express processing requirements.
12. Compare PSL/PSA to one of its predecessors, ADS. What are the similarities? What are the differences?
13. What is the relationship of SODA to PSL/PSA? Expand Figure 6 to include the functions of PSL/PSA.
14. Where would the Hoskyns system fit into the SODA scheme? Show which functions in Figure 6 are performed by the Hoskyns system.

SECTION 4

FOURTH GENERATION DEVELOPMENT TECHNIQUES FOR COMPUTER-BASED SYSTEMS

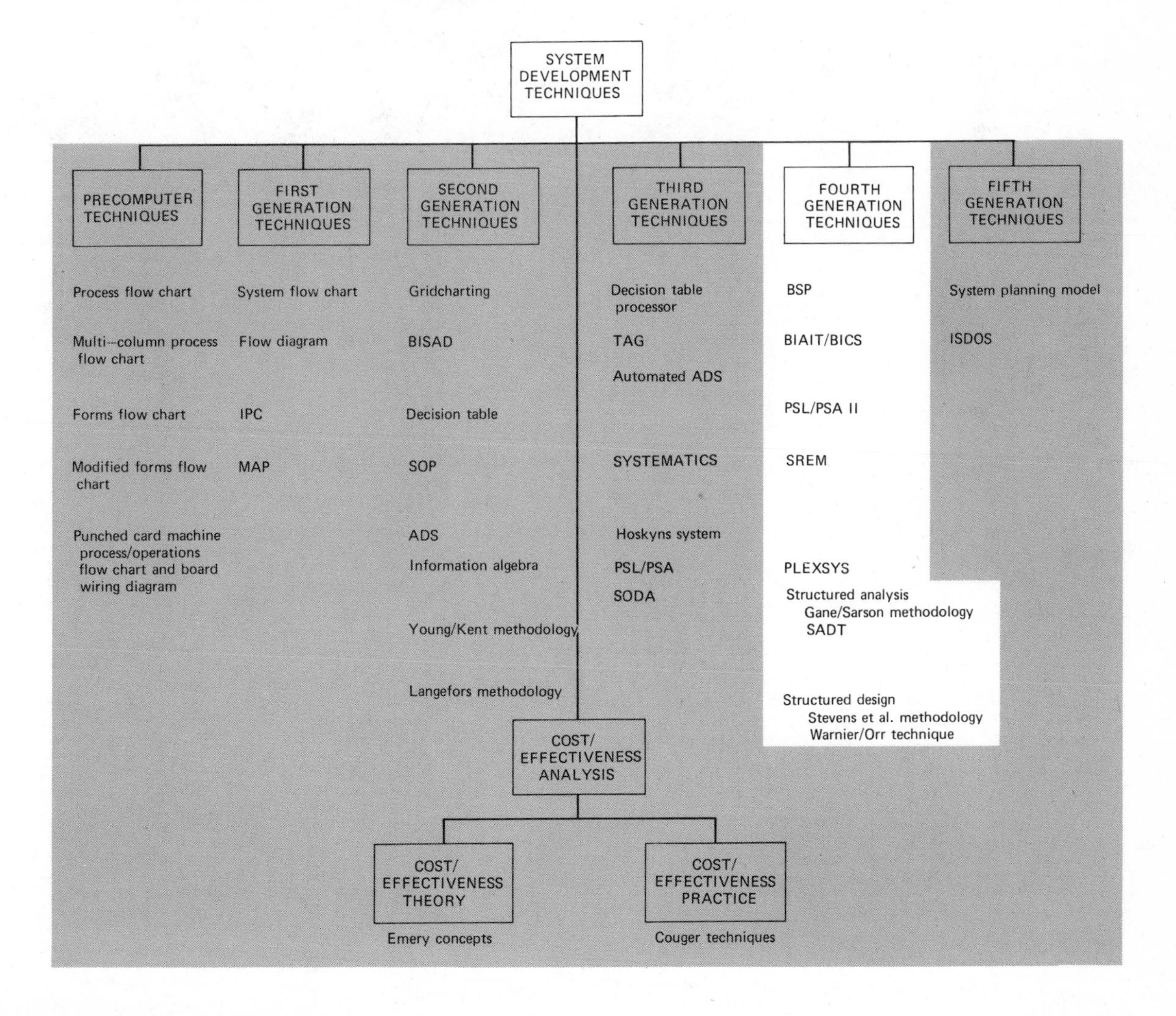

SECTION 4

FOURTH GENERATION DEVELOPMENT TECHNIQUES FOR COMPUTER-BASED SYSTEMS

J. Daniel Couger

OVERVIEW

An enigmatic situation arose during the era of the fourth generation of system development techniques. As could be expected, the computer-aided techniques continued to improve. An enhanced version of PSL/PSA was developed. The PLEXSYS methodology, an extension of the Teichroew work at the University of Michigan, emerged. A new technique, developed by TRW, SREM, also complemented the Michigan research.

Other techniques introduced during this era included Business Systems Planning (BSP), an extension of SOP, released by IBM. Complementary techniques, BIAIT (Business Information Analysis and Integration Technique), and BICS (Business Information Characterization Study), were also introduced.

The surprise was the origination of a set of techniques not designed for computer-aided system development. This was the structured methodology: structured analysis, structured design, and structured programming.

Our chart on the evolution of techniques (Figure 4 in the section entitled Evolution of System Development Techniques) shows that the structured methodology evolved from second rather than third generation techniques. They are located in the fourth generation era because of the date of emergence rather than their evolutionary pattern. If they had been available, the structured techniques would have clearly fit into the third generation methodology.

One explanation for the emergence of structured methodology is that most firms were not ready for the computer-aided techniques, despite their proven advantages. Many of these firms were still using second generation techniques and were not able to assimilate such a high level of sophistication. The new structured methodology provided an easier transition for these firms.

Unfortunately, at this stage of development there is no integration of structured techniques. The analyst must select a set of techniques from an array of structured methods and revise each to achieve an integration of techniques. Nevertheless, the structured techniques provide a basis to facilitate the move to fifth generation techniques.

Five key papers on structured methodology are included in this section. They are preceded by Mel Colter's survey paper which provides an insightful classification/integration of these tools. The literature on structured methodology has been void of such an analysis.

For the computer-aided techniques, six important papers are included: on SREM, PLEXSYS, and an enhanced version of PSL/PSA.

Finally, three significant papers are included on techniques for Phase I of the system development cycle. The other fourth generation techniques discussed above are used in Phases II and III. The new Phase I techniques (BSP, BIAIT, and BICS) could also have been classified as third generation techniques in terms of their stage of evolution as opposed to their dates of release. They were appropriate for the third generation era; however, few firms recognized a need for the degree of preplanning for systems that the BIAIT/BICS and BSP techniques required.

BICS is a logical extension of the BSP methodology, using Donald Burnstine's BIAIT theory to provide a systematic approach for analyzing the information systems of a business. The major difference between BSP and BICS is that the latter provides the study team with an information model of the business which it then verifies, whereas BSP builds a model through the discovery process.

Papers on data base optimizers and network modeling techniques are excluded from this book because they are covered in the data base/data communication course rather than in the system analysis course in most universities. The Bibliography provides references on these techniques for those who want to pursue them.

SUMMARY

One who peruses the system development literature is overwhelmed at the variety of techniques. Just keeping up with the acronyms is an imposing task. Closer analysis reveals a pattern, however. Although many techniques continue to be developed for small tasks in the system development process, few projects are devoted to the integration of tools to optimize the full development cycle. The major contributions are readily apparent. At the front end of the development cycle (Phase I), SOP evolved into BSP and BIAIT/BICS. SREM and PLEXSYS evolved from PSL/PSA and an enhanced version of PSL/PSA was produced. Development Phases II through IV were also aided by the emergence of structured methodology.

System designers now have a number of powerful tools for improving the quality of their work.

Evolution of The Structured Methodologies

Mel A. Colter

INTRODUCTION

The developmental roots of the structured methodologies can be traced to the 1960s, when the complexity of software systems began to exceed the capabilities of existing development techniques. The increasing use of on-line systems, data bases, and process control systems forced an awareness of the need to adopt new methods to ensure that reliable, maintainable systems could be produced to meet the needs of users. The history of systems failures in the software industry served both to emphasize the existing problems and to provide a framework for the understanding of the causes and possible solutions to those problems.

By 1965, there was general agreement that our ability to manage the software development process could not meet the need for increasingly complex systems. Many systems failed to meet their original design objectives; systems often failed to satisfy users' requirements; and budgets and schedules were seldom met. In addition, changes to systems often were extremely expensive and sometimes actually impossible within the framework of the existing design, and maintenance was a time-consuming and frustrating process.

These problems spanned the life cycle and crossed the boundaries of all types of systems. This wide scope of issues precluded any comprehensive solutions, but clarifying work began to appear. In 1965, Knuth and Dijkstra discussed the use of the "Go To" statement, arguing that, at the code level, this type of branching was closely related to the complexity of a program. Further work by Boehm and Jacopini and Dijkstra clarified the concept of the structure of programming logic in the late 1960s.

During this period, others began to consider software from a more macro perspective. The concept of *software engineering* was accepted as the study of processes that could be applied to ensure the orderly development of systems. In

1968 and 1969, NATO sponsored two conferences that focused interest on the broad issues of software design. This set the stage for the vast amount of work appearing in this field in the 1970s.

The work of the early 1970s offered little immediate benefit to the practicing software designer, but provided a foundation for the later development of general methodologies. The concepts of stepwise refinement and modularity were clearly delineated. The concept that programs and systems can be defined through a decomposition from general function to successive levels of more specific function has been generally accepted and is foundational to the current structured design methods. The argument that the resulting modules are best related in a hierarchical structure was also central to the discussions of this period and continues to underlie current practices.

Following this period of concept development, the primary thrust of continued efforts was toward the development of coherent methodologies. The goal was the presentation of processes that could guide the software designer and programmer toward the production of software systems and programs. It was this set of efforts that began to have impact upon the practicing professional.

The development of methodologies resulted in a series of processes being made available to practitioners. This production of methods has been of a "bottom up" nature. Early efforts began with the familiar and presented improved practices at the code level. Thus, *structured programming* emphasized the production of code in a well-defined style following a set of well-understood primitive concepts. The realization that a well-coded implementation of a poor design solved few problems supported the development and industry acceptance of *structured design methodologies*. These processes seek to produce a good design based upon a set of requirements. Understanding of the need for a carefully defined set of requirements led to the consideration of *structured analysis*. These areas will be discussed in detail later.

Concurrent to the above developments, a series of related issues received attention. Programming teams, testing, walk-throughs, and implementation techniques are all supportive of the structured techniques and will be treated also.

The Changing Emphasis in Software Development

The hardware environment existing during the 1960s was characterized by limitations of memory and execution speed. Therefore, storage and execution efficiency were critical measures of the quality of software. In the late 1960s and early 1970s, the problems experienced in the software industry clearly indicated the need for additional measures of system or program goodness. While hardware systems continued to expand in both size and speed, making storage and execution efficiency less important, the increasing complexity of systems, coupled with the increasing costs of human aspects of systems development brought new problems. Although hardware costs dropped, software costs rose dramatically, principally as a result of increased labor cost.

The software costs of systems are related to the way in which we deal with high complexity during analysis, design, construction, implementation, and maintenance. We now realize that this management of complexity is a high priority issue during the system life cycle. The structured techniques address this issue directly through the concepts of decomposition, modularity, and structured program logic. This reduction in complexity can greatly reduce costs during analysis, design, and construction, while making testing and implementation more straightforward also.

In the operational life of the system, where maintenance and modification occur, the management of complexity during design is not, by itself, enough. Although a well-structured program or system automatically supports maintenance and modification through the reduction of complexity, other issues are also important. Reliable software will perform as expected under all normal conditions and minimize maintenance. Therefore, *reliability* is a meaningful measure of goodness because reduced maintenance reduces long-term costs.

Because few programs or systems can be expected to serve their life without incurring some changes in their specifications or scope, *modifiability* is also important. Particularly in environments of changing user requirements, it is critical that we anticipate the need for future modifications. We must design system structures that support modification with minimal impact on the structural characteristics. The importance of this issue is underscored by the findings of many practitioners that even good designs often become altered so significantly by maintenance and modifications that the benefits of structured design may be lost long before the useful life of the system is over.

The current emphasis in the design of software systems is, therefore, that we first manage complexity in order to ensure that the system meets its requirements at the lowest possible cost. Next, issues of reliability and modifiability must be considered with respect to the system itself and the nature of its environment. If reliability is critical, then increased testing is indicated. If the environment is dynamic, indicating that continued modifications will be made, then the design should attempt to anticipate the changes through the creation of a general structure. Thus, we first concentrate on reducing software cost. Storage and execution efficiency are secondary considerations, and then only where such issues are truly critical.

The Direction of Structured Methods

The general goal of structured methods is the development of systems that meet user requirements through an orderly and manageable process. The management of complexity and the assurance of reliability and modifiability are central to all such methods. Thus, one view of the direction of structured concepts is that of structuring the process of analysis, design, and construction. Indeed, all such approaches do provide well-defined processes to guide the software expert in the production of systems.

Another important view of the concept of structured techniques relates to the structure of the systems that we build. At the code level, structured programming provides a set of logical primitives that are sufficient, upon proper combination, to implement any logical process. The result of the use of structured programming is therefore the creation of logic, implemented in code, which has a simple logical structure. This eases not only the construction of programs, but also the maintenance of such programs. Modifiability can also be enhanced because the logical constructs facilitate the development of code that has general logical structure rather than only a sufficient structure for specific problems. General logical structures often allow modifications to be made within the existing structure, minimizing the rewriting of sections of code.

Structured design seeks to implement a solution to a problem in such a way as to ensure that the structure of the solution parallels the structure of the problem. Such solutions tend to be of lowest complexity and of lowest cost. Finally, structured analysis seeks to discover the requirements of the system and the structural relationships of those requirements.

From this viewpoint, structured analysis, design, and programming seek first to analyze the structure of the problem, then to design a solution such that its structure most closely matches the structure of the problem, and then to implement the solution using the least complex logical structure possible.

A COMPARATIVE FRAMEWORK FOR STRUCTURED METHODS

A development process must consider analysis, design, and construction, and each organization must determine the combination of techniques and methodologies that best serves its needs. There are no techniques that are complete in the coverage of the entire set of development activities so this combination is necessary. In the area of structured analysis, very few choices exist. At this time, Structured Analysis and Design Technique (SADT) (a proprietary methodology offered by SofTech) is widely accepted as a complete and rigorous structured analysis process. In the area of structured design, several methodologies are available, and the choice between them is not easy. For structured programming, the concepts of structured logic are generally accepted and the organization need choose only the method of implementing the concepts.

As the organization evaluates structured methodologies with the intent of creating a coherent development process, three primary bases of analysis exist. These involve the consideration of:

1. The theory of system structure.
2. The specifics of the system to be designed.
3. The specifics of the development environment.

These issues will be discussed in turn.

System Structure Issues

A consideration of system structure is foundational to any methodology directed toward the development of today's systems. However, there are multiple dimensions involved in the concept of structure. For the purposes of this analysis, structure can be analyzed from the points of view of:

1. General structure.
2. Control structure.
3. Data flow.
4. Data structure.

The concept of *general structure* refers to the modularization of the system or program, usually through the analysis of functions that must be performed. The hierarchical relationship between the modules, as well as the internal characteristics of individual modules and the relationships between modules serve to define the goodness of the overall design.

Control structure defines the procedural characteristics of the system. The study of coupling and cohesion involves issues of control, as does the analysis of span of control. On the most basic level, consideration of control structure involves limiting the logical structure to include only sequence, selection, and iteration constructs.

The analysis of *data flows* clarifies the flow of data through a system. Transformations of data are identified and inputs and outputs of all processes are detailed. Delineation of *data structure* allows comparisons of the structure of the input data and output data. This analysis often has major implications on program structure.

Although it is clear that design methodologies must be based on strong consideration of structure, the multiple structural issues noted above cause problems in practice. The simultaneous treatment of all of the structural factors has so far proved to be impossible. Therefore, each methodology available today tends to place primary emphasis on one type of structure while considering the others in an iterative process. Thus, a particular approach may first analyze data flow, and next develop the hierarchical structure and control structure, while continuing to reanalyze data flows to ensure that they remain acceptable in the later stages of design. Other methodologies place the initial emphasis on different structural issues.

No methodology exists that places initial emphasis on control structures at the system level. This is probably because control is inextricably related to the other structural issues. The development of a design based, for example, on data flows will automatically create a control structure. For this reason, the analysis of control tends to be used to evaluate designs rather than to create them.

System Specific Issues

The consideration of structural issues is broad-based and general. The value of a particular methodology is, however, related to its ability to guide the software designer to the development of specific systems. Thus, one could conceivably evaluate the various methodologies in terms of their ability to address specific types or a broad range of systems. This type of comparative framework is difficult to develop and even more difficult to put into practice. However, some general comments appear to be valid.

It is unlikely that any one methodology can be expected to produce the "best" design for all types of systems. Differing types of systems require different approaches. Because no methodology can base design on all of the structural issues simultaneously, we can argue that no development process gives equal attention to all of those issues. As an example, systems that exhibit major data structure problems require a specific design approach. When the data structure on the input side is vastly different from the structure on the output side, consideration of the data structure should take priority. Methodologies that concentrate on module structure or data flow first may fail to produce the best solution to this problem.

Development Environment Issues

The fit of a methodology to an organization's needs and the demands it places on the organization are important. Most important, the emphasis of the process should match the types of problems most often faced in the shop. Secondly, the resources required by the design process must be considered. Methodologies requiring a great deal of technological support may not be feasible for the small shop. Similarly, small- to medium-sized shops may not be able to afford consultive efforts aimed at customizing a particular methodology to fit their needs. Other issues such as those of educational requirements are also of importance.

Concerns related to the development environment are organizationally specific and do not allow general or industry-wide analysis. Thus, this comparative framework is of less concern to this work than the others detailed above, although it may be critical to the individual shop.

A Comparative Framework

The preceding discussion supports the conclusion that there is no globally "best" design methodology. The multidimensionality of the concept of system structure, along with the wide variation between individual design projects, precludes a choice of a single best procedure. The concept of the goodness of specific methodologies is therefore relative.

The above statements invite critical discussion. Systems experts differ in their placement of initial emphasis, each beginning with a consideration of hierarchy—or of data flows—or of data structure!

The only resolution of this type of argument lies in the evaluation of the use of the methodologies. At this point, no single design process has been shown to consistently produce systems that are superior to those produced by other processes.

The following sections discuss structured analysis and design from the perspective of system structure.

STRUCTURED ANALYSIS

Functional decomposition was the first structured process that incorporated elements of structured analysis. Although the primary focus of functional decomposition was toward design, its actual practice merged analysis and design into a single process. The decomposition of systems and programs, generally based on functional analysis, was analytical in nature, and the resulting hierarchical structure was assumed to represent the design. Structurally, only general (hierarchical) structure was analyzed, with data flows, data structures, and control structure beyond the scope of the process.

Functional decomposition was an incomplete methodology, primarily because of the absolute merging of analysis and design, but also because of its inability to consider other structural issues. Its use as a major tool has declined, but the concept of decomposition has remained central to all of the structured methods. Therefore, this method will not be treated further here.

Structured Analysis Using Data Flow Diagrams

Data flow diagrams, also known as *bubble charts*, *DFDs*, or *entity diagrams* were first seen as a part of the structured design methodology of the Yourdon group.[1] Here, the analysis of data flows and transformations of data precedes the actual hierarchical design. The move to the hierarchy chart representing the design occurs after the DFD is completed.

Tom DeMarco and Victor Weinberg extended the use of the data flow diagram to a formal analysis process. Both use the concept of *leveling*, where high level processes are broken into more detailed diagrams. This creates a hierarchical structure during analysis, easing the transition to design. The approach used by Gane and Sarson is similar in the use of this leveling process.

The data flow diagram, although general in concept, varies greatly in application across the various methodologies. The differences, however, are due mainly to variations in graphic standards.

Structured Analysis and Design Technique (SADT)

SADT is a process developed by Douglas Ross of SofTech. The methodology is similar to those based upon data flow diagrams, but the level of detail is greater. A DFD shows only general inputs and outputs associated with a function or transformation, as shown in Figure 1. This generality, although limiting the complexity of the graphic, provides no detail as to the nature of the data flowing into the function or process.

Inputs to a process can fall into three categories:

1. Pure data.
2. Data that *may* function as control.
3. Pure control.

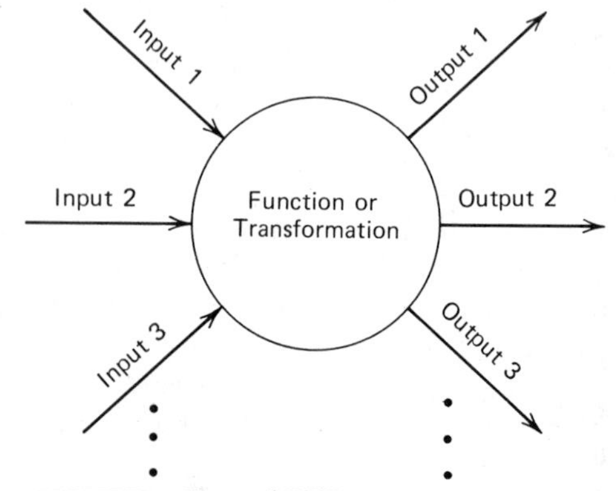

FIGURE 1 General DFD representation.

[1]The Stevens, Meyers, and Constantine paper included in this section of the book treats DFDs lightly. The Gane and Sarson paper, also included, provides more detail.

For example, the data item, "DATE," considered as input to the function, "PRINT REPORT LINE," is pure data. The data item, "TRANSACTION CODE," when input to the function, "PROCESS TRANSACTION," is pure control, because it has the single function of controlling transaction processing. However, the data item, "TRANSACTION AMOUNT," when input to "PROCESS TRANSACTION," may have both a pure data component and a control component. The data component exists because the amount is processed to create account totals. If transaction amounts over a certain level trigger special functions, then the data item may have a conditional control feature.

SADT clearly shows the difference between elements of pure data and pure control, as shown in Figure 2. This capability allows the early representation of the control structure and aids analysis from that perspective. The case of data that conditionally functions as control is often obscure at higher levels of this diagram, but becomes clear at more detailed levels.

Additionally, in Figure 2, we see the delineation of the mechanism responsible for the function listed in the box. This combination of detailed statements that accompanies each function provides information not available in the standard DFD representation. The trade-off clearly involves increased complexity in the diagram, but most users find the added information worth the increased complexity.

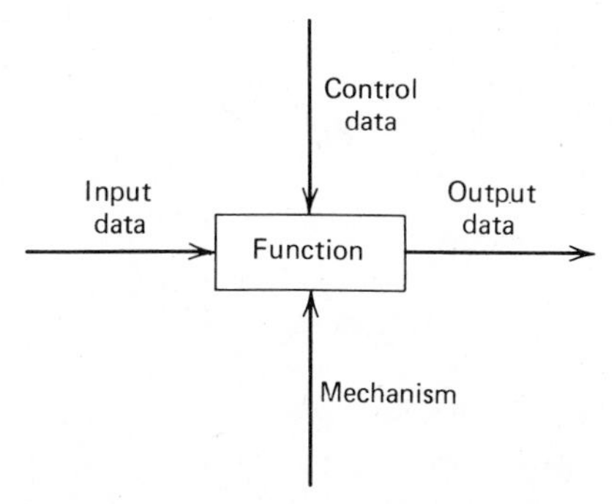

FIGURE 2 General SADT activity representation.

The diagram shown in Figure 2 is activity related and its use results in a decomposition based upon function. SADT also supports the analysis of data. Figure 3 shows the general structure of the data model. The combined analysis of activity (function) and data yields a reasonably complete picture of the system of interest. The hierarchical structure is created by decomposing each box into its component detail; data flows are analyzed through the activity diagrams; and control is viewed in both diagrams. Data structure is not specifically considered.

STRUCTURED DESIGN

The structured design methodologies in use today can be classified according to their primary focus. The major areas of emphasis are: (1) hierarchy, (2) data flow, and (3) data structure. The design methods to be discussed here are classified below as:

1. Hierarchical Focus
   (a) Functional Decomposition
   (b) Structured Analysis and Design Technique
   (c) Hierarchical Input, Process, and Output (HIPO)
2. Data Flow Focus
   (a) The Yourdon and Constantine and Myers Approaches

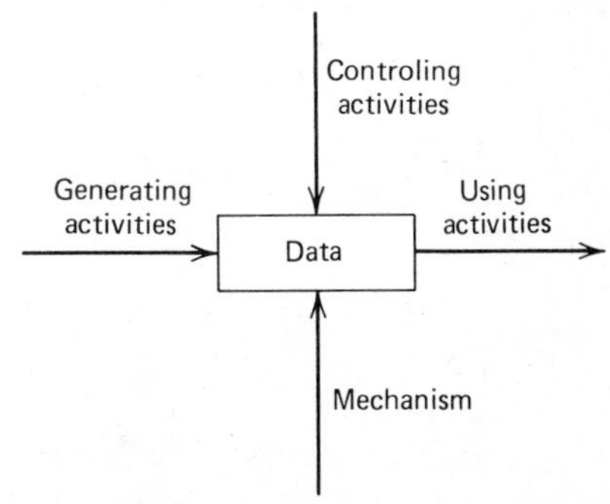

FIGURE 3 General SADT data representation.

3. Data Structure Focus
   (a) The Warnier-Orr Approach
   (b) The Jackson Methodology

Many of these approaches consider additional structural issues on a secondary level. All of them result in a hierarchical structure, although the representation of hierarchy varies.

Hierarchical Methods

The primary emphasis of methodologies concentrating on hierarchy is the process of dividing the system into increasingly smaller subunits while maintaining a hierarchical relationship between the components. In general, the definition of the subunits may be based upon considerations of procedure, function, time, or common data needs. Each of the specific methodologies discussed below may, in fact, involve decomposition using any of these perspectives. They differ primarily in scope and in the tools used to implement the process.

The earliest hierarchical process was functional decomposition which treated only hierarchical structure and did not provide for any analysis of data or control issues. This approach has been absorbed into more general methodologies.

In the early 1970s, IBM developed HIPO (Hierarchy plus Input, Process, and Output). HIPO is primarily a documentation package, supporting design more than directing it. However, many organizations have adopted HIPO as a design methodology with success. This success is due to the completeness of the HIPO documentation which encourages its users to consider a wide range of issues in order to satisfy its information requirements.

The original HIPO package had three major objectives:

1. To show the decomposed hierarchical structure of the system functions clearly.
2. To provide detail of the functions of the system or program, independent of programming language.
3. To visually describe input and output flows at the function level.

These objectives suggested that the functional decomposition should occur first, followed by a diagrammatic representation of the data flows implied by the resulting hierarchical structure. No analysis of the "goodness" of the design was involved.

Three major types of diagrams represent the basic HIPO package. The *Visual Table of Contents*, (VTOC) shows the hierarchical structure with each function named and numbered for cross-referencing purposes (Figure 4). In practice, many users modify the numbering system, but its existence is critical to allow easy movement between the VTOC and the detailed diagrams. The VTOC is often accompanied by a *Summary Description Table* (Figure 5) and an *Interfunction Communication Table* (Figure 6). The summary description table contains a brief narrative description of each function represented in the VTOC, whereas the interfunction communication table details parameter passage between the functions. This allows some specification of the data flows implied by the VTOC.

The second major component of the HIPO package is the *Overview Diagram* (Figure 7). An overview diagram exists for each function that has subordinate functions in the VTOC. The functions subordinate to the one of interest are depicted, together with their associated input and output data flows. Subordinate functions that are further decomposed are enclosed in boxes in the process portion of the diagram. The arrows representing input and output are drawn to convey the maximum information about specific contents, sources, and destinations of the data flows.

The third major component of HIPO is the *detail diagram*. Detail diagrams are at the level of specific function, showing detailed input and output and directly referencing the associated

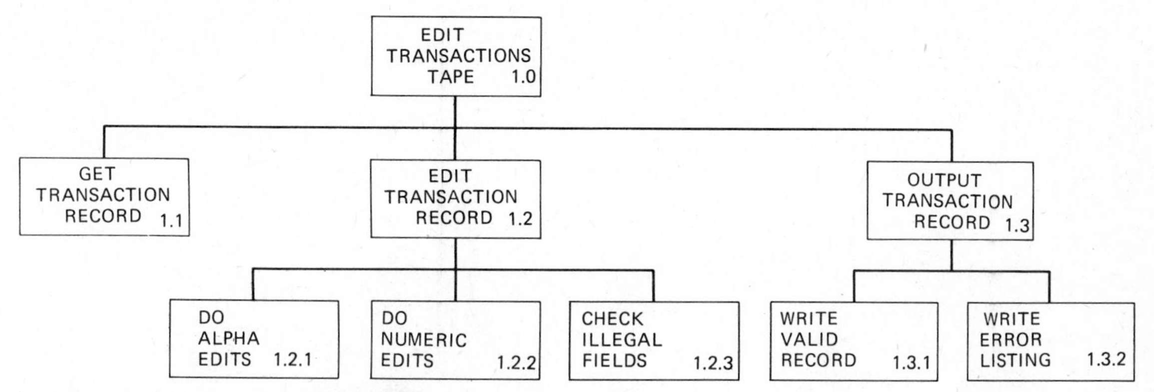

FIGURE 4 HIPO visual table of contents.

code. In practice, some organizations have omitted the detail diagram, choosing instead to add detail to the overview diagram. One common practice is to depict logical processes by inserting pseudocode into the process section of the diagram.

All of these diagrams may be accompanied by an extended description section that amplifies the primary information. In addition, data dictionaries, flowcharts, decision tables, report formats, and so on, may be appended to the appropriate diagrams to complete the package.

In summary, HIPO places the highest priority on hierarchical structure. Data flows are detailed only as they result from the VTOC, and data structure is only weakly handled in the overview and detail diagrams. Control structure is not formally considered, although some practicing professionals have chosen to make the VTOC procedural by inserting pseudocode into the diagram (Figure 8).

Design Based on Data Flow

The data flow design methodology was developed by Larry Constantine and Ed Yourdon, and by Glen Myers. It is primarily a sophisticated extension of functional decomposition. Prior to the decomposition, however, an analysis of data flows is performed, using the *Data Flow Diagram* (or Bubble Chart). The Data Flow Diagram (DFD) is a model of the inputs and outputs of the processes involved in the system. The analysis of the DFD supports the development of the hierarchical structure of the system.

This methodology is more complete than the HIPO process because it specifically precedes de-

SUMMARY DESCRIPTION TABLE	SYS: EDIT TRANS FILE
MODULE	DESCRIPTION
(1.0) EDIT TRANSACTION TAPE	EDIT ALL TRANSACTIONS ON BATCH TAPE PRODUCING A VALID TRANSACTION TAPE AND AN ERROR LISTING
(1.1) GET TRANSACTION RECORD	SEQUENTIALLY READ NEXT TRANSACTION FROM BATCH TAPE

FIGURE 5 Summary description table.

INTERFUNCTION COMMUNICATION TABLE			SYS: EDIT TRANS TAPE
FROM	TO	IN	OUT
1.0	1.1		TRANSACTION RECORD
1.0	1.2	TRANSACTION RECORD	EDIT CODES
1.2	1.2.1	ALPHA FIELDS	ALPHA EDIT CODE
1.2	1.2.2	NUMERIC FIELDS	NUMERIC EDIT CODE

FIGURE 6 Interfunction communication table.

sign with analysis. Certain weaknesses exist, however. First, the creation of the hierarchical structure chart from the data flow diagram is poorly defined, causing the design to be poorly coupled to the results of the analysis. As noted in the discussion of structured analysis, the process used to derive the DFD can affect this linkage.

The second weakness of the methodology is that it does not formally provide for an extensive documentation package. Although the DFD and structure chart are integral results of the process, other necessary details must be appended by the user. Function descriptions, processing logic detail, and output formats must be added. Program logic can be appended either in the form of pseudocode or in the form of the programs themselves. Data dictionaries can be created as supporting documentation using the ideas of Gane and Sarson. However, the existence of a formalized documentation package would strengthen the methodology.

The structure charts proposed by the Yourdon group are similar to those of HIPO. The differences reflect the choice of information to be

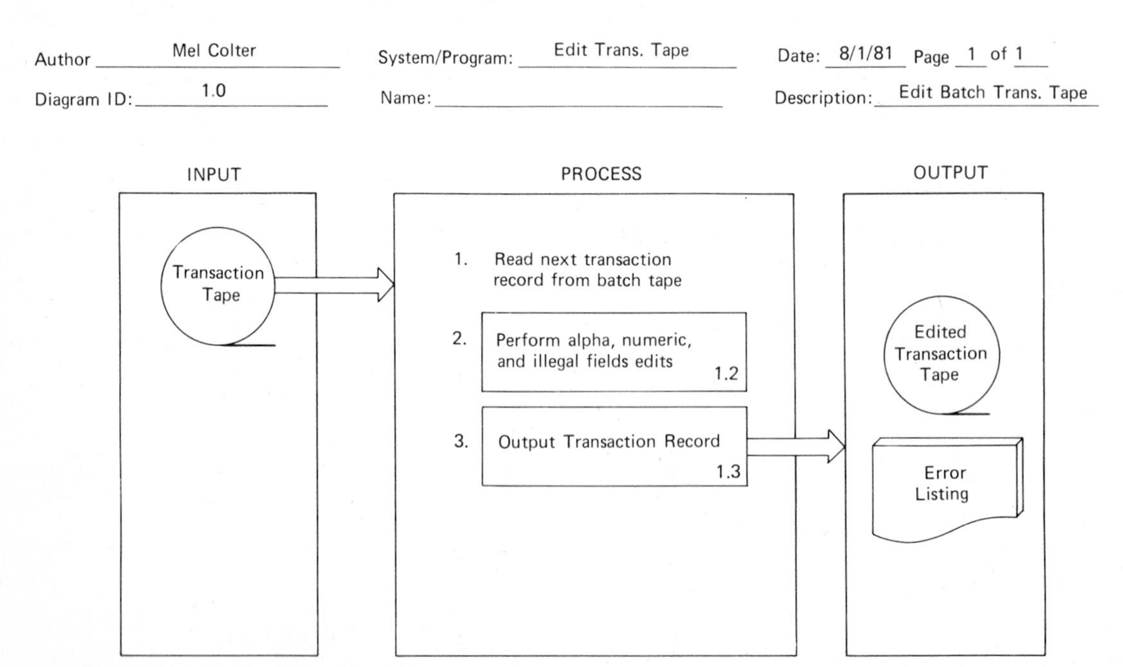

FIGURE 7 Overview diagram.

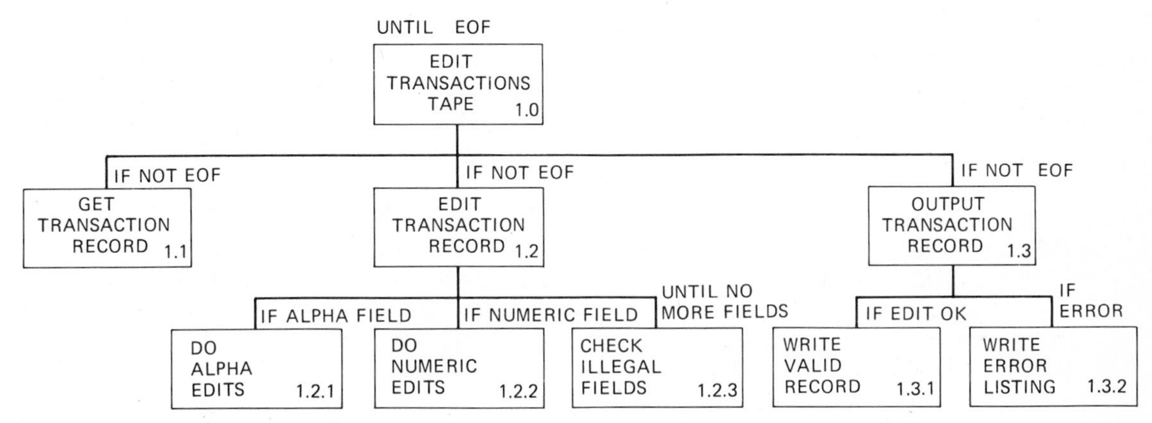

FIGURE 8 HIPO VTOC with procedural detail.

depicted rather than any foundational issues. Figure 9 shows a sample structure chart, with data flows represented as arrows on the hierarchical connectors and control shown using diamonds for decision indicators. Looping arrows are used to show iteration. Thus, the diagram allows the representation of hierarchy, data flow, and control structure. The management of complexity, however, limits the detail of the data flows and control indicators, forcing the use of supporting documentation.

In addition to their methodological development, the work of Yourdon, Meyers, Constantine, and others is particularly significant to the area of evaluating the "goodness" of a design. They suggest that any design can be analyzed to evaluate its relative quality. The most critical evaluative issues involve those of *coupling* and *cohesion*. For hierarchical modular structures, the cohesion of a module is a measure of its internal strength, whereas coupling issues address the relationships between modules. The ideal system should have modules with high cohesion and low coupling to simplify its design, construction and maintenance. Other more classical measures of goodness, including span of control and the scope of effect of a decision, are merged into this analysis framework, also.

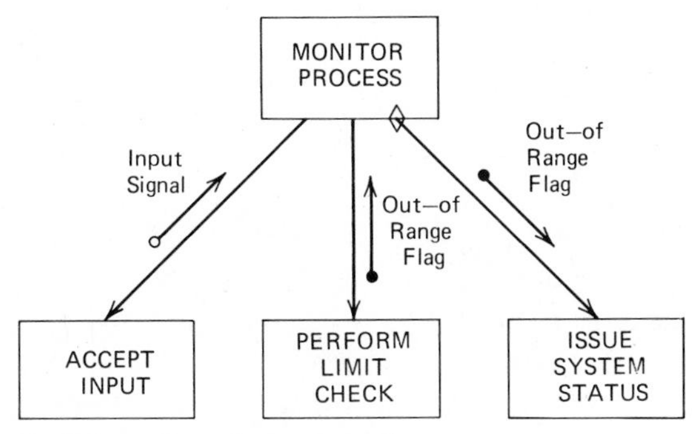

FIGURE 9 Sample structure chart.

Design Based on Data Structure

Jean-Dominique Warnier, Ken Orr, and Michael Jackson believe the relationship of the input and output data structures to be of primary importance in determining the design of systems. They argue that a proper model of data structure is essential to the development of system structure. Furthermore, they argue that programs can be judged to be "right" or "wrong," based upon such considerations. As a note of perspective, hierarchical design approaches claim only to produce good designs, whereas the data flow methods claim to be able to assess relative levels of quality in design. Only data structure advocates claim the *best* design. For systems exhibiting severe *structure clash* where input and output structures are greatly different, data structure design probably does result in the best design when compared with other methods. For systems with little or no structure clash, however, the claim is unsupportable.

The Warnier-Orr methodology begins with a detailed analysis of hierarchical data structure and then determines a program structure that parallels the data structure. The hierarchy charts used here are developed horizontally rather than vertically.[2] Control structure is shown on the program hierarchy chart at a lower level of analysis.

Michael Jackson's approach is similar to that of Warnier-Orr. He begins with an analysis of data structure, develops a parallel program structure, and then moves to the implementation of the logic. Jackson advanced the concept of structure clash and did much to foster an appreciation for data structure oriented solutions. For systems where analysis shows differing data structures on the input and output sides, the use of his techniques may be indicated. Like Warnier-Orr, however, control is considered at a low level of analysis. In fact both of these approaches are less rigorous in the analysis of the goodness of the design in the detailed manner of the Data Flow Design process.

Perspective

In summary, the methodologies differ in terms of their prioritization of basic structural issues and in terms of their scope of usefulness during the development cycle. These issues, however, are more often related to the definition of the specific methodology than to any inherent limitations. Hierarchical design methods, for example, could be combined with a preliminary evaluation of data flow and a follow-up evaluation of coupling, cohesion, and control to produce a result much like Data Flow Design. HIPO designs can be similarly evaluated. Thus, it appears that the software designer can "mix and match" portions of these methodologies to alternatively stress data flows or general structure.

The process of prioritizing data structure is less straightforward. Although one could analyze a data structure oriented design for goodness in terms of data flow and control, it is difficult to analyze a data flow design for data structure on an ex post facto basis. Such an analysis would essentially require a complete review of the problem from a data structure perspective. For problems with possible data structure clash, one might best begin with that analysis.

The scope of usefulness over the life cycle is also relative. Where HIPO was conceived as being useful from logical design through program construction, it could also be used to document and evaluate existing systems. Similarly, though SADT is aimed toward the analysis of the problem rather than the solution, its principles can be extended to the design of the solution system. Again, the knowledgeable user can customize the process to fit specific needs. The life cycle is discussed next to provide the background for this type of customization.

LIFE CYCLE CONSIDERATIONS

The optimal design methodology would be one that supports all of the necessary processes in the system development cycle. Thus, we would hope to be able to use the design process to evaluate existing systems, formulate requirements, prepare the logical and physical designs, support program construction and testing, and aid the implementation process. Figure 10 is provided to permit analysis of the design methodologies in each phase of the system life cycle.

No existing methodology fully meets this set of life cycle requirements. For example, physical design involves the constraint of the system to specific hardware and other physical configurations. This process has not been addressed by current design approaches. Similarly, although

[2] The paper by Kenneth Orr, included in this section, provides a good introduction to the use of the Warnier-Orr methodology.

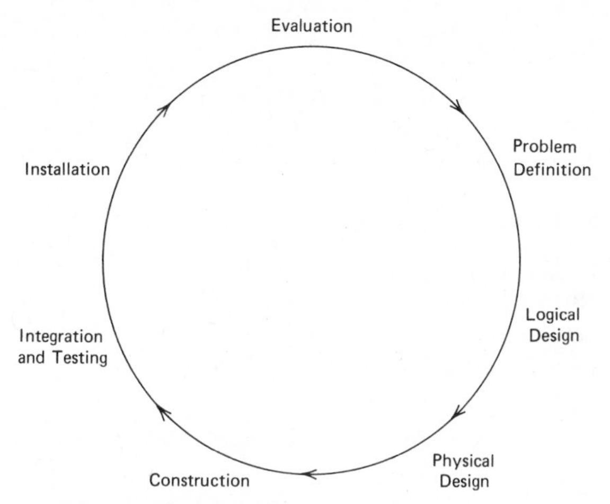

FIGURE 10 The life cycle.

there are methodologies that have addressed some implementation processes, no coherent approach exists in this phase. These gaps in the life cycle activities exist for all of the methodologies discussed here.

The treatment of other life cycle processes varies across the range of design methodologies. Each methodology was specifically designed to guide development through specific portions of the cycle. As an example, SADT is oriented primarily toward the analysis of the problem. As noted in the previous section, however, most methodologies can be extended beyond their natural bounds by customizing the processes to fit individual needs.

In Figure 11, the life cycle is shown in a linear fashion in order to depict the applicable range of each methodology. This representation is on two levels. First, if the methodology was specifically oriented toward the support of a particular life cycle process, the outline of the bar is a solid line. Processes that can be addressed if the user chooses to extend the design tool are shown with a broken outline. Second, if the methodology strongly supports a particular design process, the bar is shaded. Weak process support is shown as an unshaded portion of the bar. A discussion of the figure follows, organized by methodology.

Hierarchical Methods

These processes vary greatly in the scope of their usefulness through the life cycle. The specific methodologies will be analyzed regarding their use in the design cycle.

Functional decomposition provides a general structure at either the logical design or the physical design level. Furthermore, it can be used to represent either system structure or program structure. However, because there is no formal process for evaluating the goodness of these structures, functional decomposition cannot be said to provide strong support in any of these areas.

This technique can be used to support some definition of the problem if one considers the functions to be performed in a general sense before attempting to define the hierarchical structure. Similarly, one could examine the structure of existing systems by detailing the manner in which they are decomposed and structured. Again, only weak support is provided in these areas because no method for the evaluation of structure is provided.

SADT provides strong support for problem definition and can be easily extended to support logical design. Though this process is primarily problem oriented, the consideration of logical design is a straightforward extension. The use of SADT for physical design is, however, more difficult. An existing system could be partially evaluated through an analysis of the problems addressed by the system and a subsequent examination of the existing ability to meet those problems.

The HIPO process begins at the logical design level. But its support at that level is strong only to the extent that the designer uses other processes to evaluate the design. These additional processes would probably be best taken from the work of Yourdon and Constantine. Physical design is strongly supported by HIPO, as is the construction process. HIPO's detailed description of input, output, and process detail

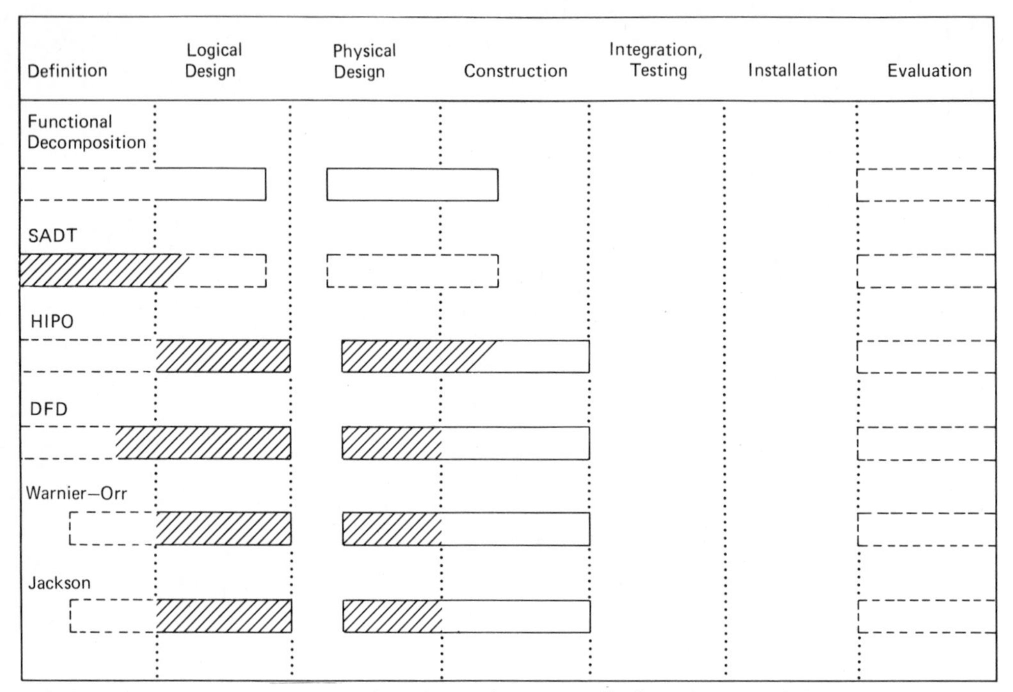

Solid Boundary: Methodology oriented towards support of indicated portion of life cycle
Dashed Boundary: Methodology can be applied to indicated portion
Shaded Interior: Methodology provides strong support
Unshaded Interior: Methodology provides weak support

FIGURE 11 Range of methodologies over the life cycle.

clearly defines the physical system. Again, formal evaluation of program design is not as strong as that of the Yourdon approach. In addition, one could use HIPO to define the structure and detail of existing systems to support their analysis.

Data Flow Methods

The data flow design methodology formally addresses the areas of HIPO's weaknesses. Evaluation of the design is an integral part of this process, although that evaluation is on a somewhat conceptual level and the iterative processes of evaluation and improvement are not tightly structured. In addition, the data flow design approach can be used to define the system through a consideration of the various ways in which it can be decomposed. The evaluation of existing systems can, of course, be addressed by the same techniques used to evaluate the new design.

Data Structure Methods

Because of their similarity in terms of the life cycle, the Warnier-Orr and Jackson methodologies may be discussed together. Neither approach completely addresses the definition phase, although an early consideration of data issues can be considered to be a part of the definition. Both methods can be used through logical and physical design, and both support the construction of the system.

Both Warnier-Orr and Jackson approach the question of the goodness of the design primarily by analyzing its ability to handle the data structures involved. Revising the techniques to handle the issues of coupling and cohesion, discussed by Yourdon and Constantine, would strengthen the processes. The use of these methodologies to evaluate existing systems is therefore limited.

Discussion

The preceding analysis clarifies the basic differences in the scope of application for the different methodologies. Functional decomposition is essentially an incomplete approach, creating structure without evaluating it. SADT, on the other hand, is designed to be problem oriented and was never intended to address the entire life cycle. The other methodologies tend to be useful in common processes with differences related primarily to the emphasis of the tool. HIPO is not heavily oriented toward the evaluation of design, whereas the inclusion of data structure issues is limited to the Warnier-Orr and Jackson methodologies.

We see that the system designer is left with a choice of methodologies with no single approach obviously superior to the others. In addition, the designer is faced with a variety of tools to represent the design and direct the design process at each stage of development. These tools are discussed next.

TOOLS FOR DESIGN

To some extent, the choice of a methodology implies a choice of tools. Each methodology is supported by a set of graphic techniques that both assists the design process and represents the design at various levels. Many of these tools, however, differ only in terms of the specific graphics used. Thus, tools are generally not methodology-specific on a conceptual level.

As an example, consider the representation of hierarchical structure with the added characterization of control structure. Warnier would represent such a process using a horizontal structure, whereas Jackson, Yourdon, and HIPO would use a vertical representation. Similarly, control is shown differently by Warnier, Yourdon, and Jackson. HIPO, on the other hand, does not specifically detail control on the hierarchy chart. However, the designer can easily adapt the HIPO representation by borrowing symbols from the other procedures and incorporating them into the graphic. Indeed, one can simply use pseudocode on the hierarchy chart to show control.

Thus, the designer is free to "mix and match" design tools to fit the needs of the process. This customization is common among practitioners and visible in the literature. As an example, Weinberg suggests the use of wavy lines to show areas of design that are inexact and need further clarification.

The real issue here is the need to characterize various design tools in terms of their ability to meet certain requirements. Any set of tools that can represent system structure, show control and data flows, and serve for communication and documentation will probably suffice for the designer. These minimal requirements are discussed next.

Tool Requirements

Any tool which is used to represent design must serve to clarify that design. Thus, we can discuss the way in which the design is shown by specific processes. First, tools should show the structure of the design. This structure exists on both a local and a global level. Here local structure refers to structure on the program level, whereas global structure is at the system level. Second, design tools must show control structure, again at both the local and global level. Third, we need to be able to depict data flows, and fourth, data structure must be included. In addition, a good tool should serve as a communication device. Tools that are strong in communication ability will serve to represent design to a wide audience, including users, analysts, and programmers. Finally, tools that become a part of the documentation of the system serve a dual (and desirable) function.

For comparative purposes, Figure 12 shows the following levels of evaluation. A rating of "1" indicates that the tool strongly supports the specific requirement. A "2" indicates that strong support can result from modification of the tool for that purpose. For example, hierarchy charts as used in HIPO can show global control well if the designer sets a standard for depicting control on the graphic. A "3" indicates only a weak natural satisfaction of the requirement; a "4" indicates a weak satisfaction if modification is made by the designer; and a "5" indicates an inability to address the requirement.

Tool Comparisons

Function descriptions, the first tool depicted in Figure 12, involve the listing of functions necessary to meet the requirements of the system and briefly describing the nature of those functions. This process is sometimes referred to as functional narratives. Function descriptions show no system structure or control structure. Data flows can be briefly discussed on a macro level at the user's discretion. This tool is primarily a communication and documentation facilitator during the early design process.

Data Flow Diagrams perform a more general function. Although they do not represent data structure, they strongly show data flow and are good communication facilitators. DFDs show global structure, but local structure is shown only at the option of the user. Control is not shown on the DFD. As a documentation process, the DFD is weaker because it represents a phase of design and not necessarily the final structure.

The *Visual Table of Contents* (VTOC) in HIPO shows both global and local structure. The addition of control notation allows this tool to show global control also. Local control structures are shown at a lower level of the HIPO process. Data flows could be shown on the VTOC by adopting, for example, Yourdon's notation, but data structure is not considered. As a communication and documentation aid, the VTOC is a strong tool.

The *IPO Chart* (Input-Process-Output) shows global structure and control weakly because it references these issues only as necessary

Tool \ Requirement	Structure		Control		Data	Data	Commu-	Documen-
	Global	Local	Global	Local	Flow	Structure	nication	tation
Function descriptions	5	5	5	5	4	5	1	1
Data flow diagrams	1	2	3	3	1	5	1	3
HIPO (VTOC)	1	1	2	5	2	5	1	1
HIPO (IPO Chart)	3	1	3	1	1	3	1	1
Structure charts (Yourdon)	1	1	3	3	1	5	1	1
Jackson charts	1	1	3	3	2	1	1	1
Warnier-Orr diagrams	1	1	3	3	2	1	2	1
Pseudocode	5	1	5	1	3	5	2	1
Chapin charts	5	1	5	1	3	5	2	1

1 = Strong support
2 = Strong support if modified
3 = Weak support
4 = Weak support if modified
5 = Tool does not address requirement

FIGURE 12 Comparison of design tools.

to place the module into perspective in the VTOC. Local structure and control are normally detailed here, however, often through the use of pseudocode. Data flow is considered explicitly, whereas data structure is implicit to the diagram. Again, this tool is a strong aid to documentation and communication.

The *structure chart* used by Yourdon and Constantine is a representation of hierarchical structure. As such, it can show both local and global structure, depending on its scale. Control is shown in terms of both the passage of control elements and the representation of sequence, selection, and iteration. However, the exact usage of specific control elements in actual control structures is weak. User modifications could be made to remedy this weakness, possibly through the use of pseudocode. But the tool cannot carry the extra detail without becoming overly complex.

Structure charts show data flow strongly, but do not represent data structure. They are strong in terms of communicational and documentation considerations. In effect, this tool "does it all" except for the representation of data structure. The only problem is that the graphic cannot support all of the information without becoming too cluttered and losing communicational ability. Thus, control detail tends to be sacrificed for overall clarity. This problem is not limited to structure charts, however. All hierarchical representations suffer from this problem.

Jackson Charts are very similar to other forms of hierarchy charts. The major differences between Jackson Charts and Structure Charts is that Jackson's approach does not show detailed data flows, but explicitly represents data structure. The data flow could, however, be added to the graphic.

Warnier-Orr Diagrams are related to Jackson Charts and the comments above will apply. However, Warnier-Orr Diagrams seem to require a higher degree of user education in order to serve as communication tools at that level.

Pseudocode, as a tool, supports any design methodology at the code level. It does not depict global structure of control, but it is very detailed in terms of local structure and control. Data flow and data structure are not considered. Pseudocode is a strong documentation tool, but some user education is necessary in order for it to be a good communicator.

Chapin Charts are an alternative representation of logic. As such, they can replace pseudocode in terms of the depiction of the logical constructs of sequence, selection, and iteration. Thus, the comments concerning pseudocode apply here.

SUPPORTING TECHNIQUES

Top-Down Concepts

There are a number of design concepts and techniques that exist independent of specific structured methodologies. Several such concepts fall under the generic term of *top-down methods*. These are:

1. Top-down analysis.
2. Top-down design.
3. Top-down construction (or coding).
4. Top-down testing.
5. Top-down (incremental) delivery.

Top-down analysis and design are similar in concept. Both involve the practice of breaking complex problems or functions into successive levels of their component parts. The terms, *stepwise refinement* and *decomposition* are synonymous here. In analysis, the result is that we first identify the primary function of the system or program and analyze the necessary subordinates. This process continues, creating a leveled system structure that is hierarchical. In design, the top-level function in the hierarchical structure is designed first, followed by successive levels of the system.

In practice, the definition of a stopping rule for this iterative decomposition is sometimes difficult. In a statistical process, for example, should the decomposition stop at the level of

"CALCULATE MEAN," or should that function be broken down to its subordinates of "CALCULATE TOTAL SUM" and "DIVIDE SUM BY NUMBER"? In both analysis and design, the general rule is to continue this process until the functions or processes are of sufficiently low complexity to allow visualization of how they are (or will be) implemented. This rule is, however, more flexible in design. If the person who is to implement the design in code is highly qualified, the design process is often stopped at a higher level than if programmer trainees will be involved.

The use of top-down analysis and design is central to all of the structured methodologies. These concepts assure an orderly and manageable execution of the complex tasks involved. The subunits of the system become visible early and their interfaces are defined clearly.

Top-down construction, testing, and *implementation* are also closely related. Here, we support the strategy of coding and testing high level modules first, followed by the construction and testing of successively lower levels of the system. In practice, high level modules may actually be coded and tested before the lower levels have been designed.

In contrast, the bottom-up approach involved the construction of the lowest level modules first, moving upward through the design. This required the creation and use of *driver modules* to exercise the low level code for tests. The common problem with the bottom-up strategy is that the latter parts of the construction and testing process involve the linkage of large blocks of the system. This causes vast increases in the complexity of the test and debug process, and is a common cause of projects staying at the 90 percent complete level for long periods of time because the system will not link and run at the high levels.

In general, the most complex and difficult module interfaces are at the top of the system structure. Top-down strategies, by building and testing the top of the structure, assure the correctness of these interfaces as early as possible.

As an example, consider the structure in Figure 13. We could code module A first and build four *stubs*, (Stubs B, C, D, and E) to test A. These stubs can simply accept the call from A and return control, or they can be used to pass test values to A or even simulate run-time events to allow timings of the system to be approximated. Next, module B can be constructed, requiring test stubs for modules F, G, and H. Of course, the testing of B includes execution of A, so the highest levels of the system are repeatedly exercised, resulting in a high degree of confidence at those levels.

Two major strategies exist for the top-down construction and testing of the structure in Figure 13. The first would involve the construction and testing of the system in the order: A-B-C-D-E-F-G-H-I-J-K-L-M-N-O. This would result in the de-

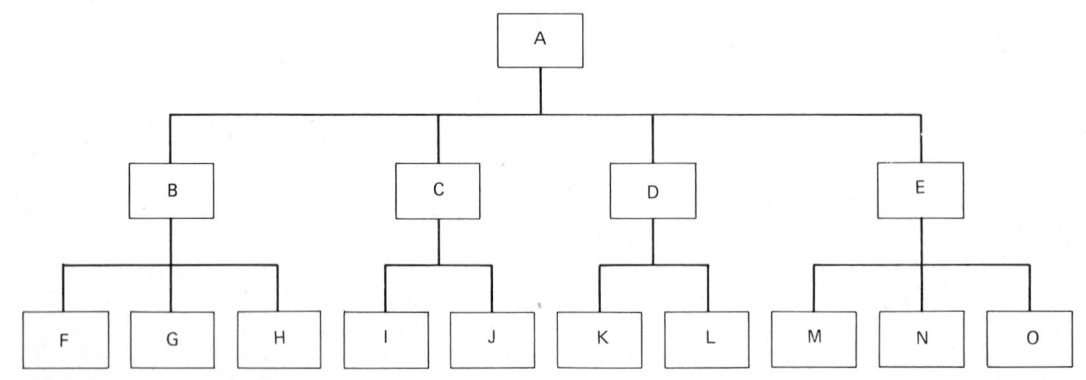

FIGURE 13 General hierarchical structure.

livery of the entire system at one time. Because of the nature of the testing process, the integration of new modules becomes less complex as the task nears completion.

Another strategy might result in construction and testing in the order: A-B-F-G-H-C-I-J. . . . This process allows the *incremental delivery* of portions of the system long before the entire package is ready. If, for example, modules B, F, G, and H supported the input, editing, and file creation for an accounting system, their early delivery could aid the file conversion process preceding full system conversion.

Walkthroughs

The testing referenced in the previous section assumed formal examination of coded modules and subsystems. Walkthroughs are less formal test procedures that can be implemented before the construction effort begins. A walkthrough involves the examination of the system, or any of its parts, by a group of people to assure design completeness and coherence. Walkthroughs can be conducted on requirement specifications, data flow diagrams, structure charts, or any other documentation that represents the system under consideration. At the structure chart level, the walkthrough would examine the hierarchical structure for functional completeness, coupling, and cohesion. Further detailed design efforts may not be allowed until the structure of the design successfully meets the approval of the walkthrough group.

At the code level, the walkthrough might concentrate on syntax and logical completeness. This type of code examination can drastically reduce machine test requirements and assure speedy development of code, particularly where test time is not readily available.

The major benefit of walkthroughs is that they allow each stage of the design to be tested before moving on to the next level of effort. In practice, organizations determine the number and timing of walkthroughs that best fit their needs, with good success.

Programming Teams

The *chief programmer team* concept, advanced in the early 1970s, was a highly formal structure, requiring a great many resources from the organization. Current teams are often less formal and almost always require fewer people. There are two primary benefits from the use of teams, and both can be obtained without the commitment of large numbers of individuals to a long-term effort. First, we seek to combine people into a team in order to assure the availability of the human resources and skills necessary to complete the project. This results in project teams in the classical sense.

A second major benefit of teams lies in their ability to maximize the productivity of systems professionals.

A *librarian*, for instance, can easily formalize diagrams, schedule meetings, maintain current listings, and record discussions. Such a person need not be of high technical skill, but he or she can perform tasks that are time-consuming for the higher paid professionals. The librarian may have a formal position, or an existing person may accept the duties while still performing other tasks.

Here we see that a part of the team concept may be very informal. An organization may choose to implement teams simply by a commitment to the merging of the individuals needed to satisfy specific requirements at a specific point in time. In this way, even small shops can benefit from this approach without formally creating organizational structures that may not be appropriate for their organization.

FUTURE DIRECTIONS

Though early applications of structured methods caused references to be made to the "structured revolution," no such revolution has occurred. In reality, both the development of structured approaches and their acceptance in industry can best be termed as evolutionary. We have seen an evolution of our understanding of these concepts from the microperspective of program logic, through concepts of design, and finally to the logical analysis process. Similarly, practitioners have adopted the new methodologies slowly, seeking positive results at low levels of commitment before making major adoptions of the entire set of techniques.

At this point in time, the current understanding of structured techniques represents a foundation upon which further improvements in the development process will stand. There are two major areas in which these improvements are likely to occur.

The first area of immediate ongoing development appears to involve the integration of the various competing methodologies in structured design. The vast utilization of the various design methodologies provides a framework for the analysis of the strengths and weaknesses of each. This supports comparative analysis of the type presented in this paper. It also pressures the proponents of specific methodologies to strengthen their techniques in handling broader classes of design issues. As this broadening occurs, the differences between the competing methodologies are lessened.

This process should, for the good of the industry, result in an integrated framework that allows users to "mix and match" elements of specific approaches subject to the specific needs of a development project. Papers that approach this integration are already appearing, and proponents of specific methodologies have begun formal discussions about the true nature of the similarities and differences among competing techniques.

Another area of development relates to the use of structured methods in the life cycle. Structured analysis, design, construction, testing, and maintenance are poorly linked, causing shops to move from methodology to methodology as they progress through the life cycle. The linkage of design and construction processes has typically been the strongest, whereas analysis and design have been poorly coupled. However, the recent work of DeMarco, Gane and Sarson, and Yourdon and Constantine has resulted in analysis processes that facilitate the movement to design. In addition, Ken Orr has now published a book on structured requirements definition to expand the scope of his approach.

The processes of testing and maintenance have been treated somewhat independently. Although some ties to structured design have been made, no unifying approaches exist.

We hope to see a continuing integration of these life-cycle processes to assure a coherence among the processes necessary during the life of a system. This type of process linkage would have major effects on the quality of our productive efforts on software.

Other areas where efforts may occur include the automation of the methodologies, the formalization of more complete and usable documentation processes, and the rigorous study of productivity implications.

CONCLUSIONS

This paper supports the argument that there is no single best methodology for the development of systems. The methodologies reviewed here vary in terms of the relative importance placed on specific design issues. They also vary with respect to their ability to support the set of processes inherent in the development cycle. This variation in terms of emphasis and scope has major implications for the software designer.

In addition, the design methodologies reviewed here fail to provide the designer with a coherent approach that can be expected to result in the automatic design of high quality systems. Although SADT does tightly structure the problem analysis, the other approaches actually provide design guidelines rather than specific design instructions. This issue is also of major importance.

System design is not a science. The development of specific methodologies has moved the field from the old "artsy" processes to improved procedural approaches to design, but much is still left to the abilities of the designer. The most important perspective that can be provided at this time is as follows: It is the designer's responsibility to choose the methodology and tools best suited to a particular design environment and to modify them as necessary to meet the needs imposed by specific design problems.

The choice of a methodology must relate to the system under development. The primary design issue must be identified and a methodology chosen which places a high priority on that issue. Other considerations must then be examined on an iterative basis after an initial design is developed. Similarly, tools must be chosen that support the design process from the necessary directions.

Although the software industry has advanced rapidly in the past decade, further advances will occur in this decade. One important direction of growth involves the intelligent choice of design strategies for software systems. This choice may vary between systems, and modification of tools and methodologies may be necessary to meet specific needs. For some systems, a customized mix of methods and tools may best serve to aid and direct the design process. It is hoped that this paper can contribute to that perspective.

REFERENCES

Baker, A. L. and Zweben, S. H., "The Use of Software Science in Evaluating Modularity Concepts," *Software Engineering*, March 1979, pp. 110–120.

Baker, F. T., "Chief Programmer Team Management of Production Programming," *IBM Systems Journal*, Vol. 11, No. 1, 1972, pp. 56–73.

Barbuto, P. F., Jr. and Geller, J., "Tools for Top-Down Testing," *Datamation*, October 1978, pp. 178–182.

Barrese, A. L. and Shapiro, S. D., "Structuring Programs for Efficient Operation in Virtual Memory Systems," *Software Engineering*, November 1979, pp. 643–652.

Bergland, G. D., "Structured Design Methodologies" in Bergland, G. D., and Bergland, R. D., *Tutorial: Software Design Strategies*, IEEE Computer Society, 1979.

Boehm, B. W., "Software Engineering," in Bergland, G. D. and Bergland, R. D., *Tutorial: Software Design Strategies*, IEEE Computer Society, 1979.

Brown, P. J., "Programming and Documenting Software Projects," *ACM Computing Surveys*, Vol. 6, No. 4, 1974, pp. 213–220.

Caine, S. H., and Gordon, E. K., "PDL—A Tool for Software Design," *Proceedings, National Computer Conference*, AFIPS Press, 1975.

Comer, D. and Halstead, M. H., "A Simple Experiment in Top-Down Design," *Software Engineering*, March 1979, pp. 105–109.

Curtis, B. et al., "Measuring the Psychological Complexity of Software Maintenance Tasks With the Halstead and McCabe Metrics," *Software Engineering*, March 1979, pp. 95–104.

DeMarco, T., *Structured Analysis and System Specification*. New York: Yourdon Press, 1978.

Elspas, B. et al., "An Assessment of Techniques for Proving Program Correctness," *ACM Computing Surveys*, Vol. 4, No. 2, 1972, pp. 97–147.

Enos, J. C. and Van Tilburg, R. L., "Software Design," *IEEE Computer*, February 1981, pp. 61–82.

Fagan, M. E., "Design and Code Inspections to Reduce Errors in Program Development," *IBM Systems Journal*, Vol. 15, No. 3, 1976, pp. 182–211.

Fagan, M. E., "Inspecting Software Design and Code," *Datamation*, October 1977, pp. 133–144.

Fitzsimmons, A. and Love, T., "A Review and Evaluation of Software Science," *ACM Computing Surveys*, Vol. 10, No. 1, 1978, pp. 3–18.

Fosdick, L. D. and Osterweil, L. J., "Data Flow Analysis in Software Reliability," *ACM Computing Surveys*, Vol. 8, No. 3, 1976, pp. 305–330.

Freeman, P., "Software Reliability and Design: A Survey," *Proceedings, 13th Design Automation Conference*, June 1976.

Gane, C. and Sarson, T., *Structured Systems Analysis: Tools and Techniques*. Englewood Cliffs, NJ: Prentice-Hall, Inc., 1979.

Gane, C. and Sarson, T., "Structured Methodologies: What Have We Learned" *Computer World/Extra*, Vol. XIV, No. 38, September 17, 1981, pp. 52–57.

Gordon, R. D., "A Qualitative Justification for a Measure of Program Clarity," *Software Engineering*, March 1979, pp. 121–127.

Gordon, R. D., "Measuring Improvements in Program Clarity," *Software Engineering*, March 1979, pp. 79–90.

Griffiths, S. N., "Design Methodologies: A Comparison," in Bergland, G. D. and Bergland, R. D., *Tutorial: Software Design Strategies*. IEEE Computer Society, 1979.

Hantler, S. L., and King, J. C., "An Introduction to Proving the Correctness of Programs," *ACM Computing Surveys*, Vol. 8, No. 3, 1976, pp. 331–353.

Higgins, D. A., *Program Design and Construction*. Englewood Cliffs, NJ: Prentice-Hall, Inc., 1979.

Higgins, D. A., "Structured Programming with Warnier-Orr Diagrams," in Bergland, G. D. and Bergland, R. D., *Tutorial: Software Design Strategies*. IEEE Computer Society, 1979.

Huang, J. C., "An Approach to Program Testing," *ACM Computing Surveys*, Vol. 7, No. 3, pp. 113–128.

IBM, *HIPO—A Design Aid and Documentation Technique*. IBM Form No. GC20-1851, White Plains, NY.

Jackson, M. A., "Constructive Methods of Program Design," in Bergland, G. D. and Bergland, R. D., *Tutorial: Software Design Strategies*. IEEE Computer Society, 1979.

Jackson, M. A., *Principles of Program Design*. New York, NY: Academic Press Inc., 1975.

Jensen, R. W., "Structured Programming," *IEEE Computer*, March 1981, pp. 31–48.

Jones, M. N., "HIPO for Developing Specifications," *Datamation*, March 1976.

Jones, T. C., "Measuring Programming Quality and Productivity," *IBM Systems Journal*, Vol. 17, No. 1, 1978, pp. 39–63.

Knuth, D. E., "Structured Programming with Go To Statements," *ACM Computing Surveys*, Vol. 6, No. 4, 1974, pp. 261–302.

Kowalski, R. A., "Algorithm = Logic + Control," *Communications of the ACM*, July 1979, pp. 424–436.

Myers, G. J., *The Art of Software Testing*. New York NY: John Wiley & Sons, Inc., 1979.

Myers, G. J., "Composite Design Facilities of Six Programming Languages," *IBM Systems Journal*, Vol. 15, No. 3, 1976, pp. 212–224.

Myers, G. J., *Composite/Structured Design*. New York, NY: Van Nostrand Reinhold, 1978.

Myers, G. J., "A Controlled Experiment in Program Testing and Code Walkthroughs/Inspections," *Communications of the ACM*, September 1978, pp. 760–768.

Naur, P. and Randall, B., eds., *Software Engineering*. Brussels, Belgium: NATO Scientific Affairs Division, 1969.

Orr, Kenneth, *Structured Requirements Definition.* Topeka, Kansas: Ken Orr and Associates, Inc., 1981.

Orr, Kenneth, *Structured Systems Development.* New York: Yourdon Press, 1977.

Parnas, D. L., "Designing Software for Ease of Extension and Contraction," *Software Engineering,* March 1979, pp. 128–136.

Peters, L. J. and Tripp, L. L., "Comparing Software Design Methodologies," in Bergland, G. D. and Bergland R. D., *Tutorial: Software Design Strategies.* IEEE Computer Society, 1979.

Randell B. et al., "Reliability Issues in Computing System Design," *ACM Computing Surveys,* Vol. 10, No. 2, 1978, pp. 123–166.

Ross, D. T., "Structured Analysis (SA): A Language for Communicating Ideas," *IEEE Transactions on Software Engineering,* January 1977.

Ross, D. T. and Schoman, K. E., "Structured Analysis for Requirements Definition," *IEEE Transactions on Software Engineering,* January 1977.

Ross, D. T. et al., "Software Engineering: Process, Principles, and Goals," *IEEE Computer,* May 1975, pp. 17–27.

Stay, J. F., "HIPO and Integrated Program Design," *IBM Systems Journal,* Vol. 15, No. 2, 1976, pp. 143–154.

Stevens, W. P. et al., "Structured Design," *IBM Systems Journal,* Vol. 13, No. 2, 1974, pp. 115–139.

Van Leer, P., "Top-Down Development Using a Program Design Language, *IBM Systems Journal,* Vol. 15, No. 2, 1976, pp. 155–170.

Warnier, J. D., *Logical Construction of Programs.* New York,: Van Nostrand Reinhold Company, 1974.

Wasserman, A. I., "Information System Design Methodology," *Journal of the American Society for Information Science,* January 1980.

Weinberg, G. M., "You Say Your Design's Inexact? Try a Wiggle," *Datamation,* August 1979, pp. 146–149.

Weinberg, V., *Structured Analysis.* Englewood Cliffs, NJ: Prentice-Hall, Inc., 1980.

Wirth, N., *Algorithms + Data Structures = Programs.* Englewood Cliffs, NJ: Prentice-Hall, Inc., 1976.

Wirth, N., "On the Composition of Well-Structured Programs," *ACM Computing Surveys,* Vol. 6, No. 4, 1974, pp. 247–260.

Woodware, M. R., Hennell, M. A., and Hedley, D., "A Measure of Control Flow Complexity in Program Text," *IEEE Transactions on Software Engineering,* Vol. SE-5, No. 1, January 1979.

Yourdon, E., "Top-Down Design and Testing" in Bergland, G. D. and Bergland, R. D., *Tutorial: Software Design Strategies.* IEEE Computer Society, 1979.

Yourdon, E. and Constantine, L., *Structured Design: Fundamentals of a Discipline of Computer Program and Systems Design.* Englewood Cliffs, NJ: Prentice-Hall, Inc., 1979.

Zelkowitz, M. V., "Perspectives on Software Engineering," *ACM Computing Surveys,* Vol. 10, No. 2, 1978, pp. 197–216.

READING QUESTIONS

1. How has the evolution of structured analysis, structured design, and structured programming strengthened the interfaces between analysis, design, and programming?
2. Briefly define the four structural points of view.
3. Relate the concept of system structure to the processes of analysis, design, and programming. Does our utilization of the concept change as we move through the life cycle? If so, how?

4. Compare the general data flow diagram (Figure 1) with the general SADT activity diagram (Figure 2) from the viewpoint of the representation of system structural information. Which structural viewpoints are represented in each?
5. Some analysts prefer the use of the DFD over the SADT diagram for analysis. What strengths and/or weaknesses do you see in this approach?
6. State the major objectives of the HIPO documentation package.
7. The Interfunction Communication Table details data flows that result from the hierarchical structure of the UTOC. How is this stronger or weaker than the use of the DFD to show data flows?
8. What structural viewpoints are represented in the modified UTOC of Figure 8? Does HIPO allow the detailing of other structural viewpoints?
9. Compare the data flow representation of Figure 9 with that of Figure 6. Which do you prefer from the point of view of:
   - clarity
   - complexity
   - completeness

   Explain your answer
10. All of the methodologies discussed in this paper are incomplete from a life-cycle perspective. Choose a set of methods and graphics that could be combined to result in a complete methodology across the life cycle. Be complete and justify your solution.
11. Analyze your solution for Question 10, above, from a tool perspective (Figure 12).

Software Engineering

Barry W. Boehm

I. INTRODUCTION

The annual cost of software in the U.S. is approximately 20 billion dollars. Its rate of growth is considerably greater than that of the economy in general. Compared to the cost of computer hardware, the cost of software is continuing to escalate along the lines predicted in Fig. 1 [1].[1] A recent SHARE study [2] indicates further that software demand over the years 1975–1985 will grow considerably faster (about 21–23 percent per year) than the growth rate in software supply at current estimated growth rates of the software labor force and its productivity per individual, which produce a combined growth rate of about 11.5–17 percent per year over the years 1975–1985.

In addition, as we continue to automate many of the processes which control our life-style—our medical equipment, air traffic control, defense system, personal records, bank accounts—we continue to trust more and more in the reliable functioning of this proliferating mass of software. *Software engineering* is the means by which we attempt to produce all of this software in a way that is both cost-effective and reliable enough to deserve our trust. Clearly, it is a

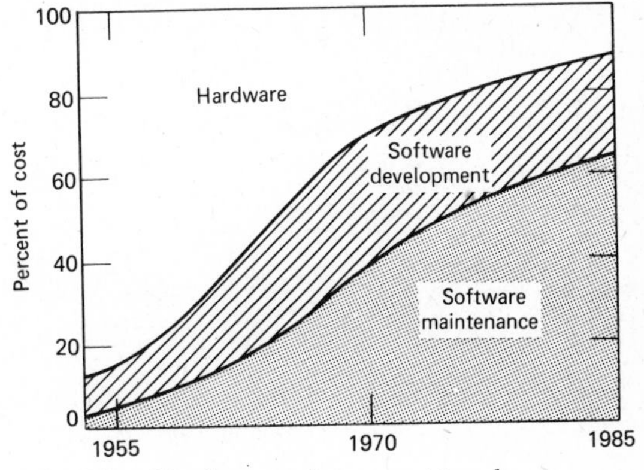

FIGURE 1 Hardware-software cost trends.

[1]Another trend has been added to Fig. 1: the growth of software maintenance, which will be discussed later.

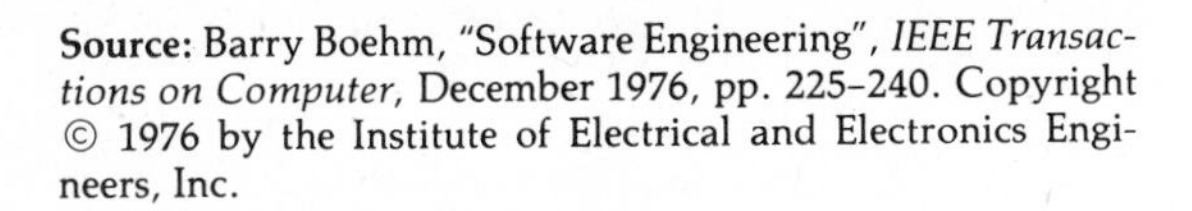

discipline which is important to establish well and to perform well.

This paper will begin with a definition of "software engineering." It will then survey the current state of the art of the discipline, and conclude with an assessment of likely future trends.

II. DEFINITIONS

Let us begin by defining "software engineering." We will define software to include not only computer programs, but also the associated documentation required to develop, operate, and maintain the programs. By defining software in this broader sense, we wish to emphasize the necessity of considering the generation of timely documentation as an integral portion of the software development process. We can then combine this with a definition of "engineering" to produce the following definition.

Software Engineering. The practical application of scientific knowledge in the design and construction of computer programs and the associated documentation required to develop, operate, and maintain them.

Three main points should be made about this definition. The first concerns the necessity of considering a broad enough interpretation of the word "design" to cover the extremely important activity of software requirements engineering. The second point is that the definition should cover the entire software life cycle, thus including those activities of redesign and modification often termed "software maintenance." (Fig. 2 indicates the overall set of activities thus encompassed in the definition.) The final point is that our store of knowledge about software which can really be called "scientific knowledge" is a rather small base upon which to build an engineering discipline. But, of course, that is what makes software engineering such a fascinating challenge at this time.

The remainder of this paper will discuss the state of the art of software engineering along the lines of the software life cycle depicted in Fig. 2. Section III contains a discussion of software requirements engineering, with some mention of the problem of determining overall system requirements. Section IV discusses both preliminary design and detailed design technology trends. Section V contains only a brief discussion of programming, as this topic is also covered in a companion article in this issue [3] Section VI covers both software testing and the overall life cycle concern with software reliability. Section VII discusses the highly important but largely ne-

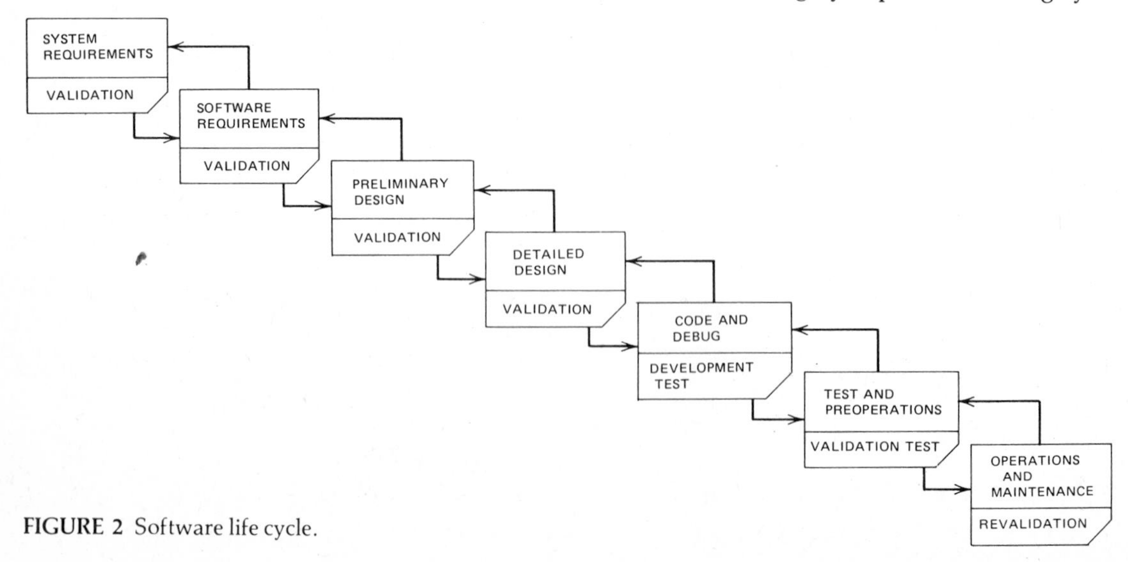

FIGURE 2 Software life cycle.

glected area of software maintenance. Section VIII surveys software management concepts and techniques, and discusses the status and trends of integrated technology-management approaches to software development. Finally, Section IX concludes with an assessment of the current state of the art of software engineering with respect to the definition above.

Each section (sometimes after an introduction) contains a short summary of current practice in the area, followed by a survey of current frontier technology, and concluding with a short summary of likely trends in the area. The survey is oriented primarily toward discussing the domain of applicability of techniques (where and when they work) rather than how they work in detail. An extensive set of references is provided for readers wishing to pursue the latter.

III. SOFTWARE REQUIREMENTS ENGINEERING

Critical Nature of Software Requirements Engineering

Software requirements engineering is the discipline for developing a complete, consistent, unambiguous specification—which can serve as a basis for common agreement among all parties concerned—describing *what* the software product will do (but *not how* it will do it; this is to be done in the design specification).

The extreme importance of such a specification is only now becoming generally recognized. Its importance derives from two main characteristics: 1) it is easy to delay or avoid doing throughly; and 2) deficiencies in it are very difficult and expensive to correct later.

Fig. 3 shows a summary of current experience at IBM [4], GTE [5], and TRW on the relative cost of correcting software errors as a function of the phase in which they are corrected. Clearly, it pays off to invest effort in finding requirements errors early and correcting them in, say, 1 man-hour rather than waiting to find the error during operations and having to spend 100 man-hours correcting it.

Besides the cost-to-fix problems, there are other critical problems stemming from a lack of a good requirements specification. These include [6]: 1) top-down designing is impossible, for lack of a well-specified "top"; 2) testing is impossible, because there is nothing to test against; 3) the user is frozen out, because there is no clear statement of what is being produced for him; and 4) management is not in control, as there is no clear statement of what the project team is producing.

Current Practice

Currently, software requirements specifications (when they exist at all) are generally expressed in free-form English. They abound with ambiguous terms ("suitable," "sufficient," "real-time," "flexible") or precise-sounding terms with unspecified definitions ("optimum," "99.9 percent reliable") which are potential seeds of dissension or lawsuits once the software is produced. They have numerous errors; one recent study [7] indicated that the first independent review of a fairly good software requirements specification will find from one to four nontrivial errors per page.

Trends

In the area of requirements statement languages, we will see further efforts either to extend the ISDOS-PSL and SREP-RSL capabilities to handle further areas of application, such as man-machine intractions, or to develop language variants specific to such areas. It is still an open question as to how general such a language can be and still retain its utility. Other open questions are those of the nature, "which representation scheme is best for describing requirements in a certain area?" BMDATC is sponsoring some work here in representing general data-processing system requirements for the BMD problem,

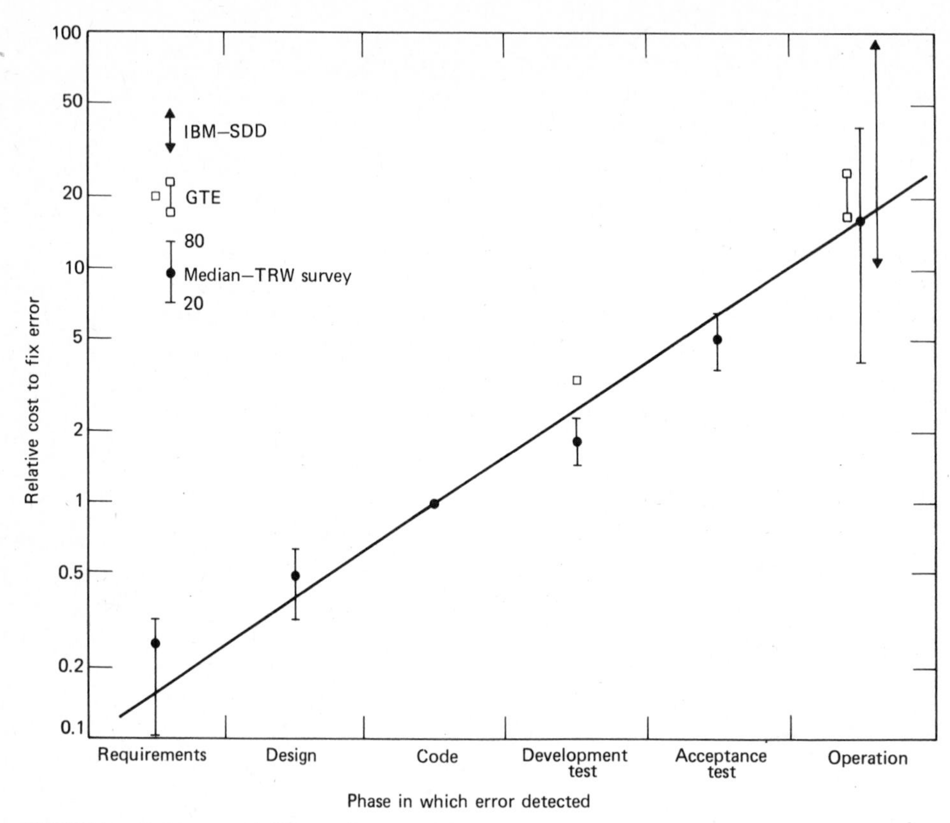

FIGURE 3 Software validation: the price of procrastination.

involving Petri nets, state transition diagrams, and predicate calculus [11]. but its outcome is still uncertain.

A good deal more can and will be done to extend the capability of requirements statement analyzers. Some extensions are fairly straightforward consistency checking; others, involving the use of relational operators to deduce derived requirements and the detection (and perhaps generation) of missing requirements are more difficult, tending toward the automatic programming work.

Other advances will involve the use of formal requirements statements to improve subsequent parts of the software life cycle. Examples include requirements-design-code consistency checking (one initial effort is underway), the automatic generation of test cases from requirements statements, and, of course, the advances in automatic programming involving the generation of code from requirements.

Progress will not necessarily be evolutionary, though. There is always a good chance of a breakthrough: some key concept which will simplify and formalize large regions of the problem space. Even then, though, there will always remain difficult regions which will require human insight and sensitivity to come up with an acceptable set of software requirements.

Another trend involves the impact of having formal, machine-analyzable requirements (and design) specifications on our overall inventory of software code. Besides improving software reliability, this will make our software much more portable; users will not be tied so much to a particular machine configuration. It is interesting to speculate on what impact this will have on hardware vendors in the future.

IV. SOFTWARE DESIGN

The Requirements/Design Dilemma

Ideally, one would like to have a complete, consistent, validated, unambiguous, machine-independent specification of software requirements before proceeding to software design. However, the requirements are not really validated until it is determined that the resulting system can be built for a reasonable *cost*—and to do so requires developing one or more software *designs* (and any associated hardware designs needed).

This dilemma is complicated by the huge number of degrees of freedom available to software/hardware system designers. In the 1950's, the designer had only a few alternatives to choose from in selecting a central processing unit (CPU), a set of peripherals, a programming language, and an ensemble of support software. In the 1970's, with rapidly evolving mini- and microcomputers, firmware, modems, smart terminals, data management systems, etc., the designer had an enormous number of alternative design components to sort out (possibilities) and to seriously choose from (likely choices). In the 1980's, as indicated by Table I, the number of possible design combinations is formidable.

TABLE 1
Design Degrees of Freedom for New Data Processing Systems (Rough Estimates)

Element	Choices or Possibilities 1960	1980
CPU	5	200
OP-Codes	Fixed	Variable
Peripherals (per function)	1	100
Programming languages	3	30 or more
Operating systems	1	10 or more
Data management systems	0	50
Communications	1	20 or more

The following are some of the implications for the designer. (1) It is easier for him to do an outstanding design job. (2) It is easier for him to do a terrible design job. (3) He needs more powerful analysis tools to help him sort out the alternatives. (4) He has more opportunities for designing-to-cost. (5) He has more opportunities to design and develop tunable systems. (6) He needs a more flexible requirements-tracking and hardware procurement mechanism to support the above flexibility (particularly in government systems). (7) Any rational standardization (e.g., in programming languages) will be a big help to him, in that it reduces the number of alternatives he must consider.

Current Practice

Software design is still almost completely a manual process. There is relatively little effort devoted to design validation and risk analysis before committing to a particular software design. Most software errors are made during the design phase. As seen in Fig. 4, which summarizes several software error analyses by IBM [4], [19] and TRW [20], [21], the ratio of design to coding errors generally exceeds 60:40. (For the TRW data, an error was called a design error if and only if the resulting fix required a change in the detailed design specification.)

Most software design is still done bottom-up, by developing software components before addressing interface and integration issues. There is, however, increasing successful use of top-down design. There is little organized knowledge of what a software designer does, how he does it, or what makes a good software designer, although some initial work along these lines has been done by Freeman [22].

Current Frontier Technology

Relatively little is available to help the designer make the overall hardware-software tradeoff analyses and decisions to appropriately narrow the large number of design degrees of freedom available to him. At the micro level, some formalisms such as LOGOS [23] have been helpful, but at the macro level, not much is available beyond general system engineering techniques. Some help is provided via improved techniques for simulating information systems, such as the Extendable Computer System Simulator (ECSS) [24], [25], which make it possible to develop a fairly thorough functional simulation of the system for design analysis in a considerably shorter time than it takes to develop the complete design itself.

Top-Down Design. Most of the helpful new techniques for software design fall into the category of "top-down" approaches, where the "top" is already assumed to be a firm, fixed requirements specification and hardware architecture. Often, it is also assumed that the data structure has also been established. (These assumptions must in many cases be considered potential pitfalls in using such top-down techniques.)

What the top-down approach does well, though, is to provide a procedure for organizing and developing the control structure of a program in a way which focuses early attention on the critical issues of integration and interface definition. It begins with a top-level expression of a hierarchical control structure (often a top level "executive" routine controlling an "input," a "process," and an "output" routine) and proceeds to iteratively refine each successive lower-level component until the entire system is specified. The successive refinements, which may be considered as "levels of abstraction" or "virtual machines" [26], provide a number of advantages in improved understanding, communication, and verification of complex designs [27], [28]. In general, though, experience shows that some degree of early attention to bottom-level design issues is necessary on most projects [29].

The technology of top-down design has centered on two main issues. One involves establishing guidelines for *how to perform* successive refinements and to group functions into modules; the other involves techniques of *representing* the design of the control structure and its interaction with data.

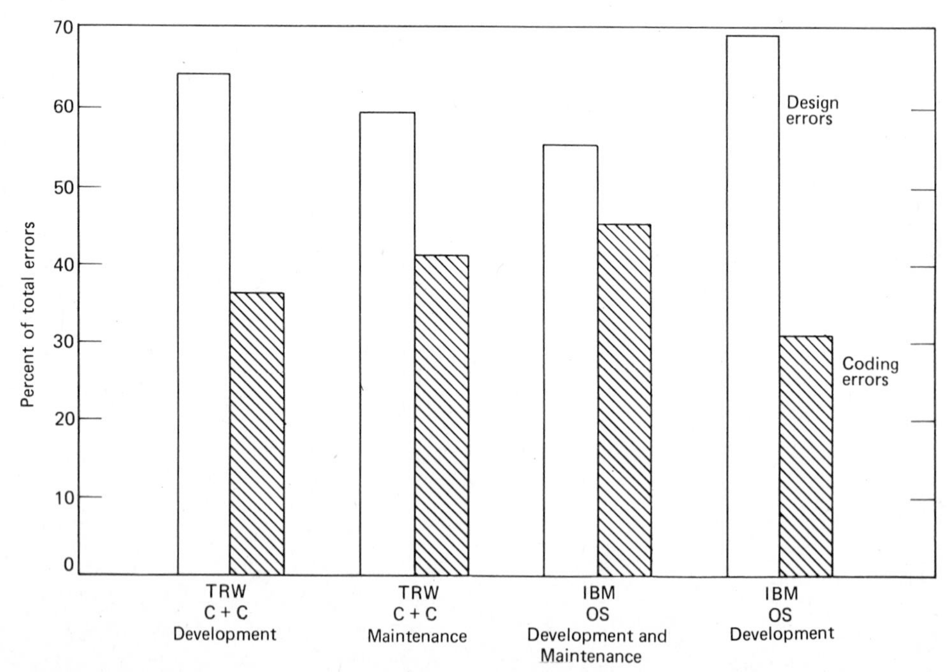

FIGURE 4 Most errors in large software systems are in early stages.

Modularization. The techniques of structured design [30] (or composite design [31]) and the modularization guidelines of Parnas [32] provide the most detailed thinking and help in the area of module definition and refinement. Structured design establishes a number of successively stronger types of binding of functions into modules (coincidental, logical, classical, procedural, communicational, informational, and functional) and provides the guideline that a function should be grouped with those functions to which its binding is the strongest. Some designers are able to use this approach quite successfully; others find it useful for reviewing designs but not for formulating them; and others simply find it too ambiguous or complex to be of help. Further experience will be needed to determine how much of this is simply a learning curve effect. In general, Parnas' modularization criteria and guidelines are more straightforward and widely used than the levels-of-binding guidelines, although they may also be becoming more complicated as they address such issues as distribution of responsibility for erroneous inputs [33]. Along these lines, Draper Labs' Higher Order Software (HOS) methodology [34] has attempted to resolve such issues via a set of six axioms covering relations between modules and data, including responsibility for erroneous inputs. For example, Axiom 5 states, "Each module controls the rejection of invalid elements of its own, and only its own, input set."[2]

Design Representation. Flow charts remain the main method currently used for design representation. They have a number of deficiencies, particularly in representing hierarchical control structures and data interactions. Also, their free-form nature makes it too easy to construct complicated, unstructured designs which are hard to understand and maintain. A number of representation schemes have been developed to avoid these deficiencies.

The hierarchical input-process-output (HIPO) technique [35] represents software in a hierarchy of modules, each of which is represented by its inputs, its outputs, and a summary of the processing which connects the inputs and outputs. Advantages of the HIPO technique are its ease of use, ease of learning, easy-to-understand graphics, and disciplined structure. Some general disadvantages are the ambiguity of the control relationships (are successive lower level modules in sequence, in a loop, or in an if/else relationship?), the lack of summary information about data, the unwieldiness of the graphics on large systems, and the manual nature of the technique. Some attempts have been made to automate the representation and generation of HIPO's such as Univac's PROVAC System [36].

The structure charts used in structured design [30], [31] remedy some of these disadvantages, although they lose the advantage of representing the processes connecting the inputs with the outputs. In doing so, though, they provide a more compact summary of a module's inputs and outputs which is less unwieldy on large problems. They also provide some extra symbology to remove at least some of the sequence/loop/branch ambiguity of the control relationships.

Several other similar conventions have been developed [37]-[39], each with different strong points, but one main difficulty of any such manual system is the difficulty of keeping the design consistent and up-to-date, especially on large problems. Thus, a number of systems have been developed which store design information in machine-readable form. This simplifies updating (and reduces update errors) and facilitates generation of selective design summaries and simple consistency checking. Experience has shown that even a simple set of automated consistency checks can catch dozens of potential problems in a large design specification [21]. Systems of this nature that have been reported include the Newcastle TOPD system [40], TRW's DACC and DEVISE systems [21], Boeing's DECA System [41],

[2]Problems can arise, however, when one furnishes such a *design choice* with the power of an axiom. Suppose, for example, the input set contains a huge table or a master file. Is the module stuck with the job of checking it, by itself, every time?

and Univac's PROVAC [36]; several more are under development.

Another machine-processable design representation is provided by Caine, Farber, and Gordon's Program Design Language (PDL) System [42]. This system accepts constructs which have the form of hierarchical structured programs, but instead of the actual code, the designer can write some English text describing what the segment of code will do. (This representation was originally called "structured pidgin" by Mills [43].) The PDL system again makes updating much easier; it also provides a number of useful formatted summaries of the design information, although it still lacks some wished-for features to support terminology control and version control. The program-like representation makes it easy for programmers to read and write PDL, albeit less easy for nonprogrammers. Initial results in using the PDL system on projects have been quite favorable.

Trends

Once a good deal of design information is in machine-readable form, there is a fair amount of pressure from users to do more with it: to generate core and time budgets, software cost estimates, first-cut data base descriptions, etc. We should continue to see such added capabilities, and generally a further evolution toward computer-aided-design systems for software. Besides improvements in determining and representing control structures, we should see progress in the more difficult area of data structuring. Some initial attempts have been made by Hoare [44] and others to provide a data analog of the basic control structures in structured programming, but with less practical impact to date. Additionally, there will be more integration and traceability between the requirements specification, the design specification, and the code—again with significant implications regarding the improved portability of a user's software.

The proliferation of minicomputers and microcomputers will continue to complicate the designer's job. It is difficult enough to derive or use principles for partitioning software jobs on single machines; additional degrees of freedom and concurrency problems just make things so much harder. Here again, though, we should expect at least some initial guidelines for decomposing information processing jobs into separate concurrent processes.

It is still not clear, however, how much one can formalize the software design process. Surveys of software designers have indicated a wide variation in their design styles and approaches, and in their receptiveness to using formal design procedures. The key to good software design still lies in getting the best out of good people, and in structuring the job so that the less-good people can still make a positive contribution.

V. PROGRAMMING

This section will be brief, because much of the material will be covered in the companion article by Wegner on "Computer Languages" [3].

Current Practice

Many organizations are moving toward using structured code [28], [43] (hierarchical, block-oriented code with a limited number of control structures—generally SEQUENCE, IFTHENELSE, CASE, DOWHILE, and DOUNTIL—and rules for formatting and limiting module size).A great deal of terribly unstructured code is still being written, though, often in assembly language and particularly for the rapidly proliferating minicomputers and microcomputers.

Current Frontier Technology

Languages are becoming available which support structured code and additional valuable features such as data typing and type checking (e.g., Pascal [45]). Extensions such as concurrent Pascal [46] have been developed to support the programming of concurrent processes. Extensions to data typing involving more explicit binding of procedures and their data have been embodied in recent languages such as ALPHARD [47] and CLU [48]. Metacompiler and compiler writing system technology continues to improve, although much more slowly in the code generation area than in the syntax analysis area.

Automated aids include support systems for top-down structured programming such as the Program Support Library [49], Process Construction [50], TOPD [40], and COLUMBUS [51]. Another novel aid is the Code Auditor program [50] for automated standards compliance checking—which guarantees that the standards are more than just words. Good programming practices are now becoming codified into style handbooks, i.e., Kernighan and Plauger [52] and Ledgard [53].

Trends

It is difficult to clean up old programming languages or to introduce new ones into widespread practice. Perhaps the strongest hope in this direction is the current Department of Defense (DoD) effort to define requirements for its future higher order programming languages [54], which may eventually lead to the development and widespread use of a cleaner programming language. Another trend will be an increasing capability for automatically generating code from design specifications.

VI. SOFTWARE TESTING AND RELIABILITY

Current Practice

Surprisingly often, software testing and reliability activities are still not considered until the code has been run the first time and found not to work. In general, the high cost of testing (still 40–50 percent of the development effort) is due to the high cost of reworking the code at this stage (see Fig. 3), and to the wasted effort resulting from the lack of an advance test plan to efficiently guide testing activities.

In addition, most testing is still a tedious manual process which is error-prone in itself. There are few effective criteria used for answering the question, "How much testing is enough?" except the usual "when the budget (or schedule) runs out." However, more and more organizations are now using disciplined test planning and some objective criteria such as "exercise every instruction" or "exercise every branch," often with the aid of automated test monitoring tools and test case planning aids. But other technologies, such as mathematical proof techniques, have barely begun to penetrate the world of production software.

Current Frontier Technology

Software Reliability Models and Phenomenology. Initially, attemps to predict software reliability (the probability of future satisfactory operation of the software) were made by applying models derived from hardware reliability analysis and fitting them to observed software error rates [55]. These models worked at times, but often were unable to explain actual experienced error phenomena. This was primarily because of fundamental differences between software phenomenology and the hardware-oriented assumptions on which the models were based. For example, software components do not degrade due to wear or fatigue; no imperfection or variations are introduced in making additional copies of a piece of software (except possibly for a class of easy-to-check copying errors); repair of a software fault generally results in a different software configuration than previously, unlike most hardware replacement repairs.

Models are now being developed which provide explanations of the previous error histories in terms of appropriate software phenomenology. They are based on a view of a software program as a mapping from a space of inputs into a space of outputs [56], of program operation as the processing of a sequence of points in the input space, distributed according to an operational profile [57], and of testing as a sampling of

MINIMUM VARIANCE UNBIASED ESTIMATOR

- PICK N(SAY, 1000) RANDOM, REPRESENTATIVE INPUTS
- PROCESS THE 1000 INPUTS. OBTAIN M(SAY, 3) FAILURES
- THEN R = PROB(NO FAILURES NEXT RUN) = $\frac{N-M}{N}$ = 0.997

Operational estimation problems

- Size of input space
- Accounting for fixes
- Ensuring random inputs
- Ensuring representative inputs

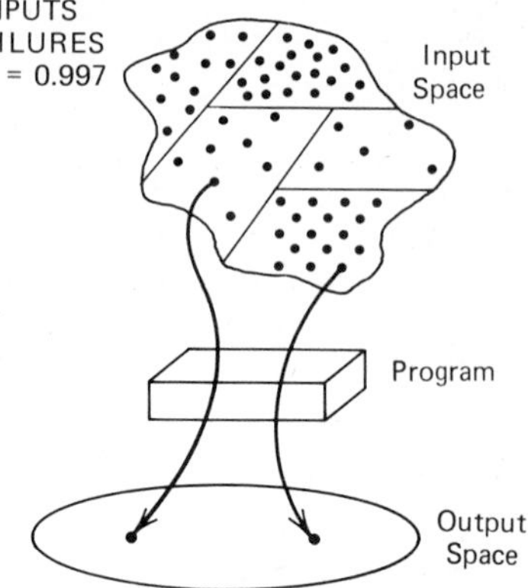

FIGURE 5 Input space sampling provides a basis for software reliability measurement.

points from the input space [56] (see Fig. 5). This approach encounters severe problems of scale on large programs, but can be used conceptually as a means of appropriately conditioning time-driven reliability models [58]. Still, we are a long way off from having truly reliable reliability-estimation methods for software.

Software Error Data. Additional insights into reliability estimation have come from analyzing the increasing data base of software errors. For example, the fact that the distributions of serious software errors are dissimilar from the distributions of minor errors [59] means that we need to define "errors" very carefully when using reliability prediction models. Further, another study [60] found that the rates of fixing serious errors and of fixing minor errors vary with management direction. ("Close out all problems quickly" generally gets minor simple errors fixed very quickly, as compared to "Get the serious problems fixed first.")

Other insights afforded by software data collection include better assessments of the relative efficacy of various software reliability techniques [4], [19], [60], identification of the requirements and design phases as key leverage points for cost savings by eliminating errors earlier (Figs. 2 and 3), and guidelines for organizing test efforts (for example, one recent analysis indicated that over half the errors were experienced when the software was handling data singularities and extreme points [60]. So far, however, the proliferation of definitions of various terms (error, design phase, logic error, validation test), still make it extremely difficult to compare error data from different sources. Some efforts to establish a unified software reliability data base and associated standards, terminology, and data collection procedures are now under way at USAF Rome Air Development Center, and within the IEEE Technical Committee on Software Engineering.

Automated Aids. Let us sketch the main steps of testing between the point the code has been written and the point it is pronounced acceptable for use, and describe for each stop the main types of automated aids which have been found helpful. More detailed discussion of these aids can be found in the surveys by Reifer [61] and Ramamoorthy and Ho [62] which in turn have references to individual contributions to the field.

Static Code Analysis. Automated aids here include the usual compiler diagnostics, plus extensions involving more detailed data-type checking. Code auditors check for standards compliance, and can also perform various type-checking functions. Control flow and reachability analysis is done by structural analysis programs (flow charters have been used for some of the elementary checks here, "structurizers" can also be helpful). Other useful static analysis tools perform set-use analysis of data elements, singularity analysis, units consistency analysis, data base consistency checking, and data-versus-code consistency checking.

Test Case Preparation. Extensions to structural analysis programs provide assistance in choosing data values which will make the program execute along a desired path. Attempts have been made to automate the generation of such data values; they can generally succeed for simple cases, but run into difficulty in handling loops or branching on complex calculated values (e.g., the results of numerical integration). Further, these programs only help generate the *inputs*; the tester must still calculate the expected outputs himself.

Another set of tools will automatically insert instrumentation to verify that a desired path has indeed been exercised in the test. A limited capability exists for automatically determining the minimum number of test cases required to exercise all the code. But, as yet, there is no tool which helps to determine the most appropriate sequence in which to run a series of tests.

Test Monitoring and Output Checking. Capabilities have been developed and used for various kinds of dynamic data-type checking and assertion checking, and for timing and performance analysis. Test output post-processing aids include output comparators and exception report capabilities, and test-oriented data reduction and report generation packages.

Fault Isolation, Debugging. Besides the traditional tools—the core dump, the trace, the snapshot, and the breakpoint—several capabilities have been developed for interactive replay or backtracking of the program's execution. This is still a difficult area, and only a relatively few advanced concepts have proved generally useful.

Retesting (once a presumed fix has been made). Test data management systems (for the code, the input data, and the comparison output data) have been shown to be most valuable here, along with comparators to check for the differences in code, inputs, and outputs between the original and the modified program and test case. A promising experimental tool performs a comparative structure analysis of the original and modified code, and indicates which test cases need to be rerun.

Integration of Routines into Systems. In general, automated aids for this process are just larger scale versions of the test data management systems above. Some additional capabilities exist for interface consistency checking, e.g., on the length and form of parameter lists or data base references. Top-down development aids are also helpful in this regard.

Stopping. Some partial criteria for thoroughness of testing can and have been automatically monitored. Tools exist which keep a cumulative tally of the number or percent of the instructions or branches which have been exercised during the test program, and indicate to the tester what branch conditions must be satisfied in order to completely exercise all the code or branches. Of course, these are far from complete criteria for determining when to stop testing; the completeness question is the subject of the next section.

Test Sufficiency and Program Proving. If a program's input space and output space are finite (where the input space includes not only all possible incoming inputs, but also all possible values in the program's data base), then one can construct a set of "black box" tests (one for each point in the input space) which can show conclusively that the program is correct (that its behavior matches its specification).

In general, though, a program's input space is infinite; for example, it must generally provide for rejecting unacceptable inputs. In this case, a finite set of black-box tests is not a sufficient demonstration of the program's correctness (since, for any input x, one must assure that the program does not wrongly treat it as a special case). Thus, the demonstration of correctness in this case involves some formal argument (e.g., a proof using induction) that the dynamic performance of the program indeed produces the static transformation of the input space indicated by the formal specification for the program. For finite portions of the input space, a successful exhaustive test of all cases can be considered as a satisfactory formal argument. Some good initial work in sorting out the conditions under which testing is equivalent to proof of a program's correctness has been done by Goodenough and Gerhart [63] and in a review of their work by Wegner [64].

Symbolic Execution. An attractive intermediate step between program testing and proving is "symbolic execution," a manual or automated procedure which operates on symbolic inputs (e.g., variable names) to produce symbolic outputs. Separate cases are generated for different execution paths. If there are a finite number of such paths, symbolic execution can be used to demonstrate correctness, using a finite symbolic input space and output space. In general, though, one cannot guarantee a finite number of paths. Even so, symbolic execution can be quite valuable as an aid to either program testing or proving. Two fairly powerful automated systems for symbolic execution exist, the EFFIGY system [65] and the SELECT system [66].

Program Proving (Program Verification). Program proving (increasingly referred to as program verification) involves expressing the program specifications as a logical proposition, expressing individual program execution statements as logical propositions, expressing program branching as an expansion into separate cases, and performing logical transformations on the propositions in a way which ends by demonstrating the equivalence of the program and its specification. Potentially infinite loops can be handled by inductive reasoning.

In general, nontrivial programs are very complicated and time-consuming to prove. In 1973, it was estimated that about one man-month of expert effort was required to prove 100 lines of code [67]. The largest program to be proved correct to date contained about 2000 statements [68]. Again, automation can help out on some of the complications. Some automated verification systems exist, notably those of London *et al.* [69] and Luckham *et al.* [70]. In general, such systems do not work on programs in the more common languages such as Fortran or Cobol. They work in languages such as Pascal [45], which has (unlike Fortran or Cobol) an axiomatic definition [71] allowing clean expression of program statements as logical propositions. An excellent survey of program verification technology has been given by London [72].

Besides size and language limitations, there are other factors which limit the utility of program proving techniques. Computations on "real" variables involving truncation and round-off errors are virtually impossible to analyze with adequate accuracy for most nontrivial programs. Programs with nonformalizable inputs (e.g., from a sensor where one has just a rough idea of its bias, signal-to-noise ratio, etc.) are impossible to handle. And, of course, programs can be proved to be consistent with a specification which is itself incorrect with respect to the system's proper functioning. Finally, there is no guarantee that the proof is correct or complete; in fact, many published "proofs" have subsequently been demonstrated to have holes in them [63].

It has been said and often repeated that "testing can be used to demonstrate the presence of errors but never their absence", [73]. Unfortunately, if we must define "errors" to include those incurred by the two limitations above (errors in specifications and errors in proofs), it must be admitted that "program proving can be used to demonstrate the presence of errors but never their absence."

Fault-Tolerance. Programs do not have to be error-free to be reliable. If one could just detect erroneous computations as they occur and compensate for them, one could achieve reliable operation. This is the rationale behind schemes for fault-tolerant software. Unfortunately, both detection and compensation are formidable problems. Some progress has been made in the case of software detection and compensation for hardware errors; see, for example, the articles by Wulf [74] and Goldberg [75]. For software errors, Randell has formulated a concept of separately-programmed, alternate "recovery blocks" [76]. It appears attractive for parts of the error compensation activity, but it is still too early to tell how well it will handle the error detection problem, or what the price will be in program slowdown.

Trends

As we continue to collect and analyze more and more data on how, when, where, and why people make software errors, we will get added insights on how to avoid making such errors, how to organize our validation strategy and tactics (not only in testing but throughout the software life cycle), how to develop or evaluate new automated aids, and how to develop useful methods for predicting software reliability. Some automated aids, particularly for static code checking, and for some dynamic-type or assertion checking, will be integrated into future programming languages and compilers. We should see some added useful criteria and associated aids for test completeness, particularly along the lines of exercising "all data elements" in some appropriate way. Symbolic execution capabilities will probably make their way into automated aids for test case generation, monitoring, and perhaps retesting.

Continuing work into the theory of software testing should provide some refined concepts of test validity, reliability, and completeness, plus a better theoretical base for supporting hybrid test/proof methods of verifying programs. Program proving techniques and aids will become more powerful in the size and range of programs they handle, and hopefully easier to use and harder to misuse. But many of their basic limitations will remain, particularly those involving real variables and nonformalizable inputs.

Unfortunately, most of these helpful capabilities will be available only to people working in higher order languages. Much of the progress in test technology will be unavailable to the increasing number of people who find themselves spending more and more time testing assembly language software written for minicomputers and microcomputers with poor test support capabilities. Powerful cross-compiler capabilities on large host machines and microprogrammed diagnostic emulation capabilities [77] should provide these people some relief after a while, but a great deal of software testing will regress back to earlier generation "dark ages."

VII. SOFTWARE MAINTENANCE

Scope of Software Maintenance

Software maintenance is an extremely important but highly neglected activity. Its importance is clear from Fig. 1: about 40 percent of the overall hardware-software dollar is going into software maintenance today, and this number is likely to grow to about 60 percent by 1985. It will continue to grow for a long time, as we continue to add to our inventory of code via development at a faster rate than we make code obsolete.

The preceding figures are only very approximate, because our only data so far are based on highly approximate definitions. It is hard to come up with an unexceptional definition of software maintenance. Here, we define it as "the process of modifying existing operational software while leaving its primary functions intact." It is useful to divide software maintenance into two categories: software *update*, which results in a changed functional specification for the software, and software *repair*, which leaves the functional specification intact. A good discussion of software repair is given in the paper by Swanson [78], who divides it into the subcategories of corrective maintenance (of processing, performance, or implementation failures), adaptive maintenance (to changes in the processing or data environment), and perfective maintenance (for enhancing performance or maintainability).

For either update or repair, three main functions are involved in software maintenance [79].

Understanding the Existing Software. This implies the need for good documentation, good traceability between requirements and code, and well-structured and well-formatted code.

Modifying the Existing Software. This implies the need for software, hardware, and data structures which are easy to expand and which minimize side effects of changes, plus easy-to-update documentation.

Revalidating the Modified Software. This implies the need for software structures which facilitate selective retest, and aids for making retest more thorough and efficient.

Following a short discussion of current practice in software maintenance, these three functions will be used below as a framework for discussing current frontier technology in software maintenance.

Current Practice

As indicated in Fig. 6, probably about 70 percent of the overall cost of software is spent in software maintenance. A paper by Elshoff [80] indicates that the figure for General Motors is about 75 percent, and that GM is fairly typical of large business software activities. Daly [5] indicates that about 60 percent of GTE's 10-year life cycle costs for real-time software are devoted to maintenance. On two Air Force command and control software systems, the maintenance portions of the 10-year life cycle costs were about 67 and 72 percent. Often, maintenance is not done very efficiently. On one aircraft computer, software development costs were roughly $75/instruction, while maintenance costs ran as high as $4000/instruction [81].

Despite its size, software maintenance is a highly neglected activity. In general, less-qualified personnel are assigned to maintenance tasks. There are few good general principles and few studies of the process, most of them inconclusive.

Further, data processing practices are usually optimized around other criteria than maintenance efficiency. Optimizing around development cost and schedule criteria generally leads to

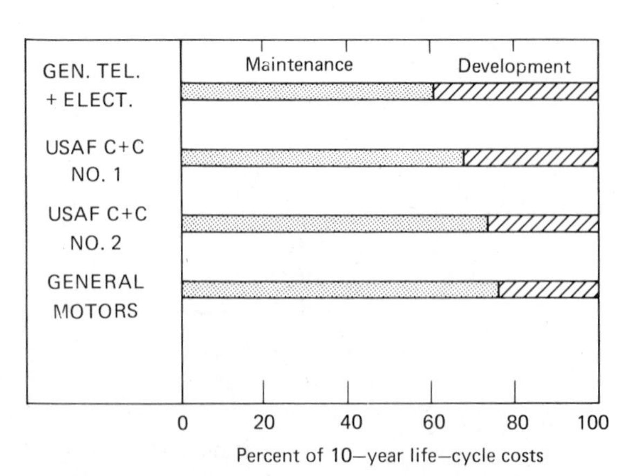

FIGURE 6 Software life-cycle cost breakdown.

compromises in documentation, testing, and structuring. Optimizing around hardware efficiency criteria generally leads to use of assembly language and skimping on hardware, both of which correlate strongly with increased software maintenance costs [1].

Current Frontier Technology

Understanding the Existing Software. Aids here have largely been discussed in previous sections: structured programming, automatic formatting, and code auditors for standards compliance checking to enhance code readability; machine-readable requirements and design languages with traceability support to and from the code. Several systems exist for automatically updating documentation by excerpting information from the revised code and comment cards.

Modifying the Existing Software. Some of Parnas' modularization guidelines [32] and the data abstractions of the CLU [48] and ALPHARD [47] languages make it easier to minimize the side effects of changes. There may be a maintenance price, however. In the past, some systems with highly coupled programs and associated data structures have had difficulties with data base updating. This may not be a problem with today's data dictionary capabilities, but the interactions have not yet been investigated. Other aids to modification are structured code, configuration management techniques, programming support libraries, and process construction systems.

Revalidating the Modified Software. Aids here were discussed earlier under testing; they include primarily test data management systems, comparator programs, and program structure analyzers with some limited capability for selective retest analysis.

General Aids. On-line interactive systems help to remove one of the main bottlenecks involved in software maintenance: the long turnaround times for retesting. In addition, many of these systems are providing helpful capabilities for text editing and software module management. They will be discussed in more detail under "Management and Integrated Approaches" below. In general, a good deal more work has been done on the maintainability aspects of data bases and data structures than for program structures; a good survey of data base technology is given in a recent special issue of *ACM Computing Surveys* [82].

Trends

The increased concern with life cycle costs, particularly within the U.S. DoD [83], will focus a good deal more attention on software maintenance. More data collection and analysis on the growth dynamics of software systems, such as the Belady-Lehman studies of OS/360 [84], will begin to point out the high-leverage areas for improvement. Explicit mechanisms for confronting maintainability issues early in the development cycle, such as the requirements-properties matrix [18] and the design inspection [4] will be refined and used more extensively. In fact, we may evolve a more general concept of software quality assurance (currently focussed largely on reliability concerns), involving such activities as independent reviews of software requirements and design specifications by experts in software maintainability. Such activities will be enhanced considerably with the advent of more powerful capabilities for analyzing machine-readable requirements and design specifications. Finally, advances in automatic programming [14], [15] should reduce or eliminate some maintenance activity, at least in some problem domains.

VIII. SOFTWARE MANAGEMENT AND INTEGRATED APPROACHES

Current Practice

There are more opportunities for improving software productivity and quality in the area of management than anywhere else. The difference between software project successes and failures has most often been traced to good or poor practices in software management. The biggest software management problems have generally been the following.

Poor Planning. Generally, this leads to large amounts of wasted effort and idle time because of tasks being unnecessarily performed, overdone, poorly synchronized, or poorly interfaced.

Poor Control. Even a good plan is useless when it is not kept up-to-date and used to manage the project.

Poor Resource Estimation. Without a firm idea of how much time and effort a task should take, the manager is in a poor position to exercise control.

Unsuitable Management Personnel. As a very general statement, software personnel tend to respond to problem situations as designers rather than as managers.

Poor Accountability Structure. Projects are generally organized and run with very diffuse delineation of responsibilities, thus exacerbating all the above problems.

Inappropriate Success Criteria. Minimizing development costs and schedules will generally yield a hard-to-maintain product. Emphasizing "percent coded" tends to get people coding early and to neglect such key activities as requirements and design validation, test planning, and draft user documentation.

Procrastination on Key Activities. This is especially prevalent when reinforced by inappropriate success criteria as above.

Current Frontier Technology

Management Guidelines. There is no lack of useful material to guide software management. In general, it takes a book-length treatment to adequately cover the issues. A number of books on the subject are now available [85]-[95], but for various reasons they have not strongly influenced software management practice. Some of the books (e.g., Brooks [85] and the collections by Horowitz [86], Weinwurm [87], and Buxton, Naur, and Randell [88]) are collections of very good advice, ideas, and experiences, but are fragmentary and lacking in a consistent, integrated, life cycle approach. Some of the books (e.g., Metzger [89], Shaw and Atkins [90], Hice *et al.* [91], Ridge and Johnson [92], and Gildersleeve [93]) are good on checklists and procedures but (except to some extent the latter two) are light on the human aspects of management, such as staffing, motivation, and conflict resolution. Weinberg [94] provides the most help on the human aspects, along with Brooks [85] and Aron [95], but in turn, these three books are light on checklists and procedures. (A second volume by Aron is intended to cover software group and project considerations.) None of the books have an adequate treatment of some items, largely because they are so poorly understood: chief among these items are software cost and resource estimation, and software maintenance.

In the area of software cost estimation, the paper by Wolverton [96] remains the most useful source of help. It is strongly based on the number of object instructions (modified by complexity, type of application, and novelty) as the determinant of software cost. This is a known weak spot, but not one for which an acceptable improvement has surfaced. One possible line of improvement might be along the "software physics" lines being investigated by Halstead [97] and others; some interesting initial results have been obtained here, but their utility for practical cost estimation remains to be demonstrated. A good review of the software cost estimation area is contained in [98].

Management-Technology Decoupling. Another difficulty of the above books is the degree to which they are decoupled from software technology. Except for the Horowitz and Aron books, they say relatively little about the use of such advanced-technology aids as formal, machine-readable requirements, top-down design approaches, structured programming, and automated aids to software testing.

Unfortunately, the management-technology decoupling works the other way, also. In the design area, for example, most treatments of top-down software design are presented as logical exercises independent of user or economic considerations. Most automated aids to software design provide little support for such management needs as configuration management, traceability to code or requirements, and resource estimation and control. Clearly, there needs to be a closer coupling between technology and management than this. Some current effort to provide integrated management-technology approaches are presented next.

Integrated Approaches. Several major integrated systems for software development are currently in operation or under development. In general, their objectives are similar: to achieve a significant boost in software development efficiency and quality through the synergism of a unified approach. Examples are the utility of having a complementary development approach (top-down, hierarchical) and set of programming standards (hierarchical, structured code); the ability to perform a software update and at the same time perform a set of timely, consistent project status updates (new version number of module, closure of software problem report, updated status logs); or simply the improvement in software system integration achieved when all participants are using the same development concept, ground rules, and support software.

The most familiar of the integrated approaches is the IBM "top-down structured programming with chief programmer teams" concept. A good short description of the concept is given by Baker [49]; an extensive treatment is available in a 15-volume series of reports done by IBM for the U.S. Army and Air Force [99]. The top-down structured approach was discussed earlier. The Chief Programmer Team centers around an individual (the Chief) who is responsible for designing, coding, and integrating the top-level control structure as well as the key components of the team's product; for managing and motivating the team personnel and personally reading and reviewing all their code; and also for performing traditional management and customer interface functions. The Chief is assisted by a Backup programmer who is prepared at anytime to take the Chief's place, a Librarian who handles job submission, configuration control, and project status accounting, and additional programmers and specialists as needed.

In general, the overall ensemble of techniques has been quite successful, but the Chief Programmer concept has had mixed results [99]. It is difficult to find individuals with enough energy and talent to perform all the above functions. If you find one, the project will do quite well; otherwise, you have concentrated most of the project risk in a single individual, without a good way of finding out whether or not he is in trouble. The Librarian and Programming Support Library concept have generally been quite useful, although to date the concept has been oriented toward a batch-processing development environment.

Another "structured" integrated approach has been developed and used at SofTech [38]. It is oriented largely around a hierarchical-decomposition design approach, guided by formalized sets of principles (modularity, abstraction, localization, hiding, uniformity, completeness, confirmability), processes (purpose, concept, mechanism, notation, usage), and goals (modularity, efficiency, reliability, understandability). Thus, it accommodates some economic considerations, although it says little about any other management considerations. It appears to work well for SofTech, but in general has not been widely assimilated elsewhere.

A more management-intensive integrated approach is the TRW software development methodology exemplified in the paper by Williams [50] and the TRW Software Development and Configuration Management Manual [100], which has been used as the basis for several recent government in-house software manuals. This approach features a coordinated set of high-level and detailed management objectives, associated automated aids—standards compliance checkers, test thoroughness checkers, process construction aids, reporting systems for cost, schedule, core and time budgets, problem identification and closure, etc.—and unified documentation and management devices such as the Unit Development Folder. Portions of the approach are still largely manual, although additional automation is underway, e.g., via the Requirements Statement Language [13].

The SDC Software Factory [101] is a highly ambitious attempt to automate and integrate software development technology. It consists of an interface control component, the Factory Access and Control Executive (FACE), which provides users access to various tools and data bases; a project planning and monitoring system, a software development data base and module management system, a top-down development support system, a set of test tools, etc. As the system is still undergoing development and preliminary evaluation, it is too early to tell what degree of success it will have.

Another factory-type approach is the System Design Laboratory (SDL) under development at the Naval Electronics Laboratory Center [102]. It currently consists primarily of a framework within which a wide range of aids to software development can be incorporated. The initial installment contains text editors, compilers, assemblers, and microprogrammed emulators. Later additions are envisioned to include design, development, and test aids, and such management aids as progress reporting, cost reporting, and user profile analysis.

SDL itself is only a part of a more ambitious integrated approach, ARPA's National Software Works (NSW) [102]. The initial objective here has been to develop a "Works Manager" which will allow a software developer at a terminal to access a wide variety of software development tools on various computers available over the ARPANET. Thus, a developer might log into the NSW, obtain his source code from one computer, text-edit it on another, and perhaps continue to hand the program to additional computers for test instrumentation, compiling, executing, and postprocessing of output data. Currently, an initial version of the Works Manager is operational, along with a few tools, but it is too early to assess the likely outcome and payoffs of the project.

Trends

In the area of management techniques, we are probably entering a consolidation period, particularly as the U.S. DoD proceeds to implement the upgrades in its standards and procedures called for in the DoD Directive 5000.29

[104]. The resulting government-industry efforts should produce a set of software management guidelines which are more consistent and up-to-date with today's technology than the ones currently in use. It is likely that they will also be more comprehensible and less encumbered with DoD jargon; this will make them more useful to the software field in general.

Efforts to develop integrated, semiautomated systems for software development will continue at a healthy clip. They will run into a number of challenges which will probably take a few years to work out. Some are technical, such as the lack of a good technological base for data structuring aids, and the formidable problem of integrating complex software support tools. Some are economic and managerial, such as the problems of pricing services, providing tool warranties, and controlling the evolution of the system. Others are environmental, such as the proliferation of minicomputers and microcomputers, which will strain the capability of any support system to keep up-to-date.

Even if the various integrated systems do not achieve all their goals, there will be a number of major benefits from the effort. One is of course that a larger number of support tools will become available to a larger number of people (another major channel of tools will still continue to expand, though: the independent software products marketplace). More importantly, those systems which achieve a degree of conceptual integration (not just a free-form tool box) will eliminate a great deal of the semantic confusion which currently slows down our group efforts throughout the software life cycle. Where we have learned how to talk to each other about our software problems, we tend to do pretty well.

IX. CONCLUSIONS

Let us now assess the current state of the art of tools and techniques which are being used to solve software development problems, in terms of our original definition of software engineering: the practical application of *scientific knowledge* in the design and construction of software. Table II presents a summary assessment of the extent to which current software engineering techniques are based on solid scientific principles (versus empirical heuristics). The summary assessment covers four dimensions: the extent to which existing scientific principles apply across the entire software life cycle, across the entire

TABLE 2
Applicability of Existing Scientific Principles

Dimension	Software Engineering	Hardware Engineering
Scope across life cycle	Some principles for component construction and detailed design, virtually none for system design and integration, e.g., algorithms, automata theory.	Many principles applicable across life cycle, e.g., communication theory, control theory.
Scope across application	Some principles for "systems" software, virtually none for applications software, e.g., discrete mathematical structures.	Many principles applicable across entire application system, e.g., control theory application.
Engineering economics	Very few principles which apply to system economics, e.g., algorithms.	Many principles apply well to system economics, e.g., strength of materials, optimization, and control theory.
Required training	Very few principles formulated for consumption by technicians, e.g., structured code, basic math packages.	Many principles formulated for consumption by technicians, e.g., handbooks for structural design, stress testing, maintainability.

range of software applications, across the range of engineering-economic analyses required for software development, and across the range of personnel available to perform software development.

For perspective, a similar summary assessment is presented in Table II for hardware engineering. It is clear from Table II that software engineering is in a very primitive state as compared to hardware engineering, with respect to its range of scientific foundations. Those scientific principles available to support software engineering address problems in an area we shall call *Area 1: detailed design and coding* of *systems software* by *experts* in a relatively *economics-independent* context. Unfortunately, the most pressing software development problems are in an area we shall call *Area 2: requirements analysis design, test, and maintenance* of *applications software by technicians*[3] in an *economics-driven* context. And in Area 2, our scientific foundations are so slight that one can seriously question whether our current techniques deserve to be called "software engineering."

Hardware engineering clearly has available a better scientific foundation for addressing its counterpart of these Area 2 problems. This should not be too surprising, since "hardware science" has been pursued for a much longer time, is easier to experiment with, and does not have to explain the performance of human beings.

What is rather surprising, and a bit disappointing, is the reluctance of the computer science field to address itself to the more difficult and diffuse problems in Area 2, as compared with the more tractable Area 1 subjects of automata theory, parsing, computability, etc. Like most explorations into the relatively unknown, the risks of addressing Area 2 research problems in the requirements analysis, design, test and maintenance of applications software are relatively higher. But the prizes, in terms of payoff to practical software development and maintenance, are likely to be far more rewarding. In fact, as software engineering begins to solve its more difficult Area 2 problems, it will begin to lead the way toward solutions to the more difficult large-systems problems which continue to beset hardware engineering.

[3]For example, one survey of 14 installations in one large organization produced the following profile of its "average coder": 2 years college-level education, 2 years software experience, familiarity with 2 programming languages and 2 applications, and generally introverted, sloppy, inflexible, "in over his head," and undermanaged. Given the continuing increase in demand for software personnel, one should not assume that this typical profile will improve much. This has strong implications for effective software engineering technology which, like effective software, must be well-matched to the people who must use it.

REFERENCES

1. B. W. Boehm, "Software and its impact: A quantitative assessment," *Datamation,* pp. 48–59. May 1973.
2. T. A. Dolotta *et. al., Data Processing in 1980–85.* New York: Wiley-Interscience, 1976.
3. P. Wegner, "Computer languages," *IEEE Trans. Comput.,* this issue, pp. 1207–1225.
4. M. E. Fagan, "Design and code inspections and process control in the development of programs," IBM, rep. IBM-SDD TR-21.572, Dec. 1974.
5. E. B. Daly, "Management of software development," *IEEE Trans. Software Eng.,* to be published.
6. W. W. Royce, "Software requirements analysis, sizing, and costing," in *Practical Strategies for the Development of Large Scale Software,* E. Horowitz, Ed. Reading, MA: Addison-Wesley, 1975.
7. T. E. Bell and T. A. Thayer, "Software requirements: Are they a problem?," *Proc. IEEE/ACM 2nd Int. Conf. Software Eng.,* Oct. 1976.
8. E. S. Quade, Ed., *Analysis for Military Decisions.* Chicago, IL: Rand-McNally, 1964.
9. J. D. Couger and R. W. Knapp, Eds., *Sys-*

tem Analysis Techniques. New York: Wiley, 1974.

10. D. Teichroew and H. Sayani, "Automation of system building," *Datamation*, pp. 25–30, Aug. 15, 1971.
11. C. G. Davis and C. R. Vick, "The software development system," in *Proc. IEEE/ACM 2nd Int. Conf. Software Eng.*, Oct. 1976.
12. M. Alford, "A requirements engineering methodology for real-time processing requirements," in *Proc. IEEE/ACM 2nd Int. Conf. Software Eng.*, Oct. 1976.
13. T. E. Bell, D. C. Bixler, and M. E. Dyer, "An extendable approach to computer-aided software requirements engineering," in *Proc. IEEE/ACM 2nd Int. Conf. Software Eng.*, Oct. 1976.
14. R. M. Balzer, "Imprecise program specification." Univ. Southern California, Los Angeles, rep. ISI/RR-75-36, Dec. 1975.
15. W. A. Martin and M. Bosyj, "Requirements derivation in automatic programming," in *Proc. MRI Symp. Comput. Software Eng.*, Apr. 1976.
16. N. P. Dooner and J. R. Lourie, "The application software engineering tool," IBM res. rep. RC 5434, May 29, 1975.
17. M. L. Wilson, "The information automat approach to design and implementation of computer-based systems," IBM, rep. IBM-FSD, June 27, 1975.
18. B. W. Boehm, "Some steps toward formal and automated aids to software requirements and design," *Proc. IFIP Cong.*, 1974, pp. 192–197.
19. A. B. Endres, "An analysis of errors and their causes in system programs," *IEEE Trans. Software Eng.*, pp. 140–149, June 1975.
20. T. A. Thayer, "Understanding software through analysis of empirical data," *Proc. Nat. Comput. Conf.*, 1975, pp. 335–341.
21. B. W. Boehm, R. L. McClean, and D. B. Urfrig, "Some experience with automated aids to the design of large-scale reliable software," *IEEE Trans. Software Eng.*, pp. 125–133, Mar. 1975.
22. P. Freeman, "Software design representation: Analysis and improvements," Univ. California, Irvine, tech. rep. 81, May 1976.
23. E. L. Glaser *et al.*, "The LOGOS project," in *Proc. IEEE COMP-CON*, 1972, pp. 175–192.
24. N. R. Nielsen, "ECSS: Extendable computer system simulator," Rand. Corp., rep. RM-6132-PR/NASA, Jan. 1970.
25. D. W. Kosy, "The ECSS II language for simulating computer systems," Rand Corp., rep. R-1895-GSA, Dec. 1975.
26. E. W. Dijkstra, "Complexity controlled by hierarchical ordering of function and variability," in *Software Engineering*, P. Naur and B. Randell, Eds. NATO, Jan. 1969.
27. H. D. Mills, "Mathematical foundations for structured programming," IBM-FSD, rep. FSC72-6012, Feb. 1972.
28. C. L. McGowan and J. R. Kelly, *Top-Down Structured Programming Techniques.* New York: Petrocelli/Charter, 1975.
29. B. W. Boehm *et al.*, "Structured programming: A quantitative assessment," *Computer*, pp. 38–54, June 1975.
30. W. P. Stevens, G. J. Myers, and L. L. Constantine, "Structured design," *IBM Syst. J.*, vol. 13, no. 2, pp. 115–139, 1974.
31. G. J. Myers, *Reliable Software Through Composite Design.* New York: Petrocelli/Charter, 1975.
32. D. L. Parnas, "On the criteria to be used in decomposing systems into modules," *CACM*, pp. 1053–1058, Dec. 1972.
33. D. L. Parnas, "The influence of software structure on reliability," in *Proc. 1975 Int. Conf. Reliable Software*, Apr. 1975, pp. 358–362, available from IEEE.

34. M. Hamilton and S. Zeldin, "Higher order software—A methodology for defining software," *IEEE Trans. Software Eng.*, pp. 9–32, Mar. 1976.
35. "HIPO—A design aid and documentation technique," IBM, rep. GC20-1851-0, Oct. 1974.
36. J. Mortison, "Tools and techniques for software development process visibility and control," in *Proc. ACM Comput. Sci. Conf.*, Feb. 1976.
37. I. Nassi and B. Schneiderman, "Flowchart techniques for structured programming," *SIGPLAN Notices*, pp. 12–26, Aug. 1973.
38. D. T. Ross, J. B. Goodenough, and C. A. Irvine, "Software engineering: Process, principles, and goals," *Computer*, pp. 17–27, May 1975.
39. M. A. Jackson, *Principles of Program Design*. New York: Academic, 1975.
40. P. Henderson and R. A. Snowden, "A tool for structured program development," in *Proc. 1974 IFIP Cong.*, pp. 204–207.
41. L. C. Carpenter and L. L. Tripp, "Software design validation tool," in *Proc. 1975 Int. Conf. Reliable Software*, Apr. 1975, pp. 395–400, available from IEEE.
42. S. H. Caine and E. K. Gordon, "PDL: A tool for software design," in *Proc. 1975 Nat. Comput. Conf.*, pp. 271–276.
43. H. D. Mills, "Structured programming in large systems," IBM-FSD, Nov. 1970.
44. C. A. R. Hoare, "Notes on data structuring," in *Structured Programming*, O. J. Dahl, E. W. Dijkstra, and C. A. R. Hoare. New York: Academic, 1972.
45. N. Wirth, "An assessment of the programming language Pascal," *IEEE Trans. Software Eng.*, pp. 192–198, June 1975.
46. P. Brinch-Hansen, "The programming language concurrent Pascal," *IEEE Trans. Software Eng.*, pp. 199–208, June 1975.
47. W. A. Wulf, *ALPHARD: Toward a language to support structured programs,"* Carnegie-Mellon Univ., Pittsburgh, PA, internal rep., Apr. 30, 1974.
48. B. H. Liskov and S. Zilles, "Programming with abstract data types," *SIGPLAN Notices*, pp. 50–59, April 1974.
49. F. T. Baker, "Structured programming in a production programming environment," *IEEE Trans. Software Eng.*, pp. 241–252, June 1975.
50. R. D. Williams, "Managing the development of reliable software," *Proc., 1975 Int. Conf. Reliable Software* April 1975, pp. 3–8, available from IEEE.
51. J. Witt, "The COLUMBUS approach," *IEEE Trans. Software Eng.*, pp. 358–363, Dec. 1975.
52. B. W. Kernighan and P. J. Plauger, *The Elements of Programming Style*, New York: McGraw-Hill, 1974.
53. H. F. Ledgard, *Programming Proverbs*. Rochelle Park, NJ: Hayden, 1975.
54. W. A. Whitaker *et al.*, "Department of Defense requirements for high order computer programming languages: 'Tinman,' " Defense Advanced Research Projects Agency, Apr. 1976.
55. *Proc. 1973 IEEE Symp. Comput. Software Reliability*, Apr.–May 1973.
56. E. C. Nelson, "A statistical basis for software reliability assessment," TRW Systems, Redondo Beach, CA, rep. TRW-SS-73-03, Mar. 1973.
57. J. R. Brown and M. Lipow, "Testing for software reliability," in *Proc. 1975 Int. Conf. Reliable Software*, Apr. 1975, pp. 518–527.
58. J. D. Musa, "Theory of software reliability and its application," *IEEE Trans. Software Eng.*, pp. 312–327, Sept. 1975.
59. R. J. Rubey, J. A. Dana, and P. W. Biche,

"Quantitative Aspects of software validation," *IEEE Trans. Software Eng.*, pp. 150–155, June 1975.

60. T. A. Thayer, M. Lipow, and E. C. Nelson, "Software reliability study," TRW Systems, Redondo Beach, CA, rep. to RADC, Contract F30602-74-C-0036, Mar. 1976.
61. D. J. Reifer, "Automated aids for reliable software," in *Proc. 1975 Int. Conf. Reliable software*, Apr. 1975, pp. 131–142.
62. C. V. Ramamoorthy and S. B. F. Ho, "Testing large software with automated software evaluation systems," *IEEE Trans. Software Eng.*, pp. 46–58, Mar. 1975.
63. J. G. Goodenough and S. L. Gerhart, "Toward a theory of test data selection," *IEEE Trans. Software Eng.*, pp. 156–173, June 1975.
64. P. Wegner, "Report on the 1975 International Conference on Reliable Software," in *Findings and Recommendations of the Joint Logistics Commanders' Software Reliability Work Group*, Vol. II, Nov. 1975, pp. 45–88.
65. J. C. King, "A new approach to program testing," in *Proc. 1975 Int. Conf. Reliable Software*, Apr. 1975, pp. 228–233.
66. R. S. Boyer, B. Elspas, and K. N. Levitt, "Select—A formal system for testing and debugging programs," in *Proc. 1975 Int. Conf. Reliable Software*, Apr. 1975, pp. 234–245.
67. J. Goldberg, Ed., *Proc. Symp. High Cost of Software*, Stanford Research Institute, Stanford, CA, Sept. 1973, p. 63.
68. L. C. Ragland, "A verified program verifier," Ph.D. dissertation, Univ. of Texas, Austin, 1973.
69. D. I. Good, R. L. London, and W. W. Bledsoe, "An interactive program verification system," *IEEE Trans. Software Eng.*, pp. 59–67, Mar. 1975.
70. F. W. von Henke and D. C. Luckham, "A methodology for verifying programs," in *Proc. 1975 Int. Conf. Reliable Software*, pp. 156–164, Apr. 1975.
71. C. A. R. Hoare and N. Wirth, "An axiomatic definition of the programming language PASCAL," *Acta Informatica*, vol. 2, pp. 335–355, 1973.
72. R. L. London, "A view of program verification," in *Proc. 1975 Int. Conf. Reliable Software*, Apr. 1975, pp. 534–545.
73. E. W. Dijkstra, "Notes on structured programming," in *Structured Programming*, O. J. Dahl, E. W. Dijkstra, and C. A. R. Hoare. New York: Academic, 1972.
74. W. A. Wulf, "Reliable hardware-software architectures," *IEEE Trans. Software Eng.*, pp. 233–240, June 1975.
75. J. Goldberg, "New problems in fault-tolerant computing," in *Proc. 1975 Int. Symp. Fault-Tolerant Computing*, Paris, France, pp. 29–36, June 1975.
76. B. Randell, "System structure for software fault-tolerance," *IEEE Trans. Software Eng.*, pp. 220–232, June 1975.
77. R. K. McClean and B. Press, "Improved techniques for reliable software using microprogrammed diagnostic emulation," in *Proc. IFAC Cong.*, Vol. IV, Aug. 1975.
78. E. B. Swanson, "The dimensions of maintenance," in *Proc. IEEE/ACM 2nd Int. Conf. Software Eng.*, Oct. 1976.
79. B. W. Boehm, J. R. Brown, and M. Lipow, "Quantitative evaluation of software quality," in *Proc. IEEE/ACM 2nd Int. Conf. Software Eng.*, Oct. 1976.
80. J. L. Elshoff, "An analysis of some commercial PL/I programs," *IEEE Trans. Software Eng.*, pp. 113–120, June 1976.
81. W. L. Trainor, "Software: From Satan to saviour," in *Proc. NAECON*, May 1973.
82. E. H. Sibley, Ed., *ACM Comput. Surveys*

(Special Issue on Data Base Management Systems), Mar. 1976.

83. *Defense Management J. (Special Issue on Software Management)*, vol. II, Oct. 1975.
84. L. A. Belady and M. M. Lehman, "The evolution dynamics of large programs," IBM Research, Sept. 1975.
85. F. P. Brooks, *The Mythical Man-Month.* Reading, MA: Addison-Wesley, 1975.
86. E. Horowitz, Ed., *Practical Strategies for Developing Large-Scale Software.* Reading, MA: Addison-Wesley, 1975.
87. G. F. Weinwurm, Ed., *On the Management of Computer Programming.* New York: Auerbach, 1970.
88. P. Naur and B. Randell, Eds., *Software Engineering,* NATO, Jan. 1969.
89. P. J. Metzger, *Managing a Programming Project.* Englewood Cliffs, NJ: Prentice-Hall, 1973.
90. J. C. Shaw and W. Atkins, *Managing Computer System Projects.* New York: McGraw-Hill, 1970.
91. G. F. Hice, W. S. Turner, and L. F. Cashwell, *System Development Methodology.* New York: American Elsevier, 1974.
92. W. J. Ridge and L. E. Johnson, *Effective Management of Computer Software.* Homewood, IL: Dow Jones-Irwin, 1973.
93. T. R. Gildersleeve, *Data Processing Project Management.* New York: Van Nostrand Reinhold, 1974.
94. G. F. Weinberg, *The Psychology of Computer Programming.* New York: Van Nostrand Reinhold, 1971.
95. J. D. Aron, *The Program Development Process: The Individual Programmer.* Reading, MA: Addison-Wesley, 1974.
96. R. W. Wolverton, "The cost of developing large-scale software," *IEEE Trans. Comput.*, 1974.
97. M. H. Halstead, "Toward a theoretical basis for estimating programming effort," in *Proc. Ass. Comput. Mach. Conf.*, Oct. 1975, pp. 222–224.
98. *Summary Notes, Government/Industry Software Sizing and Costing Workshop,* USAF Electron. Syst. Div., Oct. 1974.
99. B. S. Barry and J. J. Naughton, "Chief programmer team operations description," U.S. Air Force, rep. RADC-TR-74-300, Vol. X (of 15-volume series), pp. 1-2–1-3.
100. *Software Development and Configuration Management Manual,* TRW Systems, Redondo Beach, CA, rep. TRW-SS-73-07, Dec. 1973.
101. H. Bratman and T. Court, "The software factory," *Computer,* pp. 28–37, May 1975.
102. "Systems design laboratory: Preliminary design report," Naval Electronics Lab. Center, Preliminary Working Paper, TN-3145, Mar. 1976.
103. W. E. Carlson and S. D. Crocker, "The impact of networks on the software marketplace," in *Proc. EASCON,* Oct. 1974.
104. "Management of computer resources in major defense systems," Department of Defense, Directive 6000.29, Apr. 1976.

READING QUESTIONS

1. Define the term *software engineering.*
2. Define and contrast top-down and bottom-up design strategies.
3. Does the increasing number of possible design alternatives make the system designer's job easier or harder? Support your answer.
4. Figure 4 shows that most errors in large systems occur in the early phases of the life cycle, whereas fewer are made in coding. Which of these two classes of errors do you think would be more expensive to correct?
5. Referring again to design errors versus coding errors, and to the system life cycle, when

do you think coding errors will be found? When do you think design errors will be found?

6. What does your answer to the above question imply about which area (design or coding) you should commit more resources to?

7. The author quotes, "testing can be used to demonstrate the presence of errors, but never their absence." After reading all of the testing processes and concepts covered here, comment on this quotation.

Structured Methodology: What Have We Learned?

Chris Gane
Trish Sarson

Over the last three years, a complete set of structured techniques—structured analysis, structured design, structured coding, top-down development and structured walkthroughs—has come together into a coherent structured methodology. We asked Carl Garrison, manager of information systems for The Superior Oil Co. in Houston, what his experience with the structured methodology had been to date. "We've used the complete set of techniques over the past two years" Garrison told us. "Some projects we've gone all the way through with the structured techniques; our largest project has sections that have used structured analysis and structured design and some sections that are now in programming."

We asked Carl to tell us more about this large project.

"It's a petroleum production revenue and accounting system, with a development budget of several million dollars. It will track the production of oil and gas for Superior, from the time it comes out of the ground to the time we pay royalties on the revenue," he explained.

"I think the major impact of the structured techniques has been on the communication with the users and on the user involvement we've gotten. Every one of our users was very impressed with the data flow diagrams we showed them. To our surprise, some of them were even prepared to read the hierarchical structure design charts and tell us where we'd missed a business function!

"We've also had very good results with the changeability of our designs and the changeability of the resulting code."

STRUCTURED ANALYSIS

What are the techniques to which Carl referred?

The key to successful structured analysis, it now appears, is the building of a graphical, logical (in the sense of "not-physical") model of the required system. Such a logical model, plus a

Source: Chris Gane, and Trish Sarson, "Structured Methodology: What Have We Learned?" *Computer World/EXTRA*, Vol. XIV, No. 38, September 17, 1980, pp. 52–57.

statement of systems objectives and system constraints, constitutes an adequate requirements statement, which has two virtues.

First, it expresses what the system is required to do, while not making any commitment as to how the system should physically be implemented. Thus a systems analyst can use the logical model to express system requirements without having any detailed current knowledge of DP techniques and present the model to the systems designer—leaving the designer in no doubt as to what the system should do, but giving him the greatest possible freedom to devise a physical design which does it most cost-effectively.

Second, the logical model turns out to be a very good way of showing nontechnical users what the nature of the system is going to be and how the various parts will fit together.

LOGICAL MODEL

What does such a logical model look like? Let us consider one of the simplest of all businesses, Joe's Auto Parts (slogan: "If you want it, we'll get it"). Joe holds no inventory, but takes orders for spares from his customers, promising future delivery. When he has enough orders for spares from a given car firm to earn a bulk discount, he sends the firm a bulk order and, on receiving the bulk order, gives each customer what they want, charging them the full retail price.

Figure 1 shows a very general picture of the data flow in Joe's business. Customers send in orders, and the single process block processes them, using a store of data (in some form) about parts details—price, delivery and so on—and a store of data about customers—whether they are in good credit standing, and so on.

This very crude data flow diagram (DFD) uses just four symbols, as shown in Figure 2:

- An external entity symbol (square): a source and/or destination of data outside the system.
- A data flow symbol (arrow): a pathway along which data moves into, around and out of the system.
- A process symbol (rounded rectangle): some function which transforms data.
- A data store symbol (open-ended rectangle): a place in the system where data is stored in some way.

It turns out that these four simple symbols are the minimum required to model a commercial system and that these same symbols can be used to model at any level of detail. For instance, Figure 1 is clearly rather trivial—let us expand "Process Orders" to see more detail.

Figure 3 shows orders coming from customers into a more detailed process "Verify order is valid." This process needs data about data-parts (to check that the part exists and is available) and data about customers (to check credit).

Out of the process come valid orders—one important simplification is that in high-level DFDs, we omit error paths and exception handling. The valid orders are stored in a data store of "Pending Orders" which may be a spike on the office wall or may be an IMS data base (we do

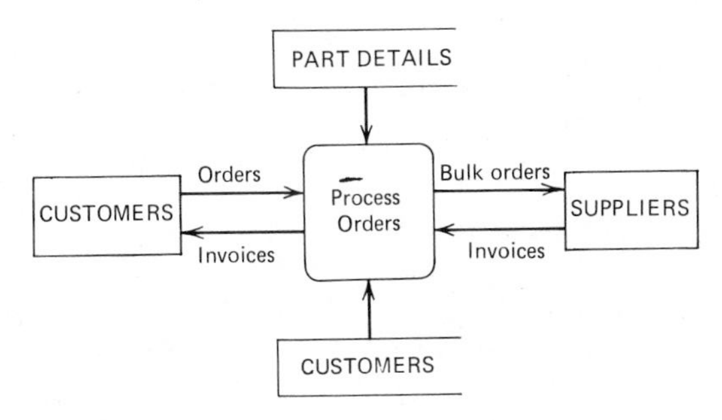

FIGURE 1

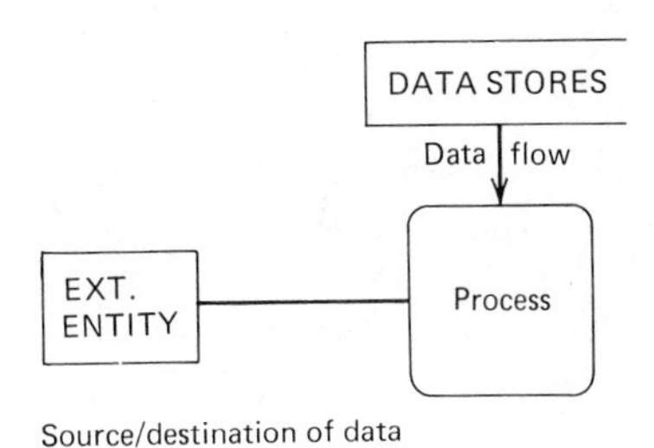

Source/destination of data

FIGURE 2

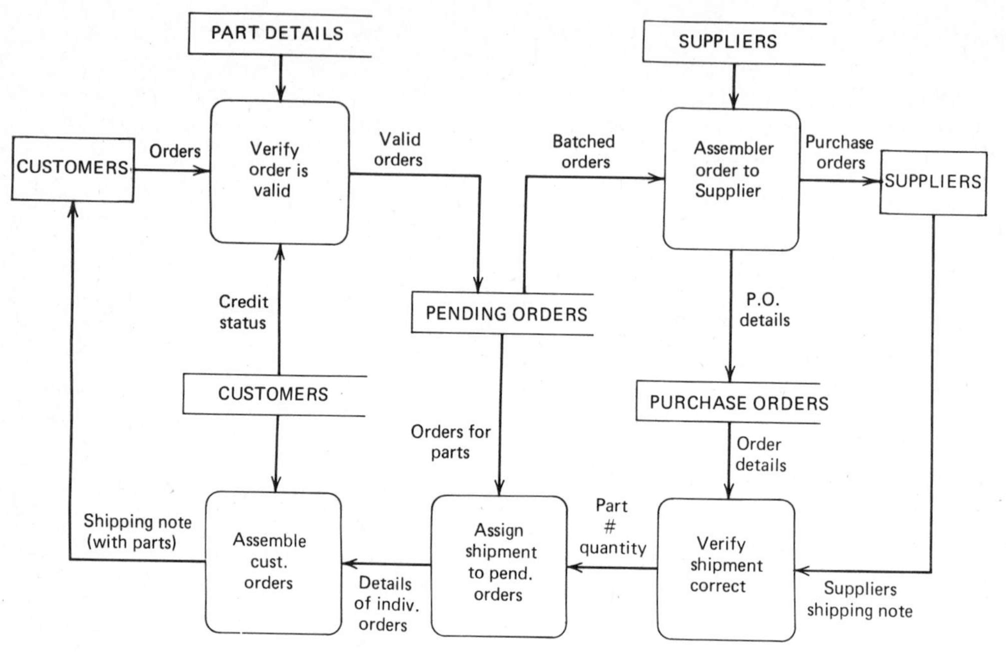

FIGURE 3

not care at this stage). We just note that there is a logical requirement to store valid orders until they can be batched to make a bulk order.

The lower stream of data in Figure 3 shows the data describing the bulk shipment from the suppliers being verified (note there are no error paths shown), assigned to individual orders and delivered to customers.

Clearly, there must be some financial flows. We must bill our customers and pay our suppliers, as shown in Figure 4.

Figure 4—a high-level logical data flow for transaction processing at Joe's Auto Parts—has several interesting features:

- It is logical, so it makes no distinction between manual processes, computer processes, tape files, disk files or any such technicality.
- Because it is logical and uses a very simple symbology, it is readily understood by non-technical users, such as Joe. User managers commonly say, when such diagrams are explained to them, "That's the first thing you computer people have ever shown me that I have understood!" Or, "Now I know what that system does!"
- Though abstract, it is precise enough to be criticized. "I see what happens to payments from my customers, but what happens if they don't pay?" Joe might ask.

We would have to admit that we are missing functions to analyze accounts receivable for delinquency and sent out dunning letters to customers. Of course, we are glad to find out that we are missing a desired function before we start to design and program. Although much more complicated for a real system, this is the type of data flow diagram with which Carl Garrison has had such good results.

GREATER DETAIL

If still more detail of a process is required, we can "explode" the process box into a more specific lower-level DFD. Figure 5 is an example

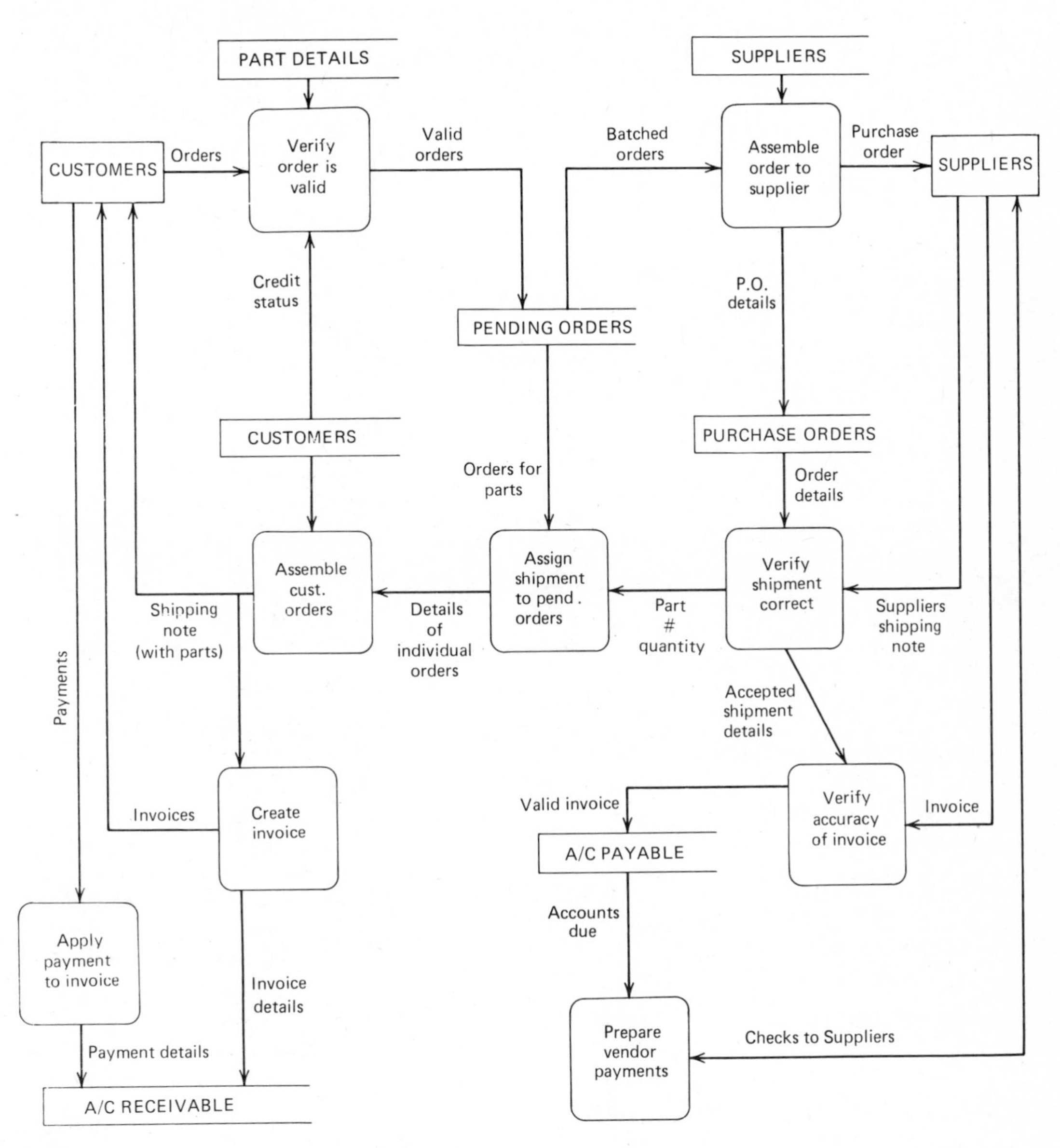

FIGURE 4

of such an "explosion" of the "Assemble Requisition to Suppliers" process from Figure 4.

If more detail of each data flow is required, we need to be able to express the logical nature of, say, "Orders." Since a data structure like this is hierarchical, we may give a series of more and more detailed answers to the question "What do you mean by orders?" First we may name the chief components:

```
ORDER
    ORDER-IDENTIFICATION
    CUSTOMER-DETAILS
    PART-DETAILS (iterated)
```

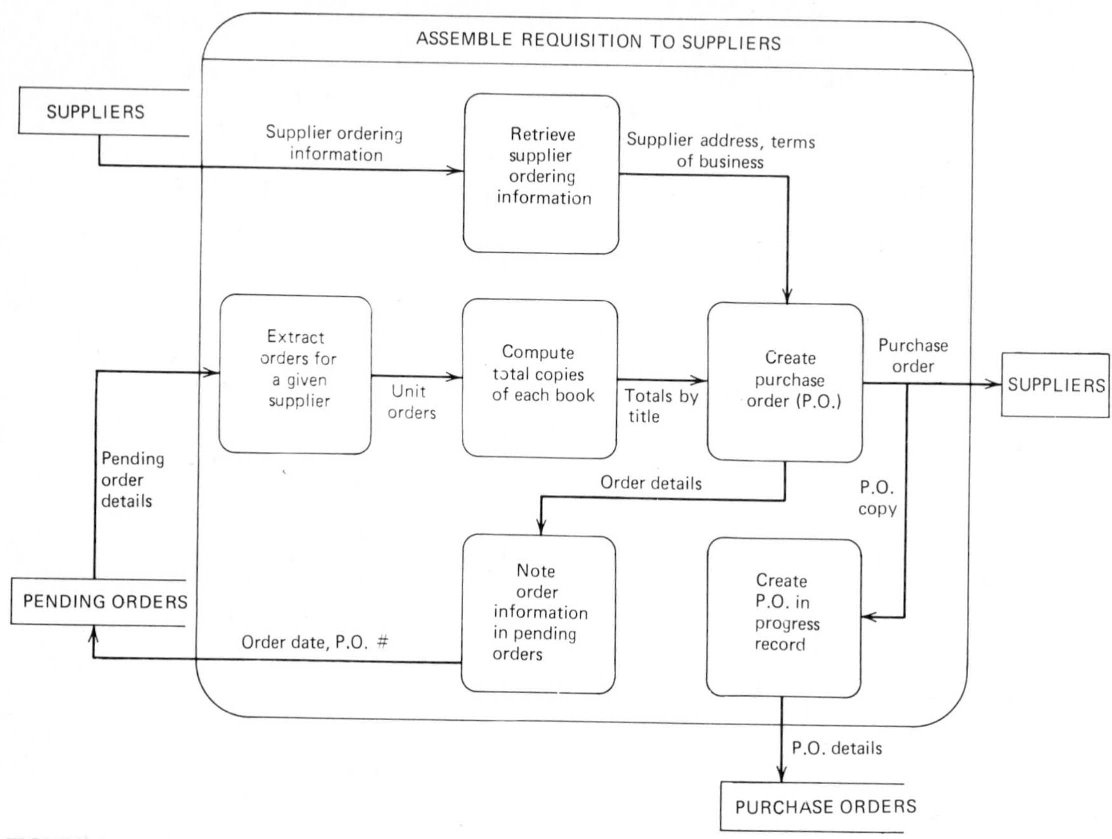

FIGURE 5

If that is not detailed enough, we can state the subcomponents of each component, as shown in Figure 6.

We express the data structures and data elements with honest, meaningful names chosen by the analyst. When we get down to the level of a data element (a piece of data that cannot be meaningfully subdivided), we must specify its logical (not-physical) nature. Thus we might express the North American telephone area codes as in Figure 7.

If all data structures are composed ultimately of data elements and if we can (a) define each data element, (b) state the way in which they are combined into data structures and (c) state which data structures move along the various data flows and which are found at rest in the data stores of our DFD (Figure 8 diagrams these relationships), then we have provided all the data objects required for a logical data dictionary.

The logical data dictionary is the place where all detailed definitions of data objects are stored.

DEFINING DATA STORES

Now we have to consider the definition of data stores. For every data store we must define its contents (in terms of the data structures defined in the data dictionary). In addition, for some data stores, we will have to show the immediate accesses that have to be made to it.

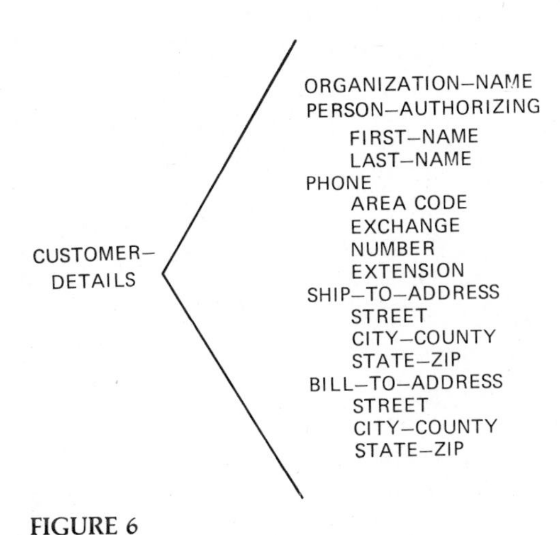

FIGURE 6

The definition of data store contents is in principle very simple: What comes out must go in—no more, no less.

So we look at the data flows coming into the data store, compare them with the data flows coming out, make sure that nothing is missing and that nothing is redundant, then derive the data store contents accordingly.

Suppose we had detailed data flows for Joe's Auto Parts and the complete flows in and out of "Pending Orders" were as shown in Figure 9. Assume that the detailed contents of each flow have already been defined in the data dictionary. We have some nonimmediate accesses to data (batched orders, orders for a given part) and some potentially immediate accesses which may physically be on-line inquiries: "What orders are

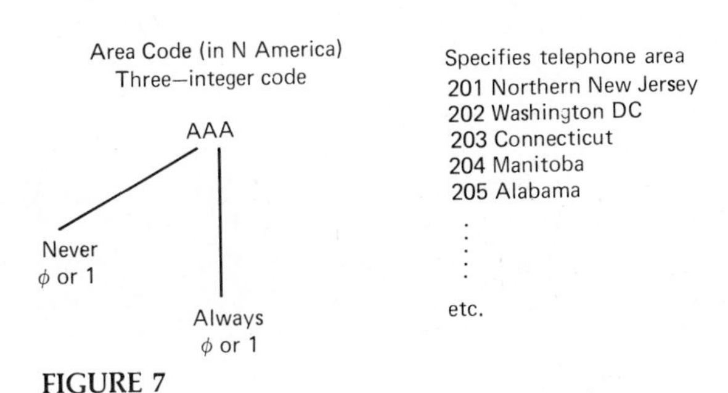

FIGURE 7

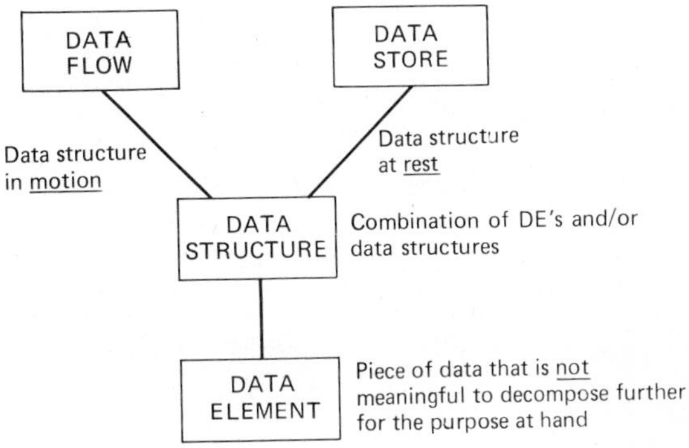

DATA FLOW — Data structure in motion

DATA STORE — Data structure at rest

DATA STRUCTURE — Combination of DE's and/or data structures

DATA ELEMENT — Piece of data that is not meaningful to decompose further for the purpose at hand

FIGURE 8

there for Supplier X?" "What is the status of Customer Y's order?" and so on.

What must the contents of the "Pending Orders" data store be? Since it is fed by a stream of orders and streams of orders come out, it surely must contain records of the data structure "Orders," defined as in our previous list.

Is that enough? Inspection shows us that it would be. Of course with a more complex data store, with say 10 inflows and 20 outflows, the derivation of contents is far from being so easy, and the detailed logical components of each flow must be compared. It is valuable to reduce the contents of each data store to several tables, or relations, in third normal form, thus removing

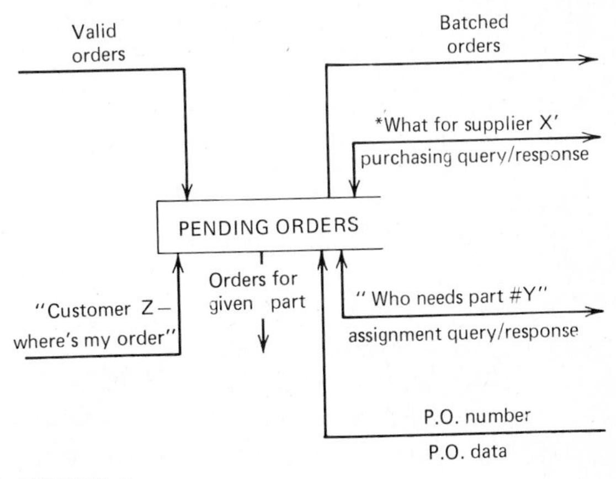

FIGURE 9

redundant data and getting the structures as simple as possible:

```
ORDER
    ORDER-IDENTIFICATION
        CUSTOMER-ORDER-NUM
        CUSTOMER-ORDER-DATE
    CUSTOMER-DETAILS
        CUSTOMER-NAME
        PERSON-AUTHORIZING
        PHONE
        SHIP-TO-ADDRESS
        BILL-TO-ADDRESS
    PART-DETAILS
        PART-NUMBER
        QUANTITY
        SUPPLIER
        PRICE
```

IMMEDIATE ACCESSES

Having defined the necessary and sufficient contents for the data store, we now have to investigate the immediate accesses required. An immediate access can be defined as one which is required faster than it is possible to search or sort the entire data store.

For example, since the supplier of each part is held as a data element in the data store, the question "What orders do we have for Supplier X?" can always be answered, given time, by scanning every record. If the question must be answered while someone is on the phone, however, we must provide for some special secondary access path (an index, a logical pointer or the equivalent).

We use an Immediate Access Diagram as shown in Figure 10 to get a consolidated view of the various immediate accesses required by different members of the user community.

The arrows connecting the boxes "Customer-Name," "Part-Number" and "Supplier-Name" to the box representing the data store are not data flow arrows—they merely represent the necessity for immediate access to a record or records given Customer-Name and so on.

When an immediate access diagram has been drawn up with volumes, file sizes, frequency of access and so on, we can use it as a data base/file design planning tool. If providing all the immediate accesses required presents no technical problem and can be done at acceptable cost, so be it.

However, if the cost of providing all the secondary access paths is unacceptable, then something has got to go. The immediate access diagram is then very useful as a tool to get the users to put some valuation on the relative importance of immediate accesses—obviously if it is very valuable to the business to be able to access pending orders by part number and only "nice-to-have" to access by customer name, then in a pinch we shall do away with the Customer-Name access.

EXPRESSION OF LOGIC

The last tool of structured analysis we need to consider is for the expression of detailed logic. We have an overall DFD for the system, with typically 20 to 50 processes in it. The logic of each of these detailed processes should be expressible in one or two pages—but one or two pages of what?

Normally we write logic in natural English. For instance, suppose we exploded "Apply payment to invoice" in Figure 4 and found one of the detailed processes was "Verify discount is correct." Upon inquiring as to the detailed logic of this process, we are given the following memo:

". . . Trade discount (to car dealers) is 20%. For private customers, 5% discount is allowed on orders for six items or more, 10% on orders for 20 items or more and 15% on orders for 50 or more. Trade orders for 20 items and over receive the 10% discount over and above the trade discount."

This is fairly clear as memos go; however carefully it is written, though, we must read all

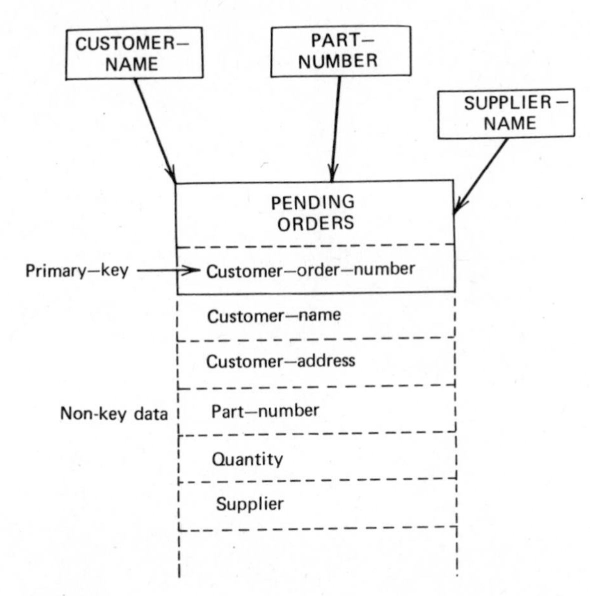

FIGURE 10

of it to process any one case. Thus, how long does it take to answer the question, "What discount for a trade customer with a 21-item order?"

We need a more graphic method of displaying the structure of detailed logic, such as a decision tree.

How long does it take you to answer the above question now? The decision tree is an ideal tool for representing the detailed logic of processes which are mainly decisions. In many cases, though, long sequences of actions must be carried out, sometimes repeatedly; then the decision tree cannot be used. We must use our second logic tool, structured English, as shown in Figure 12.

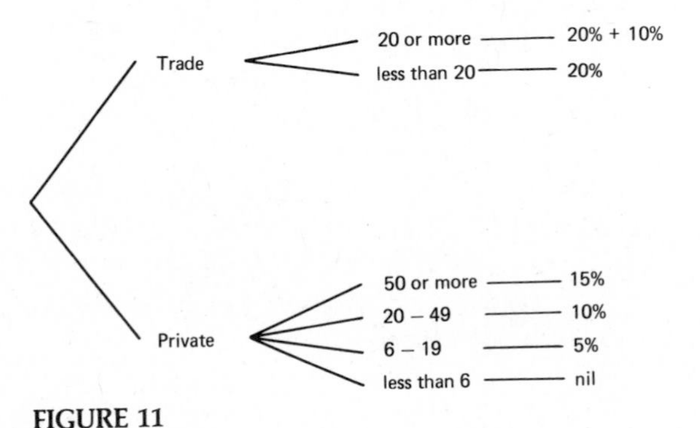

FIGURE 11

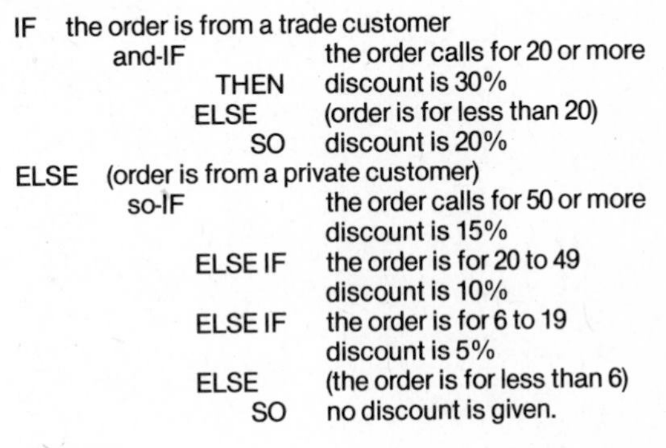

```
Add up the total number of items on the order (in the quan-
tity column)

IF    the order is from a trade customer
        and-IF                 the order calls for 20 or more
                   THEN        discount is 30%
                   ELSE        (order is for less than 20)
                         SO    discount is 20%
ELSE  (order is from a private customer)
         so-IF                 the order calls for 50 or more
                               discount is 15%
                   ELSE IF     the order is for 20 to 49
                               discount is 10%
                   ELSE IF     the order is for 6 to 19
                               discount is 5%
                   ELSE        (the order is for less than 6)
                         SO    no discount is given.
```

FIGURE 12

This is a perfectly general, strict dialect of English, using sequence, decision and repetition structures to express the detailed logic of any policy at all.

Figure 13 shows how these various tools fit together to make a logical model either of an existing or a proposed system.

Once the logical model has been developed by the analyst and the user community agrees it meets the system requirements, we can begin system design from a firm foundation. Our first concern is strategic design, sometimes known as "subsystem packaging." What will the subsystems be? What parts of the system function will be automated, and what will remain manual?

STRUCTURED DESIGN

When we have decided on the automation boundary and carved up the computer system into subsystems (creating the minimum number of intermediate files in the process), we have to design the software within each subsystem. Obviously we want a design that works, and we also want a design that is changeable—a modular piece of software whose modules have the least

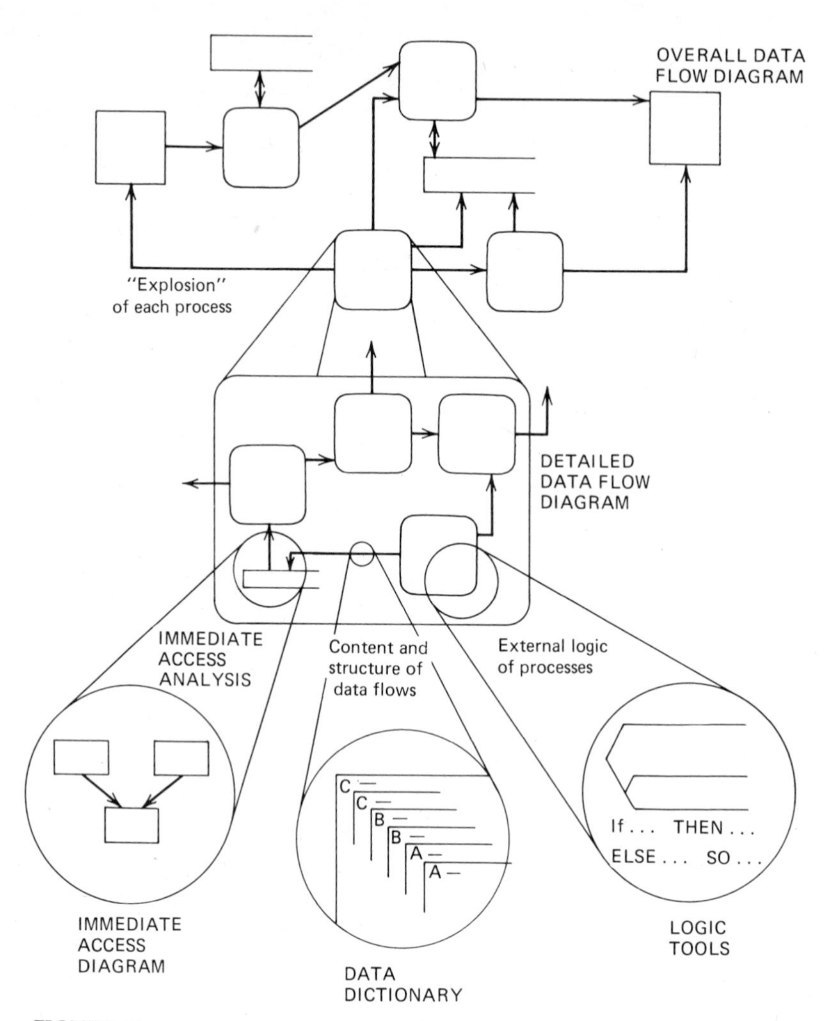

FIGURE 13

necessary coupling between them to make the system work.

Changeable modular systems frequently look like military command structures, as shown in Figure 14. The system commander gives orders to and receives information from subcommanders, which in turn give orders to worker modules that do the actual reading, writing, editing, computing and so on. Note that workers do not speak to one another—no one speaks unless spoken to, and then only to their immediate superior or subordinate.

This, of course, is exactly what we need to make a changeable system.

By artificially limiting the communication between modules in this way, we are confident that we can change, for example, the logic of the Editor module and, provided the "official" interface between the Editor and the Input Subcommander is not damaged, we know that the changed system will work. In many designs, there are hidden or nonobvious connections between modules such that, if we change one part, then we introduce a bug somewhere else; chang-

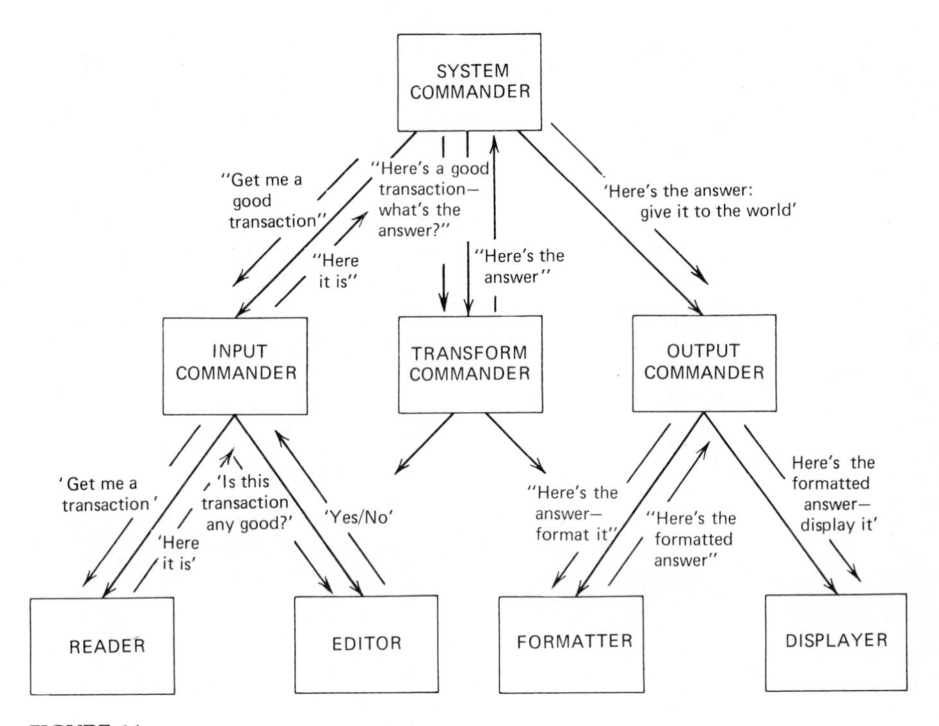

FIGURE 14

ing the second part to fix the bug creates a new bug, and so on ad infinitum.

How can we derive such a modular, changeable hierarchy for a subsystem? The most reliable method, initially conceived by Constantine, is to base it on the data flow through the subsystem. (Here of course, is the link between structured analysis and structured design.)

Figure 15 shows a very simple detailed DFD with a single stream of data going through it. All such diagrams can be divided into three parts—one part which is concerned with massaging input to get it into a clean processable form (the leftmost two functions in Figure 15); an area which transforms clean input into data ready for output; and an area which massages the output

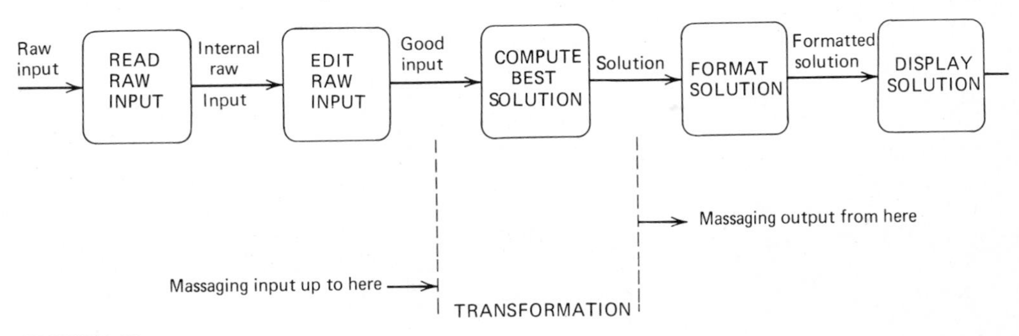

FIGURE 15

into the required form (the rightmost two functions in Figure 15).

Once we've identified these three areas, we can transform the DFD into a structure chart as in Figure 16, where the large arrows show invocation, the small arrows with white circle tails represent the data passed between modules and the small arrows with black circle tails represent control information such as a flag or a switch.

The structure chart tells us that "Produce Best Solution" invokes "Get Good Input." In turn "Get Good Input" invokes "Read Input," which returns with some Raw Input data.

"Get Good Input" then invokes "Edit Input," passing it the Raw Input data and receiving in return a flag (set to "Good" or "Bad" appropriately).

If the input is good, "Get Good Input" returns Control to "Produce Best Solution" along with the Good Input Data, and so on. Note that each function on the detailed DFD appears on the structure chart as a worker module at the bottom of the hierarchy.

The simple structure chart shown in Figure 16 is a prototype for systems with a single stream of data in the DFD; Constantine called this a "transform-centered" structure. Other structures correspond to more complicated DFDs.

By examining the detailed DFD, we can always use one of these model structures, or a

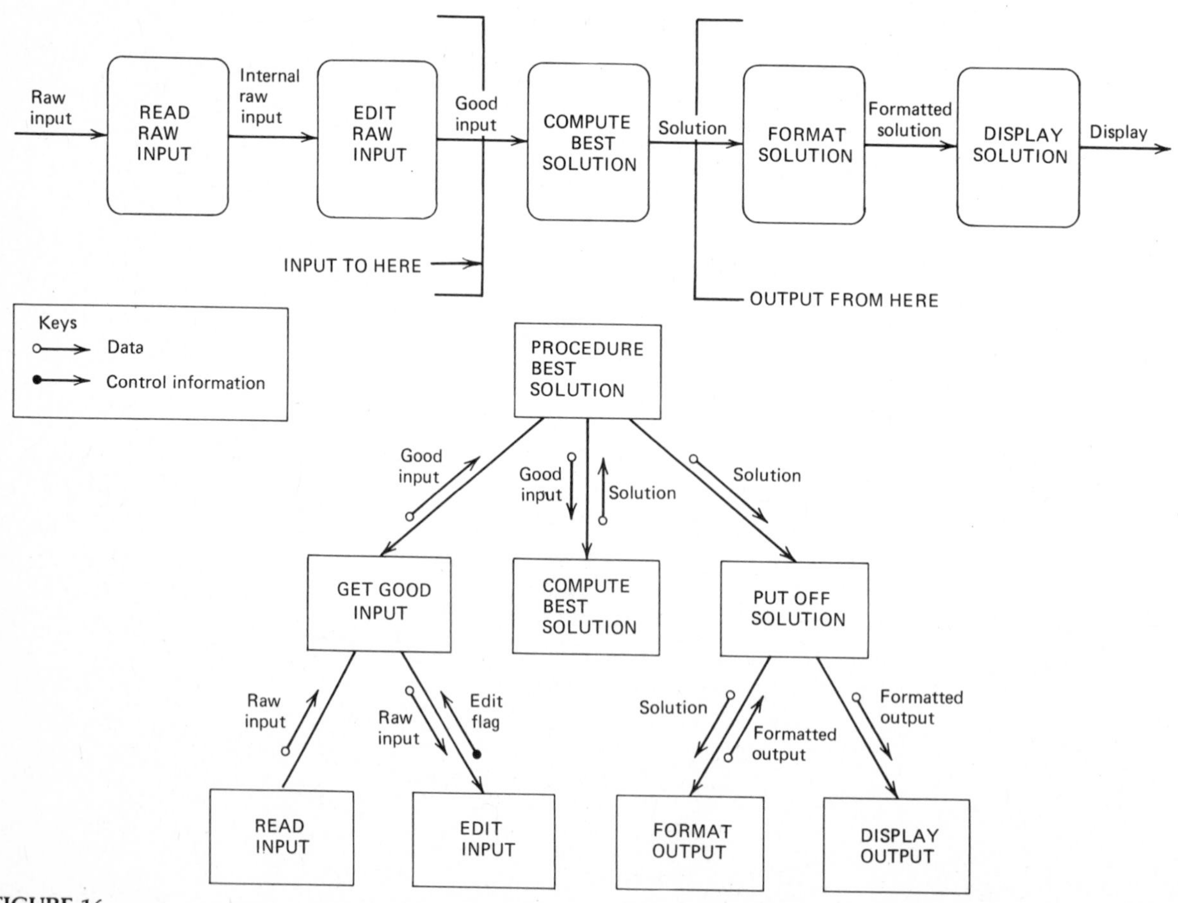

FIGURE 16

combination of them, to derive a hierarchical design. By marking the data and control interfaces on the structure chart, we can see very clearly on one sheet of paper all the significant aspects of the design, allowing us to review it and see if there is any way it can be made more changeable. This is the type of hierarchical structured design chart to which Carl Garrison referred.

We would not normally expect users to review a structure chart—although, as Garrison said, some may be prepared to do so. Normally, users review DFDs.

USER INVOLVEMENT

We asked Carl what other types of projects on which Superior Oil had used DFDs and structure charts. "We're just getting into design with a geological data base system which captures technical data about geological structures," he replied. "It's a medium-sized project. About half the project team is made up of users—geologists. They've become very enthusiastic about the data flow and the logical model of the system, to such an extent that they're now using it to train geologists who come as new hires!"

McDonnell Douglas Automation (McAuto) has had similar experiences. Jay Green, the section manager responsible for development, said, "I would summarize the benefits of structured techniques as being (a) simplified communication all the way from the final end user through to the programmer, (b) less rework due to the discovery of problems late in the game, (c) more changeable designs, (d) easier estimation of implementation costs, (e) fewer problems found during testing because the structure chart is itself testable."

To what projects has McAuto applied structured techniques? "We have quite a number in process, but we've now completed some projects where we used the structured methodology all the way through. For instance, we did this with our Property Record for Equipment Servicing and Sharing project. This is an IMS on-line project with batch reporting that we brought live in April after 25,000 man-hours of development," Green told us.

CHANGEABLE DESIGN

How does the changeability of a structured design work out in practice? Here's Carl Garrison:

"On the revenue accounting project, as we were starting into programming of one section, we were hit with a change in data base requirement. Because of the cut-and-paste nature of a structure chart, making the design change was easy."

How much longer would the same change have taken using conventional flowcharts?

"Very hard to say, but at least five times and maybe 10 times longer."

WORDS OF CAUTION

We asked Carl and Jay what words of advice they would have for people just starting to use structured techniques.

Jay: "Users must realize that we're going to spend 25% to 50% more time and money in analysis and design. Succumbing to user impatience to have programming started before these functions are complete, however, can result in more than compensating time and effort spent in difficult testing."

Carl: "Pick on a project that's large enough to be meaningful, but not so large as to be overwhelming—then use the people from the project as 'seeds' on later projects."

Carl: "It's important to develop procedures ahead of time to tie your DFDs and data structures to a data dictionary."

"We just did a labor distribution system using structured methodology. It wasn't a very large effort, three to four man-months, and it's true the people on it were good, but it was the first project recently to go in ahead of schedule and under budget—and we've had no production bugs in the last three months. A good piece of that is due to 'structured.' "

At Improved System Technologies, Inc., we are pursuing two lines of development for structured methodologies.

The first is the working out of step-by-step procedures and standards to integrate the intelligent use of data flow diagrams, logical data analysis, data dictionaries, structured design, structured walkthroughs and top-down development. This has been done in the structured analysis design and implementation of Information Systems (Stradis), whose procedures and training make up a formal system development methodology.

The second line of development is the provision of automated assistance to the analyst/designer. Several groups are experimenting with the use of automated drafting packages to assist in the drawing and maintenance of data flow diagrams, using disk storage and plotter output.

The next step, to our way of thinking, is to build a bridge between the automated data dictionary and the automated drafting package so data and processes defined on a DFD can be transferred to the data dictionary and vice versa.

If we could further integrate the data dictionary with effective text processing, we would have an "analyst's workbench" which would allow the analyst to build the logical model of a system with, say, a dedicated microprocessor, providing on-line entry or data definitions, data flows, information about system objectives, policy statements, etc. It would also permit on-line update and production of DFDs and structure charts (on a plotter) and requirements statements, policy logic, design statements, user procedures, etc., on a text printer in the project area.

While the analyst's workbench would not do any thinking for the analysts, it could speed up the clerical work of analysis and/or enable the rapid sharing of all information between members of the project team.

READING QUESTIONS

1. What do these authors feel are the major benefits of the building of a graphical, logical model?
2. Name and define the symbols used in the data flow diagram described in this paper.
3. Define the concept of the data store. Why have these authors made it an integral part of their logical model?
4. What is an immediate access, and what does it imply about the design of the data stores?
5. How does this process show hierarchical structure?
6. Relate Gane-Sarson's discussion of communication between modules and changeability to the concepts of module coupling and cohesion which are discussed in the Stevens, Myers, and Constantine paper.
7. Referring to the patient monitoring system discussed in the Stevens, Myers, and Constantine paper, identify and discuss the characteristics of the necessary data structures.

Structured Analysis (SA): A Language For Communicating Ideas

Douglas T. Ross

I. BLUEPRINT LANGUAGE

Neither Watt's steam engine nor Whitney's standardized parts really started the Industrial Revolution, although each has been awarded that claim, in the past. The real start was the awakening of scientific and technological thoughts during the Renaissance, with the idea that the lawful behavior of nature can be understood, analyzed, and manipulated to accomplish useful ends. That idea itself, alone, was not enough, however, for not until the creation and evolution of blueprints was it possible to express exactly how power and parts were to be combined for each specific task at hand.

Mechanical drawings and blueprints are not mere pictures, but a complete and rich language.

Source: Douglas T. Ross, "Structured Analysis (SA): A Language for Communicating Ideas," *IEEE Transactions on Software Engineering*, Vol. SE-3, No. 1, January 1977, pp. 16–34.

In blueprint language, scientific, mathematical, and geometric formulations, notations, mensurations, and naming do not merely describe an object or process, they actually model it. Because of broad differences in subject, purpose, roles, and the needs of the people who use them, many forms of blueprint have evolved, but all rigorously present well structured information in understandable form.

Failure to develop such a communication capability for data processing is due not merely to the diversity and complexity of the problems we tackle, but to the newness of our field. It has naturally taken time for us to escape from naive "programming by priesthood" to the more mature approaches, such as structured programming, language and database design, and software production methods. Still missing from this expanding repertoire of evidence of maturity, however, is the common thread that will allow all of the pieces to be tied together into a predictable and dependable approach.

II. STRUCTURED ANALYSIS (SA) LANGUAGE

It is the thesis of this paper that the language of structured analysis (SA), a new disciplined way of putting together old ideas, provides the evolutionary natural language appropriate to the needs of the computer field. SA is deceptively simple in its mechanics, which are few in number and have high mnemonic value, making the language easy and natural to use. Anybody can learn to read SA language with very little practice and will be able to understand the actual information content being conveyed by the graphical notation and the words of the language with ease and precision. But being a language with regorously defined semantics, SA is a tough taskmaster. Not only do well conceived and well phrased thoughts come across concisely and with precision, but poorly conceived and poorly expressed thoughts also are recognized as such. This simply *has* to be a fact for any language whose primary accomplishment is valid communication of understanding. If both the bad and the good were *not* equally recognizable, the understanding itself would be incomplete.

SA does the same for any problem chosen for analysis, for every natural language and every formal language are, by definition, included in SA. *The only function of SA is to bind up, structure, and communicate units of thought expressed in any other chosen language*. Synthesis is composition, analysis is decomposition. *SA is structured decomposition, to enable structured synthesis to achieve a given end*. The actual building-block elements of analysis or synthesis may be of any sort whatsoever. Pictures, words, expressions of any sort may be incorporated into and made a part of the structure.

The facts about Structured Analysis are as follows:

1. It incorporates any other language; its scope is universal and unrestricted.
2. It is concerned only with the orderly and well-structured decomposition of the subject matter.
3. The decomposed units are sized to suit the modes of thinking and understanding of the intended audience.
4. Those units of understanding are expressed in a way that rigorously and precisely represents their interrelation.
5. This structured decomposition may be carried out to any required degree of depth, breadth, and scope while still maintaining all of the above properties.
6. Therefore, SA greatly increases both the quantity and quality of understanding that can be effectively and precisely communicated well beyond the limitations inherently imposed by the imbedded natural or formal language used to address the chosen subject matter.

The universality and precision of SA makes it particularly effective for requirements definition for arbitrary systems problems, a subject treated in some detail in a companion paper (see [5]). Requirements definition encompasses all aspects of system development prior to actual system design, and hence is concerned with the discovery of real user needs and communicating those requirements to those who must produce an effective system solution. Structured Analysis and Design Technique (SADT™) is the name of SofTech's proprietary methodology based on SA. The method has been applied to a wide range of planning, analysis, and design problems involving men, machines, software, hardware, database, communications procedures, and finances over the last two years, and several are cited in that paper. It is recommended that that paper (see [5]) be read prior to this paper to provide motivation and insight into the features of SA language described here.

™Trademark of SofTech, Inc.

SA is not limited to requirements definition nor even problems that are easily recognized as system problems. The end product of an SA analysis is a working model of a well-structured understanding, and that can be beneficial even on a uniquely personal level—just to "think things through." Social, artistic, scientific, legal, societal, political, and even philosophic subjects, all are subject to analysis, and the resulting models can effectively communicate the ideas to others. The same methods, approach, and discipline can be used to model the problem environment, requirements, and proposed solution, as well as the project organization, operation, budget schedule, and action plan. Man thinks with language. Man communicates with language. SA structures language for communicating ideas more effectively. *The human mind can accommodate any amount of complexity as long as it is presented in easy-to-grasp chunks that are structured together to make the whole.*

III. OUTLINE OF THE DEMONSTRATION

Five years ago I said in an editorial regarding software [1]: "Tell me *why* it works, not *that* it works." That is the approach taken in this paper. This paper does not present a formal grammar for the SA language—that will come later, elsewhere. This paper also is not a user manual for either authors or readers of the language—a simple "how to" exposition. Instead, we concentrate here on the motivation *behind* the features of SA in an attempt to convey directly an appreciation for its features and power even beyond that acquired through use by most SA practitioners. SA has been heavily developed, applied, taught, and used for almost three years already, but the design rationale behind it is first set down here.

SA (both the language and the discipline of thought) derives from the way our minds work, and from the way we understand real-world situations and problems. Therefore, we start out with a summary of principles of exposition—good story-telling. This turns out to yield the familiar top-down decomposition, a key component of SA. But more than that results, for consideration of how we view our space-time world shows that we always understand anything and everything in terms of *both* things *and* happenings. This is why all of our languages have *both* nouns *and* verbs—and this, in turn, yields the means by which SA language is universal, and can absorb any other language as a component part.

SA supplies rigorous structural connections to any language whose nouns and verbs it absorbs in order to talk about things and happenings, and we will spend some time covering the basics carefully, so that the fundamentals are solid. We do this by presenting, in tabular form, some 40 basic features, and then analyzing them bit by bit, using SA diagrams as figures to guide and illustrate the discussion.

Once the basics have thus been introduced, certain important topics that would have been obscure earlier are covered in some depth because their combinations are at the heart of SA's effectiveness. These topics concern constraints, boundaries, necessity, and dominance between modular portions of subject matter being analyzed. It turns out that constraints based on purpose and viewpoint actually *make* the structure. The depth of treatment gives insight into how we understand things.

The actual output of SA is a hierarchically organized structure of separate diagrams, each of which exposes only a limited part of the subject to view, so that even very complex subjects can be understood. The structured collection of diagrams is called an *SA model*. The demonstration here concludes with several special notations to clarify presentation and facilitate the orderly organization of the material. Since actual SA diagrams (some good, some illustrating poor style) are used as figure illustrations, the reader is exposed here to the style of SA, even though the SA model represented by the collection of figures is not complete enough to be understandable by itself. Later papers will treat more advanced topics and present complete examples of SA use and practice in a wide variety of applications.

IV. PRINCIPLES OF GOOD STORYTELLING

There are certain basic, known principles about how people's minds go about the business of understanding, and communicating understanding by means of language, which have been known and used for many centuries. No matter how these principles are addressed, they always end up with hierarchic decomposition as being the heart of good storytelling. Perhaps the most relevant formulation is the familiar: "Tell 'em whatcha gonna tell 'em. Tell 'em. Tell 'em whatcha told 'em." This is a pattern of communication almost as universal and well-entrenched as Newton's laws of motion. It is the pattern found in all effective forms of communication and in all analyses of why such communication is effective. Artistic and scientific fields, in addition to journalism, all follow the same sequence, for that is the way our minds work.

Only something so obvious as not to be worth saying can be conveyed in a single stage of the communication process, however. In any worthwhile subject matter, Stage Two ("Tell 'em") requires the parallel introduction of several more instances of the same pattern starting again with Stage One. Usually a story establishes several such levels of telling, and weaves back and forth between them as the need arises, to convey understanding, staying clear of excesses in either detail (boredom) or abstraction (confusion).

V. THE SA MAXIM

This weaving together of parts with whole is the heart of SA. The natural law of good communications takes the following, quite different, form in SA:

> *Everything worth saying about anything worth saying something about must be expressed in six or fewer pieces.*

Let us analyze this maxim and see how and why it, too, yields hierarchically structured storytelling.

First of all, there must be something (anything) that is "worth saying something about." We must have some subject matter that has some value to us. We must have an interest in some aspect of it. This is called establishing the *viewpoint* for the model, in SA terminology. Then we must have in mind some audience we want to communicate with. That audience will determine what is (and is not) "worth saying" about the subject from that viewpoint. This is called establishing the *purpose* for the model, in SA terminology. As we will see, every subject has many aspects of interest to many audiences, so that there can be many viewpoints and purposes. But each SA model must have only one of each, to bound and structure its subject matter. We also will see that each model also has an established *vantage point* within the purpose-structured context of some other model's viewpoint, and this is how multiple models are interrelated so that they collectively cover the whole subject matter. But a single SA model considers only worthy thoughts about a single worthy subject.

The clincher, however, is that *every* worthy thought about that worthy subject must be included. The first word of the maxim is *everything*, and that means exactly that—absolutely nothing that fits the purpose and viewpoint can be left out. The reason is simple. By definition everything is the subject itself, for otherwise it would not *be* that subject—it would be a *lesser* subject. Then, if the subject is to be broken into six or fewer pieces, every single thing must go into exactly one of those (nonoverlapping) pieces. Only in this way can we ensure that the subject stays the same, and does not degenerate into some lesser subject as we decompose it. Also, if overlapping pieces were allowed, conflicts and confusions might arise.

A "piece" can be anything we choose it to be—the maxim merely requires that the single piece of thought about the subject be broken into several (not too few, and not too many[1]) pieces. Now, certainly if the original single piece of thought about the subject is worthy, it is very unlikely that the mere breaking of it into six-or-fewer pieces exhausts that worth. The maxim still applies so that every one of them must similarly be expressed in six-or-fewer *more* pieces—again and again—until the number of pieces has grown to suit the total worthiness. At a fine enough level of decomposition, it is not worth continuing. No further decomposition is required for completely clear understanding. Thus we see that the SA maxim must be interpreted *recursively,* and yields top-down hierarchic decomposition. The SA language allows this hierarchic structure to be expressed (see Fig. 1).

VI. EXPRESSION

In the maxim, the word "express" covers both the rigorous grammar of SA language itself, as well as the grammar (however well or ill formed) of the natural language chosen to address the subject matter. By definition, SA language includes all other languages, and regardless of what language is embedded, the decomposition discipline (expressed by the SA language component of the combined language) ensures that at each stage, the natural language (whatever it may be) is used to address and express only every worthy thought about a more and more restricted piece of the worthy subject matter. Because of this orderly zeroing-in, SA certainly cannot decrease the effectiveness of that chosen language. In effect, the SA maxim is valid by definition, for whenever the subject matter has already been broken down to such a fine level that the SA decomposition would add nothing to what already would be done (as, for example, in jokes or some poetry) the chosen language stands by itself, not decreased in effectiveness.

Most of the time the conscious practice of Structured Analysis and its thought discipline improves people's ability to think clearly and find solutions. In the cases where this does not happen, however, Structured Analysis still "works," in the sense that the bad portions stand out more clearly and are understood to be bad and needing further attention. For the next step in our demonstration we consider thoughts, and the expression of thoughts in language.

VII. THINGS AND HAPPENINGS

We live in a space-time world. Numerous philosophical and scientific studies, as well as the innate experience of every person, shows that we never have any understanding of any subject matter except in terms of our own mental constructs of "things" and "happenings" of that subject matter. It seems to be impossible for us to think about anything without having the subject automatically be bounded in our minds by a population of parts or pieces (concrete or abstract—but in any case "nominal" things, i.e., literally things to which we give names or nouns) which provide the basis for our even conceiving of the subject matter as a separate subject. Immediately, however, once we are aware of the things, we are aware of the happenings—the re-

[1]Many people have urged me to relate the magic number "six" to various psychological studies about the characteristics of the human mind. I won't. It's neither scientific nor "magic." It is simply the *right* number. (Readers who doubt my judgement are invited to read for themselves the primary source [6].) The only proper reference would be to the little bear in the Goldilocks story. His portions always were "just right," too!

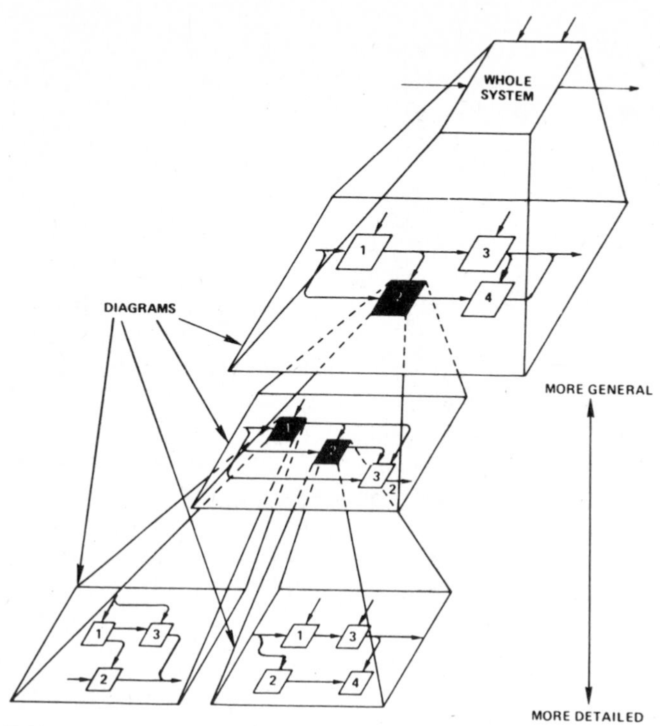

FIGURE 1 Structured decomposition.

lationships, changes, and transformations which take place between and among those things and make the subject matter meaningful or interesting (the "verbial" things, to which we give action words or verbs).

The universality of things and happenings provides the next basic step of decomposition (after the still more fundamental decomposition of recognizing and isolating the purpose and viewpoint which established the "worth" of possible things to say about the "worth" of the subject matter). Every one of our languages, whether natural or artificial, formal or informal, has those two complementary aspects—nouns and verbs, operators and operands, etc.—to permit the expression of thoughts about the subject matter. Thus the means is provided to incorporate any other language into SA. The incorporation of other languages into SA is not forced, nor awkward.

SA language provides the same graphic notation for both the things and the happenings aspects of any subject. Every SA model has two dual aspects—a *thing* aspect, called the *data decomposition,* and a *happening* aspect, called the *activity decomposition*. The model is incomplete without both decompositions.

VIII. BOUNDED SUBJECT MATTER

So we have now established the starting premises. The SA maxim forces gradual, top-down decomposition, leaving nothing out at any stage, and matching good storytelling exposition. The things and happenings (*data* and *activities,* in SA technical terms) match the nominal and verbial construction of any chosen language for directly addressing the subject, so we will never be "at a loss for words." Now we are ready to address the specifics—how SA language (mostly graphical, using boxes and arrows) actually allows well structured expression of well structured thought. We do this in stages: (1) we dump the entire body of the subject matter all at once into a table of some 40 separate items of notation and conventions—just to bound the subject itself; (2) we then start to pick our way through these topics, starting with those that define the basics of boxes and arrows; and (3) then we will use those basic expository capabilities to complete the consideration of the list.

In a prior, companion paper [2], which had its roots in the same background that led to the development of SA, we described and illustrated a univeral, standard pattern or process which appears to permeate all of software engineering and problem-solving in general. Since that pattern is so close to the natural phenomena of understanding which we are discussing here with respect to SA itself, we will use it to motivate, clarify, and structure the presentation. The idea of the pattern is captured in five words: (1) purpose; (2) concept; (3) mechanism; (4) notation; and (5) usage. Any systematic approach to a problem requires a concise purpose or objective that specifies which aspect of the problem is of concern. Within that purpose we formulate a valid conceptual structure (both things and happenings) in terms of which the problem can be analyzed and approached. We then seek out (or work out) the designs (mechanisms—concrete or abstract, but always including both data and activity aspects) which are capable of implementing the relevant concepts and of working together to achieve the stated purpose. (This combines three of the five words together.) Now, purpose, concepts, and mechanism, being a systematic approach to a *class* of problems, require a notation for expressing the capabilities of the mechanism and invoking its use for a particular problem in the class. Finally, usage rules are spelled out, explicitly or by example, to guide the use of the notation to invoke the implementation to realize the concept to achieve the specified purpose for the problem. The cited paper [2] gives numerous carefully drawn examples showing how the pattern arises over and over again throughout systematic problem solving, at both abstract and concrete levels, and with numerous hierarchic and cross-linked interconnections.

IX. THE FEATURES OF SA LANGUAGE

Figure 2 is a tabulation of some 40 features or aspects of SA which constitute the basic core of the language for communication. For each feature, the purpose, concept, mechanism, and notation are shown. Usages (for the purposes of this paper) are covered only informally by the examples which follow. The reader should scan down the "purpose" column of Fig. 2 at this time, because the collection of entries there set the objectives for the bounded subject matter which we are about to consider. Note also the heavy use of pictures in the "notation" column. These are components of graphic language. But notice that most entries mix *both* English *and* graphic language into a "phrase" of SA notation. Clearly, any other spoken language such as French, German, or Sanskrit could be translated and substituted for the English terms, for they merely aid the understanding the syntax and semantics of SA language itself.

In Fig. 2, the *name* and *label* portions of the "notation" column for rows 1 and 2, and the corresponding *noun* and *verb* indications in rows 6 and 7 are precisely the places where SA language absorbs other natural or formal languages in the sense of the preceding discussion. As the preceding sections have tried to make clear, *any* language, whether informal and natural or formal and artificial, has things and happenings aspects in the nominal and verbial components of its vocabulary. These are to be related to the *names of boxes* and *labels on arrows* in order to absorb those "foreign" languages into SA language.

Notice that it is not merely the nouns and verbs which are absorbed. Whatever richness the

	PURPOSE	CONCEPT	MECHANISM	NOTATION	NODE
1	BOUND CONTEXT	INSIDE/OUTSIDE	SA BOX	NAME	A11
2	RELATE/CONNECT	FROM/TO	SA ARROW	LABEL	A12
3	SHOW TRANSFORMATION	INPUT-OUTPUT	SA INTERFACE	INPUT --- OUTPUT	A13
4	SHOW CIRCUMSTANCE	CONTROL	SA INTERFACE	CONTROL	A14
5	SHOW MEANS	SUPPORT	SA MECHANISM	MECHANISM	A15
6	NAME APTLY	ACTIVITY: HAPPENINGS / DATA: THINGS	SA NAMES	ACTIVITY: VERB / DATA: NOUN	A211
7	LABEL APTLY	THINGS / HAPPENINGS	SA LABELS	NOUN / VERB	A212
8	SHOW NECESSITY	I-O / C-O	PATH		A213
9	SHOW DOMINANCE	C / I	CONSTRAINT		A214
10	SHOW RELEVANCE	ICO / ICO	ALL INTERFACES		A215
11	OMIT OBVIOUS	C-O / I-O	OMITTED ARROW		A216
12	BE EXPLICIT WITHOUT CLUTTER	PIPELINES, CONDUITS, WIRES	BRANCH	A A A	A221
13			JOIN	A A A	A221
14	BE CONCISE AND CLEAR	CABLES, MULTI-WIRES	BUNDLE	A B C(=AUB)	A222
15			SPREAD	C=(AUB) A B	A222
16	SHOW EXCLUSIVES	EXPLICIT ALTERNATIVES	OR BRANCH	A OR B A B	A223
17			OR JOIN	A B A OR B	A223
18	SHOW INTERFACES TO PARENT DIAGRAM	PARENT CHILD ARROWS PENETRATE	SA BOUNDARY ARROWS (ON CHILD)	NO BOX SHOWN	A231
19	SHOW EXPLICIT PARENT CONNECTION	NUMBER CONVENTION FOR PARENT, WRITE ICOM CODE ON CHILD BOUNDARY ARROWS	I C M O n	C1 C2 (ON CHILD) O2	A232
20	SHOW UNIQUE DECOMPOSITION	DETAIL REFERENCE EXPRESSION (DRE)	C-NUMBER OR PAGE NUMBER OF DETAIL DIAGRAM	BOX DRE	A233
21	SHOW SHARED OR VARIABLE DECOMPOSITION	DRE WITH (MODEL NAME)	SA CALL ON SUPPORT	BOX STUB DRE	A234

	PURPOSE	CONCEPT	MECHANISM	NOTATION	NODE
22	SHOW COOPERATION	INTERCHANGE OF SHARED RESPONSIBILITY	SA 2-WAY ARROWS		A311
23	SUPPRESS INTERCHANGE DETAILS	ALLOW 2-WAY WITHIN 1-WAY PIPELINES	2-WAY TO 1-WAY BUTTING ARROWS		A312
24	SUPPRESS "PASS-THROUGH" CLUTTER	ALLOW ARROWS TO GO OUTSIDE DIAGRAMS	SA "TUNNELING" (WITH REFERENCES)	PARENT OFFSPRING	A313
25	SUPPRESS NEEDED-ARROW CLUTTER	ALLOW TAGGED JUMPS WITHIN DIAGRAM	TO ALL or FROM ALL	TO ALL (A (A (A	A314
26	SHOW NEEDED ANNOTATION	ALLOW WORDS IN DIAGRAM	SA NOTE	NOTE:	A32
27	OVERCOME CRAMPED SPACE	ALLOW REMOTE LOCATION OF WORDS IN DIAGRAM	SA FOOTNOTE	(n=integer) n words n	A32
28	SHOW COMMENTS ABOUT DIAGRAM	ALLOW WORDS ON (NOT IN) DIAGRAM	SA META-NOTE	n (n=integer)	A32
29	ENSURE PROPER ASSOCIATION OF WORDS	TIE WORDS TO INTENDED SUBJECT	SA "SQUIGGLE"	label (TOUCH REFERENT)	A32
30	UNIQUE SHEET REFERENCE	CHRONOLOGICAL CREATION	SA C-NUMBER	AUTHOR INITS INTEGER	A41
31	UNIQUE BOX REFERENCE	PATH DOWN TREE FROM BOX NUMBERS	SA NODE NUMBER (BOX NUMBERS)	A, D, OR M · PARENT # · BOX #	A42
32	SAME FOR MULTI-MODELS	PRECEDE BY MODEL NAME	SA MODEL NAME	MODEL NAME/NODE#	A42
33	UNIQUE INTERFACE REFERENCE	ICOM WITH BOX NUMBER	SA BOX ICOM	BOX# ICOM CODE	A43
34	UNIQUE ARROW REFERENCE	FROM - TO	PAIR OF BOX ICOMs	BOX $ICOM_1$ BOX $ICOM_2$	A44
35	SHOW CONTEXT REFERENCE	SPECIFY A REFERENCE POINT	SA REF.EXP. "DOT"	A122.411 "WHICH SEE"	A45
36	ASSIST CORRECT INTERPRETATION	SHOW DOMINANCE GEOMETRICALLY (ASSIST PARSE)	STAIRCASE LAYOUT	DOMINANCE	A5
37	ASSIST UNDERSTANDING	PROSE SUMMARY OF MESSAGE	SA TEXT	NODE# · T · INTEGER	A5
38	HIGHLIGHT FEATURES	SPECIAL EFFECTS FOR EXPOSITION ONLY	SA FEOs	NODE# · F · INTEGER	A5
39	DEFINE TERMS	GLOSSARY WITH WORDS & PICTURES	SA GLOSSARY	MODEL NAME · G · INTEGER	A5
40	ORGANIZE PAGES	PROVIDE TABLE OF CONTENTS	SA NODE INDEX	NODE# ORDER	A5

FIGURE 2 SA language features.

"foreign" language may possess, the full richness of the nominal and verbial expressions, including modifiers, is available in the naming and labeling portions of SA language. As we shall see, however, normally these capabilities for richness are purposely suppressed, for simplicity and immediacy of understanding normally require brevity and conciseness.

Figure 2 has introduced our subject and has served to point out the precise way in which SA absorbs other languages, but this mode of discourse would make a long and rambling story. I therefore proceed to use SA itself to communicate the intended understanding of Fig. 2. This will not, however, be a perfect, or even a good example of SA communication in action, for the intent of this paper is to guide the reader to an understanding of SA, not to teach how to fully exploit SA diagrams and modeling. The SA diagrams presented here only as figures are incomplete and exhibit both good *and* bad examples of SA expressiveness, as well as showing all the language constructs. Our subject is too complex to treat in a small model, but the figures at least present the reader with some measure of the flexibility of the language.

The reader is forewarned that there is more information in the diagrams than is actually referenced here in text which uses them as "figures." After the paper has been read, the total model can be studied for review and for additional understanding. Everything said here about the SA language and notations applies to each diagram, and most features are illustrated more than once, frequently before they are described in the text. Therefore, on first reading, please ignore any features and notations not explicitly referenced. Non-SA "first-reading" aids are isolated by a bold outline, in the diagrams.

In practical use of SofTech's SADT, a "reader/author cycle" is rigorously adhered to in which (similar to the code-reading phase of egoless structured programming) authors, experts, and management-level personnel read and critique the efforts of each individual SADT author to achieve a fully-acceptable quality of result. (It is in fact this rigorous adherence to quality control which enables production SA models to be relied upon. So far as possible, *everything* worthy has been done to make sure that *everything* worthy has been expressed to the level required by the intended readership.)

X. PURPOSE AND VIEWPOINT

Figure 3 is an SA diagram[2] and, by definition, it is a meaningful expression in SA language. It consists of box and arrow graphical notation mixed with words and numbers. Consonant with the tutorial purpose of this paper, I will not, here, try to teach how to *read* a diagram. My tutorial approach aims only to lead to an understanding of what is *in* a diagram.

So we will just begin to examine Fig. 3. Start with the title, "Rationalize structured analysis features"—an adequate match to our understanding of the purpose and viewpoint of this whole paper. We seek to make rational the reasons behind those features. Next read the content of each of the boxes: "Define graphics; build diagram; use special notations; provide for referencing; organize material." These must be the six-or-fewer "parts" into which the titled subject matter is being broken. In this case there are five of them, and sure enough this aspect of SA follows exactly the time-tested outline approach to subject matter. Because our purpose is to have a graphics-based language (like blueprints), once we have decided upon some basics of graphic definition we will use that to build a diagram for some particular subject, adding special notations (presumably to improve clarity), and then because (as with blueprints) we know that a whole collection will be required to convey complex un-

[2]The SADT diagram form itself is © 1975, SofTech, Inc., and has various fields used in the practice of SADT methodology.

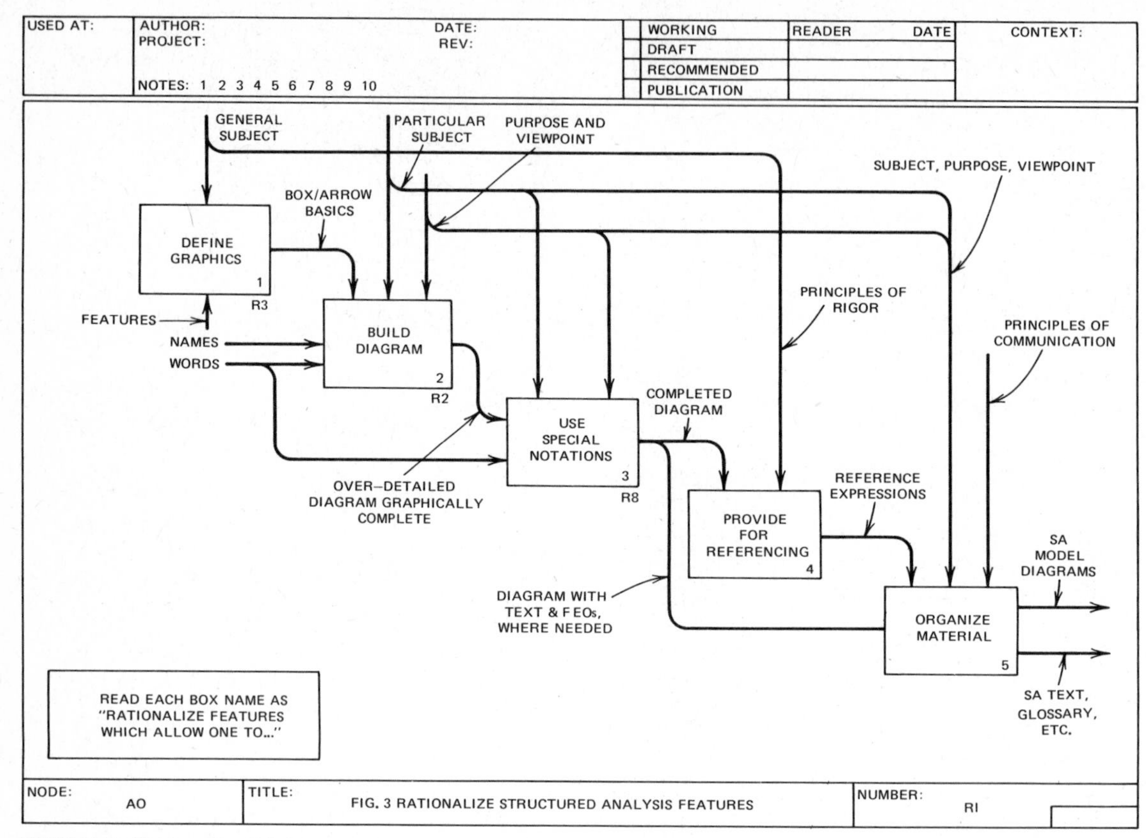

FIGURE 3 Rationalize SA features.

derstanding in easy-to-understand pieces, we must provide for a way of referencing the various pieces and organizing the resulting material into what we see as an understandable whole.

Now, I have tried to compose the preceding long sentence about Fig. 3 using natural language constructs which, if not an exact match, are very close to terms which appear directly in Fig. 3. In fact, the reader should be able to find an exact correspondence between things which show in the figure and every important syntactic and semantic point in each part of the last sentence of the preceding paragraph. Please reread that sentence now and check out this correspondence for yourself. In the process you will notice that considerable (though not exhaustive) use is made of information which is not *inside* the boxes, but instead is associated with the word-and-arrow structure of the diagram *outside* the boxes. This begins to show why, although SA in its basic backbone does follow the standard outline pattern of presentation, the box-and-arrow structure conveys a great deal more information than a simple topic outline (only the box *contents*) could possibly convey.

XI. THE FIRST DETAIL VIEW

Figure 4 is another SA diagram. Simpler in structure than the diagram of Fig. 3, but nonetheless with much the same "look." The title, "Define graphics," is identical to the name inside the first box of Fig. 3, which is here being broken

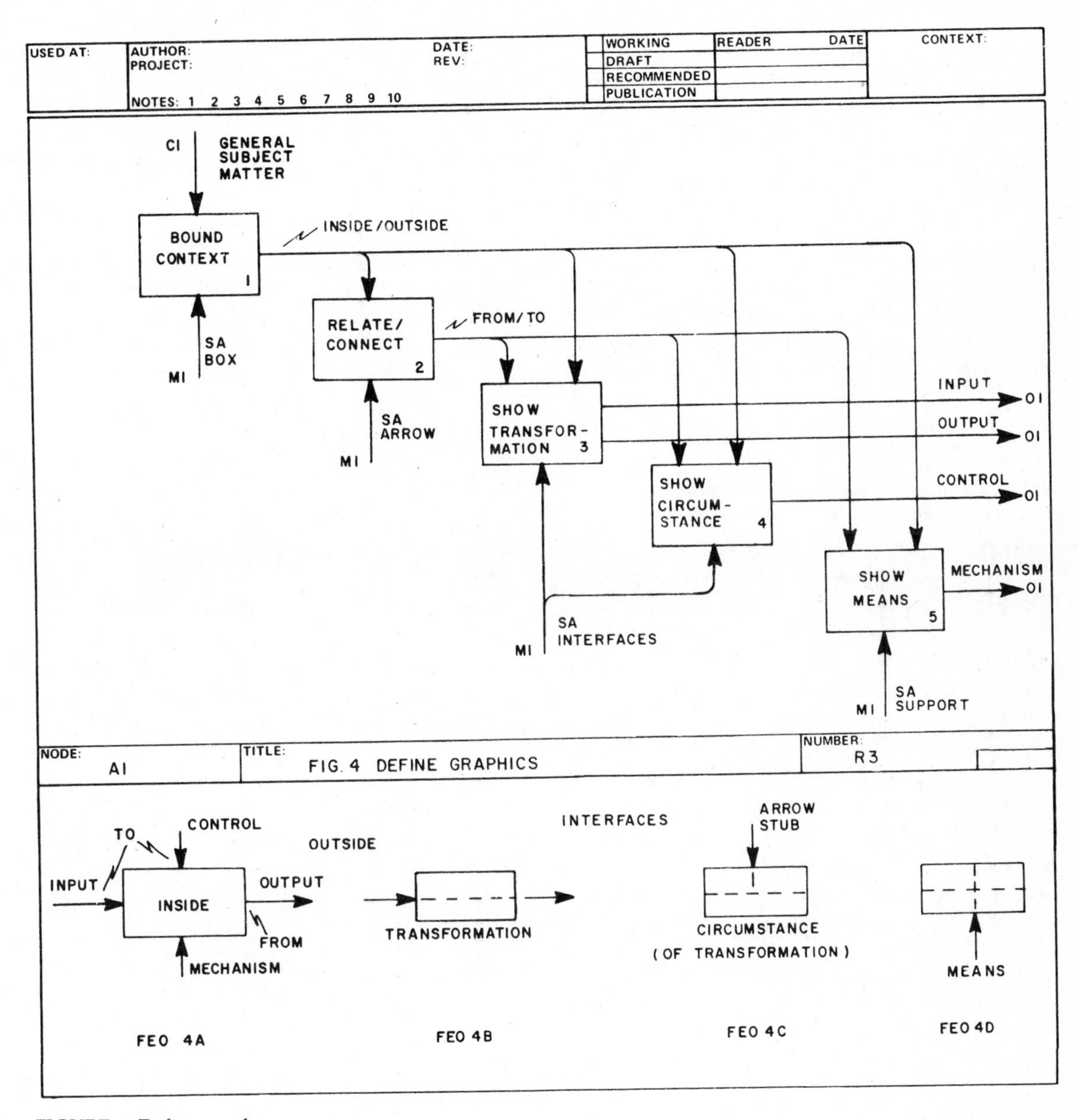

FIGURE 4 Define graphics.

into five component worthy pieces, called the *nested factors* in SA terminology. Again the words written inside the boxes are legible, but are they understandable? How can "Bound context; relate/connect; show transformation; show circumstance; show means," be considered parts of "Define graphics?" It is not very clear, is it? It would seem that something must be wrong with SA for the touted understandability turns out to be, in fact, quite obscure!

Look at Fig. 4 again and see if you don't agree that we have a problem—and see if you can supply the answer to the problem.

The problem is not with SA at all, but with our too-glib approach to it. SA is a rigorous language and thereby is unforgiving in many ways. In order for the communication of understanding to take place, we ourselves must understand and conform to the rules of that rigor. The apparent obscurity should disappear in a twinkling once the following factor is pointed out: namely, *always be sure to do your understanding in the proper context.* In this case, the proper context was established by the title of Fig. 3, "Rationalize structured analysis features," and the purpose, to define graphical concepts and notations for the purpose of representing and conveying understanding of subject matters. Now, if we have all of that firmly implanted in our minds, then surely the name in Box 1 of Fig. 4 should be amply clear. Read, now, Box 4. 1[3] for yourself, and see if that clarity and communication of intended understanding does not take place.

You see, according to the diagram, the first feature of defining graphics is to "Bound the context"—precisely the subject we have just been discussing and precisely the source of the apparent obscurity which made SA initially appear to be on shaky ground. To aid first reading of the figures, a suggested paraphrasing of the intended context is given in a bold box on each of the other diagrams (see Fig. 3).

As we can see from the section of Fig. 4 labeled FEO 4A[4] the general subject matter is isolated from the rest of all subject matter by means of the SA box which has an inside and an outside (look at the box). The only thing we are supposed to consider is the portion of that subject matter which is *inside* the box—so the boundary of the box does bound the context in which we are to consider the subject.

XII. THE SA BOX AS BUILDING BLOCK

We lack the background (at this point) to continue an actual reading of Fig. 4, because it itself defines the basic graphic notations used in it. Instead, consider only the sequence of illustrations (4A–4D) labeled FEO. FEO 4A shows that the fundamental building block of SA language notation is a box with four sides called INPUT, CONTROL, OUTPUT, and MECHANISM. As we have seen above, the bounded piece of subject matter is *inside* the box, and, as we will see, the actual boundary of the box is made by the collection of *arrow stubs* entering and leaving the box. The bounded pieces are related and connected (Box 4.2) by SA arrows which go *from* an OUTPUT of one box *to* the INPUT or CONTROL of another box, i.e., such arrow connections make the *interfaces* between subjects. The names INPUT and OUTPUT are chosen to convey the idea that (see FEO 4B and Box 4.3) the box represents a *transformation* from a "before" to an "after" state of affairs. The CONTROL interface (see FEO 4C and Box 4.4) interacts and constrains the transformation to ensure that it applies only under the appropriate circumstances. The combination of INPUT, OUTPUT, and CONTROL fully specifies the bounded piece of subject, and the interfaces related it to the other pieces. Finally, the MECHANISM *support* (not interface, see FEO 4D and Box 4.5) provide means for realizing the complete piece represented by the box.

We will see shortly why Fig. 4 contains no INPUT arrows at all, but except for that anomaly, this description should make Fig. 4 itself reasonably understandable. (Remember the context—"Rationalize the features of SA language which allow one to define graphic notation for . . . .") The diagram (with FEO's and discussion) is the desired rationalization. It fits quite well with the idea of following the maxim. We don't

[3]To shorten references to figures, "Box 4. 1" will mean "Box 1 of Fig. 4," etc. in the following discussion.

[4]This notation refers to the sequence of imbedded illustrations in Fig. 4 which are "For exposition only" (FEO).

mind breaking *everything* about a bounded piece of subject matter into pieces as long as we are sure we can express completely how all those pieces go back together to constitute the whole. Input, output, control, and mechanism provide that capability. As long as the right mechanism is provided, and the right control is applied, whatever is inside the box can be a valid transformation of input to output. We now must see how to use the "foreign" language names and labels of boxes and arrows. Then we can start putting SA to work.

XIII. USING THE BASICS FOR UNDERSTANDING

Figure 4, and especially FEO 4A, now that we have digested the meaning of the diagram itself, has presented the basic box-and-arrow-stubs-making-useful-interfaces-for-a-bounded-piece-of-subject-matter building block of SA. We now can start to use the input, output, control, and mechanism concepts to further our understanding. Knowing even this much, the power of expression of SA diagrams beyond that of simple outlining will start to become evident.

Figure 5, entitled "Build diagram," details Box 3.2. Referring back to Fig. 3 and recalling the opening discussion of its meaning (which we should do in order to establish in our minds the proper context for reading Fig. 5) we recall that the story line of Fig. 3 said that after Box 3.1 had defined the arrow and box basics, then we would build an actual diagram with words and names for a particular subject in accordance with a purpose and viewpoint chosen to convey the appropriate understanding. Looking at Box 3.2 in the light of what we have just learned about the box/arrow basics in Fig. 4, we can see that indeed the inputs are words and names, which will be transformed into a diagram (an over-detailed, but graphically complete diagram, evidently). Even though the mechanism is not specified, it is shown that this diagramming process will be controlled by (i.e., constrained by) the graphic conventions, subject, and viewpoint. Now refer to Fig. 5 with this established context and consider its three boxes:—"Build box structure; build arrow structure; build diagram structure." That matches our understanding that a diagram is a structure of boxes and arrows (with appropriate names and labels, of course). Study Fig. 5 yourself briefly, keeping in mind the points we have discussed so far. You should find little difficulty, and you will find that a number of the technical terms that were pure jargon in the tabulated form in Fig. 2 now start to take on some useful meaning. (Remember to ignore terms such as "ICOM" and "DRE," to be described later.)

If you have taken a moment or so to study Fig. 5 on your own, you probably have the impression things are working all right, but you are still not really sure that you are acquiring the intended level of understanding of Fig. 5. It seems to have too many loose ends semantically, even though it makes partial sense. If this is your reaction, you are quite right. For more detail and information is needed to make all the words and relationships take on full meaning. Figure 5 does indeed tell *everything* about "Build diagram" in its three boxes, which are themselves reasonably understandable. But we need more information for many of the labels to really snap into place. This we will find in the further detailing of the three boxes. Context *orients* for understanding (*only* orients!); details enable understanding (and strengthen context).

Figure 6 provides the detailing for Box 5.1. Especially for this diagram, it is important to keep in mind the appropriate context for reading. It is not "*Draw* an SA diagram," but to motivate the *features* of SA. Thus, when we read the title, "Build box structure," Fig. 6, we must keep in mind that the worthy piece of subject matter is not *how* to build box structure, nor even the features which *create* box structure, but motivation for *explanation* of the features which allow box structure to represent the bounded context subject matter. This actually is a very sophisticated

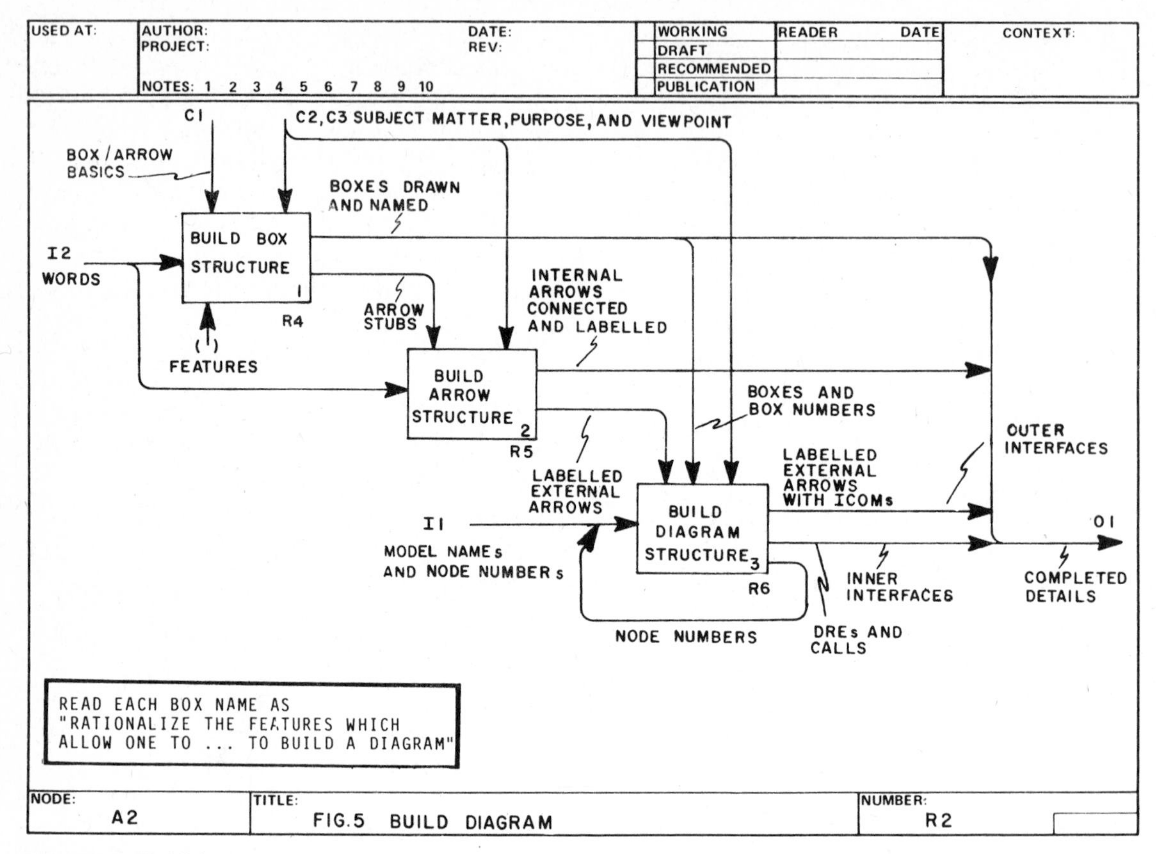

FIGURE 5 Build diagram.

subject and normally we would only be diagramming it after we had already prepared rather complete models of the "how to" of SA so that many of the terms and ideas would already be familiar. In this paper, however, the opening discussion must serve instead. The next four sections discuss Fig. 6.

XIV. DUALITY OF DATA AND ACTIVITIES

Recall that a complete SA model has to consider both the things and happenings of the subject being modeled. Happenings are represented by an activity decomposition consisting of activity diagrams and things are represented by data decomposition consisting of data diagrams. The neat thing about SA language is that both of these complementary but radically different aspects are diagrammed using exactly the same four-sided box notation. Figure 7 illustrates this fact. The happening/activity and thing/data domains are completely dual in SA. (Think of an INPUT activity on a data box as one that creates the data thing, and of OUTPUT as one that uses or references it.) Notice that mechanism is different in interpretation, but the role is the same. For a happening it is the *processor,* machine, computer, person, etc., which makes the happening happen. For a thing it is the *device,* for storage, representation, implementation, etc. (of the thing).

A quick check of Fig. 4 shows that mechanism's purpose is to show the means of realiza-

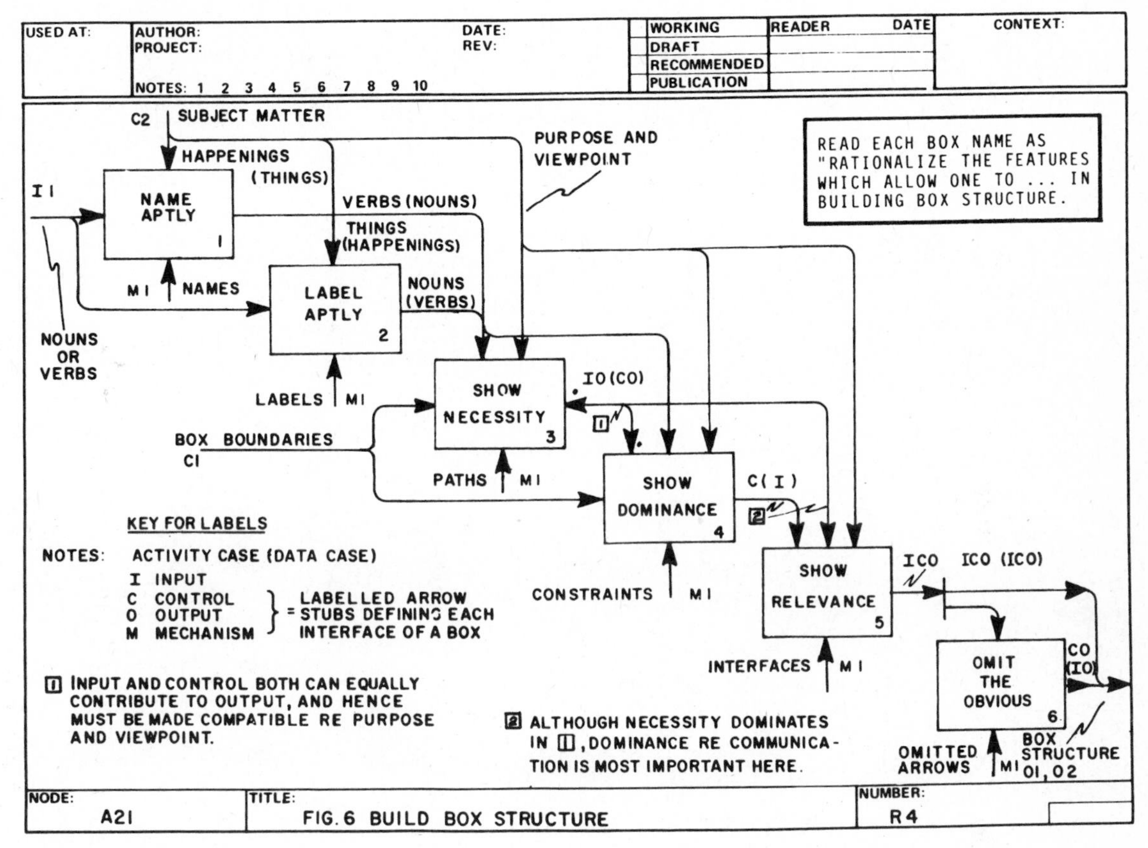

FIGURE 6 Build box structure.

tion, and that it is *not an interface* but is instead something called "support" in SA (described later in Section XXIII). For either activity or data modeling, a support mechanism is *a complete model,* with both data and activity aspects. As Fig. 7 shows, that complete "real thing" is *known by its name,* whereas things and happenings are identified by nouns and verbs (really nominal expressions and verbial expressions). With this in mind, we can see that the first two boxes in Fig. 6 motivate the naming and labeling features of SA to do or permit what Fig. 7 requires—boxes are named, and arrows are labeled, with either nouns or verbs as appropriate to the aspect of the model, and of course, in accordance with the intended purpose and viewpoint of the subject matter.

XV. CONSTRAINTS

We will consider next Boxes 6.3 and 6.4 together, and with some care, for this is one of the more subtle aspects of SA—the concept of a *constraint*—the key to well structured thought and well structured diagrams. The word *constraint* conjures up visions of opposing forces at play. *Something can be constrained only if there is something stronger upon which the constraining force can be based.* It might seem that from the ideas of SA presented so far, that the strong base will be provided by the rigorously defined bounded context of a box. Given a strong boundary, it is easy to envision forces saying either to stay inside the boundary or to stay outside the boundary. It is a pleasing thought indeed, and

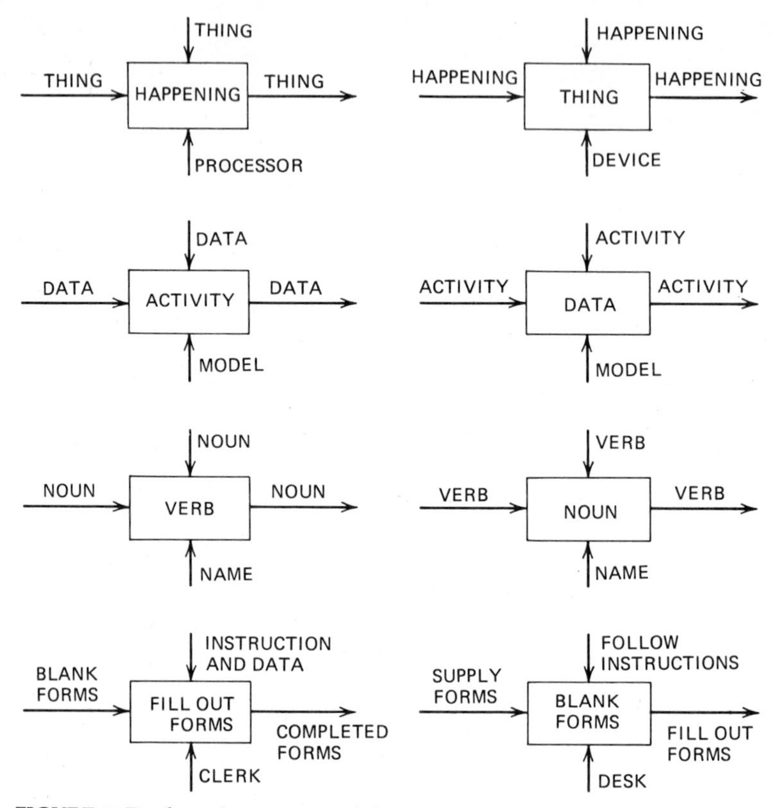

FIGURE 7 Duality of activities and data.

would certainly make strong structure in both our thinking and our diagrams. The only trouble is it does not work that way (or at least not immediately), but in fact it is just the opposite! In SA thinking *it is the constraints that make the boundaries, not the other way around.* This is a tricky point so we will approach it slowly. It is still true that a constraint, to be a constraint, has to have something to push against. If it is not the bounded-context boundary, then what is it?

The subtle answer is that *the purpose and viewpoint of the model provide the basis for all constraints* which in turn provide the strength and rigidity for all the boundaries which in turn create the inescapable structure which forces correct understanding to be communicated. This come about through the concepts of *necessity* and *dominance,* which are the subjects of Boxes 6.3 and 6.4. Dominance sounds much like constraint, and we will not be surprised to find it being the purveyor of constraint. But "necessity" has its own subtle twist in this, the very heart of SA. Therefore we must approach it, too, with some deliberation.

Figure 8 tells the story in concise form. Please read it now. Then please read it again, for all experience with SA shows that this simple argument seems to be very subtle and difficult for most people to grasp correctly. The reason is that the *everything* of the SA maxim makes the I-to-O *necessity* chain the *weakest possible structure*—akin to *no structure at all.* It merely states a fact that *must be true* for every SA box, because of the maxim. Therefore *dominant constraints,* expressed by the control arrows for activity boxes are, in fact, the *only* way possible

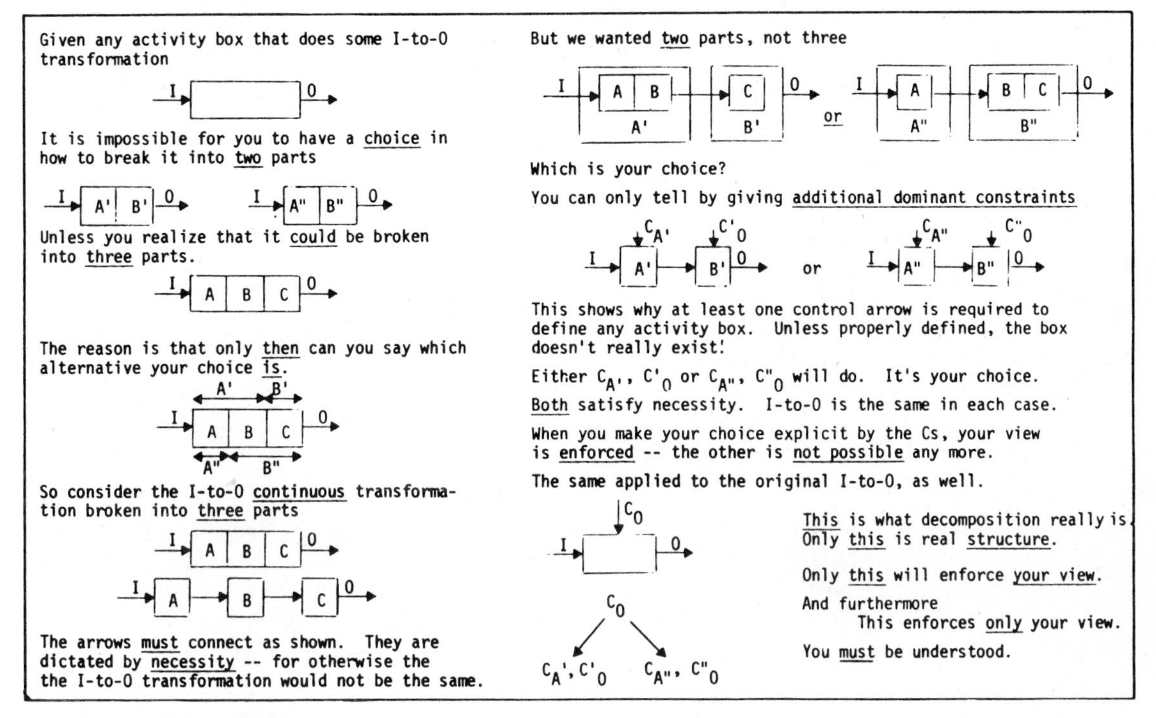

FIGURE 8 Dominance and necessity.

to impose structure. Furthermore, that enforcement of structure is unique and compelling—no other structure can be (mis-) understood in place of that intended by the SA author. This is, of course, all mediated by the effectiveness with which the SA author wields the chosen non-SA language used for names of boxes and labels on arrows, but the argument presented in this paper holds, nonetheless. This is because whenever the imprecision of the non-SA language intrudes, more SA modeling (perhaps even with new purpose and viewpoint for greater refinement, still, of objectives) is forced by the reader/author cycle of the SADT discipline.

XVI. THE RULE OF OMISSION

Now consider Boxes 6.5 and 6.6 "Show relevance" and "Omit the obvious." These two ideas follow right along with the above discussion. Namely, in the case of activity diagramming, *if inputs are relevant* (i.e., if they make a strong contribution to understandability) *then they are drawn.* But on the other hand, since the important thing is the structure imposed by the control dominance and output necessity, and inputs *must* be supplied in any case for those outputs to result, *obvious inputs can and should be omitted from* the box structure of SA *activity diagramming.* In other words, whenever an obvious input is omitted in an activity diagram, the reader knows that (because of the SA maxim) whatever is needed will be supplied in order that the control and output which *are* drawn can happen correctly. Omitting the obvious makes the understandability and meaning of the diagram much stronger, because inputs when they *are* drawn are known to be important and nonobvious. Remember that SA diagrams are not wiring diagrams, they are vehicles for communicating understanding.

Although activity and data are dual in SA and use the same four-sided box notation, they are not quite the same, for the concept of dominance and constraint in the *data* aspect centers on *input* rather than control. In the data case, the weak chain of necessity is C-O-C-O- · · · not I-O-I-O- · · · as it is in the activity case. The reason comes from a deep difference between space and time (i.e., between things and happenings). In the case of the happening, the dominant feature is the control which says when to cut the transformation to yield a desired intermediate result, because the "freedom" of happenings is in time. In the case of things, however, the "freedom" which must be constrained and dominated concerns which version of the thing (i.e., which part of the data box) is the one that is to exist, regardless of time. The input activity for a data box "creates" that thing in that form and therefore it is the dominant constraint to be specified for a data box. An unimportant control activity will happen whenever needed, and may be omitted from the diagram.

Therefore, the rules regarding the obvious in SA is that *controls may never be omitted from activity diagrams and inputs may never be omitted from data diagrams.* Figure 6 summarizes all of this discussion.

XVII. STRENGTHENING OF BOUNDARIES

Recall that a constraint does need something to push on. What is that? The answer is the one originally proposed, but rejected. *Constraints are based on the boundary of a bounded context—but of the single box of the parent diagram which the current diagram details.* (A parenthesized hint that this would turn out to be the case *does* appear in the original rejection of the view that boundaries provide the base for constraints, above.) In other words, the constraints represented by the arrows on a diagram are *not* formed by the several boundaries of the boxes of *that* diagram, but all are based on the single boundary of the corresponding box in the *parent* diagram. As was stated above in Section XV, the constraints form the box boundary interfaces and define the boxes, not the other way around. The strength of those constraints comes from the corresponding "push" passed through from the parent. As the last "C_O" portion of Figure 8 indicates, this hierarchic cascading of constraints is based entirely on the *purpose* of the model as a whole (further constrained in spread by the limited *viewpoint* of the model) as it is successfully decomposed in the hierarchic layering forced by the six-or-fewer rule of the SA maxim.

All of this is a direct consequence of the *everything* of the SA maxim, and may be inferred by considering Fig. 8 recursively. The boundary of the top-most box of the analysis is determined entirely by the subject matter, purpose, and viewpoint of the agreed-upon outset understanding. ("Tell 'em whatcha gonna tell 'em.") Then each of the subsequent constraints derives its footing only insofar as it continues to reflect that subject, purpose, and viewpoint. And each, in turn, provides the same basis for the next subdivision, etc. Inconsistencies in an original high-level interpretation are ironed out and are replaced by greater and greater precision of specific meaning.

Even though the basis for all of the constraint structure is the (perhaps ill-conceived, ambiguous, ill-defined) *outer* boundary, that boundary and the *innermost* boundary, composed of the collective class of all the boundaries of the finest subdivision taken together, are merely two representations of the *same* boundary—so that *strengthening of the inner boundary through extensive decomposition automatically strengthens the outer boundary.* It is as though the structured analyst (and each of his readers as well) were saying continually, "My outermost understanding of the problem as a whole can only make sense, now that I see all this detail, if I refine my interpretation of it in this, this, and this precise way." This is the hidden power of SA at work. This is how SA greatly amplifies the precision and understandability of any natural or formal language whose nouns and verbs are imbedded in its box and arrow structure.

XVIII. ARROW CONNECTIONS

Figure 6, which led to this discussion, detailed Box 5.1 "Build box structure"; Figure 9 decomposes Box 5.2, "Build arrow structure." From Fig. 5 we see that the controlling constraints that dominate Box 5.2 are the subject matter, purpose, and viewpoint, as we would expect, along with the "arrow stubs" which resulted from building each box separately. The outputs are to be internal arrows connected and labeled, as well as labeled external arrows. A relevant (i.e., non-obvious) input is the collection of words—nouns or verbs—for making those labels.

With this context in mind, we are now ready to look at Fig. 9, "Build arrow structure." Here is an example of the use of non-English language to label arrows. Small graphical phrases show the intended meaning of "branch" and "join" for distribution, and "bundle" and "spread" with respect to subdivision, as well as two forms of logical OR for exclusion. We have seen many examples of these in use in the diagrams already considered, so that the ideas should be quite transparent.

The little pictures as labels show how the labels attach to arrows to convey the appropriate meaning. In most good SA diagramming the

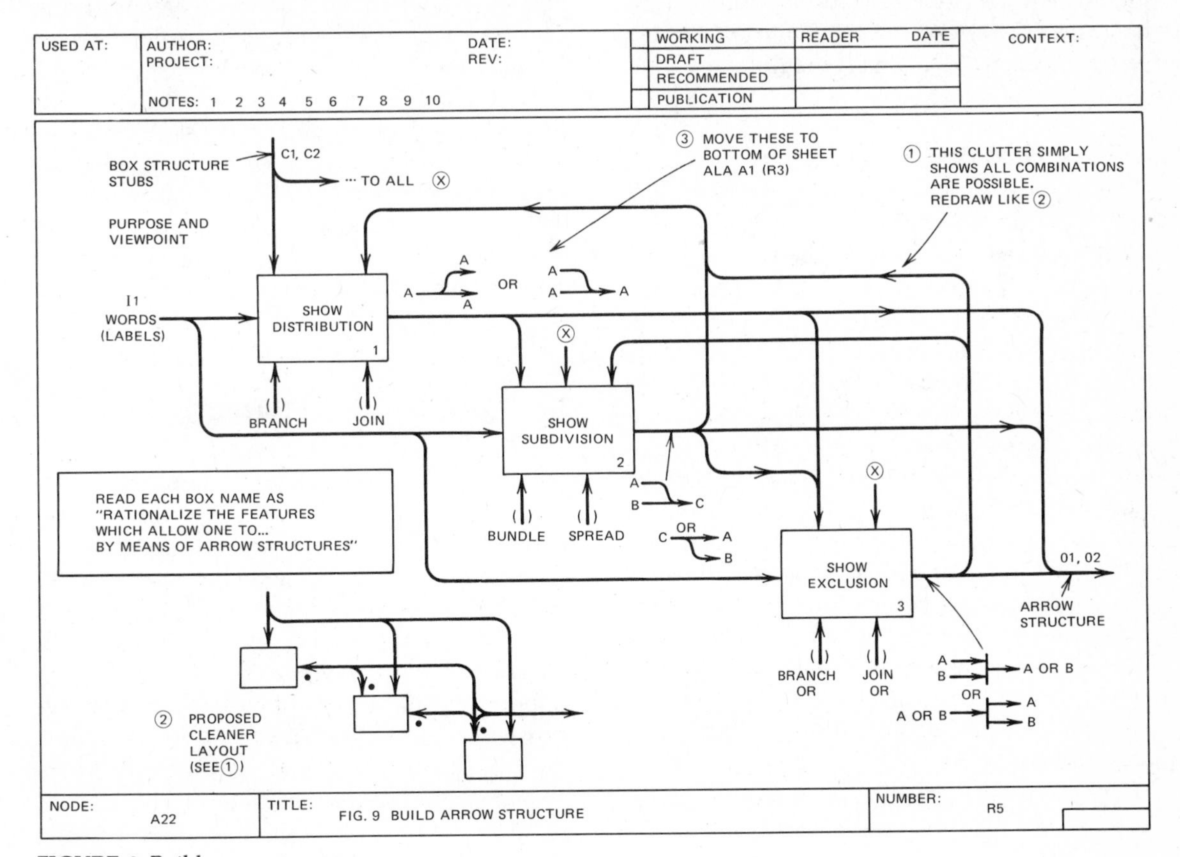

FIGURE 9 Build arrow structure.

OR's are used very sparingly—only when they materially assist understanding. In most circumstances, the fact that arrows represent constraints either of dominance or necessity supplies the required understanding in clearer form merely by topological connection. This also is the reason why there is no graphical provision for the other logical functions such as AND, for they are really out of place at the level of communication of basic SA language. In order for them to have an appropriate role, the total context of interpretation of an SA model must have been drawn down very precisely to some mathematical or logical domain at which point a language more appropriate to that domain should be chosen. Then logical terms in the nominal and verbial expressions in labels can convey the conditions. This is preferable to distorting the SA language into a detailed communication role it was not designed or intended to fulfill.

XIX. BOUNDARIES

Figure 12, "Build diagram structure," will provide detailing for the third and last box of Fig. 5. It is needed as a separate consideration of this motivation model because the building of box structure (Box 5.1, detailed in Fig. 6) and arrow structure (Box 5.2, detailed in Fig. 9) only cover arrows between boxes in a single diagram—the *internal* arrows. Box 4.2 requires that *every* arrow which relates or connects bounded contexts must participate in both a *from* and a *to* interface. Every *external* arrow (shown as the second output of Box 5.2) will be missing either its source (from) or its destination (to) because the relevant boxes do not appear on this diagram. As the relationship between Boxes 5.2 and 5.3 in Fig. 5 shows, these labeled arrows are indeed a dominant constraint controlling Box 3, "Build diagram structure."

Figure 10 helps to explain the story. This is a partial view of three levels of nesting of SA boxes, one within the other, in some model (not an SA diagram). Except for three arrows, every arrow drawn is a complete from/to connection. The middle, second-level box has four fine-level boxes within it, and it in turn is contained within the largest box drawn in the figure. If we consider the arrows in the middle, second-level box, we note that only two of them are internal arrows, all of the others being external. But notice also that every one of those external arrows (with respect to that middle-level box) are in fact *internal* with respect to the model as a whole. Each of those arrows does go from one box to another box—a lowest-level box in each case. In completing the connection, *the arrows penetrate the boundaries of the middle-level boxes* as though those boundaries were not there at all. In fact, there are only two real boundaries in all of Fig. 10—the two boundaries characteristic of every SA decomposition. These are (1) the *outer boundary* which is the outermost edge of Fig. 10, itself, and (2) the *inner boundary* which is the entire set of edges of all the lowest-level boxes drawn in Fig. 10, considered as a single boundary. As was stated above, the SA maxim requires that the outer boundary and the inner boundary must be understood to be *exactly the same* so that the subject is merely decomposed, not altered in any way.

XX. PARENTS AND CHILDREN

To understand how the structuring of Fig. 10 is expressed in SA terms, we must be clear about the relationship between boundaries and interfaces, boxes and diagrams, and the parent/child relationship. Figure 11 lays all of this out. In the upper right appears the diagram for the largest box drawn in Fig. 10, and in the lower left appears the diagram for the central middle-level box which we were discussing. The first thing to notice is that the diagrams are here drawn as though they were punched out of Fig. 10, (like cutting cookies from a sheet of cookie dough). Although the dimensions are distorted, the note

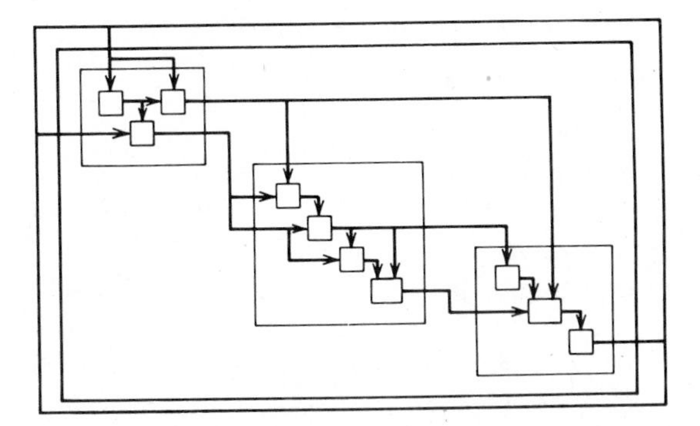

FIGURE 10 Nested factors.

in the upper left points out that, by definition, *the diagram outer boundary is actually the same as the parent box boundary* (i.e., the current child diagram is the "cookie" removed from the sheet of dough and placed to one side).

Figure 11 also points out that, just as for the hierarchic decomposition as a whole, the *inner boundary of the parent diagram is the collection of all its child box boundaries considered as a single entity.* Notice the terminology—with respect to the current child diagram, one of the boxes in the parent diagram is called the *parent box* of the child diagram. By definition of Fig. 4, that *parent box boundary* is the collection of *parent box interfaces and support* which compose it. Since we have just established that the outer boundary of the current diagram is the same as the corresponding parent box boundary, the parent box edges (interfaces *or* support) which compose the

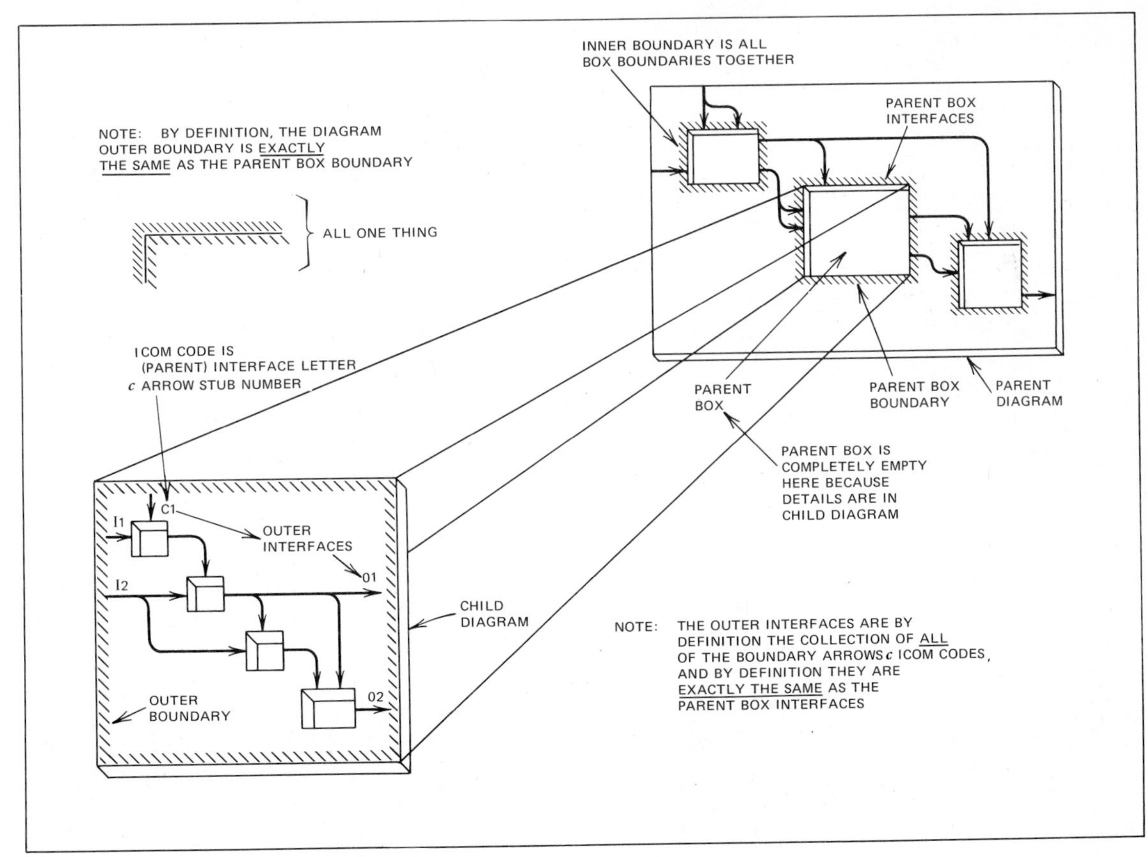

FIGURE 11 Boundaries and interfaces.

parent box boundary must somehow match the outer edges of the child diagram. This is the connection which we seek to establish rigorously.

By Fig. 10 we know that the external arrows of the child diagram penetrate through the outer boundary and are, in fact, the *same* arrows as are the stubs of the interfaces and support which compose the parent box boundary. Therefore, the connection which has to be made is clear from the definition. But for flexibility of graphic representation, *the external arrows of the current diagram need not have the same geometric layout, relationship, or labeling* as the corresponding stubs on the parent diagram, which are drawn on a completely different (parent) diagram.

In order to allow this flexibility, we construct a special code-naming scheme called *ICOM codes* as follows: An ICOM code begins with one of the letters I-C-O-M (standing for INPUT, CONTROL, OUTPUT, MECHANISM) concatenated with an integer which is obtained by considering that the stubs of the corresponding parent box edge are numbered consecutively from top to bottom or from left to right, as the case may be. With the corresponding ICOM code written beside the unattached end of the arrow in the current diagram, that arrow is no longer called "external," but is called a *boundary arrow*. Then *the four outer edges of the child diagram are, by definition, the four collections of ICOM boundary arrows which are*, by definition, exactly the same as the corresponding parent box edges, as defined by Fig. 4 and shown in Fig. 11. Thus, even though the geometric layout may be radically different, the rigor of interconnection of child and parent diagrams is complete, and the arrows are continuous and unbroken, as required. Every diagram in this paper has ICOM codes properly assigned.

The above presentation is summarized in the first two boxes of Fig. 12 and should be clear without further discussion. Boxes 12.3 and 12.4 concern the SA language notations for establishing the relationships between the child and parent diagram cookies by means of a *detail reference expression* (DRE) or an SA *call*. We will consider these shortly. For now it is sufficient to note that the topics we have considered here complete the detailing of Fig. 5, "Build diagram"—how the box structure and arrow structure for individual diagrams are built, and then how the whole collection of diagrams are linked together in a single whole so that *everything* of the top-most cookie (treated as a cookie sheet from which other cookies are cut with zero width cuts) is completely understandable. Each individual diagram itself is only a portion of the cookie dough with an outer boundary and an inner boundary formed by the decomposition operation. Nothing is either gained or lost in the process—so that the SA maxim is rigorously realized. Everything can indeed be covered for the stated purpose and viewpoint. We now complete our presentation of the remaining items in the 40 features of Fig. 2, which exploit further refinements of notation and provide orderly organization for the mass of information in a complete SA model.

XXI. WORD NOTES

In SA language, not everything is said in graphical terms. Both words and pictures are also used. If the diagram construction notations we have considered so far were to be used exclusively and exhaustively, very cluttered and nonunderstandable diagrams would result. Therefore, SA language includes further simplifying graphic notations (which also increase the expressive power of the language), as well as allowing nongraphic additional information to be incorporated into SA diagrams. This is the function of Figure 13 which details Box 3.3. Figure 13 points out that the (potentially) cluttered diagram is only graphically complete so that special *word* notations are needed. Furthermore, special arrow notations can supply more clarity,

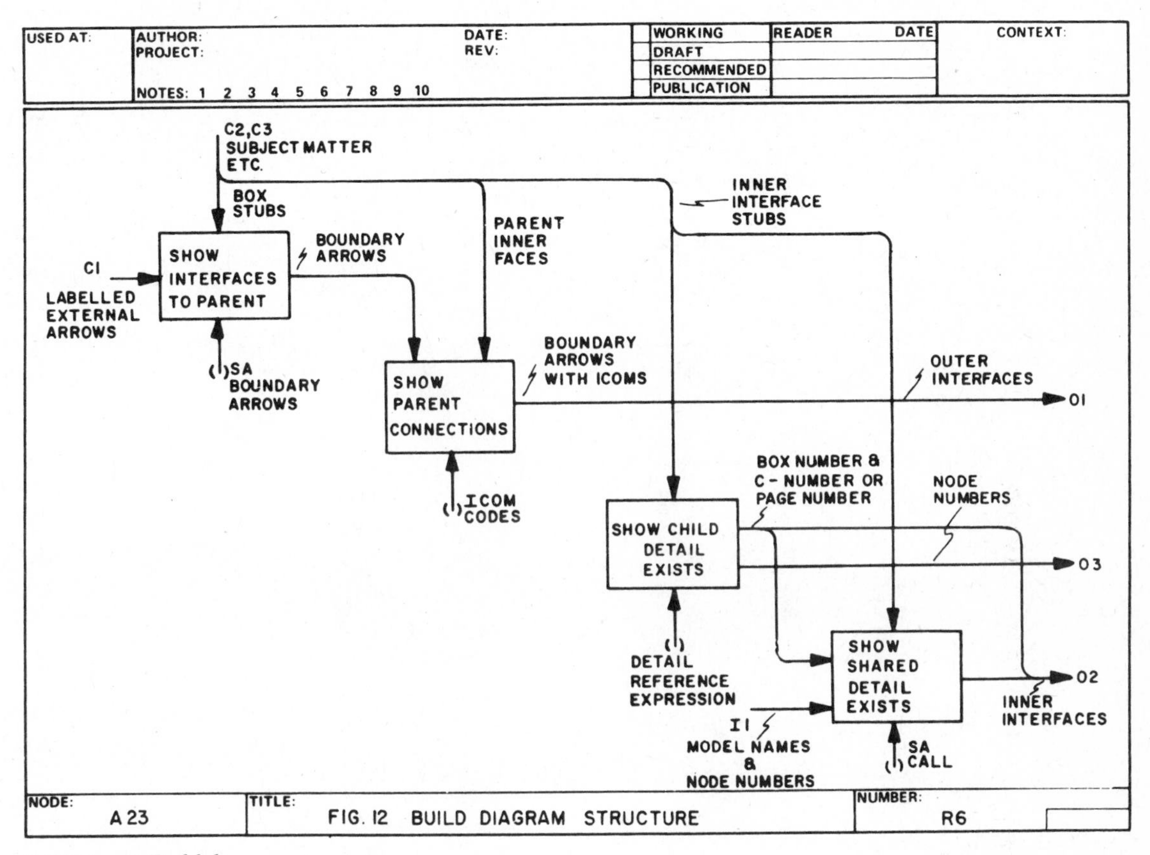

FIGURE 12 Build diagram structure.

to result in a complete *and* understandable diagram.

We will not further detail Box 13.2. We merely point out that its output consists of three forms of verbal additions to the diagram. The first two–NOTES and [n] footnotes–are actual parts of the diagram. The diagrams we have been examining have examples of each. The third category, (n) metanotes, are *not* parts of the diagram themselves, but are instead notes *about* the diagram. The (n) metanotes have only an observational or referential relation to the actual information content of these diagrams and therefore they do not in any way alter or affect the actual representational function of the SA language, either graphical or verbal. There is no way that information in (n) metanotes can participate in the information content of the diagrams, and therefore they should not be used in an attempt to affect the interpretation of the diagrams themselves, but only for mechanical operations regarding the diagram's physical format or expression. Examples are comments from a reader to an author of a diagram suggesting an improved layout for greater understandability. A few examples are included on the diagrams in this paper. The [n] footnotes are used exclusively for allowing large verbal expressions to be concisely located with respect to tight geometric layout, in addition to the normal footnoting function commonly found in textual information.

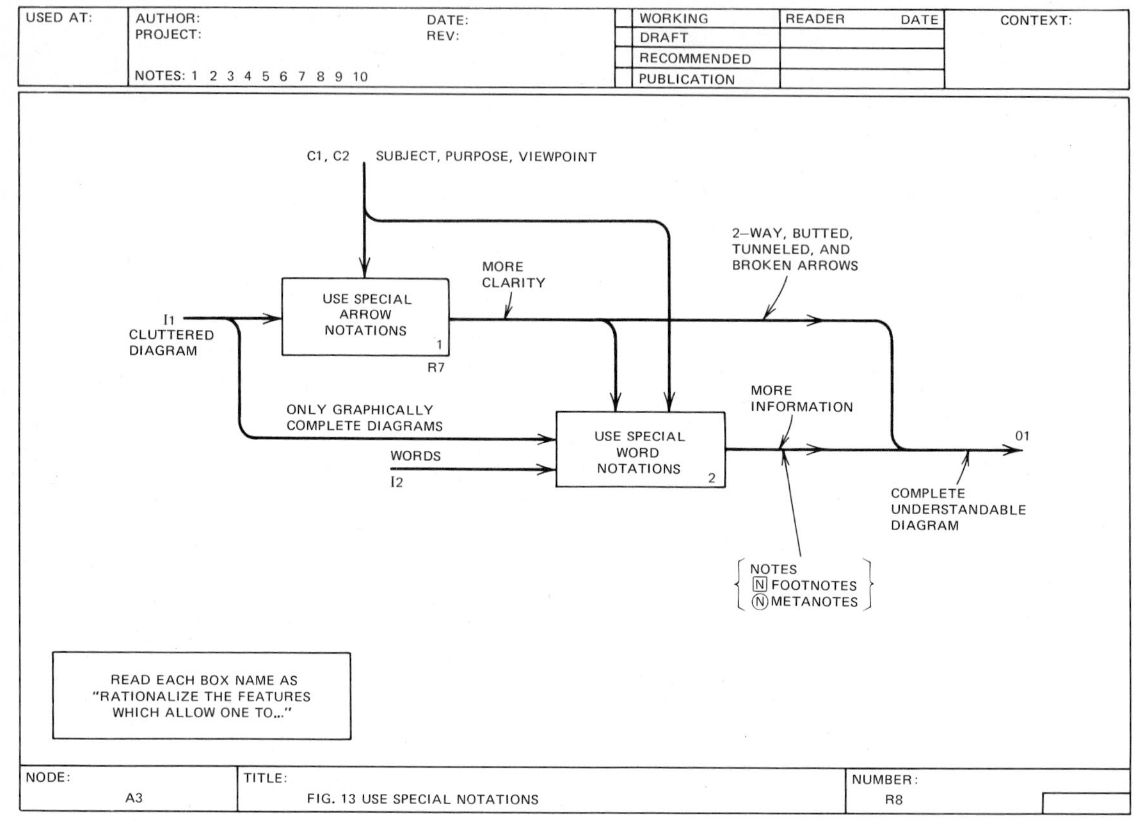

FIGURE 13 Use special notations.

XXII. SPECIAL GRAPHIC NOTATIONS

Figure 14 provides the motivation for four very simple additions to the graphic notation to improve the understandability of diagrams. With respect to a specific aspect of the subject matter, two boxes sometimes really act as one box inasmuch as each of them shares one portion of a well-defined aspect of the subject matter. In this case, arrowheads with dots above, below, or to the right of the arrowheads are added instead of drawing two separate arrows as shown in FEO 14A. Two-way arrows are a form of bundling, however, *not* a mere shorthand notation for the two separate arrows. If the subject matters represented by the two separate arrows are not sufficiently similar, they should *not* be bundled into a two-way arrow, but should be drawn separately. Many times, however, the two-way arrow *is* the appropriate semantics for the relationship between two boxes. Notice that if other considerations of diagrams are sufficiently strong, the awkward, nonstandard two-way arrow notations shown also may be used to still indicate dominance in the two-way interaction.

SA arrows should always be thought of as conduits or pipelines containing multistranded cables, each strand of which is another pipeline. Then the branching and joining is like the cabling of a telephone exchange, including trunk lines. Box 14.2 is related to both two-way arrows and pipelines, and points out that a one-way pipeline

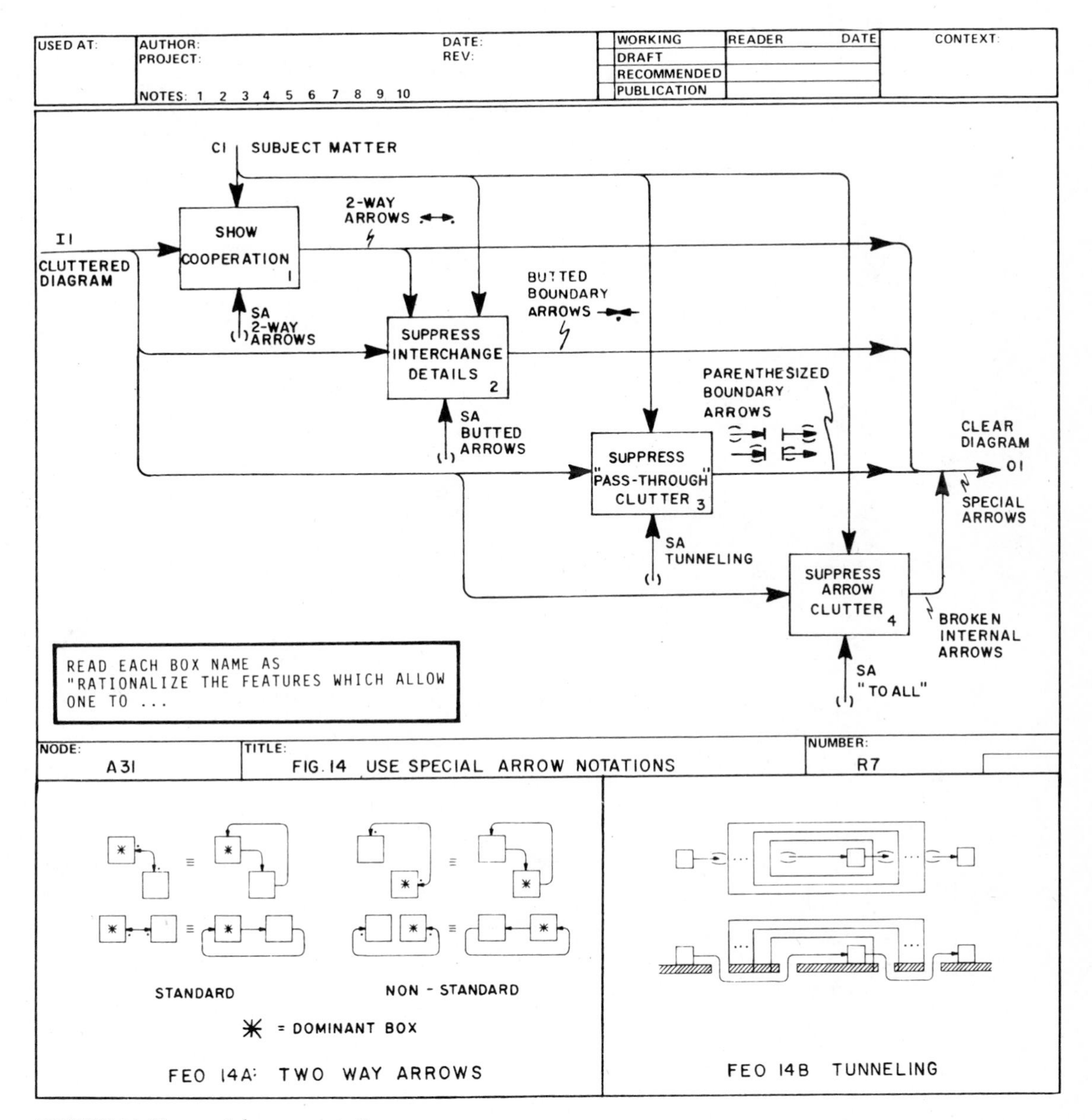

FIGURE 14 Use special arrow notations.

stub at the parent level may be shown as a two-way boundary arrow in the child. This is appropriate since, with respect to the communication of understanding at the parent level, the relationship between boxes is one-way, whereas when details are examined, two-way cooperation between the two sets of detailing boxes may be required. An example is the boss-worker relation. The boss (at parent level) provides one-way command, but (at the child diagram level) a two-way interchange between worker and boss may be needed to clarify details.

Box 14.3 motivates an additional and very useful version of Box 6.6 ("Omit the obvious"). In this case, instead of omitting the obvious, we only postpone consideration of necessary detail until the appropriate level is reached. This is done by putting parentheses around the unattached end of an external arrow, or at the interface end of a parent box stub. The notation is intended to convey the image of an arrow "tunneling" out of view as it crosses a parent box inner boundary only to emerge later, some number of levels deeper in a child's outer interface, when the information is actually required. These are known as "tunneled" or "parenthesized" boundary arrows when the sources or destinations are somewhere within the SA model, and as (proper) *external* arrows when the missing source or destination is unspecified (i.e., when the model would need to be inbedded in some context larger than the total model for the appropriate connection to be made).

Finally, Box 14.4 is a seldom-used notation which allows internal arrows themselves to be broken by an ad hoc labeling scheme merely to suppress clutter. Its use is discouraged because of its lack of geometric continuity and because its use is forced only by a diagram containing so much information already that it is likely not to be clearly understood and should be redrawn. Examples occur in Figure 9 just for illustration.

XXIII. THE REFERENCE LANGUAGE

Returning to Figure 3, we now have considered all of the aspects of basic SA language which go into the creation of diagrams themselves, with the single exception of the detail reference expressions and SA call notation of Figure 12, which were saved until this point since they relate so closely to Box 3.4, "Provide for referencing."

A complete and unique SA reference language derives very nicely from the hierarchically nested factors imposed by the SA maxim. Diagrams, boxes, interfaces, arrows, and complete contexts can be referenced by a combination of model names, *node numbers* (starting with A for activity or D for data, and derived directly from the box numbers), and ICOM codes. The insertion of a dot, meaning "which see" (i.e. "find the diagram and look at it"), can specify exactly which diagram is to be kept particularly in mind to provide the context for interpreting the SA language. Thus "A122. 4I1" means "in diagram A122, the first input of box 4"; "A1224.I1" means "in diagram A1224, the boundary arrow I4"; "A1224I1" means "the first input interface of node A1224." The SA language rules also allow such reference expressions to degenerate naturally to the minimum needed to be understandable. Thus, for example, the mechanism for showing that the child detailing exists is merely to write the corresponding chronological creation number (called a *C-number*) under the lower right-hand corner of a box on the diagram, as a DRE. (A C-number is the author's initials, concatenated with a sequential integer—assigned as the first step whenever a new diagram sheet is begun). When a model is formally published, the corresponding detail reference expression is normally converted into the page number of the appropriate detail diagram. The omission of a detail reference expression indicates that the box is not further detailed in this model. (For all the diagrams considered in this paper, the DRE's have been left in C-number form.)

The SA *call* notation consists of a detail reference expression preceded by a downward pointing arrow stub, and allows sharing of details among diagrams. It will not be covered in this paper beyond the illustration in Figure 15, which is included here more to illustrate why mechanism support is not an interface (as has repeatedly been pointed out) than to adequately describe the SA call scheme. That will be the subject of a future paper, and is merely cited here for completeness. The SA call mechanism (see also [2]) corresponds very closely to the subroutine

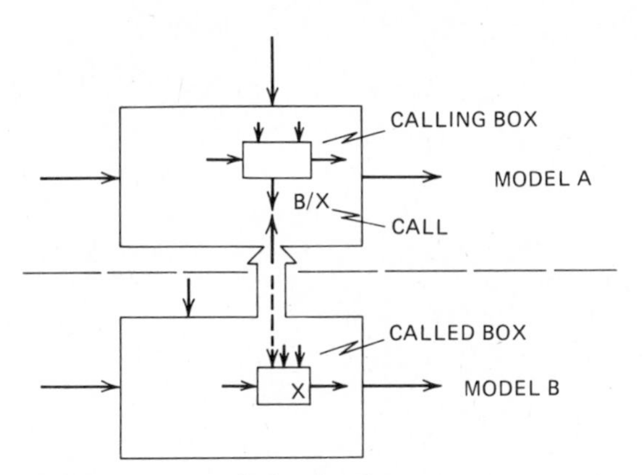

FIGURE 15 SA "call" for detailing.

call concept of programming languages, and is a key concept in combining multiple purposes and viewpoints into a single model of models.

XXIV. ORGANIZING THE MODEL

The final box of Figure 3, Box 3.5, "Organize material," is not detailed in this model. Instead, we refer the reader back to the tabulation of Figure 2 where the corresponding items are listed. In final publication form each diagram is normally accompanied by brief, carefully-structured SA *text* which, according to the reading rules, is intended to be a read *after* the diagram itself has been read and understood. The SA text supplements but does not replace the information content of the diagram. Its purpose is to "tell 'em whatcha told 'em" by giving a walk-through through the salient features of the diagram, pointing out, by using the reference language, how the story line may be seen in the diagram. Published models also include glossaries of terms used, and are preceded by a *node index,* which consists of the node-numbered box names in indented form in node number sequence. Figure 16 is the node index for the model presented in this paper, and normally would be published at the beginning to act as a table of contents.

XXV. CONCLUSION

The principle of good storytelling style (see Section IV) has been followed repeatedly in this paper. We have provided motivations for each of the 40 SA language features of structured analysis by relating each one to a need for clear and explicit exposition with no loss from an original bounded context. (The "node" column of Figure 2 maps each feature to a diagram box in the other figures.) In the process, we have seen how the successive levels of refinement strengthen the original statement of purpose and viewpoint, to enforce unambiguous understanding. The best "tell 'em watcha told 'em" for the paper as a whole is to restudy the SA model in the figures. (Space precludes even sketching the corresponding data decomposition.) The diagrams not only summarize and integrate the ideas covered in the paper, but provide further information as well.

There are more advanced features of the SA language which will be covered in subsequent papers in the context of applications. In practice, SA turns out to depend heavily on the disciplined thought processes that lead to well-structured analyses expressed in well-structured diagrams. Additional rules and supporting methodology organize the work flow, support the mechanics of the methods, and permit teams of people to work and interact as one mind attacking complex problems. These are covered in SofTech's SADT methodology. The fact that SA incorporates by definition any and all languages within its framework permits a wide variety of natural and artificial languages to be used to accomplish specific goals with respect to understanding the requirements for solution. Then those requirements can be translated, in a rigorous, organized, efficient, and above all, understandable fashion, into actual system design, system implementation, maintenance, and training. These topics also must of necessity appear in later papers, as well as a formal language definition for the ideas unfolded here.

RATIONALIZE SA FEATURES

A1 DEFINE GRAPHICS

- A11 Bound Context
- A12 Relate/Connect
- A13 Show Transformation
- A14 Show Circumstance
- A15 Show Means

A2 BUILD DIAGRAM

- A21 Build Box Structure
  - A211 Name Aptly
  - A212 Label Aptly
  - A213 Show Necessity
  - A214 Show Dominance
  - A215 Show Relevance
  - A216 Omit the Obvious
- A22 Build Arrow Structure
  - A221 Show Distribution
  - A222 Show Subdivision
  - A223 Show Exclusion
- A23 Build Diagram Structure
  - A231 Show Interfaces to Parent
  - A232 Show Parent Connections
  - A233 Show Child Detail Exists
  - A234 Show Shared Detail Exists

A3 USE SPECIAL NOTATIONS

- A31 Use Special Arrow Notations
  - A311 Show Cooperation
  - A312 Suppress Interchange Details
  - A313 Suppress "Pass-Through" Clutter
  - A314 Suppress Arrow Clutter
- A32 Use Special Word Notations

A4 PROVIDE FOR (UNIQUE) REFERENCING

- (A41 Sheet Reference)
- (A42 Box Reference)
- (A43 Interface Reference)
- (A44 Arrow Reference)
- (A45 Context Reference)

A5 ORGANIZE MATERIAL

FIGURE 16 Node index.

ACKNOWLEDGMENT

The four-sided box notation was originally inspired by the match between Hori's activity cell [3] and my own notions of Plex [4]. I have, of course, benefitted greatly from interaction with my colleagues at SofTech. J. W. Brackett, J. E. Rodriguez, and particularly J. B. Goodenough gave helpful suggestions for this paper, and C. G. Feldmann worked closely with me on early developments. Some of these ideas have earlier been presented at meetings of the IFIP Work Group 2.3 on Programming Methodology.

REFERENCES

1. D.T. Ross, "It's time to ask why?" *Software Practise Experience*, vol. 1, pp. 103-104, Jan.–Mar. 1971.
2. D. T. Ross, J. B. Goodenough, C. A. Irvine, "Software engineering: Process, principles, and goals," *Computer*, pp. 17-27, May 1975.
3. S. Hori, "Human-directed activity cell model," in *CAM-I*, long-range planning final rep., CAM-I, Inc., 1972.
4. D. T. Ross, "A generalized technique for symbol manipulation and numerical calculation," *Commun. Ass. Comput. Mach.*, Vol. 4, pp. 147-150, Mar. 1961.
5. D. T. Ross and K. E. Schoman, Jr., "Structured analysis for requirements definition," this issue, pp. 6-15.
6. G. A. Miller, "The magical number seven, plus or minus two: Some limits on our capacity for processing information," *Psychol. Rev.*, vol. 63, pp. 81-97, Mar. 1956.

READING QUESTIONS

1. How does Ross define the function of structured analysis?
2. How does structured decomposition aid the ability of humans to handle system complexity?
3. In Figure 3, what are the five activities and what is their sequence?

4. What are the two external inputs to Figure 3? What are the external outputs?
5. How well do SA diagrams show hierarchy?
6. Do SA diagrams show data flow? Do they show procedure?
7. Explain the concept of the duality of data and activities.
8. What is the rule of omission?
9. Draw a set of SA diagrams for the patient monitoring system discussed in the Stevens, Myers, and Constantine paper. Be careful that you do not simply attempt to transpose the diagrams in that paper into the SA format. Use the SA diagrams in an independent analysis and compare what you learn to what the hierarchy chart and DFD told you.

Structured Design

W. P. Stevens
G. J. Myers
L. L. Constantine

Structured design is a set of proposed general program design considerations and techniques for making coding, debugging, and modification easier, faster, and less expensive by reducing complexity.[1] The major ideas are the result of nearly ten years of research by Mr. Constantine.[2] His results are presented here, but the authors do not intend to present the theory and derivation of the results in this paper. These ideas have been called *composite design* by Mr. Myers.[3-5] The authors believe these program *design* techniques are compatible with, and enhance, the *documentation* techniques of HIPO[6] and the *coding* techniques of structured programming.[7]

These cost-saving techniques always need to be balanced with other constraints on the system. But the ability to produce simple, changeable programs will become increasingly important as the cost of the programmer's time continues to rise.

Source: W. P. Stevens, G. J. Myers, and L. L. Constantine, "Structured Design," *IBM Systems Journal*, No. 2, 1974, pp. 115–139.

GENERAL CONSIDERATIONS OF STRUCTURED DESIGN

Simplicity is the primary measurement recommended for evaluating alternative designs relative to reduced debugging and modification time. Simplicity can be enhanced by dividing the system into separate pieces in such a way that pieces can be considered, implemented, fixed, and changed with minimal consideration or effect on the other pieces of the system. Observability (the ability to easily perceive how and why actions occur) is another useful consideration that can help in designing programs that can be changed easily. Consideration of the effect of reasonable changes is also valuable for evaluating alternative designs.

Mr. Constantine has observed that programs that were the easiest to implement and change were those composed of simple, independent modules. The reason for this is that problem solving is faster and easier when the problem can be subdivided into pieces which can be considered separately. Problem solving is hardest when all aspects of the problem must be considered simultaneously.

The term *module* is used to refer to a set of one or more contiguous program statements having a name by which other parts of the system can invoke it and preferably having its own distinct set of variable names. Examples of modules are PL/I procedures, FORTRAN mainlines and subprograms and, in general, subroutines of all types. Considerations are always with relation to the program statements *as coded*, since it is the programmer's ability to understand and change the *source* program that is under consideration.

While conceptually it is useful to discuss dividing whole programs into smaller pieces, the techniques presented here are for designing simple, independent modules originally. It turns out to be difficult to divide an existing program into separate pieces without increasing the complexity because of the amount of overlapped code and other interrelationships that usually exist.

Graphical notation is a useful tool for structured design. Figure 1 illustrates a notation called a *structure chart*,[8] in which:

1. There are two modules, A and B.
2. Module A *invokes* module B. B is *subordinate* to A.
3. B receives an input parameter X (its name in module A) and returns a parameter Y (its name in module A). (It is useful to distinguish which calling parameters represent data passed *to* the called program and which are for data to be *returned* to the caller.)

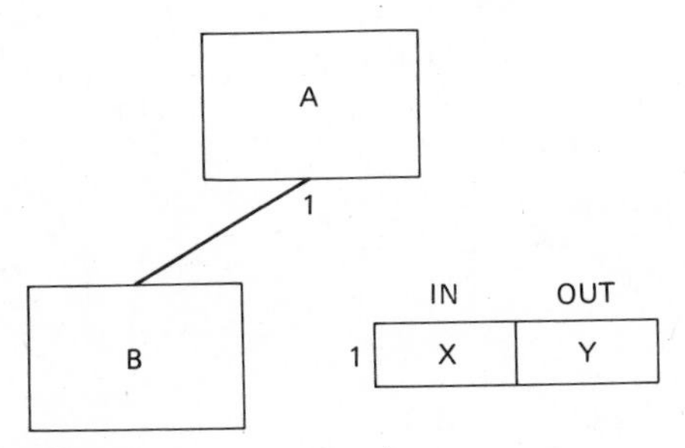

FIGURE 1 A structure chart.

COUPLING AND COMMUNICATION

To evaluate alternatives for dividing programs into modules, it becomes useful to examine and evaluate types of "connections" between modules. A connection is a reference to some label or address defined (or also defined) elsewhere.

The fewer and simpler the connections between modules, the easier it is to understand each module without reference to other modules. Minimizing connections between modules also minimizes the paths along which changes and errors can propagate into other parts of the system, thus eliminating disasterous "ripple" effects, where changes in one part cause errors in another, necessitating additional changes elsewhere, giving rise to new errors, etc. The widely used technique of using common data areas (or global variables or modules without their own distinct set of variable names) can result in an enormous number of connections between the modules of a program. The complexity of a system is affected not only by the number of connections but by the degree to which each connection couples (associates) two modules, making them interdependent rather than independent. Coupling is the measure of the strength of association established by a connection from one module to another. Strong coupling complicates a system since a module is harder to understand, change, or correct by itself if it is highly interrelated with other modules. Complexity can be reduced by designing systems with the weakest possible coupling between modules.

The degree of coupling established by a particular connection is a function of several factors, and thus it is difficult to establish a simple index of coupling. Coupling depends (1) on how complicated the connection is, (2) on whether the connection refers to the module itself or something inside it, and (3) on what is being sent or received.

Coupling increases with increasing complexity or obscurity of the interface. Coupling is lower when the connection is to the normal module interface than when the connection is to an internal component. Coupling is lower with data connections than with control connections, which are in turn lower than hybrid connections (modification of one module's code by another module). The contribution of all these factors is summarized in Table 1.

Interface Complexity

When two or more modules interface with the same area of storage, data region, or device, they share a common environment. Examples of common environments are:

- A set of data elements with the EXTERNAL attribute that is copied into PL/I modules via an INCLUDE statement, or that is found listed in each of a number of modules.
- Data elements defined in COMMON statements in FORTRAN modules.
- A centrally located "control block" or set of control blocks.
- A common overlay region of memory.
- Global variable names defined over an entire program or section.

The most important structural characteristic of a common environment is that it couples every module sharing it to every other such module without regard to their functional relationship or its absence. For example, only the two modules XVECTOR and VELOC might actually make use of data element *X* in an "included" common environment of PL/I, yet changing the length of *X* impacts *every* module making any use of the common environment, and thus necessitates recompilation.

Every element in the common environment, whether used by particular modules or not, constitutes a separate path along which errors and changes can propagate. Each element in the common environment adds to the complexity of the total system to be comprehended by an amount representing all possible pairs of modules sharing that environment. Changes to, and new uses of, the common area potentially impact all modules in unpredictable ways. Data references may become unplanned, uncontrolled, and even unknown.

A module interfacing with a common environment for some of its input or output data is, on the average, more difficult to use in varying contexts or from a variety of places or in different programs than is a module with communication restricted to parameters in calling sequences. It is somewhat clumsier to establish a new and unique data context on each call of a module when data passage is via a common environment. Without analysis of the entire set of sharing modules or careful saving and restoration of values, a new use is likely to interfere with other uses of the common environment and propagate errors into other modules. As to future growth of

TABLE 1
Contributing Factors

	Interface Complexity	Type of Connection	Type of Communication
low	simple, obvious	to module by name	data
COUPLING			control
high	complicated, obscure	to internal elements	hybrid

a given system, once the commitment is made to communication via a common environment, any new module will have to be plugged into the common environment, compounding the total complexity even more. On this point, Belady and Lehman,[9] observe that "a well-structured system, one in which communication is via passed parameters through defined interfaces, is likely to be more growable and require less effort to maintain than one making extensive use of global or shared variables."

The impact of common environments on system complexity may be quantified. Among M objects there are M ($M - 1$) ordered pairs of objects. (Ordered pairs are of interest because A and B sharing a common environment complicates both, A being coupled to B and B being coupled to A.) Thus a common environment of N elements shared by M modules results in NM ($M - 1$) first order (one level) relationships or paths along which changes and errors can propagate. This means 150 such paths in a FORTRAN program of only three modules sharing the COMMON area with just 25 variables in it.

It is possible to minimize these disadvantages of common environments by limiting access to the smallest possible subset of modules. If the total set of potentially shared elements is subdivided into groups, all of which are *required* by some subset of modules, then both the size of each common environment and the scope of modules among which it is shared is reduced. Using "named" rather than "blank" COMMON in FORTRAN is one means of accomplishing this end.

The complexity of an interface is a matter of how much information is needed to state or to understand the connection. Thus, obvious relationships result in lower coupling than obscure or inferred ones. The more syntactic units (such as parameters) in the statement of a connection, the higher the coupling. Thus, extraneous elements irrelevant to the programmer's and the modules' immediate task increase coupling unnecessarily.

Type of Connection

Connections that address or refer to a module as a whole by its name (leaving its contents unknown and irrelevant) yield lower coupling than connections referring to the internal elements of another module. In the latter case, as for example the use of a variable by direct reference from within some other module, the entire content of that module may have to be taken into account to correct an error or make a change so that it does not make an impact in some unexpected way. Modules that can be used easily without knowing anything about their insides make for simpler systems.

Consider the case depicted in Figure 2. GETCOMM is a module whose function is getting the next command from a terminal. In performing this function, GETCOMM calls the module READT, whose function is to read a line from the terminal. READT requires the address of the terminal. It gets this via an externally declared data element in GETCOMM, called TERMADDR. READT passes the line back to GETCOMM as an argument called LINE. Note the arrow extending from *inside* GETCOMM to *inside* READT. An arrow of this type is the notation for references to internal data elements of another module.

Now, suppose we wish to add a module called GETDATA, whose function is to get the

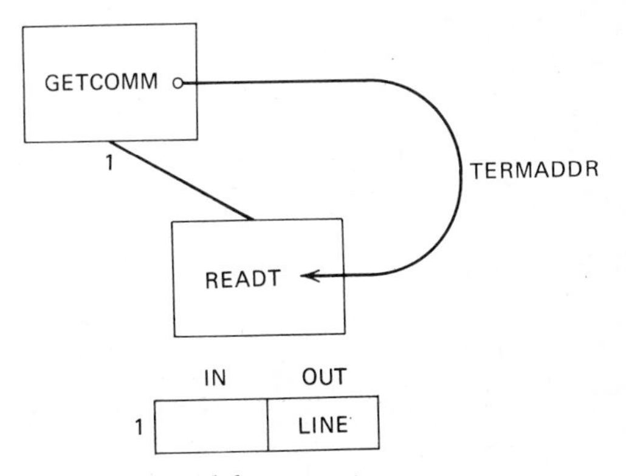

FIGURE 2 Module connections.

next data line (i.e., not a command) from a (possibly) different terminal. It would be desirable to use module READT as a subroutine of GETDATA. But if GETDATA modifies TERMADDR in GETCOMM before calling READT, it will cause GETCOMM to fail since it will "get" from the wrong terminal. Even if GETDATA restores TERMADDR after use, the error can still occur if GETDATA and GETCOMM can ever be invoked "simultaneously" in a multiprogramming environment. READT would have been more usable if TERMADDR had been made an input argument to READT instead of an externally declared data item as shown in Figure 3. This simple example shows how references to internal elements of other modules can have an adverse effect on program modification, both in terms of cost and potential bugs.

Type of Communication

Modules must at least pass data or they cannot functionally be a part of a single system. Thus connections that pass data are a necessary minimum. (Not so the communication of control. In principle, the presence or absence of requisite input data is sufficient to define the circumstances under which a module should be activated, that is, receive control. Thus the explicit passing of control by one module to another constitutes an additional, theoretically inessential form of coupling. In practice, systems that are *purely* data-coupled require special language and operating system support but have numerous attractions, not the least of which is they can be fundamentally simpler than any equivalent system with control coupling.[10])

Beyond the practical, innocuous, minimum control coupling of normal subroutine calls is the practice of passing an "element of control" such as a switch, flag, or signal from one module to another. Such a connection affects the execution of another module, and not merely the data it performs its task upon, by involving one module in the internal processing of some other module. Control arguments are an additional complication to the essential data arguments required for performance of some task, and an alternative structure that eliminates the complication always exists.

Consider the modules in Figure 4 that are control-coupled by the switch PARSE through which EXECNCOMM instructs GETCOMM whether to return a parsed or unparsed command. Separating the two distinct functions of GETCOMM results in a structure that is simpler, as shown in Figure 5.

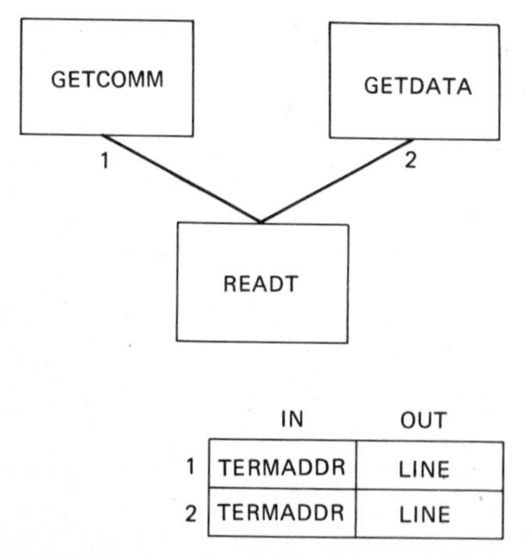

FIGURE 3 Improved module connections.

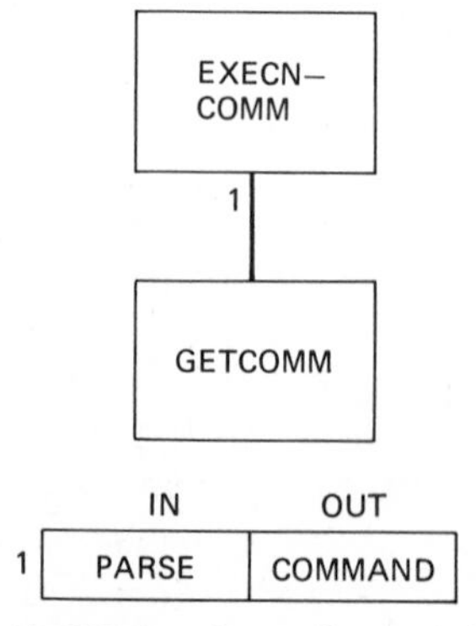

FIGURE 4 Control-coupled modules.

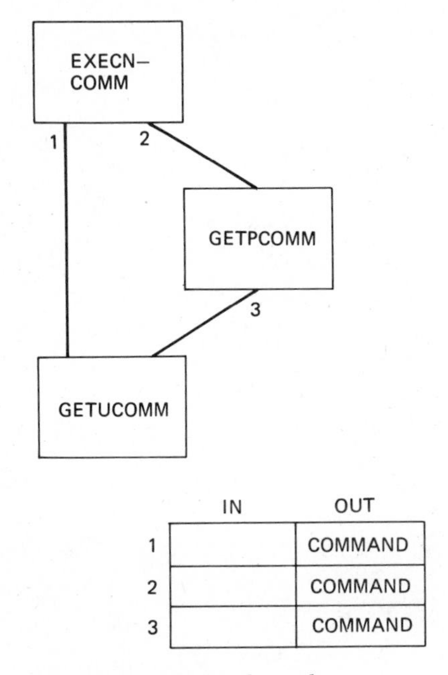

FIGURE 5 Simplified coupling.

The new EXECNCOMM is no more complicated; where once it set a switch and called, now it has two alternate calls. The sum of GETPCOMM and GFTUCOMM is (functionally) less complicated than GETCOMM was (by the amount of the switch testing). And the two small modules are likely to be easier to comprehend than the one large one. Admittedly, the immediate gains here may appear marginal, but they rise with time and the number of alternatives in the switch and the number of levels over which it is passed. Control coupling, where a called module "tells" its caller what to do, is a more severe form of coupling.

Modification of one module's code by another module may be thought of as a hybrid of data and control elements since the code is dealt with as data by the modifying module, while it acts as control to the modified module. The target module is very dependent in its behavior on the modifying module, and the latter is intimately involved in the other's internal functioning.

COHESIVENESS

Coupling is reduced when the relationships among elements *not* in the same module are minimized. There are two ways of achieving this—minimizing the relationships among modules and maximizing relationships among elements in the same module. In practice, both ways are used.

The second method is the subject of this section. "Element" in this sense means any form of a "piece" of the module, such as a statement, a segment, or a "subfunction". Binding is the measure of the cohesiveness of a module. The objective here is to reduce coupling by striving for high binding. The scale of cohesiveness, from lowest to highest, follows:

1. Coincidental.
2. Logical.
3. Temporal.
4. Communicational.
5. Sequential.
6. Functional.

The scale is not linear. Functional binding is much stronger than all the rest, and the first two are much weaker than all the rest. Also, higher-level binding classifications often include all the characteristics of one or more classifications below it *plus* additional relationships. The binding between two elements is the highest classification that applies. We will define each type of binding, give an example, and try to indicate why it is found at its particular position on the scale.

Coincidental Binding

When there is no meaningful relationship among the elements in a module, we have coincidental binding. Coincidental binding might result from either of the following situations: (1) An existing program is "modularized" by splitting it apart into modules. (2) Modules are created to consolidate "duplicate coding" in other modules.

As an example of the difficulty that can result from coincidental binding, suppose the following sequence of instructions appeared several times in a module or in several modules and was put into a separate module called X:

```
A = B + C
GET CARD
PUT OUTPUT
IF B = 4, THEN E = 0
```

Module X would probably be coincidentally bound since these four instructions have no apparent relationships among one another. Suppose in the future we have a need in one of the modules originally containing these instructions to say GET TAPERECORD instead of GET CARD. We now have a problem. If we modify the instruction in module X, it is unusable to all of the other callers of X. It may even be difficult to *find* all of the other callers of X in order to make any other compatible change.

It is only fair to admit that, independent of a module's cohesiveness, there are instances when any module can be modified in such a fashion to make it unusable to all its callers. However, the *probability* of this happening is very high if the module is coincidentally bound.

Logical Binding

Logical binding, next on the scale, implies some logical relationship between the elements of a module. Examples are a module that performs all input and output operations for the program or a module that edits all data.

The logically bound, EDIT ALL DATA module is often implemented as follows. Assume the data elements to be edited are master file records, updates, deletions, and additions. Parameters passed to the module would include the data and a special parameter indicating the type of data. The first instruction in the module is probably a four-way branch, going to four sections of code—edit master record, edit update record, edit addition record, and edit deletion record.

Often, these four functions are also intertwined in some way in the module. If the deletion record changes and requires a change to the edit deletion record function, we will have a problem if this function is intertwined with the other three. If the edits are truly independent, then the system could be simplified by putting each edit in a separate module and eliminating the need to decide which edit to do for each execution. In short, logical binding usually results in tricky or shared code, which is difficult to modify, and in the passing of unnecessary parameters.

Temporal Binding

Temporal binding is the same as logical binding, except the elements are also related in time. That is, the temporally bound elements are executed in the same time period.

The best examples of modules in this class are the traditional "initialization," "termination," "housekeeping," and "clean-up" modules. Elements in an initialization module are logically bound because initialization represents a logical class of functions. In addition, these elements are related in time (i.e., at initialization time).

Modules with temporal binding tend to exhibit the disadvantages of logically bound modules. However, temporally bound modules are higher on the scale since they tend to be simpler for the reason that *all* of the elements are executable at one time (i.e., no parameters and logic to determine which element to execute).

Communicational Binding

A module with communicational binding has elements that are related by a reference to the same set of input and/or output data. For example, "print and punch the output file" is communicationally bound. Communicational binding is higher on the scale than temporal binding since the elements in a module with communicational binding have the stronger "bond" of referring to the same data.

Sequential Binding

When the output data from an element is the input for the next element, the module is sequentially bound. Sequential binding can result from flowcharting the problem to be solved and then defining modules to represent one or more blocks in the flowchart. For example, "read next transaction and update master file" is sequentially bound.

Sequential binding, although high on the scale because of a close relationship to the problem structure, is still far from the maximum—functional binding. The reason is that the procedural processes in a program are usually distinct from the *functions* in a program. Hence, a sequentially bound module can contain several functions or just part of a function. This usually results in higher coupling and modules that are less likely to be usable from other parts of the system.

Functional Binding

Functional binding is the strongest type of binding. In a functionally bound module, all of the elements are related to the performance of a single function.

A question that often arises at this point is what is a function? In mathematics, $Y = F(X)$ is read "Y is a function F of X." The function F defines a transformation or mapping of the independent (or input) variable X into the dependent (or return) variable Y. Hence, a function describes a transformation from some input data to some return data. In terms of programming, we broaden this definition to allow functions with no input data and functions with no return data.

In practice, the above definition does not clearly describe a functionally bound module. One hint is that if the elements of the module all contribute to accomplishing a single goal, then it is probably functionally bound. Examples of functionally bound modules are "Compute Square Root" (input and return parameters), "Obtain Random Number" (no input parameter), and "Write Record to Output File" (no return parameter).

A useful technique in determining whether a module is functionally bound is writing a sentence describing the function (purpose) of the module, and then examining the sentence. The following tests can be made:

1. If the sentence *has* to be a compound sentence, contains a comma, or contains more than one verb, the module is probably performing more than one function: therefore, it probably has sequential or communicational binding.
2. If the sentence contains words relating to time, such as "first," "next," "then," "after," "when," "start," etc., then the module probably has sequential or temporal binding.
3. If the predicate of the sentence doesn't contain a single specific object following the verb, the module is probably logically bound. For example, Edit All Data has logical binding: Edit Source Statement may have functional binding.
4. Words such as "initialize," "clean-up," etc. imply temporal binding.

Functionally bound modules *can* always be described by way of their elements using a compound sentence. But if the above language is unavoidable while still completely describing the module's function, then the module is probably not functionally bound.

One unresolved problem is deciding how far to divide functionally bound subfunctions. The division has probably gone far enough if each module contains no subset of elements that could be useful alone, and if each module is small enough that its entire implementation can be grasped all at once, i.e., seldom longer than one or two pages of source code.

Observe that a module can include more than one type of binding. The binding between two elements is the highest that can be applied. The binding of a module is lowered by every element pair that does not exhibit functional binding.

PREDICTABLE MODULES

A predictable, or well-behaved, module is one that, when given the identical inputs, operates identically each time it is called. Also, a well-behaved module operates independently of its environment.

To show that dependable (free from errors) modules can still be unpredictable, consider an oscillator module that returns zero and one alternately and dependably when it is called. It might be used to facilitate double buffering. Should it have multiple users, each would be required to call it an even number of times before relinquishing control. Should any of the users have an error that prevented an even number of calls, all other users will fail. The operation of the module given the same inputs is not constant, resulting in the module not being predictable even though error-free. Modules that keep track of their own state are usually not predictable, even when error-free.

This characteristic of predictability that can be designed into modules is what we might loosely call "black-boxness." That is, the user can understand what the module does and use it without knowing what is inside it. Module "black-boxness" can even be enhanced by merely adding comments that make the module's function and use clear. Also, a descriptive name and a well-defined and visible interface enhances a module's usability and thus makes it more of a black box.

TRADEOFFS TO STRUCTURED DESIGN

The overhead involved in writing many simple modules is in the execution time and memory space used by a particular language to effect the call. The designer should realize the adverse effect on maintenance and debugging that may result from striving just for minimum execution time and/or memory. He should also remember that programmer cost is, or is rapidly becoming, the major cost of a programming system and that much of the maintenance will be in the future when the trend will be even more prominent. However, depending on the actual overhead of the language being used, it is very possible that a structured design can result in less execution and/or memory overhead rather than more due to the following considerations:

For Memory Overhead

1. Optional (error) modules may never be called into memory.
2. Structured design reduces duplicate code and the coding necessary for implementing control switches, thus reducing the amount of programmer-generated code.
3. Overlay structuring can be based on actual operating characteristics obtained by running and observing the program.
4. Having many single-function modules allows more flexible, and precise, grouping, possibly resulting in less memory needed at any one time under overlay or virtual storage constraints.

For Execution Overhead

1. Some modules may only execute a few times.
2. Optional (error) functions may never be called, resulting in zero overhead.
3. Code for control switches is reduced or elimi-

nated, reducing the total amount of code to be executed.

4. Heavily used linkage can be recompiled and calls replaced by branches.
5. "Includes" or "performs" can be used in place of calls. (However, the complexity of the system will increase by at least the extra consideration necessary to prevent duplicating data names and by the difficulty of creating the equivalent of call parameters for a well-defined interface.)
6. One way to get fast execution is to determine which parts of the system will be most used so all optimizing time can be spent on those parts. Implementing an initially structured design allows the testing of a working program for those critical modules (and yields a working program prior to any time spent optimizing). Those modules can then be optimized separately and reintegrated without introducing multitudes of errors into the rest of the program.

STRUCTURED DESIGN TECHNIQUES

It is possible to divide the design process into general program design and detailed design as follows. General program design is deciding *what* functions are needed for the program (or programming system). Detailed design is *how* to implement the functions. The considerations above and techniques that follow result in an identification of the functions, calling parameters, and the call relationships for a structure of functionally bound, simply connected modules. The information thus generated makes it easier for each module to then be separately designed, implemented, and tested.

Structure Charts

The objective of general program design is to determine what functions, calling parameters, and call relationships are needed. Since flowcharts depict *when* (in what order and under what conditions) blocks are executed, flowcharts unnecessarily complicate the general program design phase. A more useful notation is the structure chart, as described earlier and as shown in Figure 6.

To contrast a structure chart and a flowchart, consider the following for the same three modules in Figure 7—A which calls B which calls C (coding has been added to the structure chart to enable the proper flowchart to be determined: B's code will be executed first, then C's, then A's). To design A's interfaces properly, it is necessary to know that A is responsible for involving B, but this is hard to determine from the flowchart. In addition, the structure chart can show the module connections and calling parameters that are central to the consideration and techniques being presented here.

The other major difference that drastically simplifies the notation and analysis during general program design is the absence in structure charts of the decision block. Conditional calls can be so noted, but "decision designing" can be deferred until detailed module design. This is an example of where the *design* process is made simpler by having to consider only part of the design problem. Structure charts are also small enough to be worked on all at once by the designers, helping to prevent suboptimizing parts of the program at the expense of the entire problem.

Common Structures

A shortcut for arriving at simple structures is to know the general form of the result. Mr. Constantine observed that programs of the general structure in Figure 8 resulted in the lowest-cost implementations. It implements the input-process-output type of program, which applies to most programs, even if the "input" or "output" is to secondary storage or to memory.

In practice, the sink leg is often shorter than the source one. Also, source modules may pro-

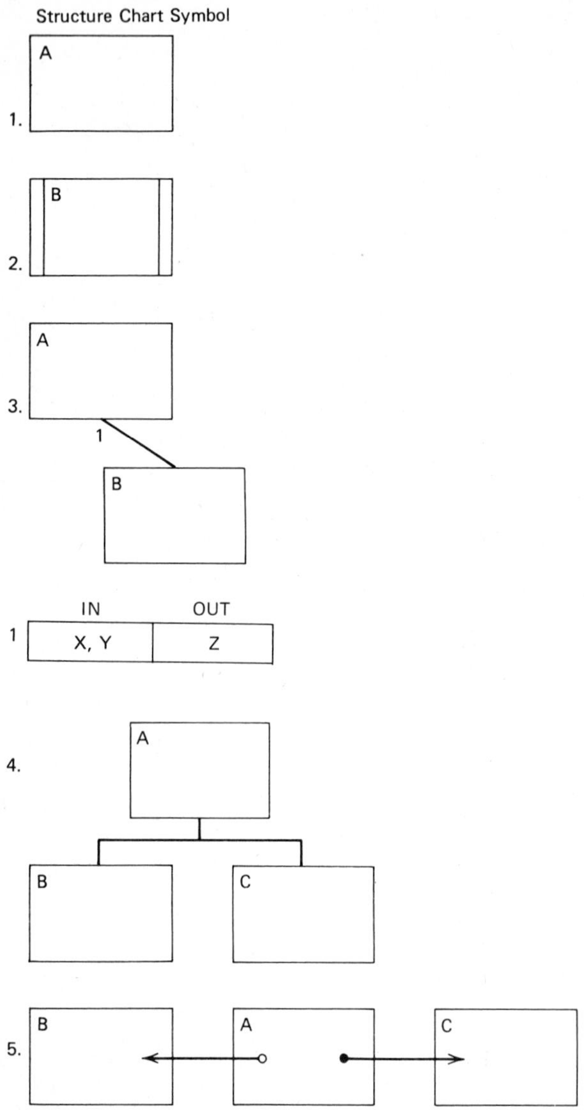

Structure Chart Symbol	Definition
1. A	Module
2. B	Predefined Module
3. A —1— B; 1: IN X, Y / OUT Z	Module A invokes Module B, and passes parameters X and Y from A to B. Module B passes parameter Z to Module A.
4. A; B, C	Module A invokes Modules B and C. Where possible, modules are placed left to right in likely order of invocation.
5. B, A, C	Module B refers to data in Module A. (Data flow from A to B.) Module A contains A branch to Module C.

The more comprehensive "proposed standard graphics for program structure." Preferred by Mr. Constantine and widely used over the past six years by his classes and clients. Uses separate arrows for each connection. Such as for the calls from A to B and from A to C to reflect structural properties for the program. The charting shown here was adopted for compatibility with the hierarchy chart of HIPO.

FIGURE 6 Definitions of symbols used in structure charts.

duce output (e.g., error messages) and sink modules may request input (e.g., execution-time format commands.)

Another structure useful for implementing parts of a design is the transaction structure depicted in Figure 9. A "transaction" here is any

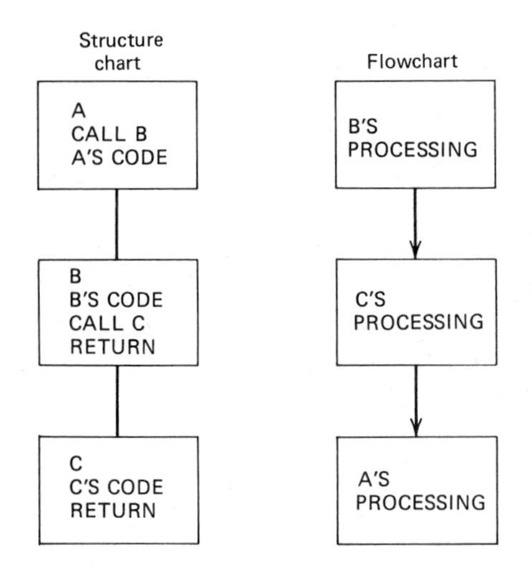

FIGURE 7 Structure chart compared to flowchart.

event, record, or input, etc. for which various actions should result. For example, a command processor has this structure. The structure may occur alone or as one or more of the source (or even sink) modules of an input-process-output structure. Analysis of the transaction modules follows that of a transform module, which is explained later.

Designing the Structure

The following procedure can be used to arrive at the input-process-output general structure shown previously.

Step One. The first step is to sketch (or mentally consider) a functional picture of the problem. As an example, consider a simulation system. The rough structure of this problem is shown in Figure 10.

Step Two. Identify the external conceptual streams of data. An *external* stream of data is one that is external to the system. A *conceptual* stream of data is a stream of related data that is independent of any physical I/O device. For instance, we may have several conceptual streams coming from one I/O device or one stream coming from several I/O devices. In our simulation system, the external conceptual streams are the input parameters, and the formatted simulation the result.

Step Three. Identify the *major* external conceptual stream of data (both input and output) in the problem. Then, using the diagram of the problem structure, determine, for this stream, the points of "highest abstraction" as in Figure 11.

The "point of highest abstraction" for an input stream of data is the point in the problem structure where that data is farthest removed from its physical input form, yet can still be viewed as coming in. Hence, in the simulation system, the most abstract form of the input transaction stream might be the built matrix. Similarly, identify the point where the data stream can first be viewed as going out—in the example, possibly the result matrix.

Admittedly, this is a subjective step. However, experience has shown that designers trained in the technique seldom differ by more than one or two blocks in their answers to the above.

Step Four. Design the structure in Figure 12 from the previous information with a source module for each conceptual input stream which exists at the point of most abstract input data; do sink modules similarly. Often only single source and sink branches are necessary. The parameters passed are dependent on the problem, but the general pattern is shown in Figure 12.

Describe the function of each module with a short, concise, and specific phrase. Describe what transformations occur when that module is called, not how the module is implemented. Evaluate the phrase relative to functional binding.

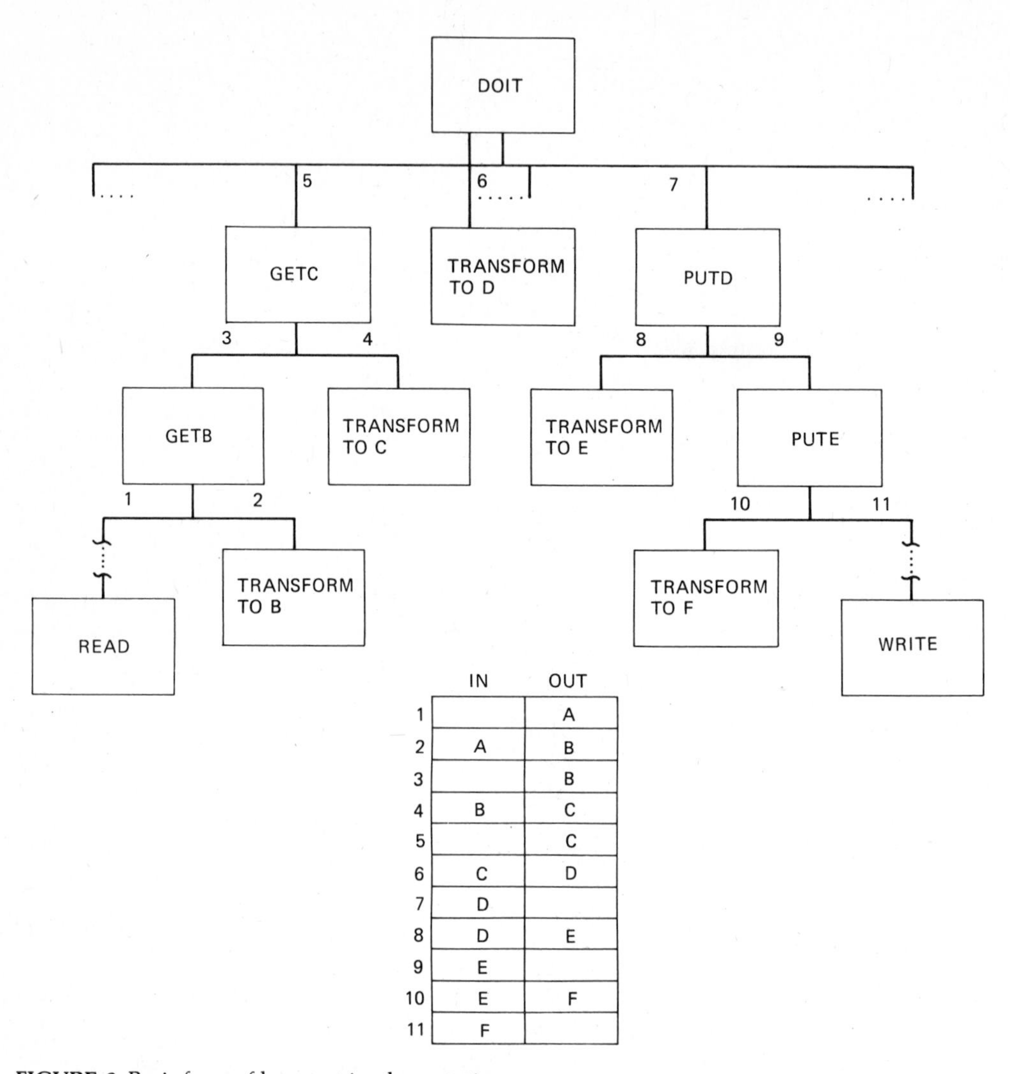

	IN	OUT
1		A
2	A	B
3		B
4	B	C
5		C
6	C	D
7	D	
8	D	E
9	E	
10	E	F
11	F	

FIGURE 8 Basic form of low-cost implementation.

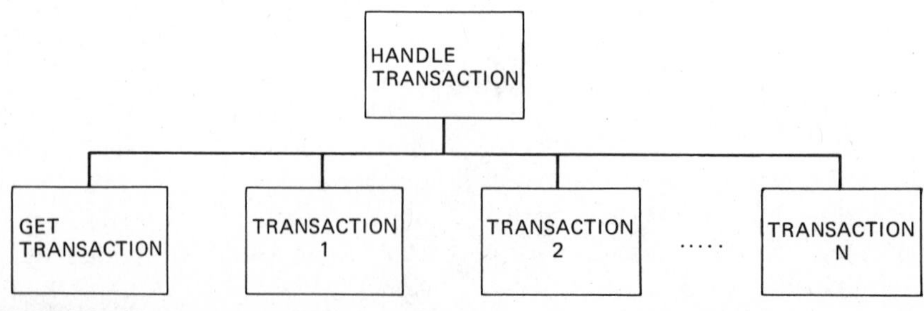

FIGURE 9 Transaction structure.

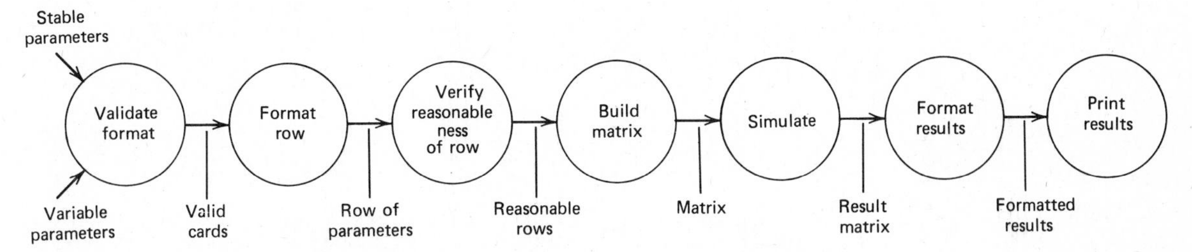

FIGURE 10 Rough structure of simulation system.

When module A is called, the program or system executes. Hence, the function of module A is equivalent to the problem being solved. If the problem is "write a FORTRAN compiler," then the function of module A is "compile FORTRAN program."

Module B's function involves obtaining the major stream of data. An example of a "typical module B" is "get next valid source statement in Polish form."

Module C's purpose is to transform the major input stream into the major output stream. Its function should be a nonprocedural description of this transformation. Examples are "convert Polish form statement to machine language statement" or "using keyword list, search abstract file for matching abstracts."

Module D's purpose is disposing of the major output stream. Examples are "produce report" or "display results of simulation."

Step Five. For each source module, identify the last transformation necessary to produce the form being returned by that module. Then identify the form of the input just prior to the last transformation. For sink modules, identify the first process necessary to get closer to the desired output and the resulting output form. This results in the portions of the structure shown in Figure 13.

Repeat Step Five on the new source and sink modules until the original source and final sink modules are reached. The modules may be analyzed in any order, but each module should be done completely before doing any of its subordinates. There are, unfortunately, no detailed guidelines available for dividing the transform modules. Use binding and coupling considerations, size (about one page of source), and usefulness (are there subfunctions that could be useful elsewhere now or in the future) as guidelines on how far to divide.

During this phase, err on the side of dividing too finely. It is always easy to recombine later in the design, but duplicate functions may not be identified if the dividing is too conservative at this point.

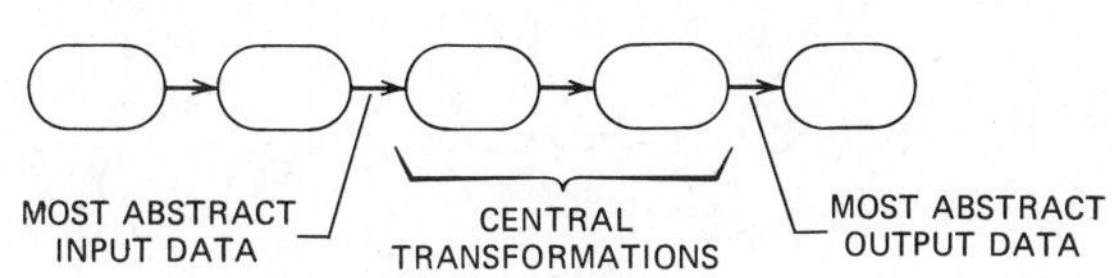

FIGURE 11 Determining points of highest abstraction.

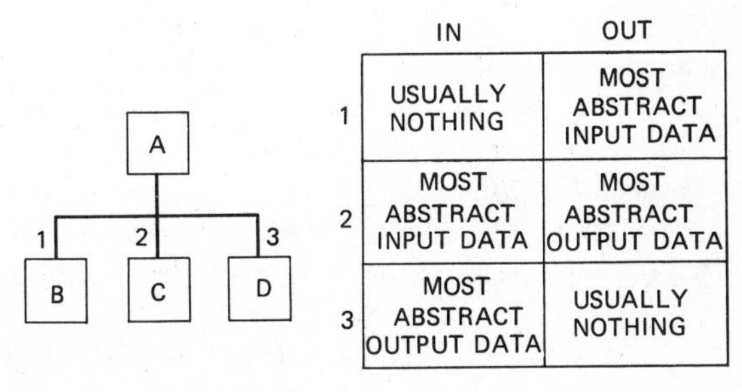

	IN	OUT
1	USUALLY NOTHING	MOST ABSTRACT INPUT DATA
2	MOST ABSTRACT INPUT DATA	MOST ABSTRACT OUTPUT DATA
3	MOST ABSTRACT OUTPUT DATA	USUALLY NOTHING

FIGURE 12 The top level.

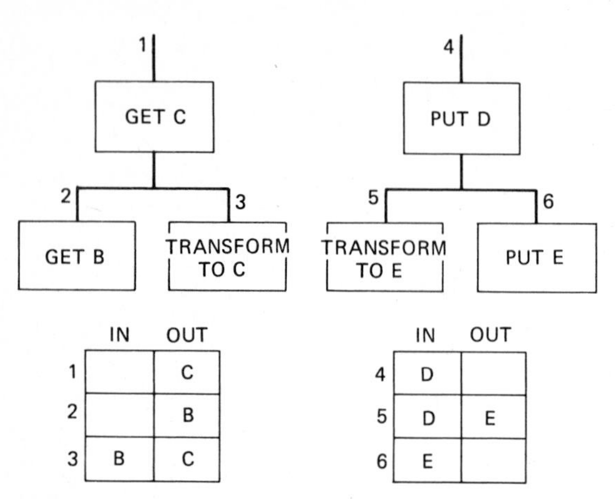

FIGURE 13 Lower levels.

DESIGN GUIDELINES

The following concepts are useful for achieving simple designs and for improving the "first-pass" structures.

Match Program to Problem

One of the most useful techniques for reducing the effect of changes on the program is to make the structure of the design match the structure of the problem, that is, form should follow function. For example, consider a module that dials a telephone and a module that receives data. If receiving immediately follows dialing, one might arrive at design A as shown in Figure 14. Consider, however, whether receiving is part of dialing. Since it is not (usually), have DIAL'S caller invoke RECEIVE as in design B.

If, in this example, design A were used, consider the effect of a new requirement to transmit immediately after dialing. The DIAL module receives first and cannot be used, or a switch must be passed, or another DIAL module has to be added.

To the extent that the design structure does match the problem structure, changes to single parts of the problem result in changes to single modules.

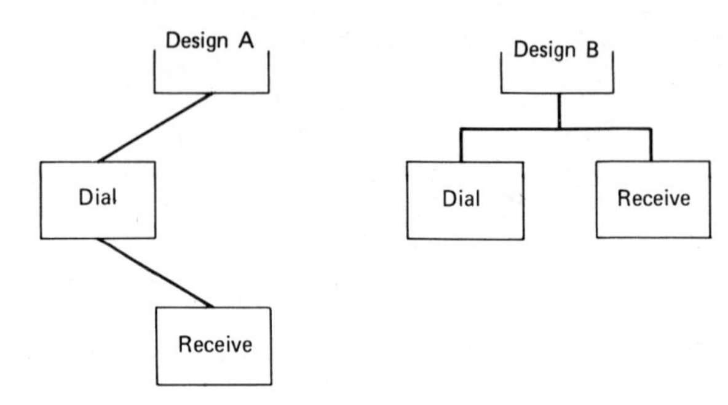

FIGURE 14 Design form should follow function.

Scopes of Effect and Control

The *scope of control* of a module is that module plus all modules that are ultimately subordinate to that module. In the example of Figure 15, the scope of control of B is B, D, and E. The *scope of effect* of a decision is the set of all modules that contain some code whose execution is based upon the outcome of the decision. The system is simpler when the scope of effect of a decision is in the scope of control of the module containing the decision. The following example illustrates why.

If the execution of some code in A is dependent on the outcome of decision X in module B,

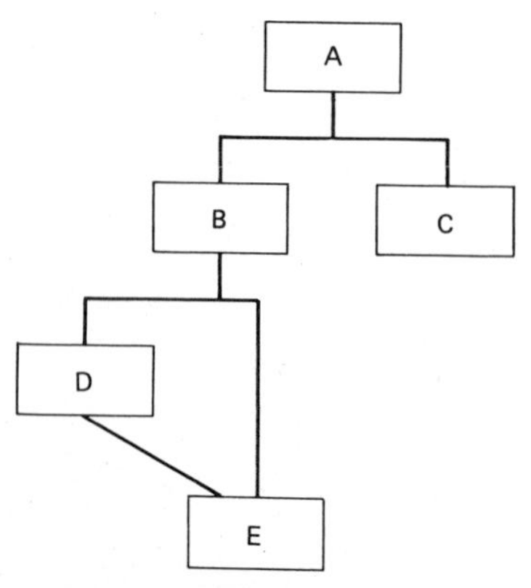

FIGURE 15 Scope of control.

then either B will have to return a flag to A or the decision will have to be repeated in A. The former approach results in added coding to implement the flag, and the latter results in some of B's function (decision X) in module A. Duplicates of decision X result in difficulties coordinating changes to both copies whenever decision X must be changed.

The scope of effect can be brought within the scope of control either by moving the decision element "up" in the structure, or by taking those modules that are in the scope of effect but not in the scope of control and moving them so that they fall within the scope of control.

Module Size

Size can be used a a signal to look for *potential* problems. Look carefully at modules with less than five or more than 100 executable source statements. Modules with a small number of statements may not perform an entire function, hence, may not have functional binding. Very small modules can be eliminated by placing their statements in the calling modules. Large modules may include more than one function. A second problem with large modules is understandability and readability. There is evidence to the fact that a group of about 30 statements is the upper limit of what can be mastered on the first reading of a module listing.[11]

Error and End-of-File

Often, part of a module's function is to notify its caller when it cannot perform its function. This is accomplished with a return error parameter (preferably binary only). A module that handles streams of data must be able to signal end-of-file (EOF), preferably also with a binary parameter. These parameters should not, however, tell the caller what to do about the error or EOF. Nevertheless, the system can be made simpler if modules can be designed without the need for error flags.

Initialization

Similarly, many modules require some initialization to be done. An initialize module will suffer from low binding but sometimes is the simplest solution. It may, however, be possible to eliminate the need for initializing without compromising "black-boxness" (the same inputs *always* produce the same outputs). For example, a read module that detects a return error of file-not-opened from the access method and recovers by opening the file and rereading, eliminates the need for initialization without maintaining an internal state.

Selecting Modules

Eliminate duplicate functions but not duplicate code. When a function changes, it is a great advantage to only have to change it in one place. But if a module's need for its own copy of a random collection of code changes slightly, it will not be necessary to change several other modules as well.

If a module seems almost, but not quite, useful from a second place in the system, try to identify and isolate the useful subfunction. The remainder of the module might be incorporated in its original caller.

Check modules that have many callers or that call many other modules. While not always a problem, it may indicate missing levels or modules.

Isolate Specifications

Isolate all dependencies on a particular data-type, record-layout, index-structure, etc. in one or a minimum of modules. This minimizes the recording necessary should that particular specification change.

Reduce Parameters

Look for ways to reduce the number of parameters passed between modules. Count every item passed as a separate parameter for this objective (independent of how it will be implemented). Do not pass whole records from module to module, but pass only the field or fields necessary for each module to accomplish its function. Otherwise, all modules will have to change if one field expands, rather than only those which directly used that field. Passing only the data being processed by the program system with necessary error and EOF parameters is the ultimate objective. Check binary switches for indications of scope-of-effect/scope-of-control inversions.

Have the designers work together and with the complete structure chart. If branches of the chart are worked on separately, common modules may be missed and incompatibilities result from design decisions made while only considering one branch.

AN EXAMPLE

The following example illustrates the use of structured design:

A patient-monitoring program is required for a hospital. Each patient is monitored by an analog device which measures factors such as pulse, temperature, blood pressure, and skin resistance. The program reads these factors on a periodic basis (specified for each patient) and stores these factors in a data base. For each patient, safe ranges for each factor are specified (e.g., patient X's valid temperature range is 98 to 99.5 degrees Fahrenheit). If a factor falls outside of a patient's safe range, or if an analog device fails, the nurse's station is notified.

In a real-life case, the problem statement would contain much more detail. However, this one is of sufficient detail to allow us to design the structure of the program.

The first step is to outline the structure of the problem as shown in Figure 16. In the second step, we identify the external conceptual streams of data. In this case, two streams are present, factors from the analog device and warnings to the nurse. These also represent the major input and output streams.

Figure 17 indicates the point of highest abstraction of the input stream, which is the point at which a patient's factors are in the form to store in the data base. The point of highest abstraction of the output stream is a list of unsafe factors (if any). We can now begin to design the program's structure as in Figure 18.

In analyzing the module "OBTAIN A PATIENT'S FACTORS," we can deduce from the problem statement that this function has three parts: (1) Determine which patient to monitor next (based on their specified periodic intervals). (2) Read the analog device. (3) Record the factors in the data base. Hence, we arrive at the structure in Figure 19. (NOTVAL is set if a valid set of factors was not available.)

Further analysis of "READ VALID SET OF FACTORS," "FIND UNSAFE FACTORS" and "NOTIFY STATION OF UNSAFE FACTORS" yields the results shown in the complete structure chart in Figure 20.

Note that the module "READ FACTORS FROM TERMINAL" contains a decision asking "did we successfully read from the terminal?" If

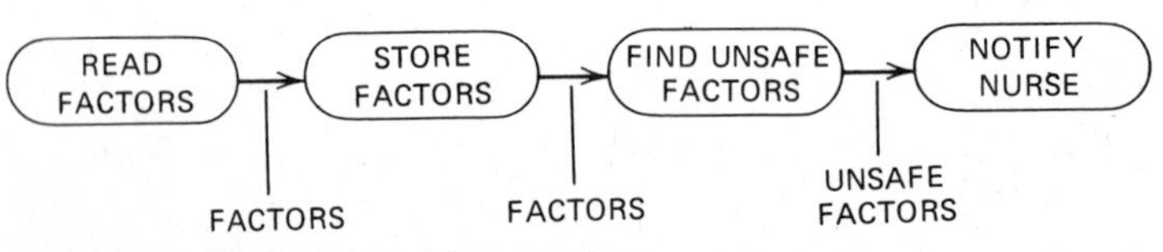

FIGURE 16 Outline of problem structure.

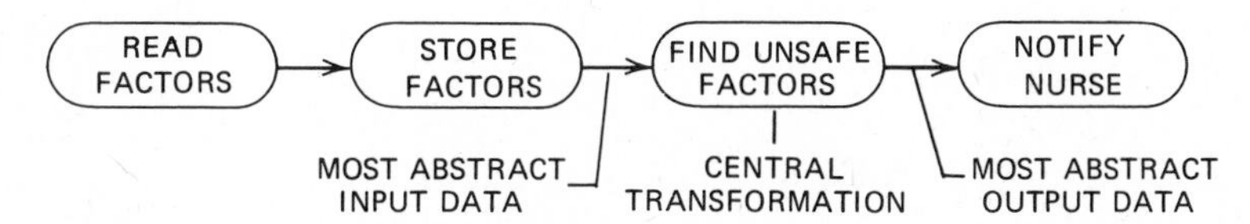

FIGURE 17 Points of highest abstraction.

the read was not successful, we have to notify the nurse's station and then find the next patient to process as depicted in Figure 21.

Modules in the scope of effect of this decision are marked with an X. Note that the scope of effect is *not* a subset of the scope of control. To correct this problem, we have to take two steps. First, we will move the decision up to "READ VALID SET OF FACTORS." We do this by merging "READ FACTORS FROM TERMINAL" into its calling module. We now make "FIND NEXT PATIENT TO MONITOR" a subordinate of "READ VALID SET OF FACTORS." Hence, we have the structure in Figure 22. Thus, by slightly altering the structure and the function of a few modules, we have completely eliminated the problem.

CONCLUDING REMARKS

The HIPO Hierarchy chart is being used as an aid during general systems design. The considerations and techniques presented here are useful for evaluating alternatives for those portions of the system that will be programmed on a computer. The charting technique used here depicts more details about the interfaces than the HIPO Hierarchy chart. This facilitates consideration during general program design of each individual connection and its associated passed parameters. The resulting design can be documented with the HIPO charts. (If the designer decides to have more than one function in any module, the structure chart should show them in the same block. However, the HIPO Hierarchy chart would still show all the functions in separate blocks.) The output of the general program design is the input for the detailed module design. The HIPO input-process-output chart is useful for describing and designing each module.

Structured design considerations could be used to review program designs in a walk-through environment.[12] These concepts are also useful for evaluating alternative ways to comply with the requirement of structured programming for one-page segments.[7]

Structured design reduces the effort needed to fix and modify programs. If all programs were written in a form where there was one module, for example, which retrieved a record from the master file given the key, then changing operat-

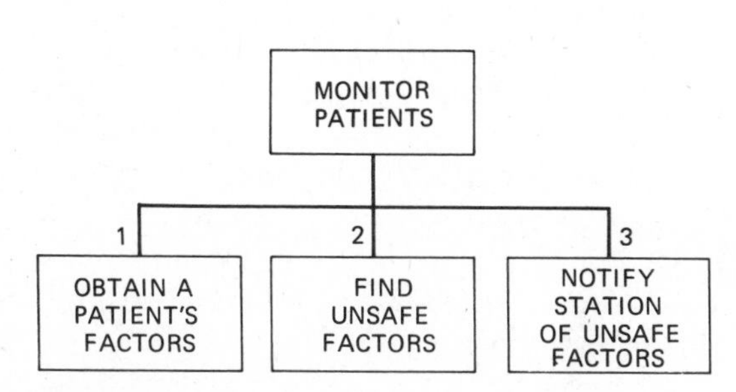

	IN	OUT
1	NOTHING	TEMP, PULSE, BP SKINR, PATIENTNUM
2	TEMP, PULSE, BP SKINR, PATIENTNUM	LIST OF UNSAFE FACTOR NAMES AND VALUES
3	PATIENTNUM AND LIST OF UNSAFE FACTOR NAMES AND VALUES	NOTHING

FIGURE 18 Structure of the top level.

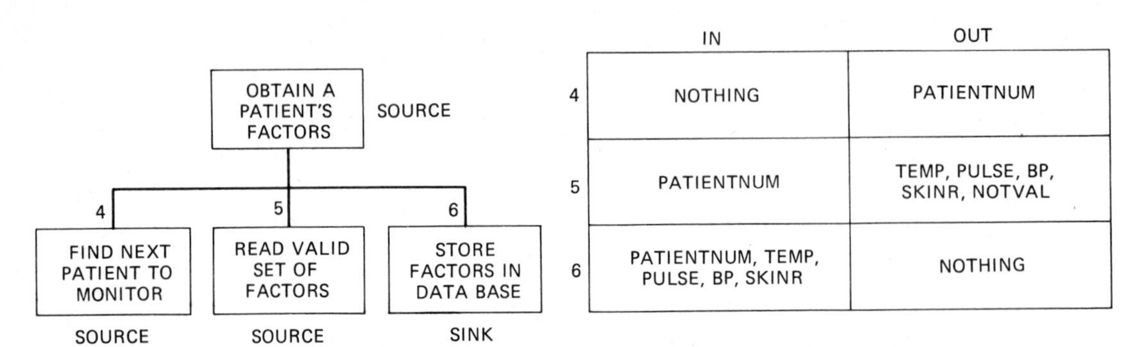

FIGURE 19 Structure of next level.

ing systems, file access techniques, file blocking, or I/O devices would be greatly simplified. And if *all* programs in the installation retrieved from a given file with the same module, then one properly rewritten module would have *all* the installation's programs working with the new constraints for that file.

However, there are other advantages. Original errors are reduced when the problem at hand is simpler. Each module is self-contained and to some extent may be programmed independently of the others in location, programmer, time, and language. Modules can be tested before all programming is done by supplying simple "stub" modules that merely return preformatted results rather than calculating them. Modules critical to memory or execution overhead can be optimized separately and reintegrated with little or no impact. An entry or return trace-module becomes very feasible, yielding a very useful debugging tool.

Independent of all the advantages previously mentioned, structured design would *still* be valuable to solve the following problem alone. Programming can be considered as an art where each programmer usually starts with a blank canvas—techniques, yes, but still a blank canvas. Previous coding is often not used because previous modules usually contain, for example, *at least* GET and EDIT. If the EDIT is not the one needed, the GET will have to be recoded also.

Programming can be brought closer to a science where current work is built on the results of earlier work. Once a module is written to get a record from the master file given a key, it can be used by all users of the file and need not be rewritten into each succeeding program. Once a module has been written to do a table search, anyone can use it. And, as the module library grows, less and less new code needs to be written to implement increasingly sophisticated systems.

Structured design concepts are not new. The whole assembly-line idea is one of isolating simple functions in a way that still produces a complete, complex result. Circuits are designed by connecting isolatable, functional stages together, not by designing one big, interrelated circuit. Page numbering is being increasingly sectionalized (e.g., 4—101) to minimize the "connections" between written sections, so that expanding one section does not require renumbering other sections. Automobile manufacturers, who have the most to gain from shared system elements, finally abandoned even the coupling of the windshield wipers to the engine vacuum due to effects of the engine load on the performance of the wiping function. Most other industries know well the advantage of isolating functions.

It is becoming increasingly important to the data-processing industry to be able to produce more programming systems and produce them with fewer errors, at a faster rate, and in a way

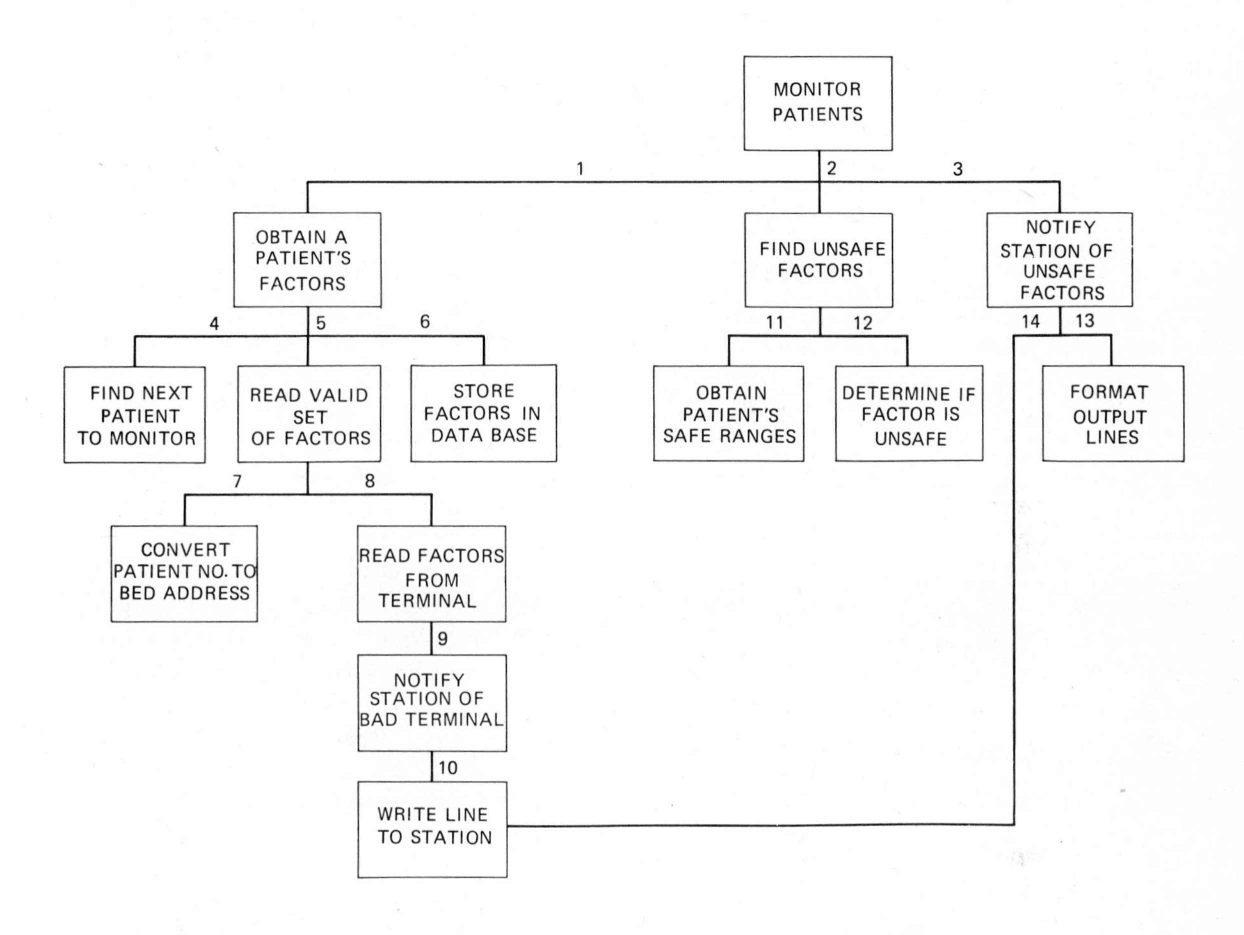

	IN	OUT
1	————	TEMP, PULSE, BP, SKINR, PATIENTNUM
2	TEMP, PULSE, BP, SKINR, PATIENTNUM	LIST OF UNSAFE FACTOR NAMES AND VALUES
3	PATIENTNUM & LIST OF UNSAFE FACTOR NAMES & VALUES	————
4	————	PATIENTNUM
5	PATIENTNUM	TEMP, PULSE, BP, SKINR, NOTVAL
6	PATIENTNUM, TEMP, PULSE, BP, SKINR	————
7	PATIENTNUM	BEDNUM
8	BEDNUM	TEMP, PULSE, BP, SKINR, NOTVAL
9	BEDNUM	————
10.14	LINE	————
11	PATIENTNUM	TEMPR, PULSER, BPR, SKINRR
12	FACTOR RANGE	UNSAFE
13	LIST OF UNSAFE FACTOR NAMES AND VALUES	LIST OF LINES

FIGURE 20 Complete structure chart.

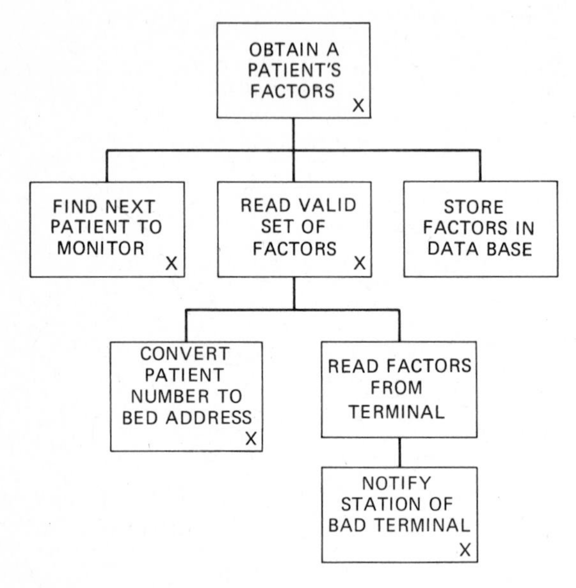

FIGURE 21 Structure as designed.

that modifications can be accomplished easily and quickly. Structured design considerations can help achieve this goal.

CITED REFERENCES AND FOOTNOTES

1. This method has not been submitted to any formal IBM test. Potential users should evaluate its usefulness in their own environment prior to implementation.
2. L. L. Constantine, *Fundamentals of Program Design,* in preparation for publication by Prentice-Hall, Englewood Cliffs, New Jersey.
3. G. J. Myers, *Composite Design: The Design of Modular Programs,* Technical Report TR00.2406, IBM, Poughkeepsie, New York (January 29, 1973).
4. G. J. Myers, "Characteristics of composite design," *Datamation* **19**, No. 9, 100–102 (September 1973).
5. G. J. Myers, *Reliable Software through Composite Design,* Petrocelli/Charter, New York, New York (1975).
6. HIPO—Hierarchical Input-Process-Output documentation technique. Audio education package, Form No. SR20-9413, available through any IBM Branch Office.
7. F. T. Baker, "Chief programmer team management of production programming," *IBM Systems Journal* **11**, No. 1.56–73 (1972).
8. The use of the HIPO Hierarchy charting format is further illustrated in Figure 6, and its use in this paper was initiated by R. Ballow of the IBM Programming Productivity Techniques Department.
9. L. A. Belady and M. M. Lehman, *Programming System Dynamics or the Metadynamics of Systems in Maintenance and Growth",* RC 3546, IBM Thomas J. Watson Research Center, Yorktown Heights, New York (1971).
10. L. L. Constantine, "Control of sequence and parallelism in modular programs," *AFIPS Conference Proceedings, Spring Joint Computer Conference* **32**, 409 (1968).
11. G. M. Weinberg, *PL/1 Programming: A Manual of Style,* McGraw-Hill, New York, New York (1970).
12. *Improved Programming Technologies: Management Overview,* IBM Corporation, Data

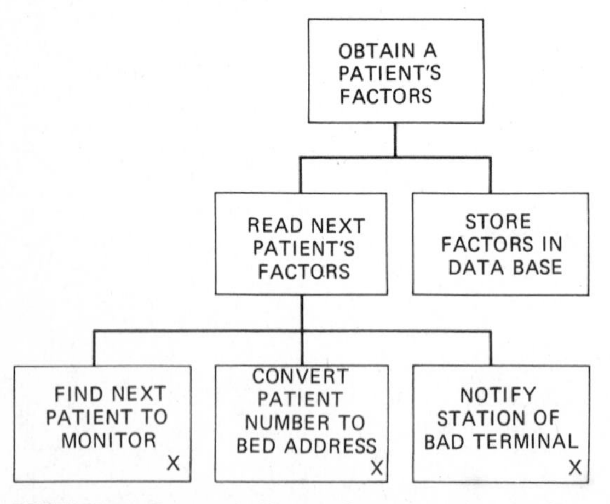

FIGURE 22 Scope of effect within scope of control.

Processing Division, White Plains, New York (August 1973).

READING QUESTIONS

1. Define the term *module*. How is this concept basic to structured design?
2. Define the term *coupling*. How does module coupling affect the quality of a design?
3. Define the term *cohesion*. How does cohesion affect system reliability and maintainability?
4. Draw a structure chart for a small payroll system which must, for each hourly employee, accept the current hours worked along with basic employee information including pay rate, year-to-date FICA, and so on. It must calculate overtime and gross pay, calculate federal, state, and FICA deductions to obtain net pay, and issue a check.
5. How do the structure charts discussed here compare to the UTOC from HIPO (discussed in the paper by Colter in this book)? How are data flows depicted? Is data structure considered?
6. Draw a data flow diagram (see Figure 10) of the payroll problem detailed above. Identify the point of highest abstraction and the central transform.
7. Compare the information contained in the structure chart with that of the data flow diagram. Use the four basic structural points of view which were presented in Colter's paper as the framework for your analysis.
8. Define and discuss the relationship between scope of effect and scope of control.
9. The authors of this paper presented methods for improved design processes. Write a short, nontechnical summary of their suggestions which would serve to encourage its readers to adopt or ignore these methods.

Structured Systems Design

Kenneth T. Orr

INTRODUCTION

A great deal has been written about information systems and their uses. Despite this wealth of data, there is very little good literature on how to put a system together. The existing systems literature is largely concerned with some tool or other that can be used to simplify the process of building a system or with some foolproof method of documenting a systems job to speed up systems development and thereby reduce maintenance costs.

Clearly, systems building is still an art, and it is unlikely that this situation will change very much until we develop new methods for training systems designers and architects in the building of complex systems. This article is aimed at doing just that: teaching systems professionals new tools and approaches for analyzing, designing, developing, and installing those things we call systems. Many of the approaches and philosophies discussed here are refinements of techniques that have been around for decades, but others, especially the portion that has to do with structured systems design, are quite new. Taken together, they represent a body of knowledge that can be extremely useful. Although there is a quantity of material about programming and systems and their development, not a lot has been written from the standpoint of the person who does not receive a well-defined problem to solve.

Systems analysis and systems development are simply other terms to describe the process of problem solving. One great benefit of structured systems design is that it provides a number of significant new theories and tools for solving problems, especially complex, logical problems. By any definition, and in any disguise, problem

Source: This paper is condensed from Chapters 1 through 5 of *Structured Systems Development,* by Kenneth T. Orr (Yourdon Press, NY, 1977) with the author's permission. It should be noted that Kenneth Orr is largely responsible for making Jean-Dominique Warnier's work known in America and that Mr. Orr has made significant developments to the Warnier methodology. Current terminology accepts the use of the term *Warnier/Orr* when referencing both the methodology and the diagrams which are noted here-in as "Warnier diagrams."

solving is a heuristic process—a process of trial and error. Anyone who fails to recognize this is doomed from the outset to outright failure or, at the very least, extreme frustration.

Today, we must recognize that *we simply don't know how to teach someone to design a system very well.* While we can teach the various features of a programming language or of a data base management system, we have tremendous difficulties teaching someone how to put it all together. That is what systems development is all about—putting together all the pieces: the user requirements, the technology, and the management.

Where do we start? First, we have to recognize that although many people know how to design a system, for the most part they don't know how to tell anyone else how to do it.

That problem is not unique to systems analysis; most young professions have the same problem. This is particularly true when there is no strong underlying theory to guide the professionals in their field. This same situation existed in programming for many years; only with the advent of such concepts as structured programming and top-down development has it become possible to put programming on a more professional basis. We must now begin to do the same for systems analysis and design.

THE SYSTEMS MODEL

In structured systems design, we take advantage of a number of tools and concepts that have been developed specifically to aid us in thinking about information handling. Most of these tools are simple to understand, but they are not always simple to use, at least not initially. As you come to understand the underlying theory, you will find their application more and more natural.

Structured systems design is an approach founded upon a rather simple systems model. That model is predicated on the fact that any (information) system can be considered as consisting of three basic parts: (1) *output,* (2) a *black box* that produces the output, and (3) *input* from which the black box produces the output. (This might seem to be a peculiar way of saying that a system is made up of inputs, a black box, and outputs; however, you will see the point later.) A graphic version of the concept is shown in Figure 1.

In structured systems design, output and input (along with certain aspects of the black box) fall into a category called *systems requirements.* While dealing with a number of elements, *systems requirements are primarily concerned with outputs and the logical rules for their derivation.* Somewhat surprisingly, from a requirements standpoint, we are far less concerned with inputs than with outputs. The reasons for this will become more clear later.

If a change is made to a system and if it affects the output, then it is a change to the systems requirements. If, on the other hand, you convert a system from one computer operating system to another, or change the sequence of operations without affecting the output produced by the system, then that change is involved with the *systems architecture.* Systems architecture is involved with the way output is produced from input. While systems requirements primarily involve output and its derivation, systems architecture largely concerns the makeup of the

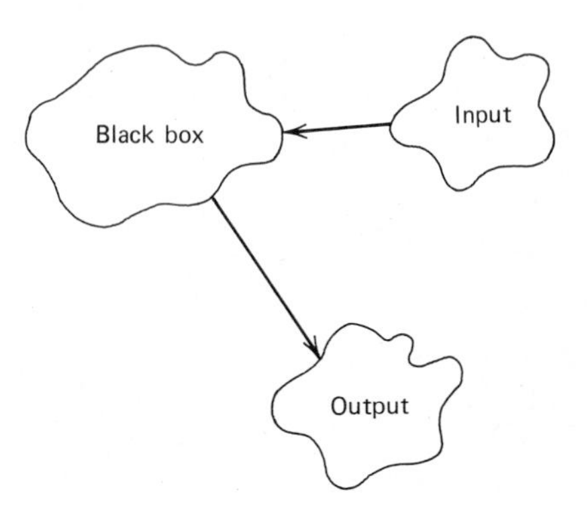

FIGURE 1 Structured systems design based upon a three-part model.

black box, that is, the organization of the pieces of the system. Information is entered, processed, stored, and put out. Systems architecture is concerned with how this process will produce the correct results.

Let's look at a system from another slightly different standpoint. Outputs and inputs can be thought of as "sets of data." For example, a payroll report can be considered a set of data about employees. Now, the black box in our model contains sets of data; but it also contains something else: "sets of actions" (or operations). *A correct system, then, is one in which the right sets of actions are executed on the right sets of data at the right times.* If the correct set of monthly actions (called report programs) are done on the right sets of data (called payroll files) at the right time (called accounting month end), then you will produce the right output and have a correct system. *A correct system is the minimum criterion for doing systems work.* In addition to producing the right output, though, you typically will also have to produce that output at a certain cost and within a given time frame prescribed by management.

At each stage of refinement, we learn more about the system's sets of data and sets of actions. Intellectually, to deal with all this data, the analyst has to summarize or classify this information into structures that can be manipulated easily. For example, the analyst might begin to talk about weekly reports or monthly reports as a group. Or he may categorize them into mandatory and optional, or federal and company. Each method of grouping may be useful, because it helps the analyst deal with the complexity of thousands of details. In fact, without some way of organizing information to find the big picture, it's highly unlikely that you'll ever get started. Many systems efforts fail simply because the analyst drowns in a sea of details. We must strive to avoid this pitfall. Again, our model can help. We said earlier that the black box part of our system can be considered as either sets of actions or sets of data. Let's show that graphically.

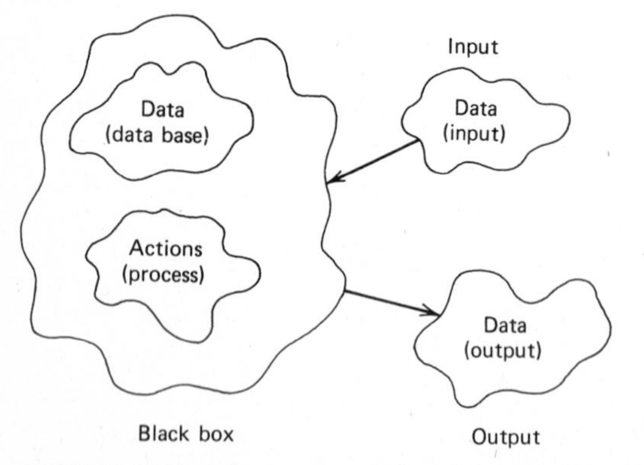

FIGURE 2 Black box part considered as sets of data or sets of actions.

This refinement of our systems model has added additional structure, for it shows how the basic pieces actually relate to one another. Now, we can think of a system in terms of the relation between sets of actions and sets of data. We also can give names to the major elements: output, process, data base, and input.

To design a system, start with the output and work backward through the system until you finally arrive at the processing steps that are required, the data base that you need, and finally the inputs you have to collect. (At first glance, working backward in systems design seems unnatural. The reason is simply that starting at the beginning, i.e., with the inputs, and going forward doesn't work very well. It's a great deal like going on a trip without first having decided where you're going. How do you know what to pack? Since you don't know where you're going, you don't know what the weather will be, and, therefore, you'll probably take too much clothing, just in case. Your next problem is encountered the moment you leave the driveway. Which way you go?)

A number of benefits result from working backward. One major benefit is that in the ideal case (which will probably never happen), you won't collect or store data that you don't use—a factor certain to have a major effect on both the cost and the complexity of the end product, not to mention on questions of data security and privacy. Other benefits will be discussed as we uncover them.

Structured systems design aims at producing minimal systems. In this age of flexible, general, powerful systems, we want to design something that is minimal. What makes up a minimal system? How do you define it? *A minimal system does the absolute minimum to produce the desired outputs.* Such a system only captures data that appear as output or that affect something that comes out. Another way to look at our definition is to think of yourself as a mathematician who has derived a perfect formula for producing a correct systems design every time. In such a mathematical formula, if given a set of values for the independent variables, then all the dependent variables can be mechanically computed. But what exactly are the independent variables in systems design? They would be those elements that determine everything else, wouldn't they? What category of things that we deal with in systems work fits that description? The output, of course! If we change the output requirements, everything else in the systems design has to change to accommodate it. If the output changes minimally, we must change the processing; and if we don't have the data from which the new output is derived, then the change will affect the data base and the input as well. Therefore, *the output requirements can be thought of as independent variables of the systems process.*

In theory, working backward from the output seems reasonable; and the more you think about it, the more you will realize that it's the *only* reasonable approach. In practice, however, working backward is difficult to apply. One of the major reasons is that thinking about outputs is hard work. It requires that you set up goals and objectives and that you make hard decisions such as, Why do I need this information? How do I want my information ordered? On the other hand, talking about input is much easier, for the hard questions always can be left until later; you don't have to be too specific about how you will use the information—you can simply say, We ought to collect it, because someday it may be needed. Most approaches in the past have indicated that one of the first systems tasks (right after specifying the outputs) is to define the inputs with the user. There are a number of strong pressures to consider the definition of input out of sequence.

ANALYSIS, SYNTHESIS, AND PROCESS FLOW

In the last section, we began to develop an intellectual model for an information system. We will return to that model after we have looked at how some traditional approaches to systems analysis and design are used. The purpose in introducing traditional approaches at this point is to give us practice in systems work before we must use our knowledge in a structured environment.

The model in the last section served a very important function: It gave us a conceptual framework or theory. We will pick up more theory as we proceed. Many of us go through our entire lives without recognizing the importance of theory in everything we do, primarily because theories seem too abstract and remote to be of much use. Often people in new professions are too busy *doing* to reflect much about *what* they are doing or *how* it relates to other things.

Take computer programming as an example. For years, people developed program after program, using a variety of machines and languages, convinced that programming was at best an art. Most programmers felt that while you could learn programming, there weren't any scientific ways of writing a correct program. A good program was simply a program that worked, and the only way to know whether a program actually worked was to test and test and test. Even then, you couldn't be absolutely sure it would always work. Today, we know there are better ways of developing correct programs. We know, for example, that if we understand logically what we want to do, we can build a correct program to do it. Moreover, we are quite certain that we can make our programs work, if not the first time, at least within a few tries. How has this transition from doubt to certainty in programming taken place in such a short time? The answer simply is that we now have a theory to guide our activities—a theory for programming. We now know that if we work out all the logical possibilities in detail and construct our program in a hierarchical logical fashion, it will work. We also know that we can use the structure of the data as a guide for the construction of our program, thereby eliminating many tough decisions once facing us in organizing or structuring our programs.

If we apply this same technique of hierarchical logical analysis to an entire system, we can become increasingly certain that we will develop a correct system, i.e., one that performs the right sets of actions on the right sets of data at the right times. There are a number of differences, however, between the design and construction of a program and the design and construction of an entire system. For one thing, the programmer ordinarily has fewer worries than has the systems analyst. Often an analyst or someone else already has defined the outputs and inputs and the basic processing for the programmer. So while the programmer has to figure out many details, he doesn't have to define basic parameters.

The analyst, on the other hand, is in a different situation: He not only has to figure out what needs to be done, he has to do it as well. *Nothing except the problem is given to the analyst.* He has to add his own ideas to the wishes of the various people who must interact with the system to define the scope of the problem. This becomes the *why* and *what* of the system. Then he has to figure out a way in which to do these things—the *how* and *when* of the system.

Working out this relationship is integral to the process of design. Clearly, the determination of why and what and of how and when are related. In a classical sense, we are talking about two activities: analysis and synthesis. Therefore, we will look at *systems analysis* and *systems synthesis* (also called design or integration) and see how they relate to the real process of designing a system.

Analysis is an old word meaning to break something into its fundamental pieces. That definition is true whether referring to chemical analysis, linguistic analysis, or systems analysis. *In systems analysis, we are concerned with breaking a system into pieces and with the tools for doing that.* In general, there is a hierarchical approach characteristic of all types of analysis; that is, break something big into successively smaller pieces, until it is no longer possible to further subdivide any of the pieces. Each new piece is subordinate to the piece above.

How does the process of hierarchical analysis actually work? Let's take an example: Suppose you are assigned the problem of developing a system to produce a monthly statistical report, as represented in Figure 3.

In hierarchical analysis we ask, How can we break this job into a series of simpler ones? One way is to think of some of the problem's basic parts that might expectedly meet the original problem definition. In this case, our goal is to develop a system that produces a monthly statistical report. From past experience, we might expect (at least on the computer side) to produce that monthly report in a number of steps: For example, one part of the system would produce the report, one part would update the master file (or data base) from which the reports will b pro-

FIGURE 3 Problem to produce monthly report.

duced, and one part would edit the input data initially to exclude any erroneous values. The typical method of showing the relation is to draw something resembling an organizational chart (Figure 4).

A similar hierarchical analysis is conducted in nearly every systems job and, in fact, is so common and natural that many analysts don't even bother to write it down. They simply assume that everyone understands how they arrived at their design. This is often not the case and it is important for those who will implement, maintain, or operate a system also to understand how it was constructed.

The analysis of a problem into parts is a useful process, and it is most effective if you continue systematically to apply the same process of divide and conquer to each part of the problem. For example, we might subdivide (or analyze) the part called PRODUCE MONTHLY STATISTICAL REPORT into several parts: one to read the master file, one to summarize the data, and one to format and print the report (Figure 5).

In a complete analysis, this process continues until each of the pieces is so simple that there is no need to break them down further. If carried on systematically, analysis is an extremely powerful tool. When finished, you will have a complete idea of what needs to be done, and how to do it.

Conceptually, analysis is a simple technique that should work all the time. Over the years, nearly every successful systems analyst has employed some form or other of hierarchical analysis. Currently, analysis is enjoying new popularity under the name of top-down design, which recommends that each problem be analyzed step-by-step until it is completely resolved. The HIPO (Hierarchical-Input-Process-Output) methodology utilizes hierarchical analysis quite extensively.

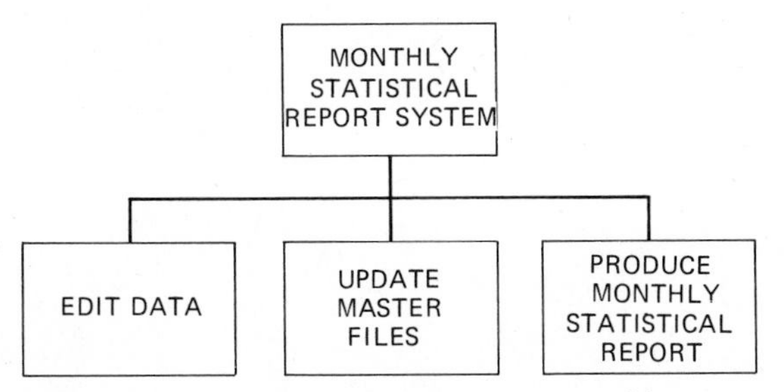

FIGURE 4 Hierarchical analysis requires problem breakdown.

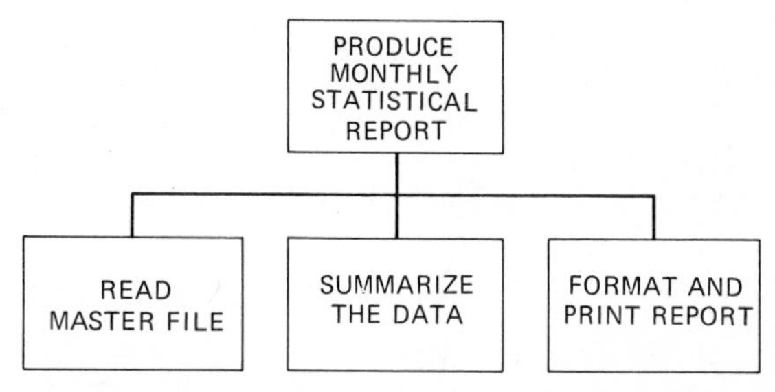

FIGURE 5 More detailed breakdown of problem into steps.

You might wonder why, if hierarchical analysis is so natural and so popular, we haven't used it more in the past. There are a number of reasons. One is that it seems like a lot of work, especially at the lower levels. Further, it requires that you have some idea of the top of the system and of the major pieces at each step. Finally, hierarchical analysis poses the question: How do we know what pieces to break the system into in the first place?

This problem is a tricky one. The example above of a monthly reporting system was uncomplicated, but suppose we had to build a total management information system? Certainly the problem would be more complex, and the analysis normally would be very difficult to do from the top down. For that reason, hierarchical analysis has not been recognized as a fundamental tool in systems work until only very recently, despite its common use.

Analysis has an important by-product: making systems that are both hierarchical and modular. This is extremely important, for our experience suggests that all good systems are both hierarchical and modular. However, analy-

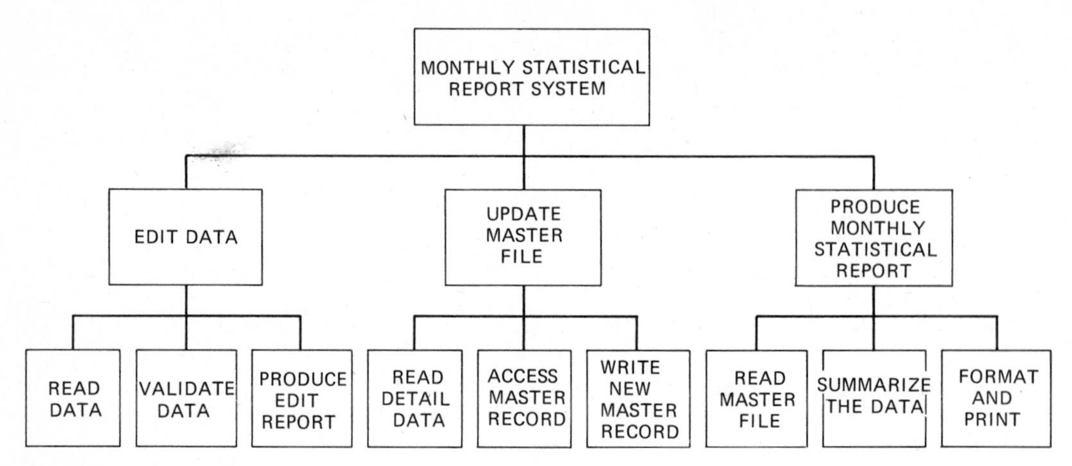

FIGURE 6 The process continues.

sis does not deal easily with one fundamental design question: How are the pieces (modules) of the hierarchy related to one another? This question is a principal concern of synthesis.

If analysis is involved with breaking a problem into its pieces, then *synthesis is involved with putting the pieces of a system together.* In fact, a term often used to mean synthesis is "integration," for synthesis is concerned with order and with time. If analysis is concerned with why and what, then synthesis is concerned with how and when.

In design we see two phases: breaking down and fitting together. In our example, the system was broken into a number of parts: one that edited the transactions, one that updated the master file, and one that produced the monthly report. But we can arrange those pieces derived from the analysis stage in a number of different ways. For example, both of the systems in Figure 7 have the same pieces. The experienced systems analyst will recognize that Figure 7b is a more realistic solution than is 7a. Although the major pieces are the same, the questions of when and how have been considered in more detail.

Also, an additional piece has been discovered that we missed in our first analysis (or in-

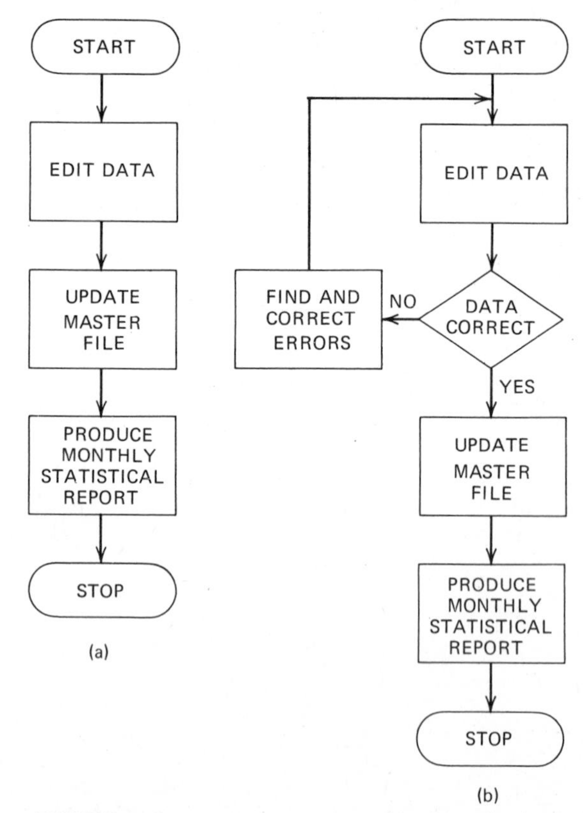

FIGURE 7 Comparing two systems *(a, b)* with identical parts.

cluded in our definition of EDIT DATA). In fitting the pieces together again, we often find that we have omitted some basic consideration or included something extraneous. When this occurs, we have to repeat the process of analysis. In our example, if we decide to include a piece to FIND AND CORRECT ERRORS (and it seems unlikely that we wouldn't), we will have to analyze that statement into its appropriate pieces (see Figure 8).

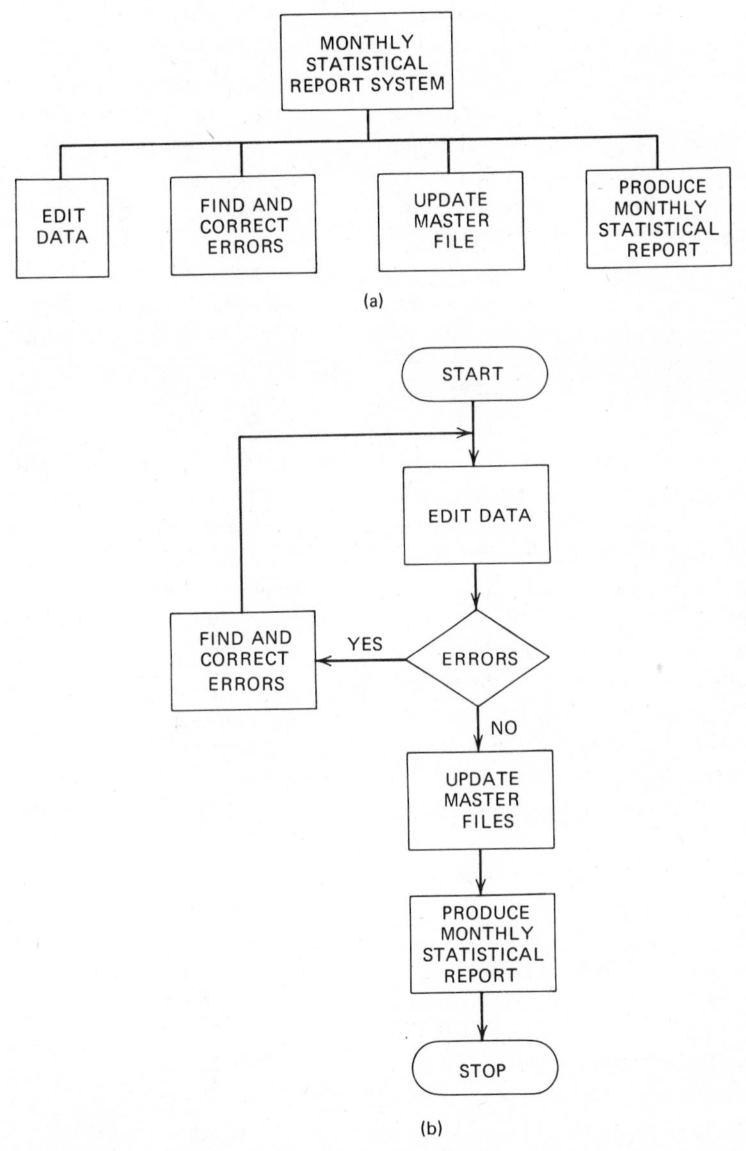

FIGURE 8 Structure chart with additional command *(a)* and analysis of that piece *(b)*.

Earlier, we mentioned briefly the difference between the design and the process of designing. Now, our words become clearer. On the one hand, after putting the pieces together again we have a finished design, or at least a better one than we had initially. Unless the reader of such a diagram has a copy of all the intermediate analysis steps and charts used to develop this design, he is apt to think that we came up with the final design the first time. Many times, beginning analysts are frustrated by the process of analyzing, integrating, analyzing, integrating, and so on. They feel they must be doing something wrong, because they look at other people's finished designs and assume that they didn't have to go through any periods of trial and error. These beginning analysts simply confuse the design with the process of designing. Making mistakes is no sin—failing to recognize and deal with them, however, can be fatal.

Recently, we have seen the process of analysis and integration (or synthesis) combined in a single approach called top-down development. As distinguished from top-down design, top-down development requires not only that your problem be analyzed into pieces, but also that those pieces be built and tested together as a structure of the solution before any of the pieces can be analyzed in great detail. By forcing continuous integration, top-down development seeks to avoid many of the more serious problems that have limited past use of hierarchical analysis.

Thus far in this section, we have tried purposely to avoid discussions of structuring the systems process or of using the model explained in the previous section. We wanted to concentrate on the ways we design systems today, for structured systems design has a great deal in common with traditional methods. The difference is largely in terms of emphasis and focus. Therefore, before introducing any formal structuring, I wanted the reader to understand the traditional concepts of analysis and synthesis as applied to systems building.

Traditionally, systems analysis and synthesis have been used in conjunction with yet another technique called "process flow." In looking at things from the process flow standpoint, we usually attempt to find out what happens when. To analyze a problem, we use current processing steps as a clue to tell us to break our problem, we use current processing steps as a clue to tell us to break our problem into pieces. In synthesis or integration, we visualize what the process steps will have to be to make the whole thing work.

Programmers have used the process flow approach as the primary means of designing their programs. In fact, the systems or program flowchart is still the most common documentation tool in use anywhere. Figure 9 is perhaps the most common programming model expressed in terms of a simple flowchart.

Process flow is a natural and compelling design approach because you start at the beginning and proceed to the end. Unfortunately, it also has many shortcomings. One thing seems to lead to another, but proceeding from a general process flow diagram to a more detailed one may lead to design problems. For example, in the simple program given above, there is an infinite loop, or

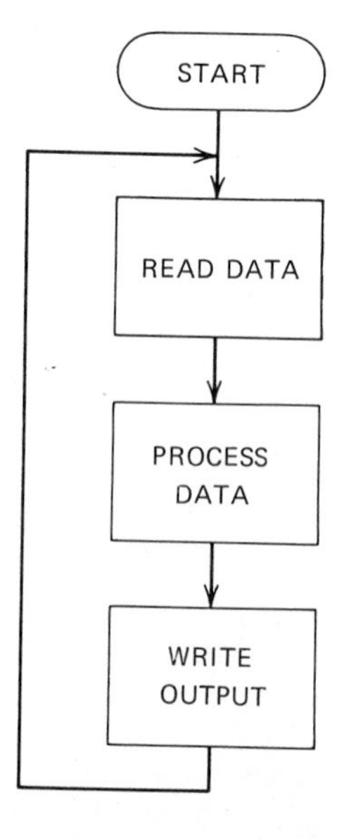

FIGURE 9 Basic programming model.

error, since the program fails to provide a means for stopping: i.e., it appears simply to ignore the signal that it is out of data. Making this model workable then requires the modification shown in Figure 10.

It is unavoidably characteristic of process flow designs that as new logical conditions crop up (call them exceptions), the basic overall process flow becomes harder to recognize. As we learn more about program development, the less attractive we find process flow as a method of designing either a program or a system. Because process flow designs habitually become extremely complex, and since designing with process flow is fundamentally a trial-and-error method, we usually end up with far more error than is acceptable.

But no matter what our feelings are about process flow designs, any finished system must assure that the proper processes happen at the right times within a system. Fortunately, hierarchical structuring makes this possible in ways never suspected a few years ago. At the same time, it eliminates much of the confusion and complexity of earlier process flow systems designs.

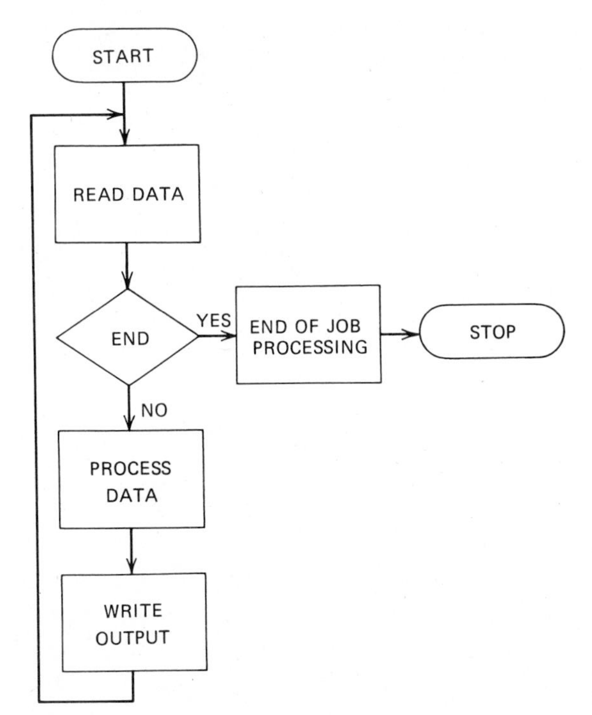

FIGURE 10 Corrected basic programming model.

We've now seen a few of the principal tools of traditional systems design: systems analysis, systems synthesis (or integration), and process flow. Let's put these elements together in a new way I referred to earlier as structured systems design. This method allows us to combine the tools of analysis, integration, and process flow in a logical fashion and, moreover, to take advantage of our basic model of a system.

STRUCTURED TOOLS: WARNIER-ORR DIAGRAMS

Some methods for doing systems work are more productive than others, and each method has advantages and disadvantages. We want tools that both maximize the advantages and minimize the disadvantages. As stated before, structured systems design techniques seem to meet those requirements better than anything else we've seen.

An important structured design tool is the *Warnier Diagram*. Warnier Diagrams are named for Jean-Dominique Warnier (pronounced Warn-yeh) who was the first person, to my knowledge, to systematically apply hierarchical logical methodology to building systems. A Warnier Diagram is fundamentally a series of brackets used with a small number of other symbols to portray a problem.

The best way to appreciate the power of Warnier Diagrams is to use them. Suppose we take the earlier problem and show how to employ these Warnier Diagrams to express the same thoughts, (see Figures 11, 12, and 13).

A Warnier chart can be regarded simply as an organizational chart (or tree diagram) laid on its side. However, we can do things with Warnier Diagrams that are not possible with other hierar-

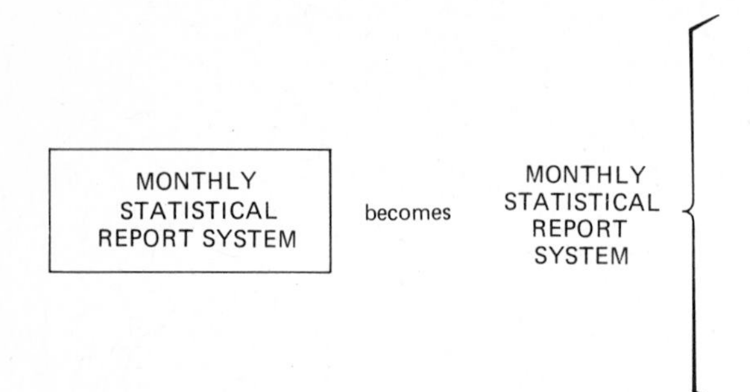

FIGURE 11 Example of a Warnier Diagram.

chical forms of organization—for example, dealing with logic, repetition, and sequencing.

Remember that in the first analysis of our simple reporting system, we had some problems; and, during the synthesis step, we had to add another piece (module) and some logic to the system. Warnier charts can be used to show this easily (see Figure 14).

You will recall that the hierarchical organization chart we used failed to show the emerging relationship between the various modules (see Figure 15). The Warnier Diagram at left in Figure 14 on the other hand, was modified to contain all aspects of both process flow and hierarchical analysis. How did we do that? The symbols in the parentheses below the names of each piece in the diagram at the right represent the number of times that each piece occurs. Figure 16 illustrates this.

In Warnier Diagrams *the sequence of activity is presumed to be left to right and top to bottom*. So, a Warnier Diagram always is concerned with order and sequence, whereas a typical hierarchical organization chart is not. The following description is roughly what the Warnier chart above expresses:

> *The system is defined as an edit cycle that is performed some variable number of times (until no more errors) followed by a set of actions to update the master file, followed by a set of actions to produce the monthly statistical report. The edit cycle is defined as a set of actions to edit the input or a mutually exclusive set of actions if there are no errors in the input. If there are errors in the input, then a set of actions to find and correct the errors is executed.*

Now, you will have noticed that we have again returned to talking about sets of actions. Indeed, in a Warnier Diagram, the names of the pieces describe a set of actions, and the bracket to the right of the name explains in detail how that piece (module) is accomplished. Consider the example below (Figure 17).

In this case, the pieces READ PHYSICAL RECORD and MOVE "TRUE" TO END-DATA INDICATOR probably can be executed directly

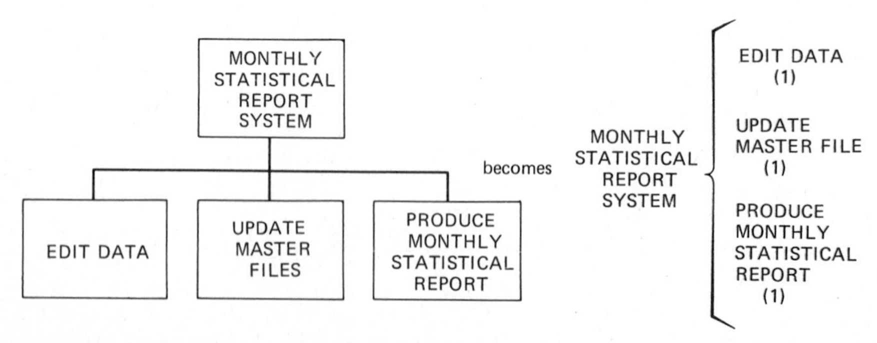

FIGURE 12 Warnier Diagram as a structured tool.

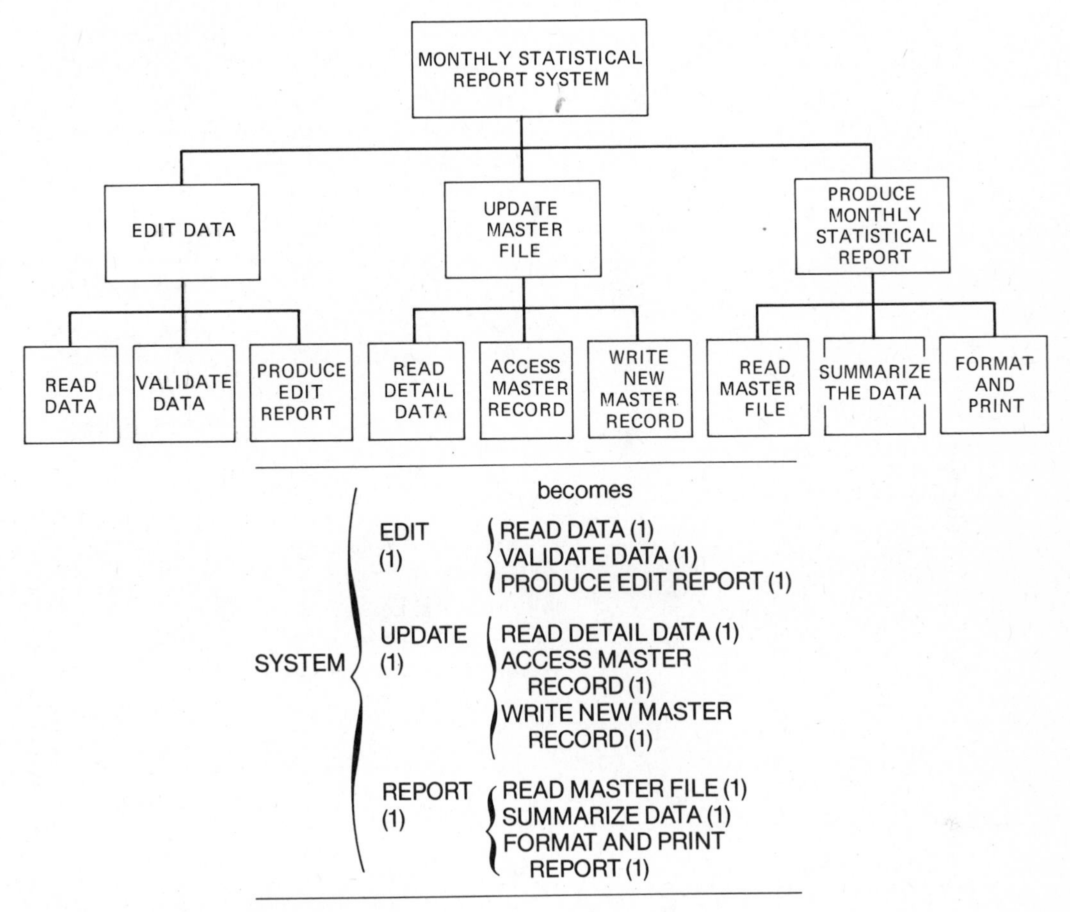

FIGURE 13 Using a Warnier Diagram.

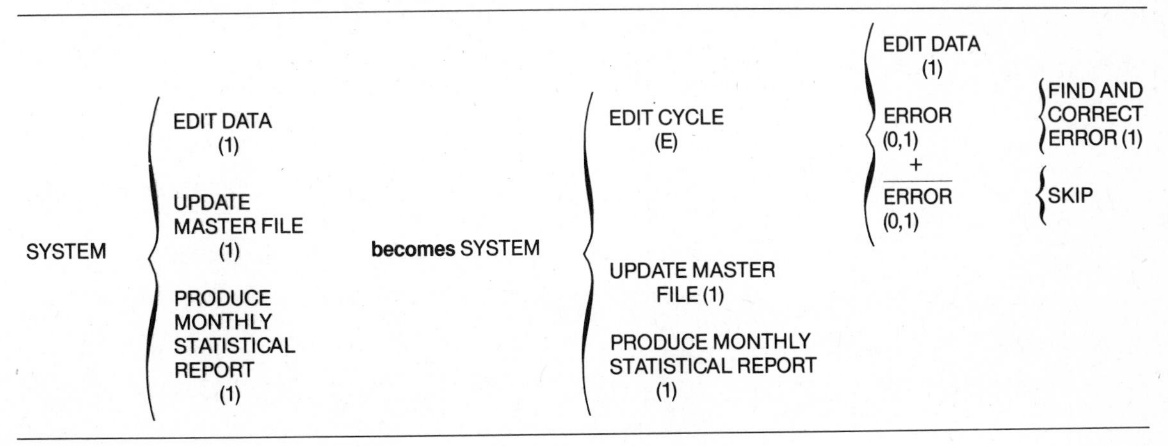

FIGURE 14 Warnier charts showing changes to the system.

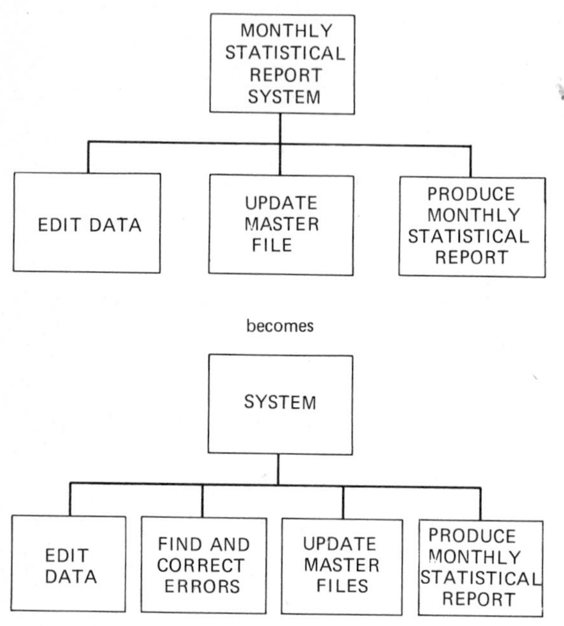

FIGURE 15 Hierarchical chart with additional module.

and, therefore, do not require any additional definition.

Structured systems design ultimately involves developing a hierarchical Warnier Diagram from the top for the entire system. Unfortunately, the top of a Warnier Diagram is on the left side, and top-down implementation means left to right (a small problem in mental gymnastics that you will get used to).

When you've finished a structured systems design, you will have broken your problem into sequences of things to do, hierarchically organized, using only alternation, repetition, and sequence as a means of structuring. This seems simple. On the other hand, it is not all that intuitively obvious, but, of course, nothing worth having is ever that simple. Otherwise, people would have discovered it a long time ago. Warnier Diagrams not only allow us to break down problems hierarchically, logically, and in a readily understandable graphic method, but they also present us with some new, unexpected approaches. Let's see what those are.

When we initially employed analysis to break our system into pieces, we did so more or less arbitrarily. We chose three pieces—EDIT, UPDATE, and REPORT—because they seemed natural. At the next level, we broke each of the major pieces into another set of three pieces, characterized as READ, PROCESS, and WRITE.

UPDATE MASTER FILE (1)	**means**	The module Update Master File is executed 1 time at this point.
EDIT CYCLE (E)	**means**	The module Edit Cycle is executed E times at this point, where E is a variable.
ERROR (0,1)	**means**	The module Error is executed 0 or 1 times depending upon whether there was an error.
$\overline{\text{ERROR}}$ (0,1)	**means**	The module Not Error is executed 0 or 1 times depending upon whether there was an error.
{ ERROR (0,1) + $\overline{\text{ERROR}}$ (0,1)	**means**	The modules Error and Nor Error are mutually exclusive, and only one of them will be executed.

FIGURE 16 Explanation of process flow and hierarchical analysis in a Warnier Diagram.

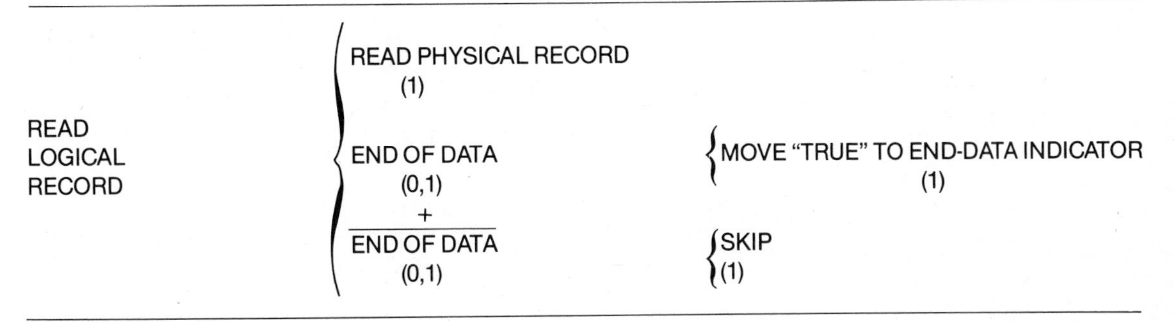

FIGURE 17 Example of a Warnier Diagram.

These forms of program structures are characteristic of most systems. But the very obviousness of these modules hides the fact that they are not necessarily the only or even the most logical means of constructing a program or a system.

Let's return to our model and see if it can help us. Discussing the model, we said that correct systems design was a process of working backward from the output to the input. Moreover, we said that correct systems design was involved in getting the right sets of actions performed on the right sets of data at the right times. How can we apply the model here? Suppose we put the information gained about our system into the model itself, as shown in Figure 18.

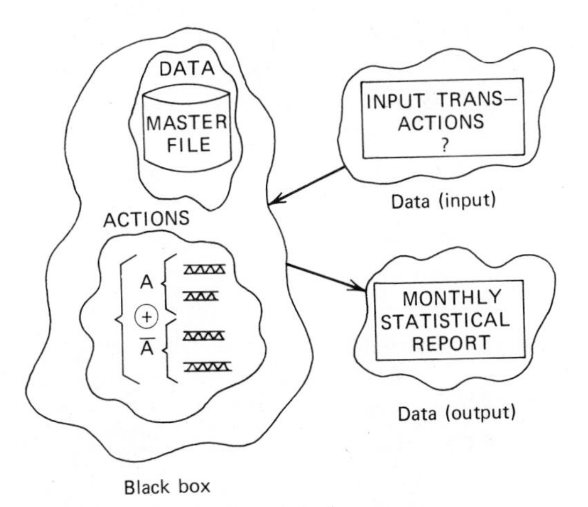

FIGURE 18 Revised model of our system.

We can see in the figure that there is a place for all of the information accumulated about the system, but we would like to be more specific. We need a better means of showing this information while preserving our systems model, say, with a modified type of Warnier chart (these modifications, by the way, are not Warnier's but my own).

The modifications shown in Figure 19 seem to work pretty well. In fact, if we only wanted to talk about systems requirements, leaving out consideration of systems architecture for the moment, we could do so very simply, just by eliminating everything within the system except for the major inputs and outputs as shown in Figure 20.

So far, so good. Our model seems to be working out as a means of describing all the various aspects of our problem, simple as well as complex. Let's see now if it will help us to understand even more about the problem.

As stated before, a structured systems designer focuses initially on output and works backward from there. However, the method of systems analysis discussed in the previous section didn't do that, did it? As a matter of fact, experience suggests that traditional techniques don't incorporate output-oriented thinking unless you concentrate on doing so. Since we have only one output in our example system, it is relatively simple to see what part of our system should be examined in detail—the monthly statistical report.

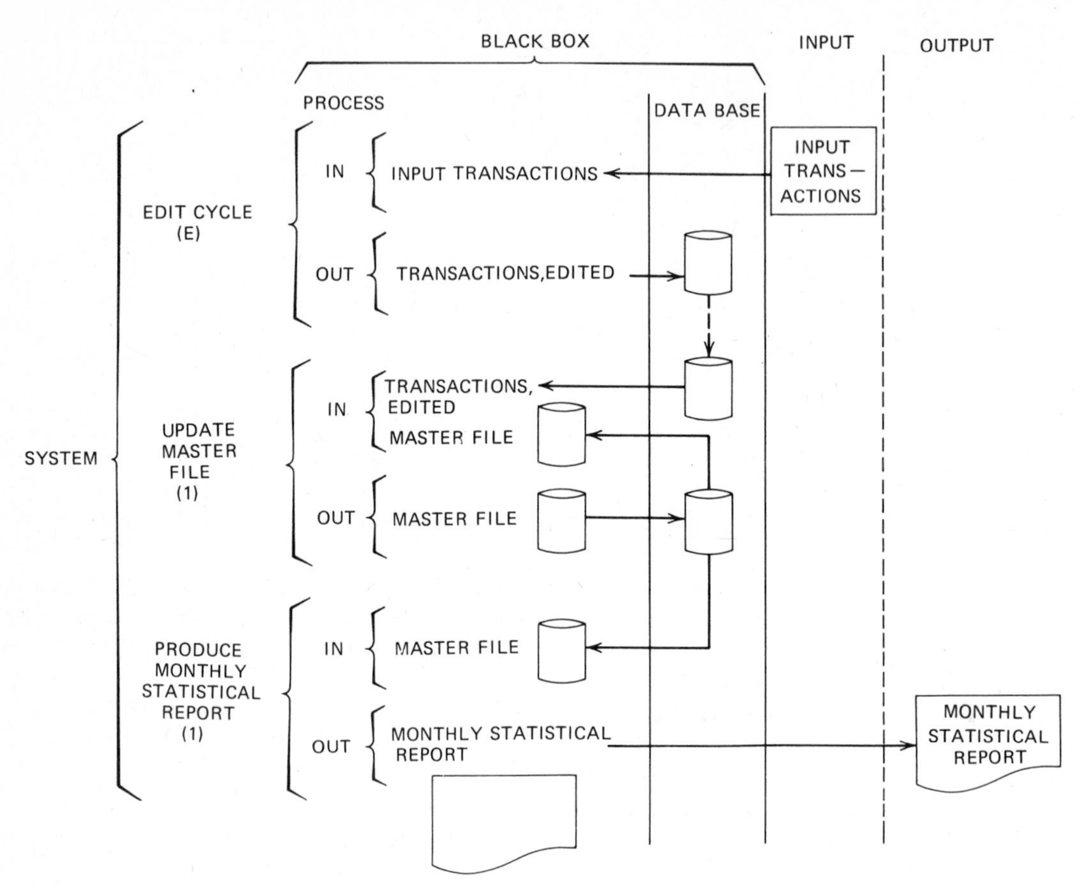

FIGURE 19 Modified Warnier/Orr chart.

But what questions do we ask? The traditional method of discussing output has been for the analyst to ask the user for his rough information needs (and his intended uses) and then to develop a detailed layout showing all the appropriate fields. Although eventually we also will want more of the same information, in a structured systems design we are interested initially in other things, particularly in knowing the logical structure of the report.

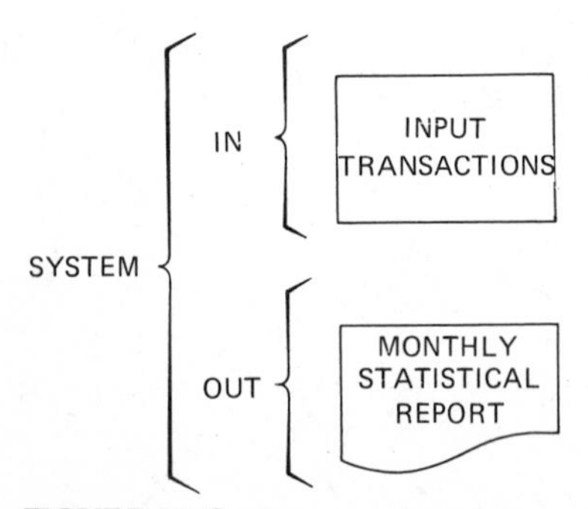

FIGURE 20 Systems requirements model.

What do we mean by that? The logical structure is more or less the hierarchy, or sequence, of the report. When the user says, "I'd like to see customer sales data by salesman, by district, and by region," he's talking about the data structure of the report. If our monthly statistical report was designed precisely in the sequence, that would be the logical structure as shown in Figure 21.

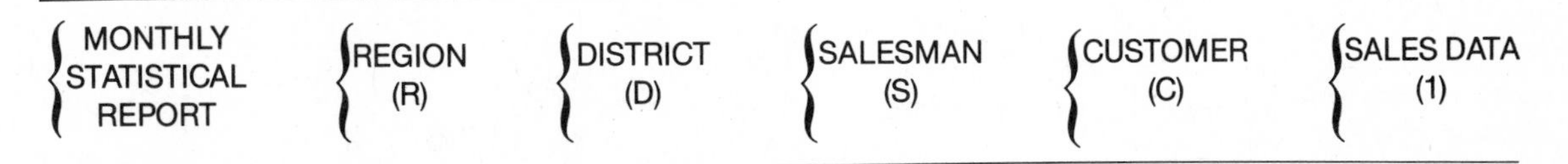

FIGURE 21 Report's logical structure.

Historically, the sequence of a report always has been considered significant in program design. Since the earliest days of data processing, input files to produce reports have been sorted into the same hierarchical sequences as the report to save space (and time) on machines, especially those with limited capacity. Before computers, we had accounting machines, and before that we had tabulating machines. These machines in conjunction with sorting machines solved complex problems using only a few primitive operations.

The hierarchical structure of the sorted file made all this possible. Now, that same idea of hierarchically structured files is a natural means of organization using structured design. Having defined the structure of our report, we can be more specific about the monthly statistical report and its layout.

Following closely, you'll notice that we slipped in a number of new requirements between the original discussion and the final specification for the report. Where did this new information originate? As we talked to our hypothetical user, he thought of more and more things he would like to see in the report, a typical reaction. As he did so, however, the problem changed somewhat. He added a new field to the output, labeled PROFIT; and he introduced a COMPANY GRAND TOTAL to the report. These are natural expected modifications. In fact, don't be surprised if the user makes many additional changes before completion.

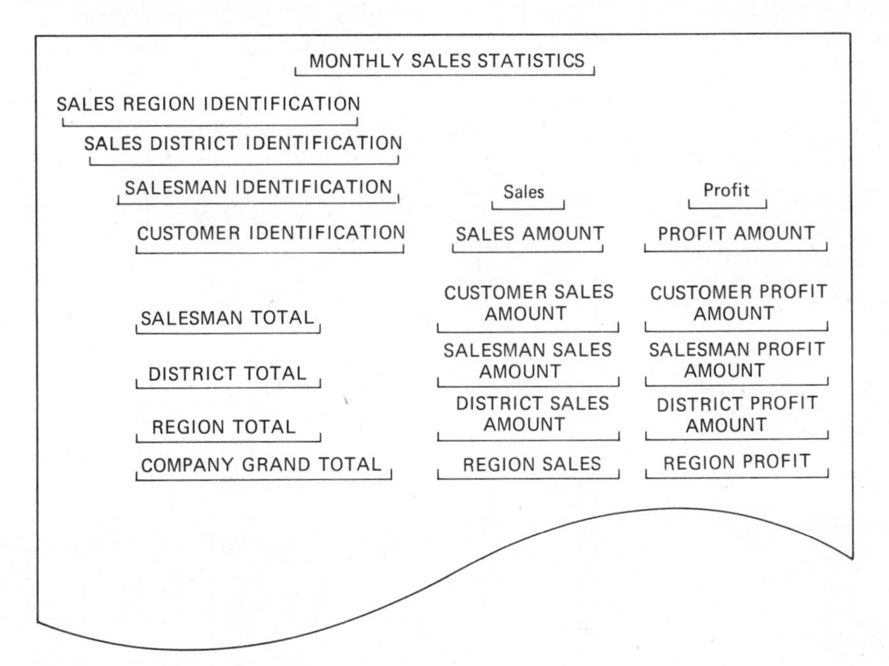

FIGURE 22 Report layout.

PRODUCE MONTHLY COMPANY SALES STATISTICS REPORT (1) { REGION (R) { DISTRICT (D) { SALESMAN (S) { CUSTOMER (C) { SALES (1) { PROFIT (1)

FIGURE 23 Revised Warnier Diagram.

With the revised specifications, let's redo our Warnier Diagrams to reflect these changes (Figure 23).

Remember that each bracket represents a set of actions and each level represents a set of actions within a set of actions. From our example, we can say:

For the company do the following:
For each region within the company do the following:
For each district within the region do the following:
For each salesman within the district do the following:
For each customer of the salesman do the following:
Print customer sales and profit.

Roughly, we would be correct, but we need to be more specific than that with structured systems design. We'll come back to this report later. First, let's fit this new information into our old diagram (Figure 24).

We didn't have any trouble working the output structure into our Warnier Diagram. But it is certainly different from the original model. Our new structure had a lot more resemblance to the problem than the original one, and that should prove helpful if the user later changes his mind regarding the desired output.

Returning to our original discussions about correctness, we said that correct systems design involves getting the right sets of actions done on the right sets of data at the right times. We've been largely concerned with the right sets of output data and with the right actions. Suppose we

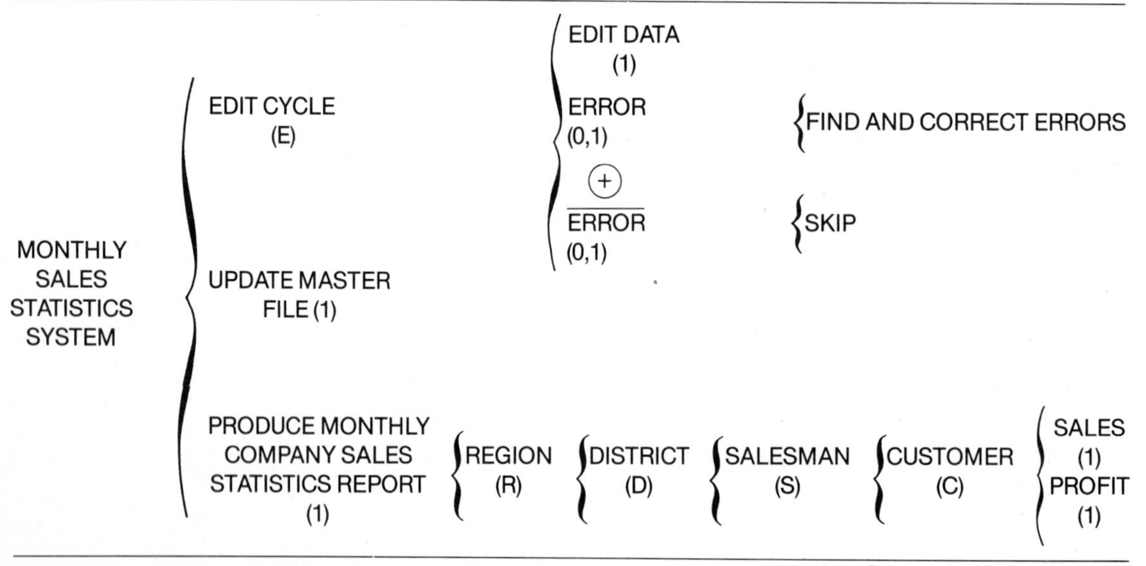

FIGURE 24 Warnier Diagram with additional output structure.

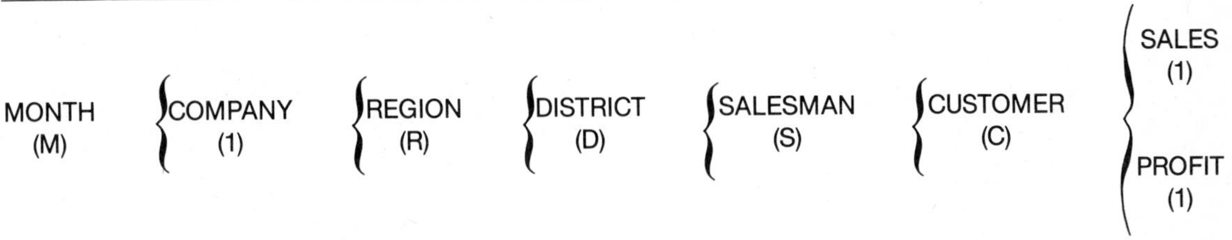

MONTH (M) { COMPANY (1) { REGION (R) { DISTRICT (D) { SALESMAN (S) { CUSTOMER (C) { SALES (1) / PROFIT (1)

FIGURE 25 Warnier Diagram of data structure.

draw a Warnier Diagram of the structure of the data needed to produce a correct report (Figure 25).

Where did the MONTH level come from? We picked it up from the title (often things like the frequency or period of a report are omitted from the definition of requirements). But we don't have any monthly considerations in the Warnier Diagram of our system! Let's see if we can insert them in the diagram in Figure 26.

Our revised diagram now means we're going to do certain things over again every month. As a matter of fact, that change has made our system into a continuous operation rather than a one-time shot, something the users always wanted even if they couldn't say so. Many times, we must add a new level to the systems diagram after integrating some new information into the systems architecture. Adding this new level often solves many problems and the technique has been so common in our work that we have given it a name: "finding the hidden hierarchy."

In the last two sections, we've seen only the tip of the structured systems design iceberg. Using a number of traditional techniques for analyzing systems in traditional ways, we've come

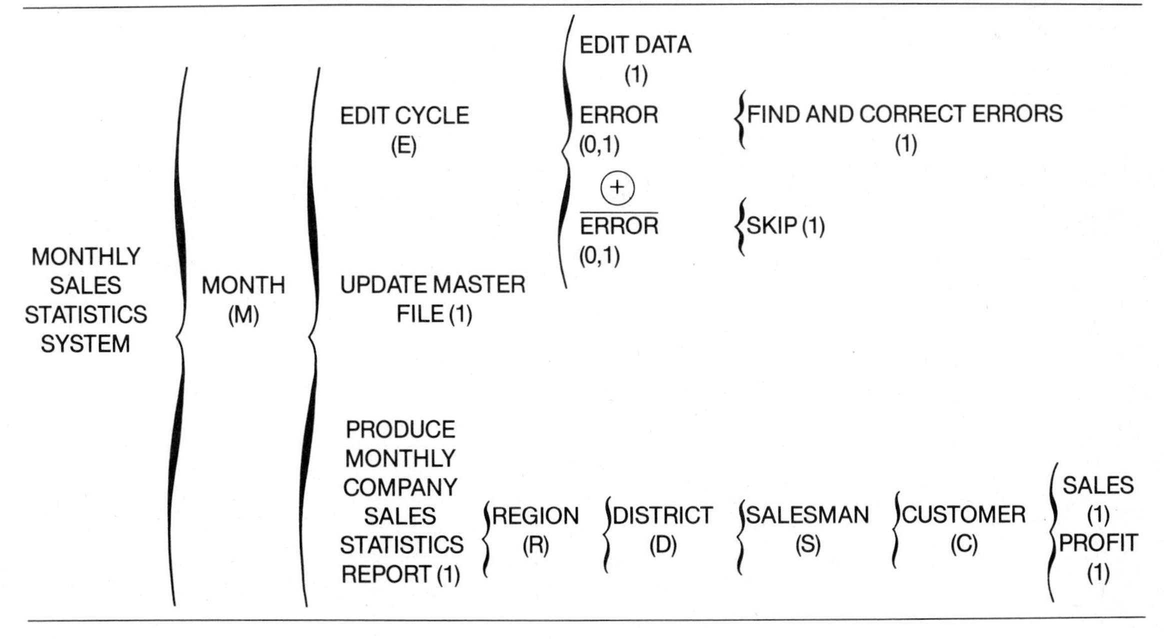

FIGURE 26 Warnier Diagram showing continuous system.

up with traditional systems designs. We've also begun to use some structured techniques such that our example system, after its usual start, now has begun to look increasingly structured. However, at this point, we shouldn't be at all surprised that our approach contains elements of both traditional and structured approaches. Many structured techniques have grown out of traditional ones, and traditional techniques are easy to grasp, accounting in part for their widespread use.

By introducing Warnier Diagrams and by applying some rules drawn from our model, we have been able to show both the hierarchical and the process flow (including logic) of our system. As we use these charts more in our work, we find that they are of tremendous value. Warnier charts are rapidly becoming the flowchart of the future, for they show a complete problem solution in an easy-to-grasp graphic form.

SETS OF ACTIONS

The Warnier Diagrams we've seen have been used primarily to describe sets of actions. By subdividing each piece into simpler pieces, and by providing a simple means for dealing with order and logic, the Warnier Diagram provides a major benefit over other structured documentation tools, such as HIPO charts or pseudocode. Other uses of a Warnier Diagram follow:

> *Problem—We wish to obtain statistics from a file of employees of a university. If a staff member is part of the teaching faculty, then count him as such. If the staff member is part of the nonteaching faculty, count him as such. If he is not on the faculty, count him simply as nonfaculty.*

In this case, our Warnier Diagram helps to express a systems problem simply, completely, and logically. It is, then, a relatively small matter to turn this problem statement into a problem solution. In fact, from one viewpoint, *in structured systems design the proper definition of the problem is the solution.*

Warnier Diagrams allow us to concentrate on the logical requirements of a problem instead of on the physical ones. This is particularly beneficial in dealing with users. Indeed, because users are normally put off by jargon and technical

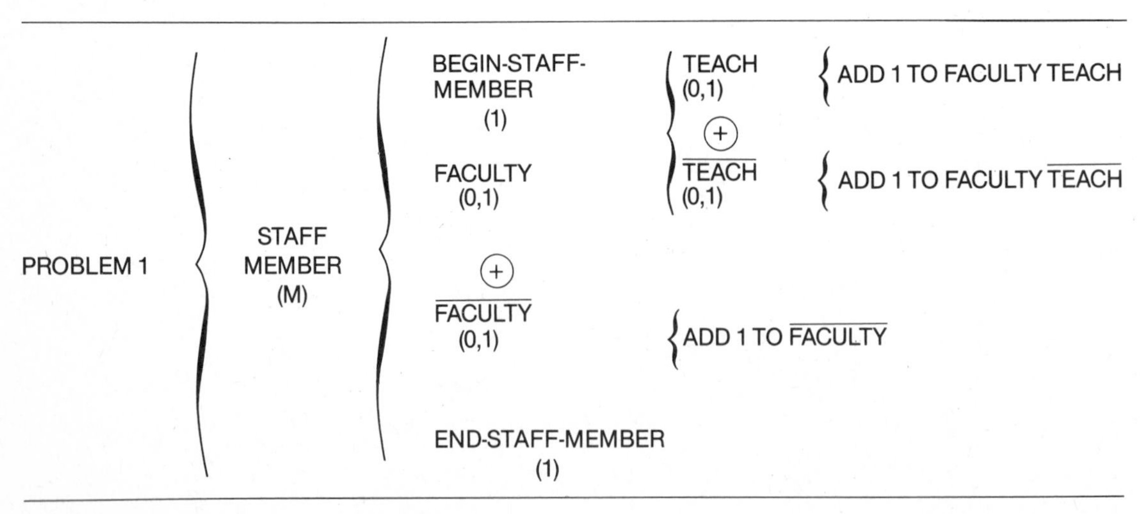

FIGURE 27 Warnier Diagram of problem statement.

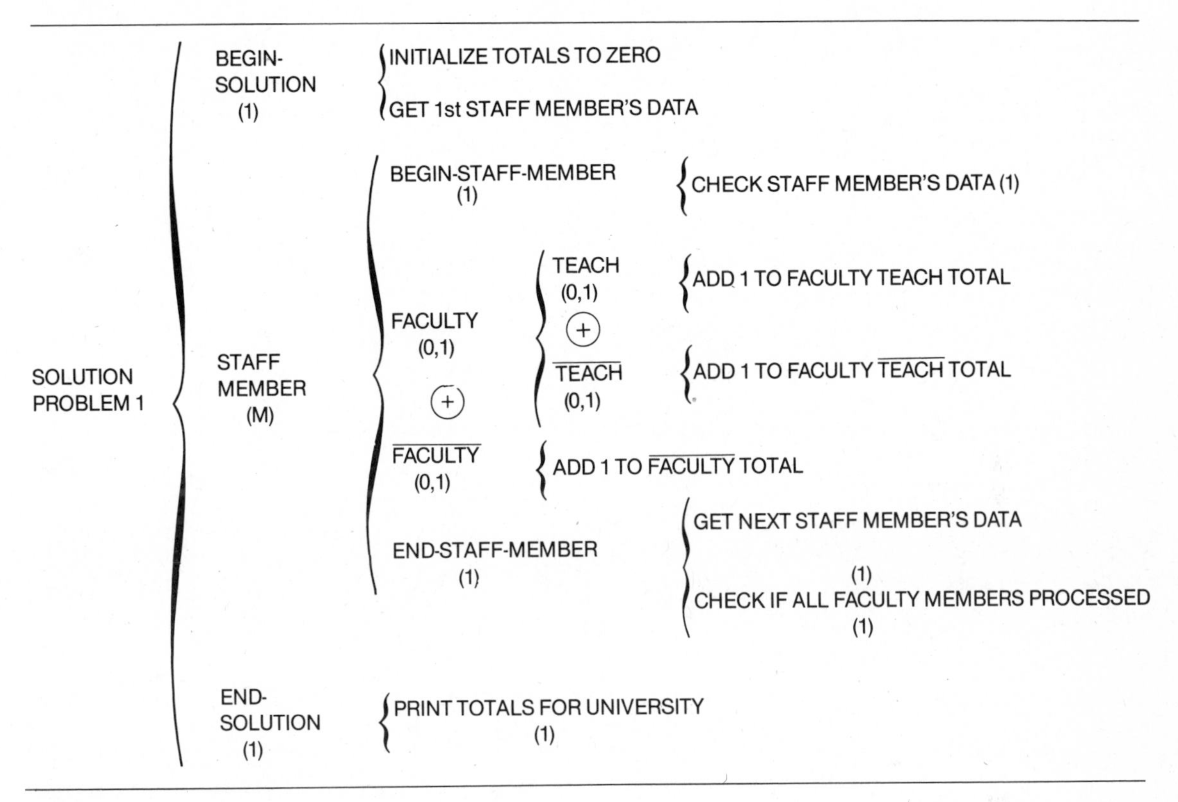

FIGURE 28 Warnier Diagram of problem solution.

terms, the ability of stating problems in a fashion understandable to a user helps significantly to overcome many of the serious communications problems that always have kept us away from the problem. Structured systems design restricts us to the use of *operations* (or sets of actions), *sequence of operations, repetition of operations*, and *alternation* (or selection) between operations. Only a few years ago, most of us would have found this list of possible operations impossibly restrictive. Today, however, based on theoretical and practical experiences with structured programming and structured systems design, we know that these forms of representation are enough. Moreover, we now know that there is a fundamental relationship between these forms and logic, the foundation of all mathematics.

SETS OF DATA

Warnier Diagrams are useful for describing not only logical sets of operations but logical sets of data as well. Suppose we had to describe the following data set:

> *A customer master file is composed of records. For each customer there is one customer header record containing customer number, record type, and customer name. This is followed by a variable number (possibly zero) of customer orders, which are followed by a variable number of customer accounts receivable transactions (possibly zero).*

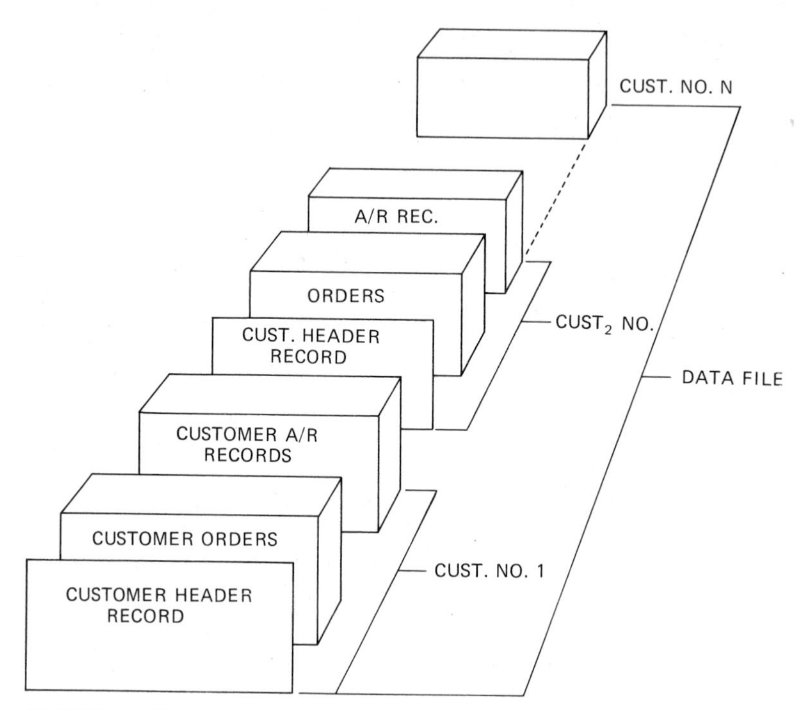

FIGURE 29 Data structure of customer master file.

In the past, we would have shown this file schematically as shown in Figure 29.

For the most part, when thinking of data, we have thought of files as being made up of records, and records made up of fields. In fact, our experience with structured systems data base designs indicates that we're dealing with sets and subsets of data. The Warnier Diagram of the above data structure is shown in Figure 30.

The use of Warnier Diagrams to describe, graphically, complex data structures is of particular value to a systems designer. As our problems have become more complex, their data structures often have grown more complex with them. This tends to lead to serious confusion. We need new methods of documenting exactly how data files are structured. Data base diagrams, pioneered by Bachman[1] and others, have helped to show that logical hierarchical organization chart methods are not as explicit as Warnier Diagrams when describing all relevant logical facts. Take the following example:

> *Problem—Describe a file that is made up of two types of customers: individuals and businesses. For individuals, only basic information and records of purchases are kept. For businesses, in addition to information about the firm, data is kept with respect to purchases by each of its locations. Customers are either individuals or businesses.*

The data base diagram for this problem only approximates the problem statement (see Figure 31).

By contrast, the Warnier Diagram describing the same file (see Figure 32) follows the problem exactly. As we will see later, this exactness

[1]C. W. Bachman, "Data Structure Diagrams," *Data Base*, Vol. 1, No. 2 (Summer 1969).

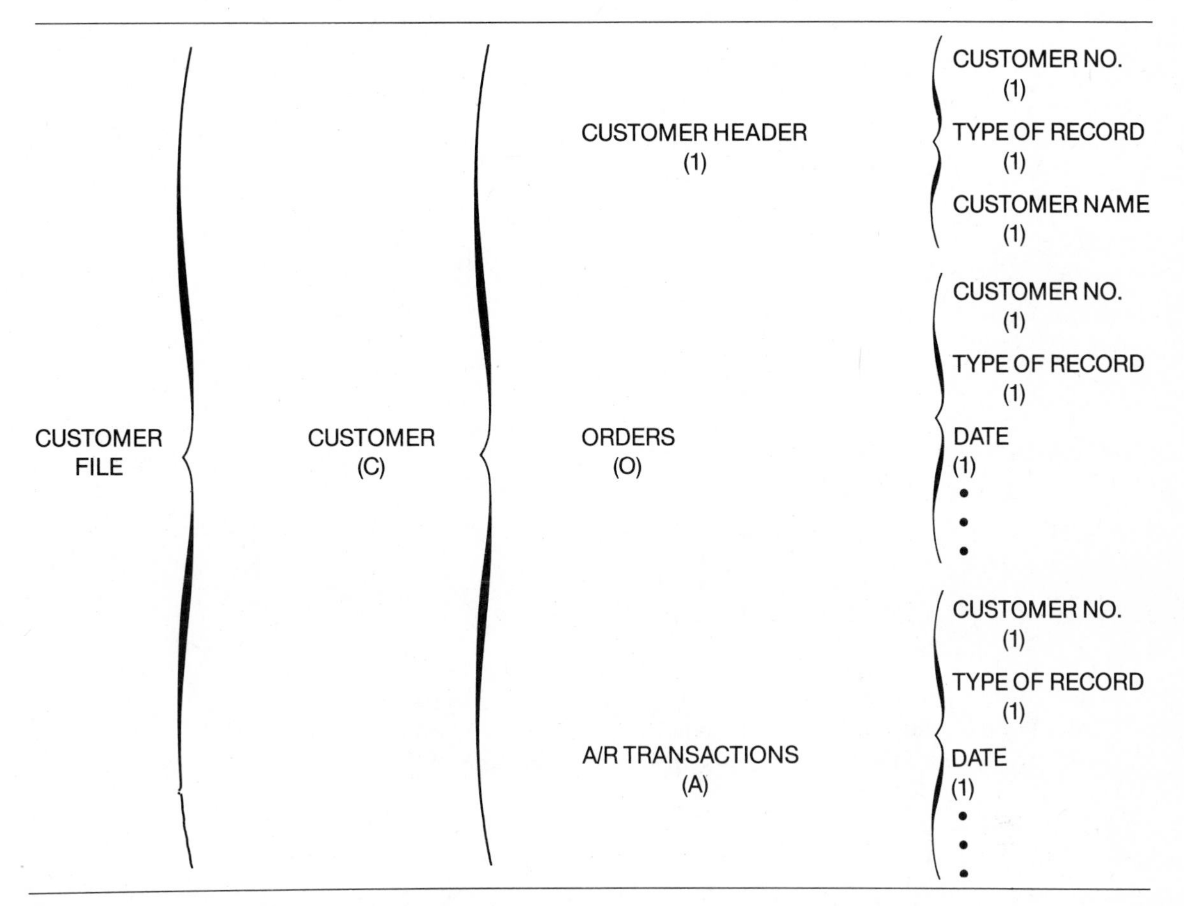

FIGURE 30 Warnier Diagram of data structure.

makes it possible to develop data structured programs to process structured data files.

The use of Warnier Diagrams to define data structures makes many things possible, not the least of which is improved maintenance of systems. In the past, people often have defined very complex data structures with the (expressed) aim of improving the efficiency of some aspect of their processing. But more often than not, as the people on the original design team moved to other presumably better jobs, and as new people took their places, maintenance problems arose due to misunderstandings about the data structure. As a consequence, old programs were difficult to maintain and new programs difficult to debug. Invariably, since new people assigned to those systems took such a long time to understand how the system worked, a small number of unfortunates were eventually stuck with maintenance of the system—that is, until they quit.

Clear, understandable documentation is essential in any system. The larger and/or more complex the system is, the greater is the need for exact documentation. Whenever we allow ambiguity to slip into a systems definition, we open a veritable Pandora's box of evils.

In general, when describing data structures, especially complex ones, remember that a *Warnier Diagram always describes a logical data access path for sets and subsets of data.* Thinking

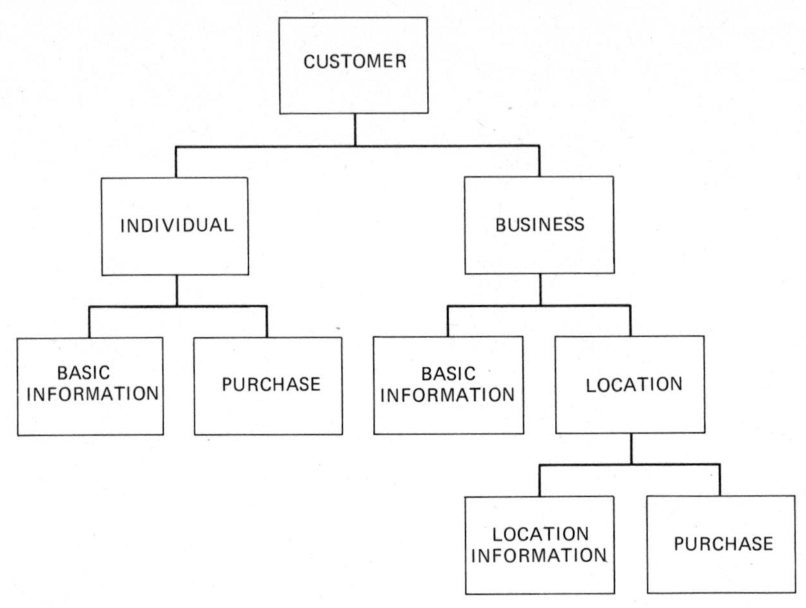

FIGURE 31 Data base diagram.

of data in terms of access paths at first may be difficult. Indeed, until the advent of data base management systems, we were so accustomed to thinking of data in physical terms that most of us still have great difficulty in breaking away from that mode of thinking.

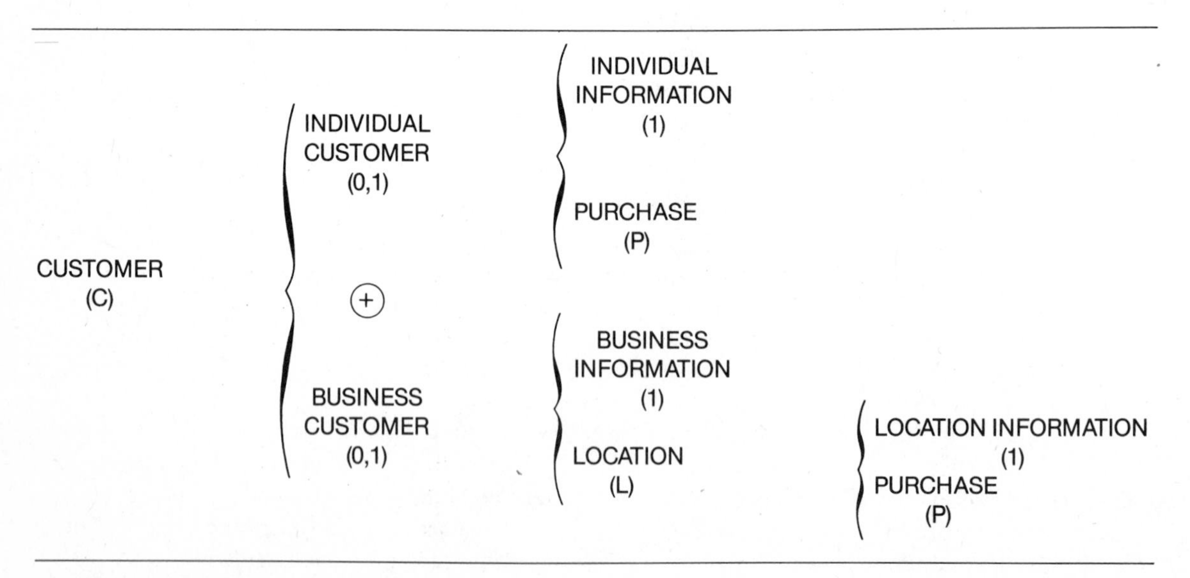

FIGURE 32 Warnier Diagram of same problem.

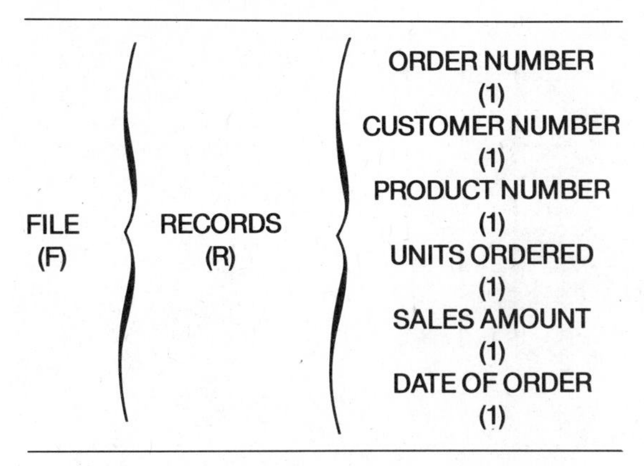

FIGURE 33 Warnier Diagram of file and records.

An example that illustrates our inability to see clearly the differences between logical and physical data structure is our treatment of sequential files. Such files, we have been taught, are inadequate, rigid, and out-of-date. On the other hand, newer file structures such as hierarchical data bases, are capable of wonderful and miraculous things. In fact, while both statements are true to some extent, they both are misleading. Why? Because most of the files that we have traditionally called sequential are, in fact, hierarchically sequential files, i.e., sorted. Moreover, many of the advantages that are ascribed to hierarchical data base structures also apply, if you think about it, to hierarchically sequential files. For example:

> *A data file is made up of detailed records of line item purchases, sorted by product number, customer number, and order number. An order number, customer number, product number, units ordered, sales amount, and date are kept for each line item.*

In the traditional view, a file is a file is a file; and the only important thing to consider is that a file is made up of records, and records are made up of fields. We could diagram it as such (see Figure 33).

However, this would not be a complete logical description of the file in question, because we have failed to treat the hierarchical structure sequence of data access. If we thought of the file only in terms of the previous description, we would miss the fact that the data on this file have been ordered into sets, subsets, and sub-subsets of data (Figure 34).

Now our ugly duckling sequential file suddenly has become a beautiful swan-like hierarchical structure. We could even draw a data base diagram just to be in step with the times (Figure 35).

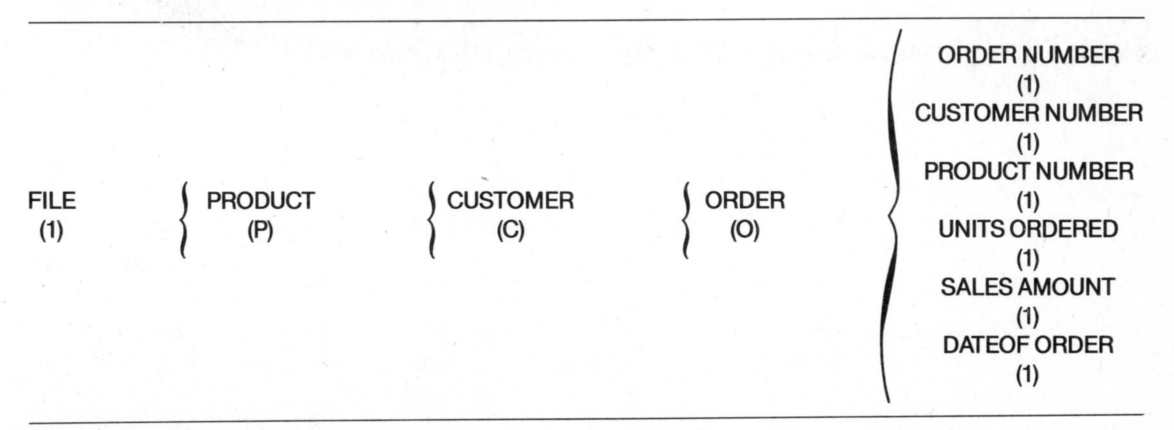

FIGURE 34 More complete diagram of data.

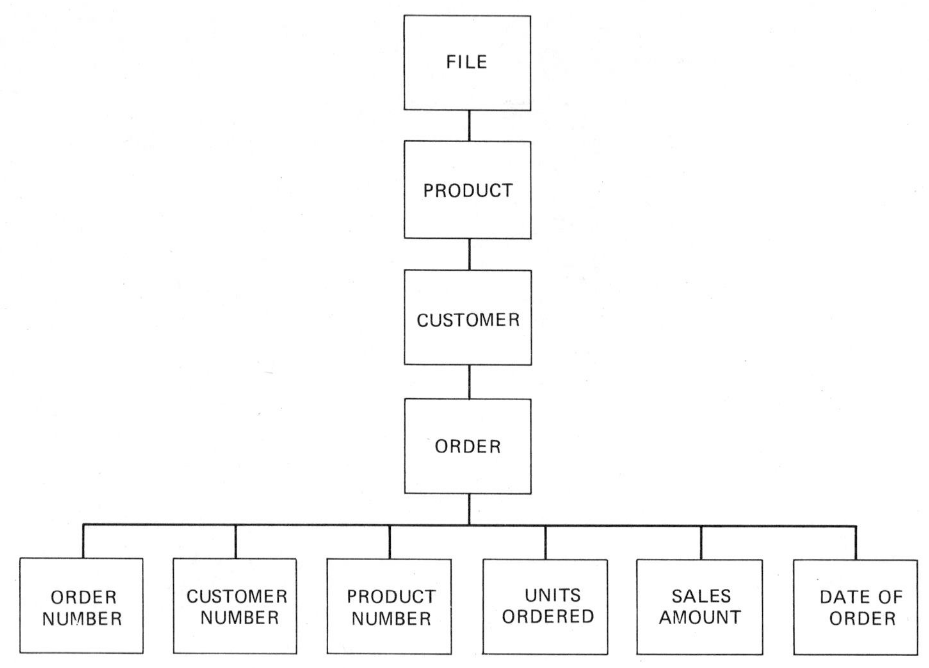

FIGURE 35 Data base diagram.

We could actually reorient our data according to use and availability, as shown in Figure 36.

It turns out that it is not only possible to describe complex data structures in a better fashion using Warnier Diagrams, but we can also learn more about data structures once considered self-evident.

BIBLIOGRAPHY

Bachman, C. W. "Data Structure Diagrams," *Data Base*, Vol. 1, No. 2 (Summer 1969).

Bertini, M. T., and Y. Tallineau. *Le COBOL Structure: Un Modéle de Programmation*. Paris: Editions d'Informatique, 1974.

A description of coding Logical Construction of Programs (L.C.P.)-designed solutions in a structured COBOL.

Caine, S., and E. K. Gordon "P.D.L.—A tool for Software Design," *AFIPS Proceedings of 1975 National Computer Conference*. Vol. 44, pp. 271-276. Montvale, N.J. AFIPS Press, 1975.

A description of the use of a program design language (P.D.L.) to aid in the systematic development of structured programs.

Jackson, M. A. *Principles of Program Design*. New York: Academic Press, 1975.

A discussion of a programming methodology very similar in many respects to Warnier's L.C.P.

Kernighan, B. W., and P. J. Plauger. *Elements of Programming Style*. New York: McGraw-Hill, 1974.

Neely, P. M. *Fundamentals of Programming*. Lawrence, Kan.: University of Kansas Computation Center, 1973.

This was one of the first books to use Warnier's methodology in conjunction with structured programming.

_______, and K. T. Orr. "A Home-Handyman's Guide to Structured Programming in COBOL"

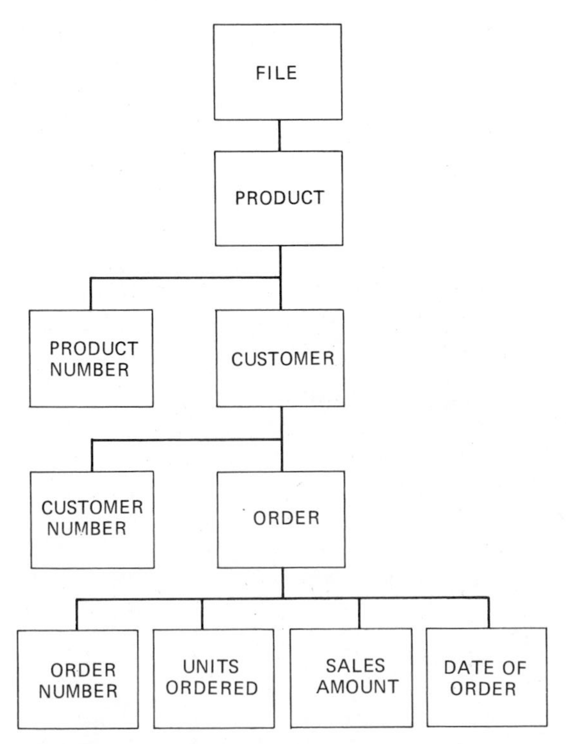

FIGURE 36 Reorientation of data base diagram.

Paper delivered for Langston, Kitch & Associates, Inc., Topeka, 1975.

Orr, K. T. "Structured Systems Design." Student text for Ken Orr and Associates, Inc., Topeka, 1975.

This text shows the development of structured systems technology. Included are a variety of examples of the use of Warnier Diagrams, HIPO/DB charts, structured narratives, and structured COBOL.

Taylor, B., and S. C. Lloyd. "DUCHESS—A High Level Information System" *AFIPS Proceedings of 1974 National Computer Conference*. Vol. 43, pp. 35-41. Montvale, N.J.: AFIPS Press, 1974.

One of the first articles on the development of a structured data base management system. The paper describes the methodology of structured organization and utilization of data.

_______. "Implementation of the DUCHESS Data Base Structure." Paper delivered at Duke University, Durham, N.C., 1975.

A further description of the use of structured data base methodology employed in the DUCHESS system.

Warnier, J. D. *L'Organization des Données d'Un Système*. Paris: Les Editions d'Organization, 1974.

This book on the application of L.C.P. to the construction of a systems data base is a start toward the building of a systems science.

_______. *Logical Construction of Programs*. 3rd ed., trans. B. M. Flanagan. New York: Van Nostrand Reinhold Co., 1976.

The only volume of Warnier's work in English. This book expounds Warnier's L.C.P. methodology in a very careful manner. While the book does not adopt structured programming directly, the programs developed are clearly well structured.

_______. *La Transformation des Programmes*. Paris: Les Editions d'Organization, 1975.

In this book, Warnier develops a method of modifying and maintaining well-structured L.C.P. programs.

_______, and B. Flanagan. *Entrainment á la Programmation*. Vols. I-II. Paris: Les Editions d'Organization, 1972.

These two volumes represent a statement of Warnier's L.C.P. methodology. This methodology, relating the development of logical data structured programs, is one keystone of structured systems design.

Yourdon, E. *How to Manage Structured Programming*. New York: YOURDON inc., 1976.

A discussion of the management problems encountered by introducing structured techniques to existing organizations. The book describes methods for beginning in a manner that is likely to prove successful.

_______, and L. Constantine. *Structured Design*. New York: YOURDON inc., 1975.

READING QUESTIONS

1. What is Orr's definition of correct system?
2. Orr argues that system design should work backward from the required outputs. What are the benefits of this approach?
3. Assume that you require a simple payroll system which accepts edited current hours worked for hourly employees along with basic employee information including pay rate, year-to-date FICA, wages, and so on. You must calculate gross pay, including overtime, calculate federal, state, and FICA deductions, and print paychecks for net pay. (This is the same as Question 4 from the Stevens, Myers, and Constantine article.)
   a. Represent this system using a hierarchy chart.
   b. Represent the system with a Warnier-Orr Diagram, including information on sequence and frequency.
   c. Compare the information content of the two diagrams. Do they show the same hierarchical structure? Do they show data flow, data structure, or control?
4. Refer to Figure 25. Assume that the company has chosen to include inactive accounts (customer has no sales or profit) in the report. Also, they recognize that a single customer may have multiple transactions in a given month, so the report must include all sales and profits with totals for each customer. Expand or alter Figure 25 to reflect these new requirements.
5. Which type of diagram more directly supports the implementation of logic into code—the hierarchy chart or the Warnier-Orr Diagram? Explain your answer.
6. A company had D1 divisions, D2 departments, E employees in each department, and D3 dependents per employee. Design an output report for the company that lists employees and their dependents by department and represent the structure of the report with a Warnier-Orr Diagram.
7. The Warnier-Orr approach does not consider the evaluation of design through the examination of coupling and cohesion. Could you merge these concepts into this process? Explain.
8. This technique explicitly considers data structure, where most others do not. How well does it represent hierarchy, control, and data flow?

Business Information Analysis and Integration Technique (BIAIT)—the New Horizon

Walter M. Carlson

One of the deep-seated problems plaguing the computing profession is poor communication between top executives and data processing managers about effective use of new information technologies. Some research in IBM over the past several years has focused on this problem. This paper and its companion by David Kerner present an overview of this research. The papers are being published at this stage to solicit suggestions on directions which further studies and experiments might take.

I have not conducted any of the research and development personally. I have, however, helped guide the effort, and the results to date respond to a goal that I set nearly 25 years ago when I first became responsible for DuPont's computer activity.

The goal was to establish a "business calculus" which could describe and analyze information-handling processes and which could integrate solutions with the same rigor and discipline that a traditional "calculus" deals with mathematical, engineering or scientific processes. We tested a wide range of approaches in the late '50s, but none yielded useful results.

Now, however, the work being described has those sought-after virtues of simplicity, rigor, disciplined reproducibility and universal applicability. It may not be the *final* solution, but it comes closest by far to meeting the goal set and pursued for more than two decades.

For nearly three decades, computer professionals have been trying to create methodologies which express business information processes in terms that are easily understood by business executives. To date, the efforts have had marginal success, at best.

Some recent research has pointed the way to a new horizon in the search for a generalized solution. In fact, *the* new horizon may be in view. It consists of a simple set of questions which lead to a crisp definition of an organization's business objectives and the related information-handling disciplines required to manage the organization's

Source: Walter M. Carlson, "Business Information Analysis and Integration Technique (BIAIT)—The New Horizon," *DATA BASE*, Vol. 10, No. 4, Spring 1979, pp. 3–9.

resources toward those objectives. Experience to date indicates that this new methodology is effective without regard to size of the organizational unit or the product or service it provides. Most importantly, the description of the information system is presented in terms that are understood by both the organization's top management and the data processing management.

The data processing literature is full of articles, proceedings and reports which extol information as a resource to be managed in its own right, in the same sense that men, money, materials and so on are managed as resources. This author has had his name on such exhortations for a long time.

But is this correct? No. It is deadly wrong and misleading. The fundamental error is that the accounting profession has no way to deal with information in precise ways comparable to the acquisition, use and disposal of the other traditional resources. Consider, as one example, how the accounting records can deal with the fact that data or information is transferred from one location to another at a price, yet no change whatsoever has been made to the data or information at the originating source!

The way out of this dilemma was provided by Dr. John Richardson, U.S. Department of Commerce, at the ACM 70 conference in New York City. In one of those spontaneous (and rare) flashes of insight, he observed, "Information conserves other resources through improved decisions."

In a single sentence (or equation, if you will) the proposition is completely bounded.

This paper will focus on the "information" portion of the proposition. The companion paper will begin to address the "resource" aspects of the proposition. Continuing work, not reported on here, is addressing the "decision" part of the proposition. It is merely sufficient to say here that nothing, so far, has led the work to consider factors which clearly reside outside this elegant proposition.

The greatest persuasion, of course, is that top management knows which resources it plans, controls, and operates with, and that the accountants are able to report systematically on the acquisition, use, and disposition of those resources.

When we look closely, there are two worlds in our organizations.

The Chief Executive Officer (CEO) finds a wide and deep chasm between his world and the world of the DP department. The research reported on here has been explicitly directed toward bridging this chasm. It may be eliminating it altogether. Of greatest importance, the research results apply whether manual, mechanical or electronic means are used to serve the organization's decisions.

The reality is that only one world really exists in the sense shown in Figure 1. It is only our perceptions, biases and priorities that cause it to be fragmented.

The one world is the "enterprise".

The "enterprise" may be industrial, commercial, academic or governmental. It can be a total "enterprise," a location (establishment), a department or a single occupation (or group of single skills).

Every such enterprise, establishment, department or occupation constructs a business

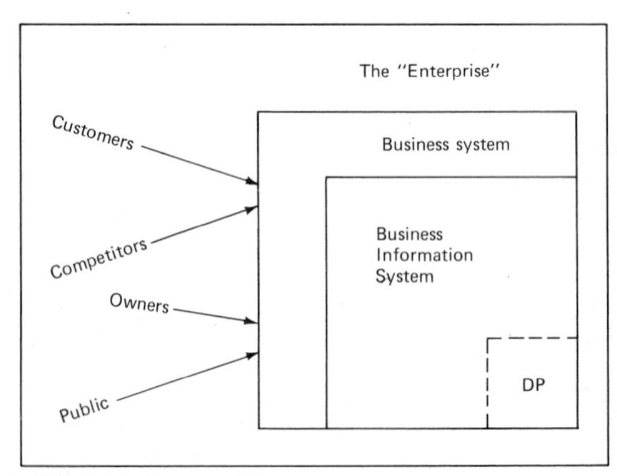

FIGURE 1 One world.

system that is tailored to respond to external forces, which are rich in variety and priority and which have the insidious habit of changing without warning.

Every such organization constructs a business system designed to respond to these external forces and, at the same time, keep the organization alive, healthy and respected.

Every such organization that is large enough, or aggressive enough, has set up a data processing facility to play a part in the business information system. On the average, the DP expenditures reflect from 3 to 7 percent of the total business information expenses.

The DP manager sees the world from one perspective. The CEO sees it from another perspective.

The DP version sees the boundaries of the enterprise as fixed and unchanged over time. The DP objective is to expand its participation in the business information processes by judicious selection of target areas for applications.

The CEO version is that the organization is under many stresses that change over time, sometimes abruptly. He sees his role as adjusting the business system and the supporting information system to meet and overcome these stresses as they occur. Meanwhile, he sees the DP department as a black box of technology which remains fixed and unchanged within the time frames available for the organization to react to threats and stresses.

As stresses on the enterprise cause adjustments in the business system and its supporting business information system, the DP department will adjust its priorities, too. The DP services will be geared directly to the points of greatest stress. And they will be changed with time constants that help management respond to the stresses with better decisions. While that is the goal, there is a gulf of misunderstanding yet to be bridged.

How do we get the CEO and the DP department on the same wave length, using the same language and synchronized in their priorities? Fortunately, this question has yielded to some research that began early in the 1970s. The result of the research is a theory known as BIAIT, an acronym for Business Information Analysis and Integration Technique.

The research was conducted by Donald C. Burnstine. His initial goal was to see if some better way could be found to perform product planning. Specifically, he was looking for a way to describe a customer's needs for computer products and services in terms of information-handling functions rather than in terms of the products or services made or provided by the customer.

Don sought out and researched every question he could find that seemed to relate to how information requirements emerge in an organization. After checking out 300 or 400 questions and their implications, he found eight that uniquely and systematically characterized the way an organization uses information—independent of its size and independent of the products or services it provides.

After another year or so of research, he found that one question was redundant and was not needed. So BIAIT now has seven questions—a prime and lucky number—which provide the systems analyst with a complete boundary around the information needs of the organization.

Before reviewing the seven questions, it is important to recognize the simple ground rules that have evolved along with the BIAIT theory. They are listed in Table 1.

The first ground rule introduces the generic concept of an "order." An order can take many forms, some formal and some informal. It can be a purchase order, for example. It can be a request, or it can be merely a question. An order can arise from any source external to the organization under study. The net of it is that an order is anything that requires a response from a supplier.

- Classify *"ORDER"*
- Define ordered entity
- View the supplier—*not "customer"*
- Multiple orders per supplier can occur
- All questions have one of only two answers

TABLE 1. Ground Rules of BIAIT

The next ground rule requires that the ordered entity be identified and defined. The ordered entity must be either a thing, a space or a skill. These three categories exhaust the possible classifications.

The idea of a thing is fairly easy to grasp. A chair, a shoe, a house, a building, a machine or a report—they are all things.

The concept of a space includes something as temporary as a seat for a movie or as permanent as a grave. Also, the space can move about, as in a truck or an airplane.

A skill may consist of formal schooling or a craft as physical as ditch digging. A skill may be provided by anyone, male or female, young or old.

The third ground rule resolves the question of whether the BIAIT questions apply to the supplier or to the customer.

The view taken by the analyst is directed only toward the supplier of the ordered entity. Whenever a customer places an order, the supplier responds with a thing, a space or a skill. So, from this point forward, we shall be looking at the organization or person who responds to an order. A simple way to keep this viewpoint in mind is to recall that if a business or organization receives no orders, it has no reason for existing.

The next ground rule clarifies the matter of how many orders a supplier can receive. A supplier can receive many kinds of orders under the ground rules of BIAIT.

For example, a department store may bill some of its customers as well as taking cash from others; these are two different orders. Similarly, a vendor may rent equipment to one customer and sell it to another, again, two different orders.

A rigorous application of the BIAIT questions to establishments defined in the Standard Industrial Classification Manual published by the U.S. government has shown that U.S. establishments receive between four and five types of orders, on the average. Some receive 12 or more orders.

The final ground rule specifies that each of seven BIAIT questions can be answered only one of two ways. As will be seen, some of the questions require only a simple "yes" or "no." Others provide a choice, as in the "bill" or "take cash" example used above.

Table 2 shows the four questions which deal with the supplier.

Table 3 shows the three questions which relate to the ordered entity.

There are three levels of organizational elements shown in each table to illustrate that the phrasing of the question needs to be tailored to maintain the basic concept within the decision environment that is actually involved. While managers of enterprises may think their decisions through in terminology that is quite different from the terminology used by a single professional or small group of professionals, the information-handling implications are the same, whatever the terminology used.

At the enterprise or establishment level, the semantics are traditional and readily understood. Does the supplier bill the customer or take cash?

At the department level, the phrasing of the question has to fit. Is the department a cost center which accounts for each service provided or is it operating on a budget basis that requires no accounting of individual service actions?

At the occupation level, is a white collar worker on commission or straight salary? Or as suggested earlier, some combination of each? Is the blue collar worker or service tradesman paid by piecework or strictly on an hourly rate?

TABLE 2
Four Questions About the Supplier

SUPPLIER QUESTIONS → ORGANIZATION LEVEL ↓	1. BILLING?	2. DELIVER LATER?	4. PROFILE CUSTOMERS?	8. NEGOTIATE PRICE?
ENTERPRISE/ ESTABLISHMENT	BILL OR TAKE CASH	LATER OR NOW	RECORD PREVIOUS ORDERS FROM SOURCE OR NO PROFILE	NEGOTIATE OR FIXED
DEPARTMENT	COST CENTER OR BUDGET	PLAN WORK OR FIRE CALL	RECORD PREVIOUS ORDERS FROM SOURCE OR NO PROFILE	COSTED WORK ORDER OR STANDARD RATE
OCCUPATION	COMMISSION OR SALARY PIECE WORK OR HOURLY WAGE	SELF-SCHEDULED OR PRIORITY SET BY OTHERS	RECORD PREVIOUS ORDERS FROM SOURCE OR NO PROFILE	COSTED WORK ORDER OR STANDARD RATE

TABLE 3
Three Questions About the Ordered Entity

ORDERED ENTITY QUESTIONS → ORGANIZATION LEVEL ↓	16. RENTED?	32. TRACKED?	64. MADE TO ORDER
ENTERPRISE/ ESTABLISHMENT	RENTED OR SOLD	RECORD WHO RECEIVED OR NO RECORD	MADE/ASSEMBLED TO ORDER OR FROM STOCK
DEPARTMENT	LOANED OR GIVEN	RECORD WHO RECEIVED OR NO RECORD	ASSEMBLE/CREATE OR PROVIDE FROM FILES
OCCUPATION	LOANED OR GIVEN	RECORD WHO RECEIVED OR NO RECORD	ASSEMBLE/CREATE OR PROVIDE FROM FILES

It becomes apparent that wide differences appear in the information-handling requirements, depending on which of the two answers is correct for the specific situation being analyzed.

For the enterprise or establishment, the question is deliver later or now? The department has to decide whether to plan the work involved or drop everything and go into a "fire call" mode. At the occupation level, the question becomes whether we as individuals (or teams) can schedule the work for ourselves or whether our job priority is set by others.

Whether or not the supplier keeps a profile of its customers can be phrased the same way at all three levels. The profile mechanics are independent of the nature of the order or the definition of the ordered entity. If a profile is kept, there are specific information requirements. If no profile is kept, those requirements disappear.

By the way, the numbers in the headings are going up by the powers of 2. Each question can only be answered one of two ways, hence the question's value can be assessed in a sequence that increases by powers of 2. In this instance, the answer has a value of 4 if a profile is kept. If not, the value is zero.

An enterprise or establishment either negotiates the price or satisfies the order at a fixed price. Or, as we have noted, it may do some of each, depending on the customer, the quantity or the ordered entity. At the department or occupation level, the question needs to be phrased in terms of costed work orders or some standard, nonnegotiable rate.

These are the items we need to know about the supplier. They are all we need to know about the supplier to understand his business system in terms of his information-handling needs. The next three questions relate to the ordered entity.

If the supplier rents the ordered entity, he retains ownership with all the record-keeping implied by such ownership. If he sells it, the deed is done, and records of the transaction are quite simple, relatively speaking.

A department or occupation either lends things (like reports) and thus keeps records or gives things in response to an order and keeps no records of the transaction itself.

The questions are still going up in value by the power of 2. This indicates that the information-handling complexities inherent in the questions are rising as product/service considerations take over.

An enterprise or establishment tracks products for such reasons as warranty or recall purposes. A department tracks a report to be able to update it, when necessary. You and I keep track of a number we gave our management in the event that later investigation requires that it be adjusted. The complexity of tracking products through successive locations and owners can be enormous.

There is no information-handling involved if there is no tracking of the ordered entity. As some firms and government agencies have discovered, however, there can be severe business or social problems if no tracking is done.

The enterprise or establishment can chose to wait and make the entity to order or it can build an inventory and provide from stock. Assembly from stock parts upon receipt of an order is equivalent to "made to order," incidentally.

A department or occupation also responds to an order by assembling or creating the response. On the other hand, the response may be handled by simply pulling something from a file.

These are the seven questions and the 14 possible answers they provoke in the various situations we might need to analyze at any level of the organization.

What is done with the answers?

The values attached to the 14 answers leads to adding them up. Each order analyzed by BIAIT then lands in one of 128 cells on the basis of accumulating values from the seven questions. Each of the BIAIT cells has associated with it two kinds of analytical and integrative information, as indicated in Figure 2.

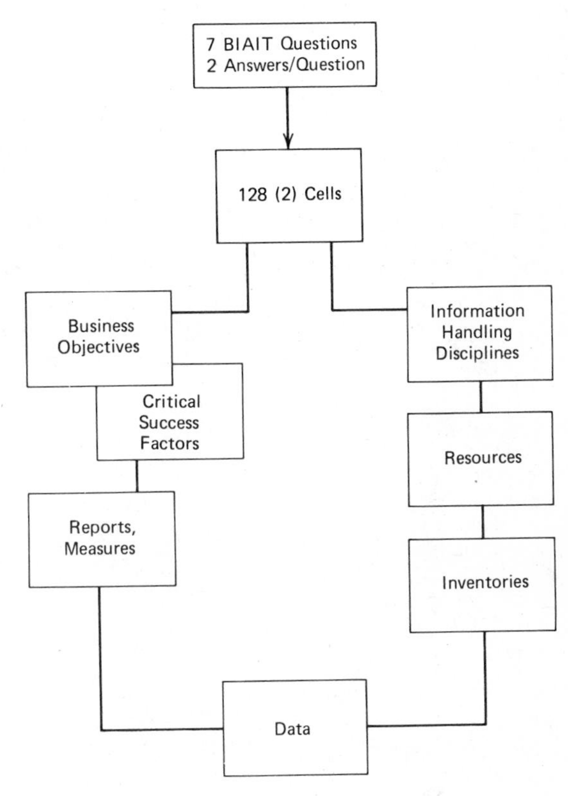

FIGURE 2 Two kinds of analytical and integrative information.

Each cell has associated with it a set of business objectives that normally come into play when a supplier receives and responds to that kind of order. Also, each cell has identified with it a set of information-handling disciplines that are required to process the order.

The work done by John Rockart at MIT on what he calls "critical success factors" is a currently useful elucidation of business objectives. He shows the way in which these factors lead to specific reports and measures needed to run the organization. The measures, in turn, define the specific data requirements to support the relevant decision processes.

Similarly, the information-handling disciplines have to be executed to manage specific resources, whose inventories define the specific items of data to be collected and managed.

One of the pleasant surprises of the BIAIT theory is that it points directly toward who the data owners and data users should be. This may be one of its most powerful results in helping to design distributed systems, for example.

If an order-processing system were to be analyzed, we would have a number of data elements (or fields) to consider. As indicated in Figure 3, the data would be used or not used according to answers to the seven questions. For example, customer I.D. is used if you bill but not if you take cash; used if you profile the customer and used if you negotiate price. But customer I.D. is owned by the function that handles future delivery, since the knowledge of who the customer is has its greatest and most critical value there.

The other data items can be reviewed the same way, as shown. It seems that about 1,000 data items would appear in a complete study at the enterprise or establishment-level. There would be fewer at the lower levels.

Nothing has been said up to now about using computers to carry out the information processing. That is exactly the point.

	Cust. I.D.	Order No.	Item No.	Item Description	Price	Quantity	$ Total
Bill	U	U	U	U	U	U	O
Cash							
Future	O	O	U	U	U	O	
Immed.			U	U	U		
Profile	U	U					
None							
Negot.	U	U			O		
Fixed					O		
Make		U	O	O			
Stock		U	O	O			

FIGURE 3 Data owner and data users.

The BIAIT process is designed to get full agreement between the end-user management and the analyst before anyone writes code or even installs a manual system. Both parties speak the same language. Their agreements are readily recorded. The data processing tasks and procedures, when undertaken, are directly relatable to the business objectives through BIAIT.

As currently envisioned, the process takes the seven questions about the order, locates the resultant cell and goes through four stages to reach the next application in priority and then recycles back to set the next priority. The overall process is diagrammed in Figure 4.

Stage I is entirely in the hands of the analyst. From the answers to seven questions, the BIAIT data base would give him a generic model of the organization or occupation being analyzed. He can read from his terminal what to expect in the way of other Standard Industrial Classification code establishments in the same cell. Logically expected objectives (or critical success factors) and reports of measures to be expected are provided him.

The information-handling disciplines to look for are automatically provided. The expected disciplines needed to manage the resources and their inventories are listed. And finally, a listing can be provided of the data elements needed to support operations within that particular cell, the function that should own each and the users of each data element.

The target is to do Stage I in less than four hours. The product of Stage I is a generic business information model that needs to be tailored to the actual organization in Stage II. Stage II involves getting out to interview the decision-makers in the organization to see how closely their operations fit the model.

It appears that use of the generic model in developing the interview procedures can save lots of time. The analyst and the executive do not have to invent things to talk about. The model gives them a wealth of detail that they can quickly confirm or modify to fit what shows up

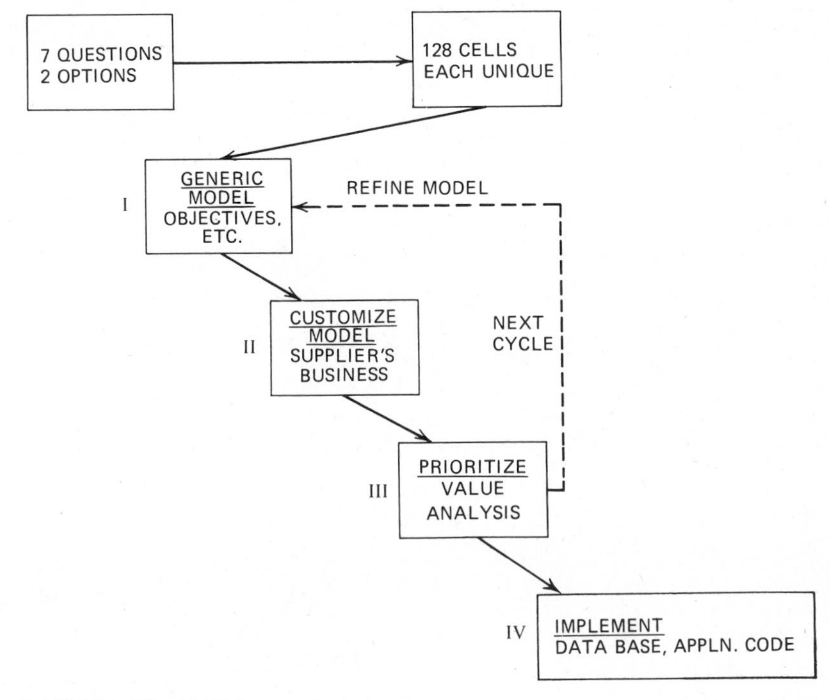

FIGURE 4 The BIAIT process.

on actual forms and reports. The decision points (or lack of them, sometimes) become quickly identified.

The target for this stage is less than six weeks for an enterprise or major division; much less time is needed for departments. The product is an agreed-upon model of the actual business information flow in the organization.

Stage III involves the prioritization and value analysis needed to select the next application which is most important to top management. The analyst in his office and in a dialog with key executives works with a large data base. In addition to the model of the business information obtained from Stage II, he has confirmed which objectives and measures actually run the organization. He knows where the forms and reports fit in.

He also takes the time to map current applications into the model to see where overlap exists, where holes exist and even where existing applications serve no useful purpose. Perhaps of greatest importance, he and the management have a set of conflicts and issues to resolve between elements of the organization before work can be started on new applications. The target is one to four weeks to develop the specifications for the next application.

Stage IV reduces all of the study work to a practical application. The analyst and programming assistants at their terminals convert the specifications to running code.

Presumably, their computer-based tools and aids will be managed for them by a data base management system. By this time, the organization's BIAIT model and the forms and reports supporting it will be in machine-readable form. So will the matrix defining data owners and users for each data element.

Compilers and report generators will be invoked, as needed, to convert the source code derived from the specifications into running code.

The target duration is four to 10 weeks, after which the analyst is ready to recycle back, refine the model on the basis of experience and pick the next application.

While BIAIT is still in a late research status and development is only recently under way, there has been some experience to support targets discussed for each of the four stages.

Don Burnstine tried his theory and some of his early assignments of objectives, information-handling disciplines and so on on some of his friends running their own businesses. He was able to provide them a complete business information system tailored on their needs in two days. He did not have to write computer code, of course, since such small businesses do not have computers—yet.

Don also trained some summer employees from graduate schools of business, and they were turned loose on analyzing some fairly large business units. Using manual methods only, they were able to complete building a Stage I generic model in each case in 17 manhours or less.

David Kerner has been experimenting with the BIAIT process to develop efficient methodologies for application of the theory. His paper describes, in general, the approach taken in an early application. More recent uses of BIAIT by Kerner have demonstrated that Stage I and Stage II can be completed inside one week, using strictly manual methods.

BIAIT is a formal, yet simple analytical tool (see Table 4). The results are reproducible. Different analysts can check each other against formal rules. It is a theory which is reducible to practice. There are not yet many theories in the information systems business that have the property of being usable in day-to-day business. It is a communication vehicle that bridges the gulf between the front office and the DP department. Of great significance, the DP manager can tell the top executives what the impact will be of the changes proposed.

BIAIT works in a way that is independent of the size of the organization, the products or services it provides, or the structure of the organization with respect to type or level. It serves as an attractive foundation for revitalizing the business analyst profession.

BIAIT truly is *the* new horizon for all of us concerned with helping management conserve resources through better decisions.

- Formal, simple analytical tool
- Theory reducible to practice
- Communication vehicle front office to DP and back
- Independent of size, product/service, structure
- Foundation for revitalizing business analyst profession
- *The* new horizon for all of us

TABLE 4. The Net of BIAIT

READING QUESTIONS

1. Define BIAIT. What functions does it perform?
2. What is accomplished through BIAIT that was missing with the use of prior techniques?
3. What are the advantages of BIAIT? What are its shortcomings?
4. Record the steps in the BIAIT procedure.
5. Explain Table II—how it is developed and what it means.
6. Explain Table III—how it is developed and what it means.
7. Explain Figure 2—how it is developed and what it means.
8. Explain Figure 3—how it is developed and what it means.
9. Compare BIAIT to SOP. What are the differences? Similarities?

Business Information Characterization Study

David V. Kerner

In the late 1960s, IBM recognized the need for new information systems that would improve planning and control for some of the business problems facing the company at that time. A small organization, created by the company's Data Processing Group Headquarters, was assigned the task of defining the required information systems. From this group, Business Systems Planning (BSP) was later formed in the Data Processing Division to carry out this type of definition in customer locations.[1]

This article describes the Business Information Characterization Study (BICS) which is a logical extension to the BSP methodology via use of Donald Burnstine's BIAIT theory[2] to provide a structured approach for understanding business information systems. The Business Information Analysis and Integration Technique (BIAIT) is described in the companion paper by Walter Carlson. The major difference between the two methodologies is that BICS provides the study team with an information model of the business which it then verifies, while BSP builds a model through a discovery process.

Data processing addresses the application of computers to such activities as inventory control, truck scheduling, invoicing, shipping, reserving space, banking and the preparation of reports to help management measure results. These processes permit the manager to conduct business as he or she wishes and in a disciplined way. The manager, for example, may wish to supply a customer with a bill, may wish to deliver a product later, may wish to keep historical records about a customer or may want to make a product to order.

Executives know what is necessary to do billing, invoicing and so on, but they often do not understand these activities in terms of the data necessary to do them. Data processing people, on the other hand, understand data organization but not necessarily all the functions which need to be performed, both manually and electronically.

Source: David V. Kerner, "Business Information Characterization Study," *DATA BASE*, Vol. 10, No. 4, Spring 1979, pp. 10–17.

Through the use of a generic model, BIAIT Business Data Classification (BBDC), BICS bridges the communication gap between management and the information systems department. BBDC is a business model that affords an abstract view of the information activity and data usage of businesses.

Using the techniques described in this article, end users (managers) can develop an information-handling definition of the business (or businesses) they are in. In this paper, business is defined as any establishment or organization that exists to provide some product or service. The term includes for-profit organizations, not-for-profit organizations and public institutions (schools, government agencies and so on).

R. N. Anthony has identified three distinct levels of planning and control which overlay the operational processes of a business.[3] This study examines those three levels of planning and control with emphasis on improving the effectiveness and efficiency of information needed in strategic planning and management control. Figure 1 shows the levels of planning and control in a business. In broad terms, Anthony defines the three planning and control levels as follows:

1. Strategic Planning—the activity of deciding on objectives, changes to objectives and establishing policies that govern acquisition, use and disposition of resources.
2. Management Control—the management activity which assures that resources are obtained and used effectively and efficiently to accomplish objectives.
3. Operational Control—the activity of assuring that specific tasks are carried out effectively and efficiently.[4]

The four major tasks in the BICS methodology are constructing the business model, verification of the business model, examination of the current information system and analysis and implementation specifications. The following procedures have been applied in a number of actual studies.

The business model is generated from the generic model BBDC. The components of the generic model are BIAIT variables, Unique Inventories (UI), Data Groups (DG) and Information-Handling Disciplines (IHD).

BIAIT consists of seven basic questions (variables) that can be answered only by a "yes" or "no." These questions, listed in Figure 2, are sufficient to describe all businesses through their orders.

Unique Inventory (UI) is a basic resource the business identifies to control (manage or track). The UIs fall into two categories: those which identify basic resources and those which identify relationships of the basic resources. For example, a sales order provides the relationship between basic resources called "product" and "customer." There are 12 UIs in the generic business model. Tests to date indicate the UIs are adequate to characterize the basic data of a business.

Data Group (DG) is a further breakdown of the Unique Inventories. The Data Groups fall into two categories: plan data and actual data. Plan data is divided into plan descriptive data and plan value data. Actual data is divided into actual descriptive data and actual value data. Note value always implies dollars. Figure 3 lists the 12 Unique Inventories and four Data Groups. Figure 4 is an example of a UI with some of the attributes which describe the entity called "product." The Data Groups are indicated in the example.

The IHD is an information activity of the business. The IHD manages a Data Group within a Unique Inventory. Manage is defined as having responsibility for definition and maintenance (update) of the Data Group within a Unique Inventory. There is a one-to-one relationship of an IHD to a UI and DG by BIAIT variable.

Some IHDs are common to all businesses because all businesses have common resources such as people, space and so on. These IHDs are always needed and are independent of the seven questions. They tend to support functional areas such as marketing, finance, facilities, personnel, procurement and general management.

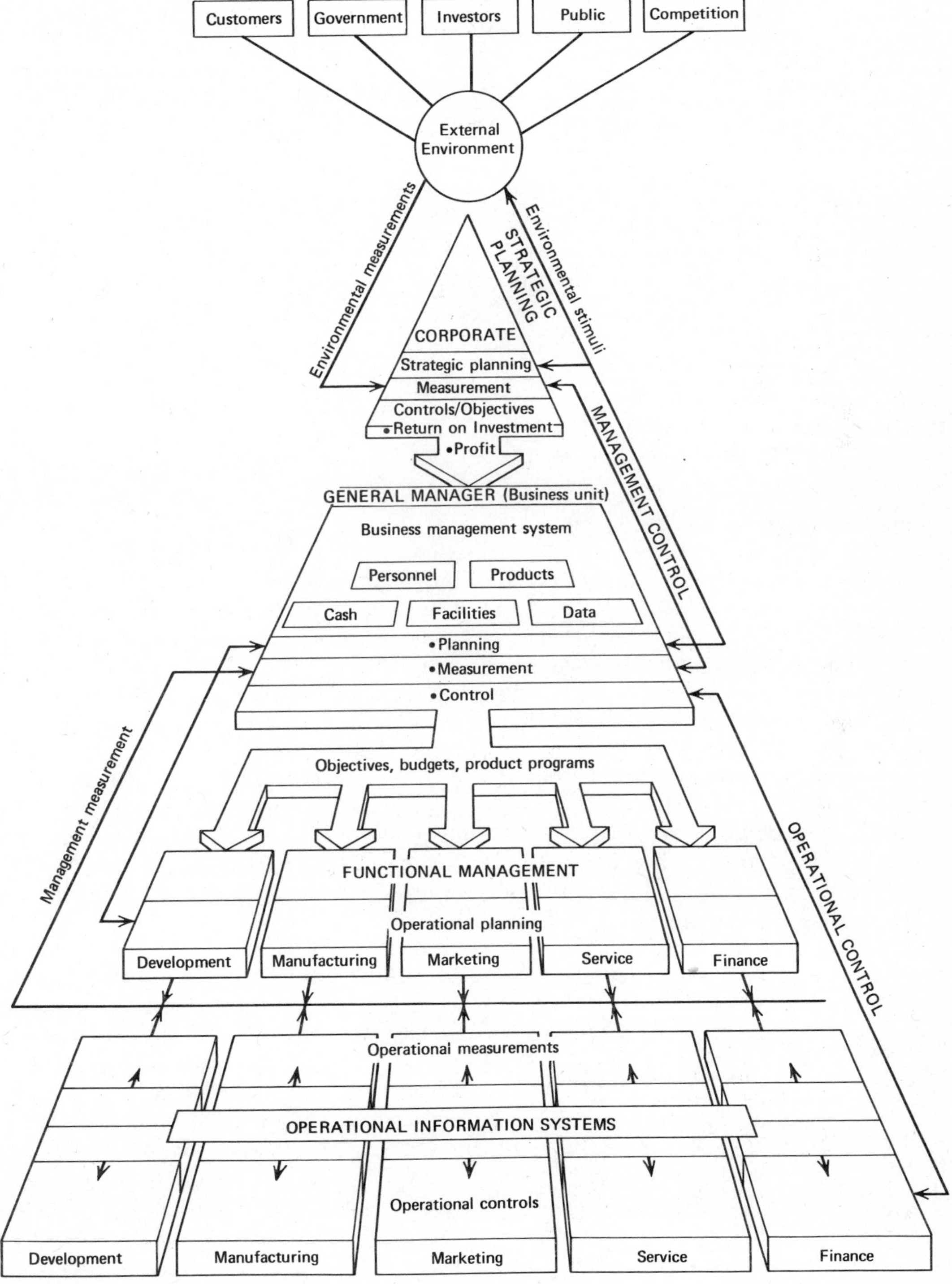

FIGURE 1 Levels of planning and control.

YES	*QUESTIONS*	NO
The supplier bills the consumer.	BILL	The order is paid for with cash or cash equivalent (check or charge card).
The supplier delivers the product or service to the consumer at some time in the future.	FUTURE	The consumer takes what is ordered with him.
The supplier keeps records by individual customer about prior transactions (for purposes other than billing).	PROFILE	The supplier does not keep records about prior transactions with individual customers (every customer is a surprise).
The consumer and supplier negotiate price.	NEGOTIATE	The transaction is fixed price (to any given set of customers).
The supplier rents the product or service to the consumer; supplier retains title.	RENT	Consumer buys the product and takes title to it.
The supplier keeps track of the product or service for subsequent recall or change.	TRACK	Supplier does not track the product or service.
The product is made to the consumer's specifications	MAKE TO ORDER	The product or service is provided from stock.

FIGURE 2 The seven questions of BIAIT.

To construct a tailored model—that is, a model which represents the actual business to be studied—the orders the business receives from its customers and the ordered entities (products or services) it produces must be identified. When the orders have been determined, the BIAIT questions are applied to each order.

Orders are the basic driving force of business: no orders, no business. The term "order" is used here in the broadest possible sense as a request for a product or service. There are fundamentally three classes of items ordered from suppliers: space, skill and things. Everything that is ordered from a supplier fits into one of these three categories.[2]

RESOURCE UIs	*RELATION UIs*
PRODUCT	SALES ORDER
CUSTOMER	TRACK
VENDOR	BILL OF MATERIAL
FACILITIES	ROUTING
EMPLOYEE	PURCHASE ORDER
MONEY	ACTIVITY
DATA GROUPS	
PLAN	*ACTUAL*
PLAN DESCRIPTIVE	ACTUAL DESCRIPTIVE
PLAN VALUE	ACTUAL VALUE

FIGURE 3 Unique inventories and data groups.

Each class of order received by an establishment must have Information-Handling Disciplines (IHDs) to manage and control the data about the order and ordered entity as it moves through the business. BIAIT identifies 14 potential groups of Information-Handling Disciplines and their related data.[2] Figure 5 shows a sample of IHDs and their relationship to the seven questions.

If an establishment has different types of orders, it will have different businesses for managing and controlling those orders. For example, an establishment that manufactures products will have orders for those products. If the same establishment provides education courses to instruct customers on the use of the products, it will receive orders for courses. If it also publishes documentation about its products or courses, it will receive orders for the documentation. Therefore, the establishment is in at least the manufacturing, education and publication businesses.

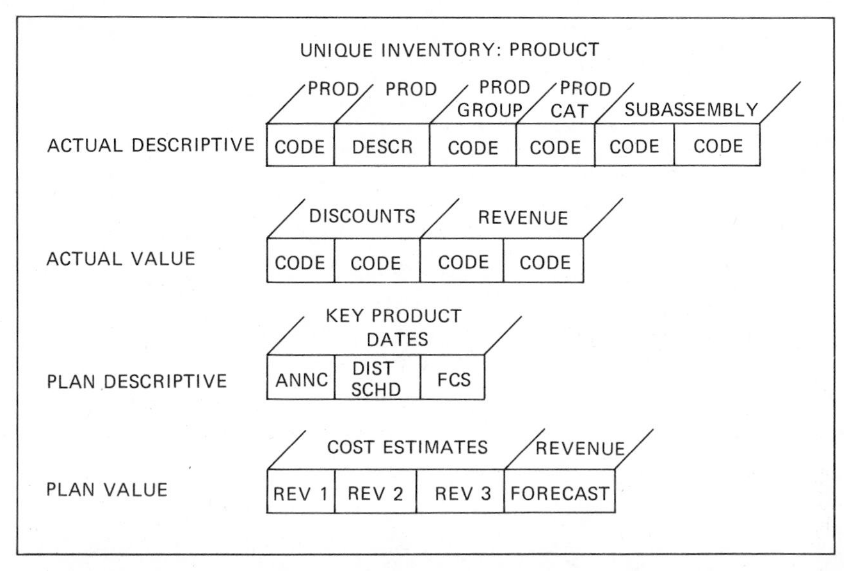

FIGURE 4 Unique inventory and attributes by data group.

From the answers to the questions, a list of IHDs and their related Unique Inventories (UIs) is produced. The relationship of Information-Handling Disciplines to Unique Inventories can be shown by building a matrix which has IHDs along one axis and UIs along the other. Relationship of IHDs to UIs are represented wherever interactions exist. Figure 6 shows IHDs and associated UIs in matrix form.

This matrix shows how the information-related activity and the basic data of a business correlate to each other, and it shows where the responsibility lies (in information-handling terms) for creating and maintaining the accuracy and timeliness of the data. Work to date indicates that any business would have a matrix comprising Information-Handling Disciplines selected from 56 candidates and Unique Inventories from 12 candidates.

Using the IHDs generated from answering the BIAIT questions, the organization can be mapped to the IHD to show where the functional responsibility lays for managing the IHDs or for participating in their operation. Figure 7 represents a typical relationship of an organization to Information-Handling Disciplines. Organizational numbers are used to make the illustration general.

The following relationships of IHDs to organization are determined by the study team:

1. Ownership of IHD (definition and update capability).
2. Update responsibility (can maintain current copy of data).
3. Major user of the data.
4. Minor user of the data.

Figure 8 is a combination of the previous two matrices, showing the relationship of IHDs to UIs and the organization. This pictorial model represents the business information-handling activity and its associated data by organizational element. The model in Figure 8 permits management to determine by way of their organizational elements:

1. How each element interacts with the information activity throughout the business.
2. Which pieces of data are essential to operation of each organizational element.

INFORMATION HANDLING DISCIPLINES

BILL	FUTURE	PROFILE	NEGOTIATE	RENT	TRACK	MAKE TO ORDER
• CREDIT RECORDS • CUSTOMER ACCOUNTING	• CUSTOMER BASIC RECORDS • INVENTORY COMMITMENT • SALES ORDER ACCOUNTING • SALES ANALYSIS • SALES ORDER MANAGEMENT • SHIPPING RECEIVING	• ACCOUNT HISTORY MAINTENANCE • ACCOUNT PLANNING • CUSTOMER DATA GATHERING • CUSTOMER FORECASTING	• CONTRACT NEGOTIATION	• RENTAL MANAGEMENT	• FIELD E.C. PLANNING • FIELD SERVICE ACCOUNTING • FIELD INVENTORY MANAGEMENT	• ENGINEERING BOM MANAGEMENT • MANUFACTURING ENGINEERING • MANUFACTURING BOM MANAGEMENT • PRODUCT COST ESTIMATING
CASH	IMMEDIATE	NO PROFILE	FIXED PRICE	SELL	NO TRACK	PROVIDE FROM STOCK
NO IHDs	NO IHDs	NO IHDs	• PRODUCT FORECASTING • PRODUCT PRICING	NO IHDs	NO IHDs	• FINISHED GOODS INVENTORY • FINISHED GOODS INVENTORY PLANNING • FINISHED GOODS INVENTORY RECORDS

FIGURE 5 Seven BIAIT questions and information-handling disciplines.

INFORMATION HANDLING DISCIPLINE / UNIQUE INVENTORY	BILLING	YEAR END CLOSING	SERVICE COUNT ACCOUNTING	FACILITY ACCOUNTING	FUNDS FLOW PLANNING	SERVICE REVENUE ACCOUNTING	PROJECT ACCOUNTING	VENDOR ACCOUNTING	CUSTOMER BASIC RECORDS	SITE SERVICE SUMMARY	SALES ORDER MANAGEMENT	ACCOUNT HISTORY MAINTENANCE	ACCOUNT PLANNING	CUSTOMER DATA GATHERING	CUSTOMER REVENUE PROJECTION	SERVICE FORECASTING	SERVICE PRICING	BUDGETING	ORGANIZATION PLANNING
PRODUCT						AV										PD	PV		
CUSTOMER	AV								AD			AV	PD	AD	PV				
SALES ORDER		AV								AD	PD								
FACILITIES				AV															
BILL OF MATERIAL																			
JOB FLOW			AV																
VENDOR								AV											
PURCHASE ORDER																			
EMPLOYEE																			
FUNDS					PV														
DEPARTMENT							AV											PV	PD
TRACK																			

UI/IHD

PD = PLAN DESCRIPTIVE
PV = PLAN VALUE
AD = ACTUAL DESCRIPTIVE
AV = ACTUAL VALUE

FIGURE 6 Information-handling disciplines and associated unique inventories matrix.

INFORMATION HANDLING DISCIPLINE / ORGANIZATION ELEMENTS	BILLING	YEAR END CLOSING	SERVICE COUNT ACCOUNTING	FACILITY ACCOUNTING	FUNDS FLOW PLANNING	SERVICE REVENUE ACCOUNTING	PROJECT ACCOUNTING	VENDOR ACCOUNTING	CUSTOMER BASIC RECORDS	SITE SERVICE SUMMARY	SALES ORDER MANAGEMENT	ACCOUNT HISTORY MAINTENANCE	ACCOUNT PLANNING	CUSTOMER DATA GATHERING	CUSTOMER REVENUE PROJECTION	SERVICE FORECASTING	SERVICE PRICING	BUDGETING	ORGANIZATION PLANNING
1000	o	○	○			o	x		o	○		o		●	○	x		x	
2000		x	x	o	o		o		/								●	o	
3000			x				x		x									x	o
4000		x	o	x	o	x	x		/	x			○	o	○	○	o	x	
5000			x				x		/									x	
6000			x				x		/									x	
7000			x			x	x		○	/	o						x	x	o
8000			x	x			x		x		○							x	o
9000		x	x		●	●	●		/	x					x		●	●	●
10000								○											

ORGANIZATION / HD

● = OWNER (DEFINITION AUTHORITY)
o = UPDATE
○ = OWNER & UPDATE
x = MAJOR USER
/ = OCCASIONAL USER

FIGURE 7 Information-handling disciplines and associated organization elements.

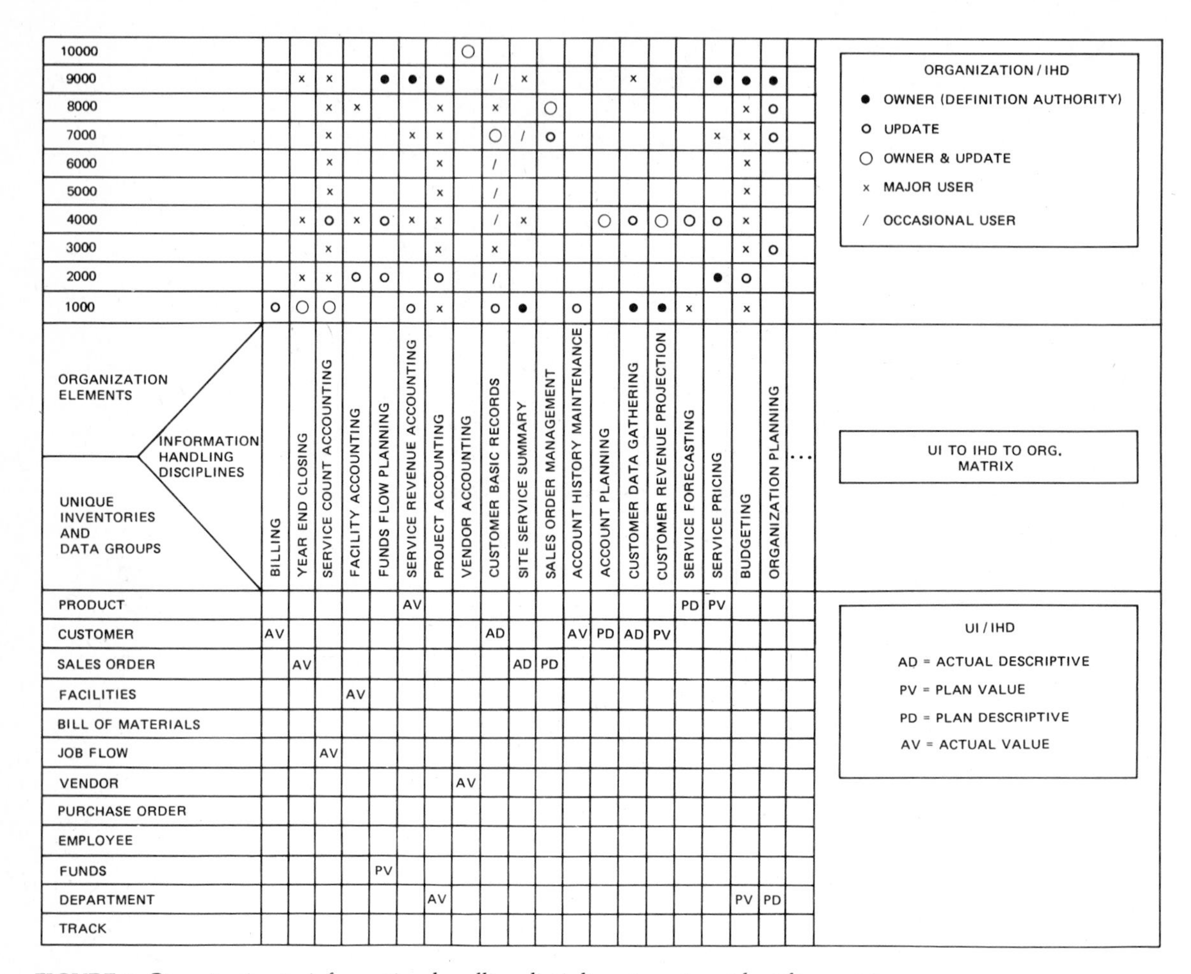

FIGURE 8 Organization to information-handling disciplines to unique identifier matrix.

Conversely, the model in Figure 8 permits the Information Systems department to determine by way of the unique inventories:

1. Which organizational element has an interest in the creation or use of the data.
2. Which data-handling processes (Information-Handling Disciplines) involve use of the data for which parts of the organization.

Also, the model in Figure 8 is used by the BICS team to expedite and simplify the interviewing of the organization's executives by identifying:

1. Which organizational element has key responsibility for planning and controlling specific items of information.
2. Which organizational elements are directly affected by a particular information item.

At this point, management has a model of its business in terms that provides direct communication with the information systems department. The model provides a common language for the information-handling activity and basic data of the business.

This model can be developed by a study team in a matter of a few days. The seven BIAIT questions and the common resource inventories provide an initial model which needs only to be verified and adapted to the local situation. It must be noted that the model reflects data usage and not the actual physical storage of data.

The next step is to verify the model and to gain insight into actual information problems of the business. This is accomplished by interviewing senior management to verify the IHDs and UIs of the business. The interviews will also determine the objectives and measures of the business and problems, if any, in meeting those objectives. This information is added to the business model.

The model will then reflect the verified IHDs and UIs and the relationship of objectives/measures and problems to the IHDs and UIs. The completed model becomes the communication vehicle whereby the business analyst and the data processing analyst plan how the data of the business is used in planning, control and operations.

The study must also examine the current information system to discover where the data of the business resides. This examination is at a high level. Its purpose is to identify the major data bases, both manual and automated. Figure 9 is a representation of this examination.

The rectangles are major data base groups within a functional area of the business. The numbers inside the rectangles represent individual data bases used and owned by the functional area of the business. The UIs for each data base are also indicated in the rectangle between parentheses. The numbers outside the boxes represent the input and output reports and forms. The diagram also shows manual entry and/or machines processing. The arrows show the information flow across data base groups.

Information systems can be improved in two areas:

1. Efficiency of information processing. Analysis of the current data bases and their input/output can produce recommendations to eliminate duplicate data, manual entry and duplicate reporting. While the elimination of redundant data and manual entry will also reduce the losses caused by inconsistent and inaccurate data, the most visible benefit is the reduction in processing and/or people time.
2. Effectiveness of the information usage. A benefit that is harder to justify (or even recognize) is the improvement of the information content or effectiveness of the information in the decisions made in the course of operating the business. For each IHD and UI, the following questions must be asked:

   - Does one UI exist for every IHD?
   - Does each entity have a single UI?
   - Does an objective exist for every IHD?
   - Does a measure exist for each objective?
   - Is there a data processing system to generate the measurement reports?
   - Is the cost of reporting worth the value of tracking?

If the answer to any of the above questions is no, does management have problems that can be solved by correcting the situation so the answer would be yes?

IHDs that do not correlate to measureable objectives are *potentially out of control.* Entities that are not uniquely identified are probably not being managed in an organized manner. If there are single entities, such as a product, that are being identified with different identifiers, there will be problems tracking the entity across functions.

Benefits of the analysis are:

1. Identification of information bottlenecks caused by manual or duplicate entry of data.

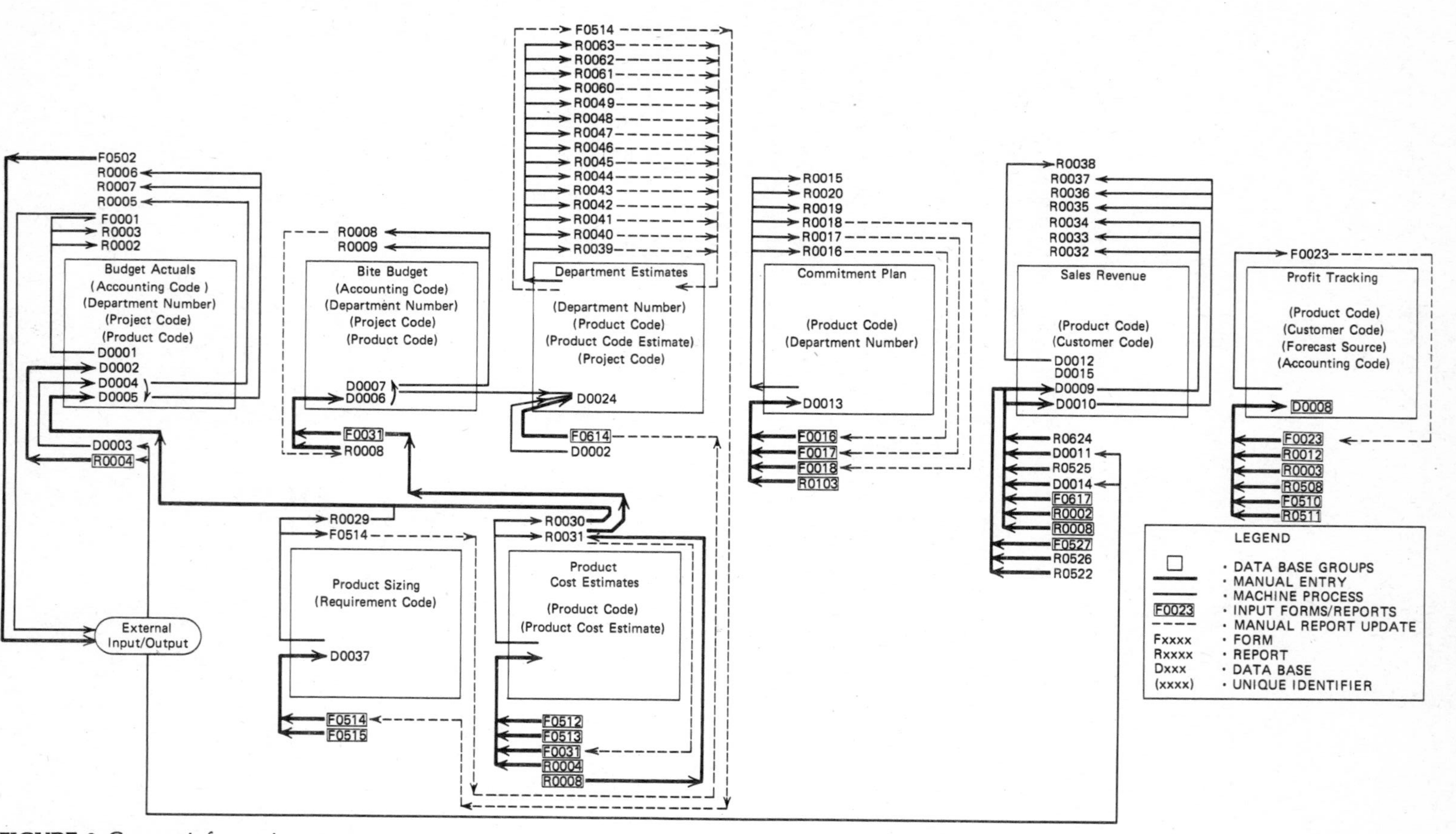

FIGURE 9 Current information system.

2. Identification of entities of the business which are not uniquely identified.
3. Identification of entities of the business which have multiple identifiers.
4. Identification of IHDs that have no associated measures.
5. Identification of IHDs that have no associated objectives. Note that to develop business information systems, one must first define the overall business strategy and business objectives. This is necessary for setting direction and standards for information systems.[5]
6. Identification of relationships of IHDs to each other. These relationships are shown in Figure 10. The functional view of an information activity of a business in Figure 10 explicitly ignores organization lines. Presented in this format, the BBDC model has developed the information flow pattern that is necessary to meet objectives set by management. It is this flow that remains stable even while management alters the organization chart to balance available personnel resources to the maximum extent.

A Business Information Characterization Study as described here results in the creation of a conceptual model of the business. After the model has been verified by management, it is stable and will change only when the orders received by the business change. The model reflects the objectives/measures or the lack thereof to the Information-Handling Disciplines (IHDs). The basic data of the business Unique Inventories (UIs) are represented in the model. The usage of data and the relationship of IHDs to UIs across the organization is represented and maintained in the model.

Having identified the IHDs, UIs, objectives and measures, management is in a position to:

1. Communicate effectively with the information systems organization regarding the role that data plays in business operations.
2. Participate in decisions as to which function of the organization must be responsible for acquiring and controlling data and which function will be using the data.

Since the data, procedures and tasks that make up computer programs have been linked via the BBDC model to the business objectives and the key decision variables, modifications of direction or decisions at the top of the organization can be promptly translated into the necessary changes in how data is processed to satisfy the new requirements.

The techniques described in this paper have been applied in a "live" environment and found to work. A complex business can be studied and its business information systems defined in a matter of weeks. It is extremely important to note that the output of the study is directly relatable to data organization and data base design.

The methodology will work whether applied to large or small businesses or to a subset of a business.

REFERENCES

1. "Business Systems Planning Information Systems Planning Guide," (IBM 1978) GE20-0527
2. "BIAIT Business Information Analysis and Integration Technique," Donald C. Burnstine, Fox Hollow, Petersburg, N.Y. 1978
3. *Planning and Control Systems A Framework For Analysis,* Robert N. Anthony, Harvard University Press, 1965
4. "A Concept for Information Systems Control & Planning," J.A. Zachman, IBM Western Region, 3424 Wilshire Blvd., Los Angeles, Calif. 90010
5. "Designing the General Manager's Information System," S.D. Catalano, P.D. Walker (IBM 1973)

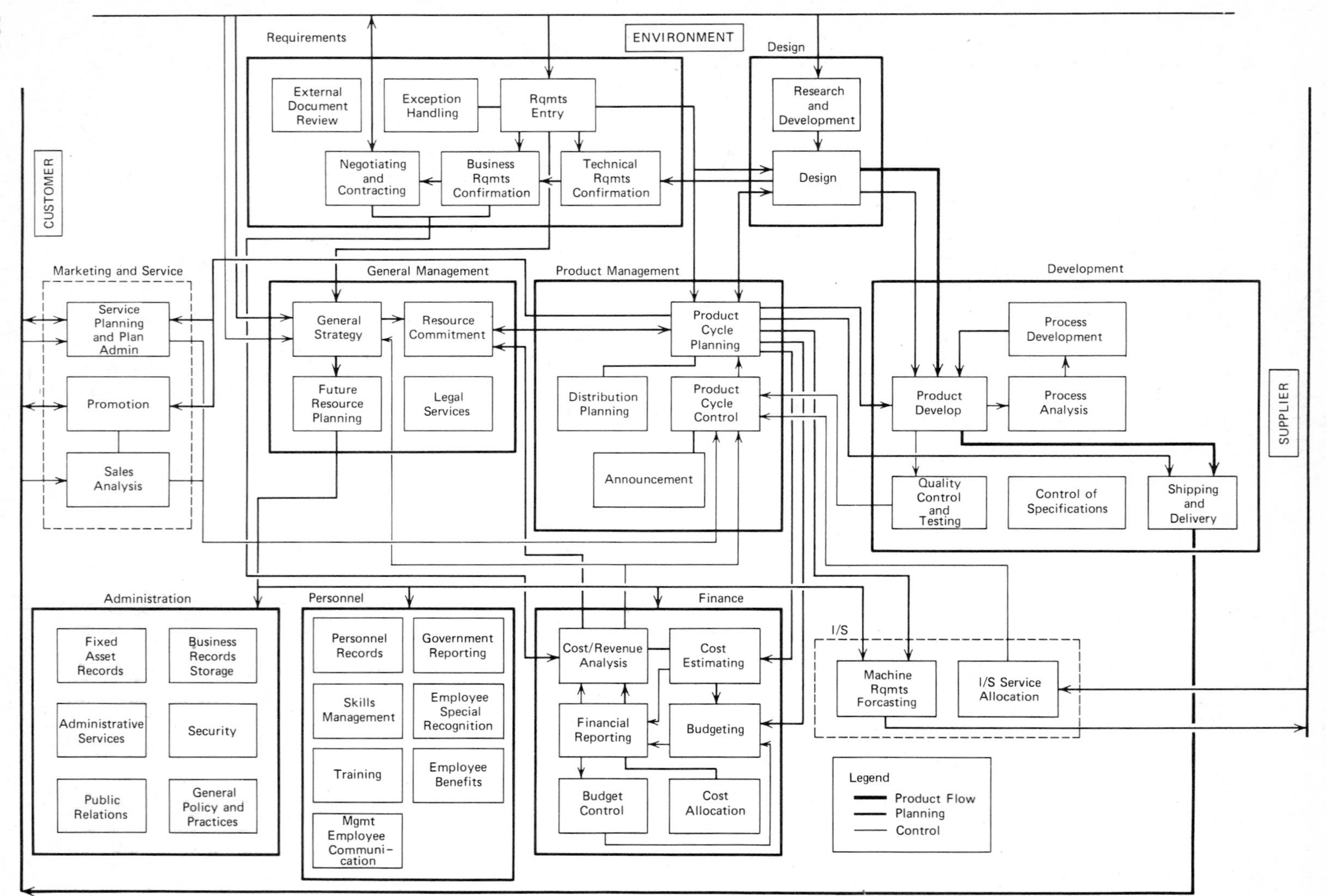

FIGURE 10 Information-handling discipline network.

READING QUESTIONS

1. Define BICS. What functions does it perform?
2. What is accomplished through BIAIT/BICS that was missing with the use of prior techniques?
3. Record the steps in the BICS procedure.
4. What are the advantages of BIAIT/BICS? What are its shortcomings?
5. Explain Figure 5—how it is developed and what it means.
6. Explain Figure 6—how it is developed and what it means.
7. Explain Figure 7—how it is developed and what it means.
8. Explain Figure 8—how it is developed and what it means.
9. How does BIAIT/BICS relate to the system life cycle? How would an analyst use it in conjunction with PSL/PSA?
10. Compare the features of BIAIT/BICS with those of BSP. Show how these approaches might be used in conjunction.

Business Systems Planning

IBM

INTRODUCTION

BACKGROUND

To survive, organizations must change as the environment changes. Today the environment is changing more rapidly and dramatically than at any time since the beginning of the industrial revolution. Technological, sociological, political and psychological changes—all combine to make business decisions increasingly complex.

To cope with these changes and improve productivity, executives need to base their decisions on a more complete view of business activities across functional and operational lines. Such an overall view helps management understand how an action in one area may impact other areas.

Executive decisions are usually as good as the information on which they are based. Information is required to operate any business, and its value can be judged by how well it contributes to the achievements of the business. For total-business decision making, information should be available throughout the business rather than only to individual functions or departments.

Since information is used to control resources such as cash, personnel, materials, and facilities, it is itself a valuable resource and should be managed with as much care as other more visible resources. To obtain information of maximum value with the least cost in the long run, a network of information systems should be build around the business processes essential to achieving business objectives.

Recognizing the need for such management information, some companies have attempted to develop management information systems. Many of these efforts have fallen short of their objectives because the planners:

- Attempted to implement information systems without first understanding the business from general management's viewpoint

Source: IBM, Business Systems Planning (GE 20-0527-1), 1975, pp. 1–92.

- Attempted to implement a totally new management information system rather than one planned to evolve from current systems
- Lacked sufficient management acceptance of the role of information systems in helping them achieve business objectives. Hence, management support and direction were not adequate.
- Lacked real management understanding of the standards and disciplines necessary to implement information systems across the entire business rather than simply to individual operating units

BUSINESS SYSTEMS PLANNING

A fundamental tenet underlying BSP is that an information systems plan for a business must be integrated with the business plan and should be developed from the point of view of top management and with their active participation.

The primary objective of the *identification phase* is to understand the business. After acquiring this knowledge, the planner can identify the information systems that are needed to logically support the business and group them into an information systems network. Understanding management and the value it places on the component subsystems within the network leads to a recommendation for the first, or most needed, subsystems to be implemented. An action plan can then be developed to guide the activities in the follow-on *definition phase*.

The definition phase validates the information systems network identified in the first phase and defines and justifies the first and most needed subsystems necessary to implement the network. This phase culminates with the development of an information systems plan for managing the network implementation, including definition of the major actions, schedules, and resources required. The activities required to complete both of these phases are described in this guide.

INFORMATION SYSTEMS PLANNING

In an article entitled "Blueprint for MIS,"* Dr. William Zani states, "Traditionally, management information systems have not really been designed at all. They have been spun off as byproducts while improving existing systems within a company. No tool has proved so disappointing in use. I trace this disappointment to the fact that most MIS's have been developed in the 'bottom up' fashion . . . an effective system, under normal conditions, can only be born of a carefully planned, rational design that looks down from the top, the natural vantage point of the managers who will use it."

A basic framework is needed for approaching the subject of management planning and control systems. Usually, planning and control are discussed in theory as separate management functions. In practice, however, these are not separate activities carried on at different times, or by different people. To be effective, planning and control must be integrated; furthermore, they must be integrated vertically to include the functions of strategic planning, management control, and operational control. These functions can be defined as follows:

- *Strategic planning.* Deciding on the objectives of the business, changes in these objectives, the resources required to attain these objectives, and the policies that are to govern the acquisition, use, and disposition of these resources. Strategic planning is the process of forming plans and policies that determine or change the charter or direction of the business. Both long- and short-range plans are formed as required. A considerable amount of external data is used at this level.
- *Management control.* The process by which managers ensure that resources are obtained and used effectively and efficiently in the accomplishment of the business' objectives. Management control reflects the general

**Harvard Business Review,* November/December 1970.

management view of the total business, where a balancing of resources among the various processes must be achieved to maximize output with minimum resources. Information is needed about each part. In the past, information tended to have a financial structure, with money as the common denominator. However, money is not the only basis for measurement; other quantitative measurements (market share, yield, productivity, industry averages and volumes, etc.) and qualitative measurements are also important. Supporting information systems need to be coordinated and integrated so as to reflect this data in a form that can be reconciled within and across the business. Management control is therefore heavily oriented toward line management and holds the key to the success and survival of the business.

- *Operational control.* The process ensuring that specific activities and tasks are carried out effectively and efficiently. This function focuses on the execution of specific tasks. It takes place within a context of decisions made and rules formulated in the management control process, and to some extent in the strategic planning process. Overall performance in activities where operational control is applicable is reviewed as part of the management control process. Examples of tasks are scheduling and controlling individual jobs, procuring specific items for inventory, and specific personnel actions. Some but not all of the tasks can be predefined and programmed. This focus contrasts with the view of management control over the whole stream of ongoing activities, using summaries, aggregates, and totals of information to monitor progress. Whereas a management control system is usually built on finances, operational control data is frequently nonmonetary and appropriate to the specific area of application.

Just as business planning activities must be integrated with control, so must the information systems planning and development activities be integrated with the business objectives. This aim is accomplished by information systems that support the business processes independent of organizational structure. An overall information systems strategy is necessary, and a plan that identifies the systems, their interrelationships, priorities, and migration from current systems.

In most instances, the current operational systems were developed for the right reasons and have been performing effectively for their specific intent. BSP does not result in a radical redo of the operational systems environment; although the BSP approach identifies long-range requirements from the top down, it defines the systems from the bottom up using most of the operational systems and data as they currently exist.

An information systems plan should facilitate a modular approach to implementation, providing confidence that each of the modules will fit and function properly in an integrated network. The plan should also allow for better decisions concerning the efficient and effective commitment of information systems development resources. This control is particularly necessary when moving toward a data base environment, which requires common data to be provided from and to multiple sources.

THE BSP PROGRAM

Although the approach and methodology for BSP originated primarily within IBM, work with customers across many industries has proved the approach to be effective regardless of industry type and size. The BSP approach is founded on a set of basic tenets:

- Businesses should change in response to changes in their environment.
- The success of a business depends to a great extent on the effectiveness of management decisions.
- The effectiveness of management decisions is

closely related to the information on which they are based.

- Resource optimization is a key to increased profitability.
- Management control is a key to resource optimization.
- Information systems requirements should be identified from the top down, with design and implementation from the bottom up.
- Because organizational changes generally occur more frequently than changes to business activities, information systems designed to support the business activities will survive longer and be easier to maintain.

The planning approach forces managers to get involved and to contribute their perspective in planning and systems definition. The management personnel of the business must assume this role and apply their leadership and business knowledge to achieve the desired output. IBM can assist and act as a consultant, but the plan will not succeed without top management understanding and commitment.

The effectiveness of the BSP methodology can be attributed to two components:

- *Principles and concepts.* The fundamental, unvarying ideas and logic that form the basis for BSP. Included are the standards upon which the procedures are based.
- *Procedures.* The sequenced activities, techniques, disciplines, time, outputs, planning team composition, etc., established to fill a particular organization's need and situation. Although the procedures will be consistent with the BSP's basic principles and concepts, they are flexible and will vary with the particular environment.

VALUE OF BSP

The rapidly changing environment and the need for businesses to adjust quickly to these changes make it necessary for executive management to have useful information available when needed. The information should be structured and processed as an integrated system. It should give executives the ability to make meaningful resource allocation tradeoffs. With a plan that leads to the implementation of information systems to support the processes of the business as a whole, such data can be more readily obtained.

Application of the approach and methodology contained in this planning guide offers many potential benefits.

Executive management can gain:

- An evaluation of the effectiveness of current information systems
- A defined, logical approach to aid in management control problems from a business perspective
- An assessment of future information systems needs based on business-related impacts and priorities
- A planned approach that will allow an earlier return on the company's information systems investment
- Information systems that are relatively independent of organization structure
- Confidence that information systems direction and adequate management attention exist to implement the proposed network

Functional and operational management can expect:

- A defined, logical approach to management control and operational control problems
- Consistent data to be used and shared by all users
- Top management involvement to establish organizational objectives and direction, as well as agreed-upon system priorities
- Systems that are management- and user-oriented rather than data processing-oriented

Information system management can take advantage of:

- Top management communication and awareness
- Agreed-upon system priorities

- A better long-range base for data processing resources and funding
- Better trained and experienced personnel in data processing planning to respond to business needs

The plan that results from a BSP study should not be considered immutable; it simply represents the best thinking at a certain point in time. The real value of the BSP approach is to create an environment and an initial plan of action that will enable a business to react to future changes in priorities and direction without radical disruptions in systems design.

PRINCIPLES OF BSP

Systems that are planned and implemented to support the needs of a business and that treat data as a business resource tend to reap significant payback; they can also take better advantage of technology and the business capabilities. A strategic information systems plan that is consistent with the objectives of a business requires development by and for the management of that business.

To achieve such a plan, BSP follows three principles:

1. Establishment of a business-wide perspective
2. Top-down analysis, bottom-up implementation
3. Systems and data independence

ESTABLISHMENT OF A BUSINESS-WIDE PERSPECTIVE

In defining information systems to deal with business decisions and problems, establishment of a general management vantage point is important. This high-level, wide-scope perspective forms one of the major distinctions between the BSP approach and more traditional data processing-oriented studies.

The level of perspective refers to the organizational segment for which the planning definitions will be done. BSP is not intended to address the information needs of a single area of the business—manufacturing, marketing, distribution, etc. It is intended to take the perspective level of general management—corporation, group, division, agency—where multiple functional areas are involved. The selected organizational segment is referred to as the business unit of BSP.

The BSP business unit may comprise an entire parent organization or some subcomponent. The selection of business units is illustrated in Figure 1, which depicts three sample options. Option 1, the least inclusive unit, takes in one division of general management at a level immediately above functional management. Option 2 includes that level plus a group at a higher level. Option 3 comprises the entire organization. In general, the selected unit should not be below a general manager level, where the potential exists for common systems and data sharing. The scope of the information planning within the selected organizational segment may exclude particular functions or departments (such as legal or personnel) as long as the major thrust is still cross-functional.

Once the organizational entity is agreed upon, the systems definition perspective within that segment needs to be established. Traditionally, data processing systems have evolved naturally in accordance with the type and level of problem to be addressed and the technological capability to solve it. To solve higher level problems, the business must be considered from the perspective of higher levels of management. Figure 2 depicts the possible differences in system characteristics as the perspective level rises from operational management, to functional management, and finally to general management. In a real environment, combinations of system types do exist.

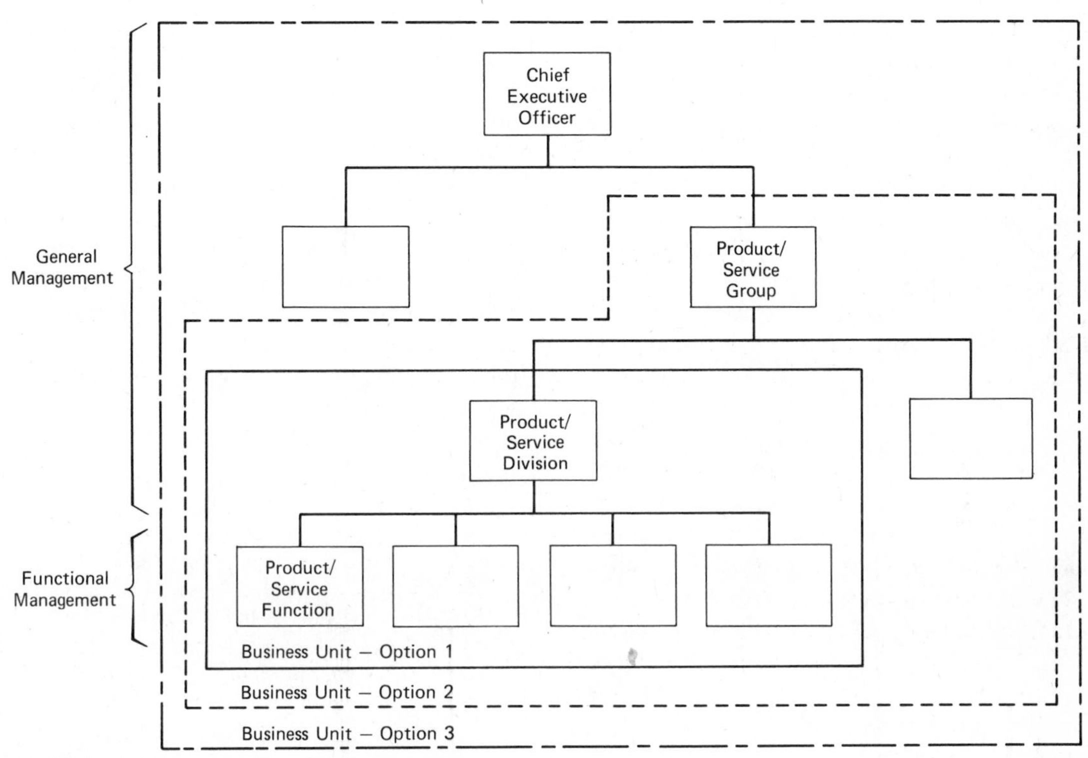

FIGURE 1 BSP organizational business unit options.

Definition Perspective	System Type	Problems Addressed	Planning Approach	Justification	Data
General Mgmt.	Business Information Systems	• Business-oriented • Mgmt. ctrl. • Decision-oriented	• Top-down planning • Bottom-up implementation	• Business results • Policy making	• Integrated data bases across business unit
Functional Mgmt.	Functional Information Systems	• Function-oriented • Mgmt. control	• Organization chart • Bottom-up implementation	• Avoidable costs • Productivity	• Integrated files per system (organization)
Operational Mgmt.	Computer Applications	• Operational control • Transaction-oriented	• Prioritized applications • Bottom-up implementation	• Displaceable costs	• Separate file per application

FIGURE 2 Definition of information systems characteristics from different perspectives.

Applications defined and implemented for operational management were the starting point for nearly all data processing activity and still represent the bulk of computer usage. Various problems surface as a data processing facility matures and attempts to provide real management information beyond that supplied as a by-product of operational transaction-processing. Technical problems develop, largely because of the difficulty of managing large numbers of data files that evolved with redundancies and processing dependencies. The provision of timely and accurate management information becomes more difficult, hampered by inconsistencies in data definition, file maintenance updating requirements, and scheduled processing times. Gaps in applications and data crop up because of a lack of past user demand for mechanization in particular areas or an inability to cost-justify a particular activity. As a result, resystematizing (redesign and integration of existing computer programs and data) has become advisable or necessary to deal with larger or higher level problems or to increase data processing efficiency.

Functional-level information systems often result from resystematizing to address the problems from the perspective of a functional manager. Examples of this type of system are a manufacturing information system or a marketing information system. Such systems are normally larger, with information and processing advantages over smaller, unrelated applications. However, this approach has some drawbacks. The system definition scope normally parallels organizational boundaries; the resulting vertical orientation may restrict system modularity, particularly when data needs to be shared across functional organization lines. General management may also encounter inconsistencies in information from a multiplicity of vertically oriented systems, much as the functional manager encounters inconsistencies from operational-level applications.

Systems defined from the perspective of the general manager, referred to in Figure 2 as "Business Information Systems," are free to take on horizontal support characteristics. Systems defined in this way should relate more directly to the objectives and problems of the business as a whole.

Management activities and decisions become more planning- and control-oriented at progressively higher levels of management. BSP is specifically oriented toward identifying and defining the planning and control of the business problems of general management. A framework for classifying management planning and control activity is shown in Figure 3.

Strategic Planning

Planning / Control & Measurement / Execution		
	Business	Goals and objectives, image, marketplace, product lines, ventures, acquisitions, organization, management system.
	Resources	Policy relative to personnel, facilities, material, vendors, customers
	Financial	Targets and policy relative to revenue, expense, profit, ROI, investments, dividends

Management Control

Planning / Control & Measurement / Execution		
	Product	Definition, selection, forecasting, pricing, sales objectives, volumes, supply/demand balancing
	Resources	Requirements definition, allocation, and retirement decisions relative to resources
	Financial	Budgeting, revenue-cost-profit by product, organization-fixed/variable-direct/indirect

Operational Control

Planning / Control & Measurement / Execution		
	Product	Design, make-or-buy, production, inventory, distribution, maintenance, sales, order-entry, advertising
	Resources	Acquisition (recruitment, hiring, vendor analysis, purchasing) and use (record-keeping by quantity project, product) of resources
	Financial	Cost accounting, receivables, payables, billings, disbursements, payroll, general ledger

FIGURE 3 Representative processes by planning and control levels.

TOP-DOWN ANALYSIS, BOTTOM-UP IMPLEMENTATION

BSP uses top-down analysis during the identification and definition phases of the study. With this approach, top management can define business needs and priorities to help assure that their perspective will predominate in the initial system definition.

The study team's first task is to gain a broad understanding of the business, or of the significant portion of it under study. The team must identify the key elements necessary to analyze the information requirements of the business. These key elements and their relationship are shown in Figure 4.

In its top-down analysis of the business and of the information necessary to support the business, the study team first requires a knowledge of the objectives and the problems faced by the business in meeting these objectives. Certain processes must be performed by the business to enable it to meet its objectives; the study team's second step is to identify these processes, or define them if necessary. The team can then relate the current organization with its assigned responsibilities and activities to the various processes. To function, the organization requires certain information. Some of this information is provided through computer applications that rely on data files containing classes of data—personnel, financial, product, etc. At this stage the study

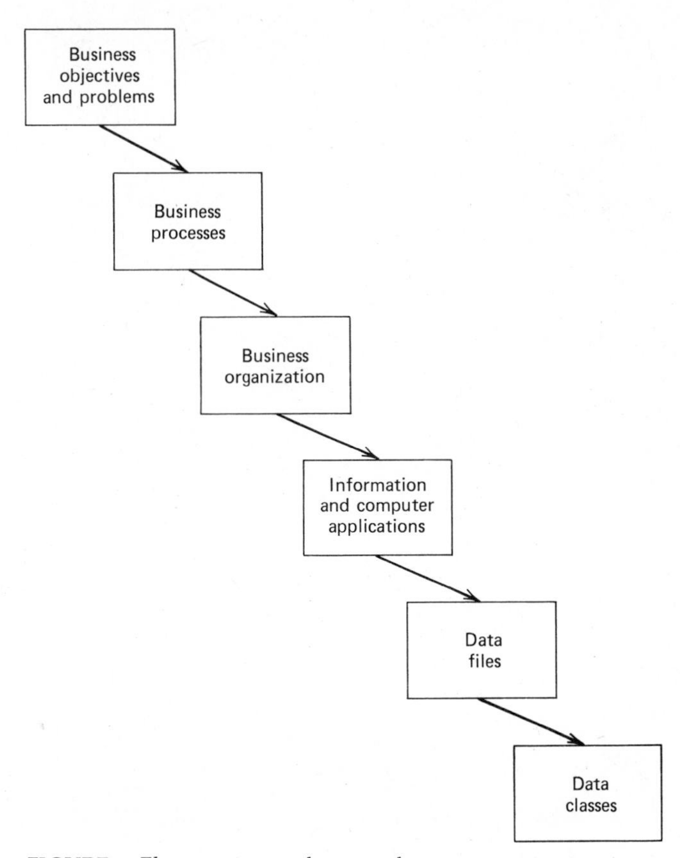

FIGURE 4 Elements in top-down analysis.

team is interested in identifying broad classes of data, not the actual data elements, which are defined later during information systems design.

The BSP approach becomes bottom-up oriented during the follow-on design and implementation phases; however, bottom-up definition of systems is addressed to some extent during the identification and definition phases. The systems definition should provide the direction for more modular and evolutionary development by taking into account the current systems and establishing practical ways of setting long-term priorities.

After the study team has gained an understanding of the business, the problems associated with meeting its objectives, and the information needs of management for operation and control, it should be able to define a business information system to meet these needs. The key elements of this implementation process are pictured in Figure 5.

The bottom-up approach for business information systems implementation incorporates all the key elements of top-down analysis with two notable exceptions. One is the exclusion of the business organization; instead, the data provided by the information systems network is related directly to the business processes. Since this network is specifically defined to support the processes necessary to achieve the objectives of the business, it should help ensure the availability of the information to the various parts of the organization responsible for performing the processes.

The second exception is the data files, which are replaced in information systems design by a data base. Separate data files tend to proliferate when applications are developed to satisfy the particular needs of individual functions or de-

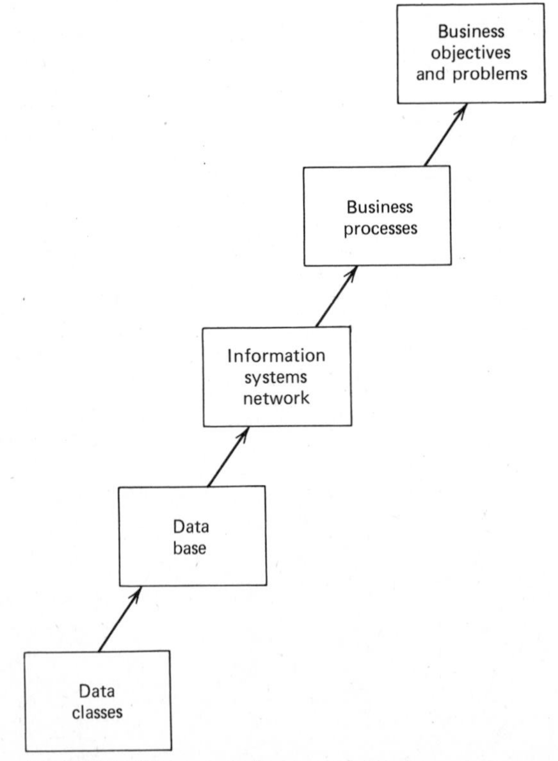

FIGURE 5 Elements in bottom-up implementation.

partments. Operation of such systems may result in redundancy, inaccuracy, and untimeliness of information. Therefore, an important potential benefit from the design of an information systems network is the collection of similar and related data into a data base, which can then supply this data to computer applications as needed. By capturing data once and making it available to all with a need to know, the data base satisfies the requirement of Business Systems Planning for data independence and organizational flexibility. This design overcomes many of the inherent problems of multiple files.

INFORMATION SYSTEMS INDEPENDENCE

A principle of BSP is to define information systems so that data is independent of the organizational structure of the business.

Organizational structures, both vertical and horizontal, can change with time as a result of growth, cutbacks, new management, external pressure, product life cycle, etc. When systems are designed to support specific organizations (and people), they often become obsolete and must be redone when the organization changes. Conversely, desirable organizational changes may be restricted because of certain major systems in place. The key, however, is to have systems address problems from a total business perspective rather than from the various perspectives of suborganizational units. A highly desirable aim, therefore, is to define systems to be as independent of a specific organization as possible. This principle does not imply that specific reports from a system should not be designed for a specific organization's use; it does mean that system support boundaries should be relatively free of organizational constraint and that multiple organizations should share in the use of consistently defined data in the solution of business-oriented problems.

As long as the principal mission of a business (that is, the general product or service area) remains constant, the basic activities and key decisions for doing business should remain essentially the same, regardless of the organizational structure. These essential activities and decision areas are referred to as business processes. The key to providing organizational independence lies in the identification and definition of the business processes. These activities constitute the two primary phases of the BSP approach.

Proper definition of business processes can provide the basis for:

- Achieving organizational independence
- Understanding how the total business unit works
- Defining the scope of the planning effort (the planning scope may be defined relative to processes as well as to a selected organizational segment)
- Defining, modularizing, and setting priorities for planned systems support
- Defining key data requirements
- Relating and understanding current systems support
- Understanding and separating management planning and control processes from operational control processes

BSP APPROACH

The key to implementing business information systems is the unified systems approach. Since a business is considered as a unified whole, information systems planning should also be treated initially in broad terms. The information systems planning approach should therefore be oriented toward the goals of the business and the activities associated with achieving those objectives rather than toward personnel, products, or organization structure. The purpose of this chapter is to describe briefly the basic BSP approach in terms of what is to be done. The next two chapters contain detailed instructions on the identification and definition phases.

The evolution of a network of unified systems requires a sequence of several phases leading to the operation of any system. Although the phases may be described differently by many businesses, the following list illustrates how the BSP identification and definition phases fit into the overall information systems implementation process.

Phase 1—Identification of Requirements
Phase 2—Definition of Requirements
Phase 3—General Design
Phase 4—Detailed Design
Phase 5—Development and Test
Phase 6—Installation
Phase 7—Operation

Figure 6 shows the time sequences of the phases for systems implementation. The arrows between the phases indicate that knowledge acquired and decisions made in later phases are fed back to earlier phases. The feedback loops reflect the iterative problem-solving process, and the overlapping of the phases in the figure illustrates concurrent implementation activities.

Throughout systems implementation, old requirements may undergo refinement, and more detailed requirements may be generated. When a system first becomes operational, actual experience with it may give rise to new requirements. Changes in the system's environment or in technology may also result in the creation of new requirements.

BSP addresses the first two phases of the seven-phase implementation cycle in an information management system: the identification and definition of the requirements. These phases culminate in the development of a plan that defines the systems and subsystems and establishes the requirements for their successful implementation.

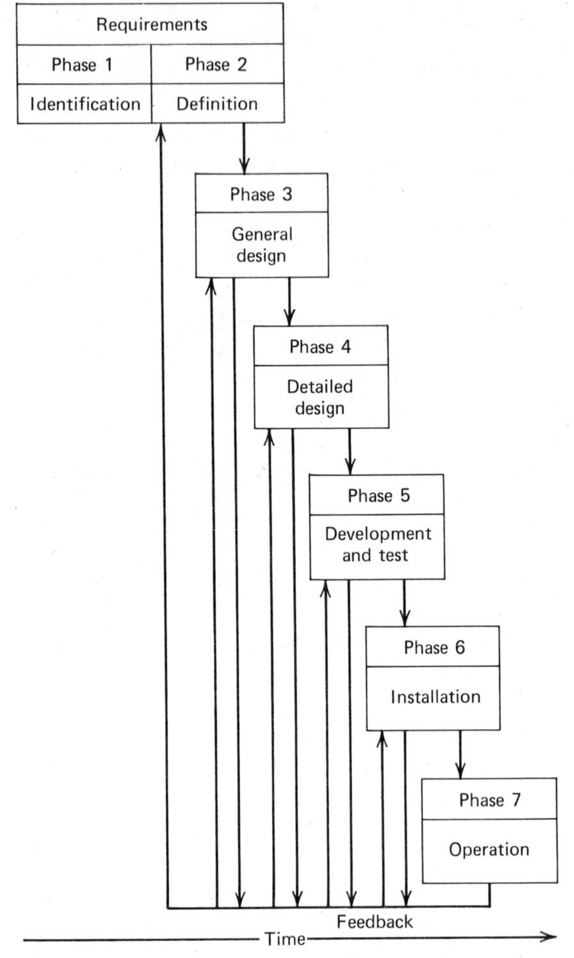

FIGURE 6 BSP information systems implementation phases.

IDENTIFICATION PHASE

The identification phase is the first of several phases of information systems development leading to the implementation of a network of data base/data communications (DB/DC) systems. The identification phase itself does not result in a design, but rather identifies the relationships and relative values of an information systems network.

Study Team

The success of BSP in any business depends from the very beginning upon executive involvement. Thus, a fundamental requirement is that a top executive of the business sponsor the BSP study. The executive sponsor appoints the team leader, a businessman with broad perspective who commands respect from management. In addition, a manager or senior systems analyst from the data processing organization is needed on the study team to supply first-hand knowledge of data processing as it is currently being performed. The team may include other members of the organization with particular attributes and knowledge within the level and scope of the business being studied.

Since the identification phase is expected to encompass all (or a very significant portion) of a business in a relatively short time, the team should be kept small and manageable. IBM Business Systems Planning consultants may assist the team and provide the added dimension of suggestions and ideas learned from other studies.

The first task of the study team is to establish the identification phase study objectives. After the executive sponsor has reviewed and approved these objectives, an action plan should be established as to how the team intends to proceed. Basically, during the identification phase the study team should expect to:

- Develop an overall understanding of the business.
- Understand how data processing currently supports the business.
- Identify a gross network of information systems that will support the business.
- Identify the first or most needed subsystems to be implemented within the network.
- Develop an action plan for the definition phase.

Data Gathering

After the study team has agreed on its objectives and obtained executive sponsor approval, data gathering assignments can be made to the appropriate members of the team. The data required is of the type that will aid the team in understanding the business from the perspective of the general manager rather than of any particular organizational unit. The team should keep this key factor in mind throughout the course of the study. The initial data gathering is intended to provide a general understanding of the business—its environment, objectives, organization, products, markets, etc.—so that team members will be better conditioned to recognize business problems and recommend appropriate business solutions.

To provide a foundation for understanding and to avoid delay when the team is assembled to actually carry out the identification phase, it is recommended that the team leader and the data processing representative prepare an orientation session. In this meeting they should identify and summarize data to show the business and data processing conditions from their perspective. Information that can help provide this basic understanding includes:

- Organization structure
- Environmental analysis
- Current plans
- Financial and product reviews
- Market analysis
- Profile of information systems
- Results of previous related studies

This information should be prepared and presented to the study team on easel charts so that it will be available for reference during the study. The initial presentation is intended to provide a summary as a starting point for the team. Additional information will be identified and obtained by the team as the study proceeds.

Business Processes

A business process is an essential decision and activity area required to manage and administer the resources and operations of a business. Business processes are defined independent of organization. Examples of major business processes are marketing, engineering, manufacturing, and distribution. A major business process area encompasses several subprocesses; for example, manufacturing includes production scheduling, raw material ordering, vendor selection, cost estimating, etc. The data needed by management should be structured and processed in such a way as to support the business processes.

Business Process/Organization Matrix

Within a business, executive management sets up an organization in order to assign responsibility for making the decisions and carrying out the activities that constitute the various business processes. However, the BSP study group seldom looks at the information needs of the business from the organizational viewpoint. Instead, it groups similar activity and decision areas (business processes) into functional units that require similar information to help them perform their assigned activities. To relate the organization to the areas where the information systems support will be directed, the study group constructs a process/organization matrix.

By completing this matrix, the BSP team can identify the business structure and the assignment of responsibilities with business processes. Initially, the study team draws up the relationships as its members observe the organization working, and they can then confirm or modify the matrix after interviewing the executives of the functional areas.

Executive Interviews

Up to this point in the study, the BSP team has only the information collected from existing documentation or through observation of the business. The next step is to interview top executives and key administrators. These high-level interviews are intended to broaden the team's understanding of the business, validate its conclusions so far, and gain knowledge of management's views, values, priorities, and information needs. The team asks questions concerning:

- Objectives and responsibilities
- Methods of measurement
- Anticipated changes
- Major problems and their impact
- Level of satisfaction with current information
- Information requirements and value

While some teams members are carrying out the interviews, other members should be compiling information regarding current data processing support.

Current Data Processing Support

The implementation of an effective network of DB/DC systems is an evolutionary process. As a starting point, the BSP team requires a thorough knowledge of the present applications and activities of the data processing organization within the business. With the proper information, the study team can identify current strengths and weaknesses, establish priorities, and recommend the subsystem areas that should be addressed first.

In most businesses, well documented applications are rare. Information that briefly describes an application (what it does, for whom, when, what data files it uses, etc.) seldom exists. The information is needed, however, to understand what services are currently available. The study team has the task of gathering this information and organizing it so that it can be readily correlated with the information satisfaction and requirements comments obtained through the executive interviews. The team can then begin to assign priorities for future development activities.

Analysis of Findings

The study team has now assembled data from documents, from executives, and from its members' own experiences within the business. This data needs to be gathered, summarized, compiled, and interrelated in such a way that it becomes meaningful rather than simply a voluminous collection of information. The study team should adopt techniques that will ease rather than complicate the analysis. Matrices showing the interrelationship between processes/organization/data and information systems have proved to be most beneficial. Additionally, well summarized interviews that are structured the same for all interviewees can aid greatly in compiling problems and needs, analyzing them for similarity, and determining group solutions. The analysis, which leads to conclusions, recommendations, and a management presentation, is one of the creative portions of the study.

Conclusions and Recommendations

While the team is developing its conclusions and recommendations, the members should constantly keep in mind the objectives of this study phase: to develop an understanding of the business and determine its needs for information to support the effective operation and management of its processes. The various types of needed information can be grouped into systems and related to one another in terms of a network. Identification of the systems to be incorporated into this network is a further objective of this study phase. Since not all of the systems in the network can be developed at once, the next step is to identify the subsystems within the network where work should start. The following criteria are useful in determining these first subsystems:

- Will the subsystem provide significant near-term savings?
- Will it remedy a major problem in the organization?
- Will it establish a data base structure for subsequent systems in the network?
- Will it provide a substantial long-term return on investment?
- Can it be achieved with the existing resources?
- Is its success reasonably assured?

The conclusions and recommendations developed by the team are packaged, together with recommended courses of action—what needs to be done, why, and by whom. The information package is then presented to the executive sponsor for approval.

Action Plan and Presentation

Although presentation of the action plan is the last identification phase activity to be performed by the BSP team, the team should not wait until the end of this phase to decide the type of information to be presented, to whom, and through what media. It is advisable that these questions be considered when the team first prepares its plan for the study.

Usually a presentation is made to the executive sponsor of the study and the top executives of the business. The media used should be consistent with presentation methods to executives in that business. A documented study report is desirable to supplement this presentation. The report will also be useful as a reference for others in the business who will be called upon to implement the approved recommendations.

This material constitutes a brief introduction to the Business Systems Planning approach and its identification phase. Under "Identification Phase Activities," details are provided on who is responsible for the various steps of this phase and how the activities should be performed.

DEFINITION PHASE

The objective of the definition phase is to develop a long-range plan for the design, development, and implementation of a network of information systems based upon data base/data communications (DB/DC) design concepts. The plan should incorporate more detail for development of the first subsystems as recommended in the identification phase, and serve as a road map for implementation of the other systems contained in the information systems network.

The potential benefits of a long-range systems plan are considerable. To assist in the planning process, the identification phase study team provides a preliminary plan for the definition phase based on its findings, conclusions, and recommendations. The approach for the definition phase is outlined below, and details of the activities required for its completion are provided on pages 279, 280.

Study Team

A fundamental requirement of the definition phase, as of the identification phase, is that a top executive of the business sponsor the study. To lend continuity and help ensure a smooth transition between the two phases, some of the identification phase team members should be assigned to the definition phase study team. In addition, one or two user personnel knowledgeable in the areas recommended for definition of the first subsystems should be assigned to the study. User participation in the study helps to assure that the systems to be further defined will actually meet the user requirements. Assignment of a systems analyst who is familiar with the present applications that support the initially recommended subsystems is also advantageous. This analyst can help with the definition of the first subsystems and can determine what applications already installed may be usable in the new subsystems.

Action Plan

It is essential that the study team develop a detailed action plan for the definition phase, either in the interval between the two phases or at the start of the definition phase. The level of detail in the action plan will depend on how the study team leader wishes to control the activities and report progress to top management.

Interviews

One of the key activities during the identification phase is a series of interviews with top executives and key administrators, designed to give the team a broader understanding of the business and the information needs of management.

Interviewing is also a key activity in the definition phase. However, the level of management to be interviewed changes, and the types of questions vary. At least two levels of management below the identification phase interviewees should be interviewed to complete the definition of the major systems in the network. The questions are more detailed and deal more in operational aspects. User personnel directly concerned with the recommended first subsystems should be interviewed at length to help the team understand what is needed and the general design required to satisfy the needs.

Network Definition

One important objective of this phase is to further define the major systems of the network and how they interrelate, and to determine the data flow between and among the systems. The steps to be taken are:

- Assess the user satisfaction with present and planned information systems supporting the business processes.
- Identify the processes and users that can share data.
- Determine the potential for common systems across organizational entities.
- Analyze the potential for consolidated acquisition and parallel distribution of data.

Definition of First Subsystems

The next step in the definition of the network is to define a generalized business information structure to identify the data characteristics within the network that match the business needs as established in the identification phase. The definition should describe in general terms the subprocesses and data bases that compose the first subsystems and how these subprocesses and data bases relate to each other.

During the definition phase the validity of the criteria used in the identification phase to select the first subsystems (see "Conclusions and Recommendations" above) should be reconfirmed. The following additional criteria should be considered in this phase:

- Will the subsystem provide a stable environment to test new technological approaches?
- Is it cost-effective?
- What benefits does it offer, both tangible and intangible?

This step results in an in-depth, top-down definition of the processes and users involved in the scope of the first systems. The method for defining the first subsystems is the same as that for the network except that the first subsystems definitions must include consideration of all levels of management in the process, whereas the network definition requires consideration of only selected higher levels. In addition, projections of resource requirements, including equipment, must be considered for the first subsystems.

Information Systems Development Control

An important activity during the definition phase is to identify the areas of major concern in the development, control, and modification of the information systems network. The network is a very complex combination of projects, many of which may occur simultaneously. Without a systematic plan to control and measure the activities, success is unlikely. The requirements for such a plan are discussed in detail under "Network Management System."

Information Systems Plan

The information systems plan is compiled at the conclusion of the study and consists of two major elements:

1. A report on the study team's findings, conclusions and recommendations. This report should illustrate the study team's understanding of the business and present the information support requirements for the business. It should contain an analysis of existing and planned information systems, the mechanics by which they are managed, and the modifications necessary to fit them into the overall network of information systems.

2. A plan that identifies the actions required to develop an information systems network as well as the requirements to design, develop, and implement the systems in the network, with strong emphasis on the first subsystems to be developed. The plan should also contain a definition of what is necessary to manage the network. The main objectives of the plan are to:

- Get top executives of the business committed to a plan of action, supported with adequate resources.

- Provide a basis for planning, allocating, and procuring resources to implement the plan.
- Provide a discipline for measuring planned versus actual performance and keeping commitments on schedule.
- Highlight problem risks and opportunities for management action.

IDENTIFICATION PHASE ACTIVITIES

Four major activities should be performed to meet the objectives of the identification phase. These activities, and the key tasks associated with each, are listed in Figure 7. This section contains detailed guides and suggestions for performing each of them.

PREPARATION

Before the formal start of the identification phase, certain activities are necessary to help ensure continuity and success of the study and accomplishment of its objectives in a reasonable time span. Many of these activities can be done before the study to save the study team valuable time.

Study Action Plan

An action plan should be developed that spells out "who does what and when." The action plan formally identifies, defines, and sequences the major tasks and subtasks required to complete the identification phase. An action plan checklist similar to that in Figure 8 can be used in developing the action plan.

The action plan can be a Gantt chart similar to the example in Figure 9, depicting responsibilities and schedules. This plan should be developed by the team leader and agreed upon by the team members. Its main purpose is to gain better control of the study with emphasis on visual communication.

Study Team Orientation

Before the start of the identification phase the team leader should conduct an orientation for the study team, including such items as:

- BSP background
- Administrative details
- Specific objectives and expected output
- Action plan
- Project control
- Business environment
- Activities required before identification phase
- Study team assignments

A similar orientation should be held for the key executives that control the major functions under the study. If the executives have not been exposed to the basic BSP approach and principles, these should be included. In addition, the executives should be introduced to the type of questions they will be asked (see "Executive Interviews" later in this chapter). The executive orientation is normally conducted following the study announcement to executives.

Announcement to Executives

A formal announcement of the program should be made to emphasize the support that top management is giving to the study. The study sponsor prepares this announcement in the form of a letter, preferably signed by the chief executive officer. Distributed to all executives who control the major functions under study, the announcement states the objectives of the program, points out the potential value to the business, and stresses the desired candor and cooperation.

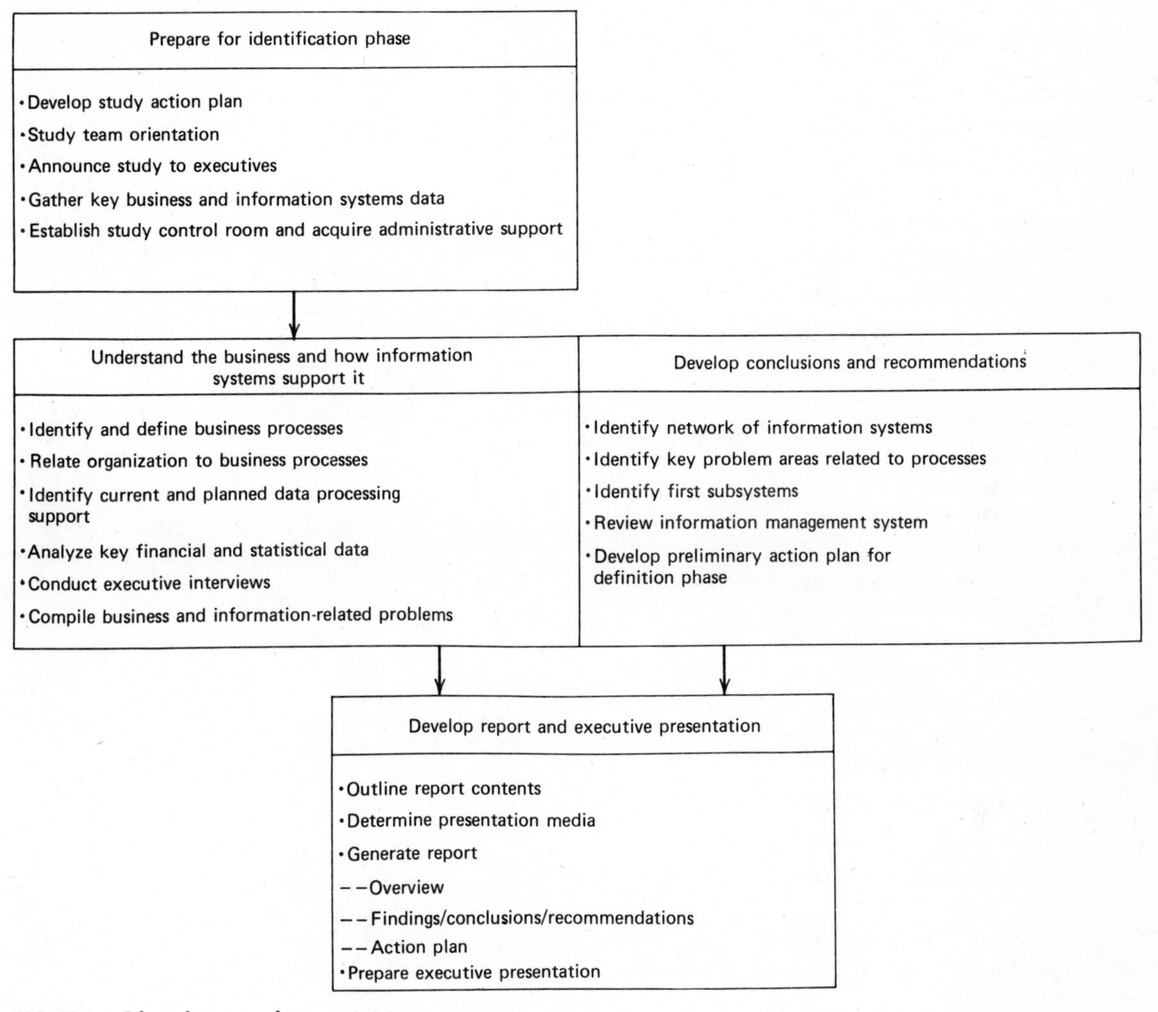

FIGURE 7 Identification phase activities.

Preliminary Data Gathering

The key business and information systems data listed in the table on the following page should be collected before the identification phase to save the study team valuable time. The items indicated with X's should be provided by the business-oriented team leader and the information systems representative or by people they designate outside the team.

As the identification phase begins, the team leader and information systems representative should be prepared to present this information in summary form to the entire team. Hard copy of the summary information should be made available to team members. This presentation will help all members understand the business, how data processing is presently supporting the business, and in what direction the business is heading. Although all the data listed above may not

	Responsible	Days	Date
• Orientation —Understand business —Understand DP environment			
• Administrative Detail —Study team control room —Clerical assignments			
• Define Processes/Subprocesses Of Business			
• Relate Processes To Organization			
• Develop Current DP Applications Profile —Description —Output —Operating environment			
• Relate Current DP Appliations To: —Organization —Data files —Processes			
• Interview Executives and Key Administrators —Determine list of interviewees —Prepare questions —Prepare and send letter to interviewees —Perform interviews —Summarize interviews			
• Outline Report Contents			
• Draft Basic Sections Of Report			
• Identify Major Problems And Impacts			
• Define Logical Information Systems Network And Data Bases			
• Relate Systems And Subsystems To: —Data Bases —Processes			
• Develop Conclusions And Recommendations			
• Identify First Subsystems			
• Develop Benefit Statements			
• Develop Action Plan For Definition Phase			
• Prepare Report			
• Prepare Presentation			
• Make Presentation			

FIGURE 8 Identification phase action plan checklist.

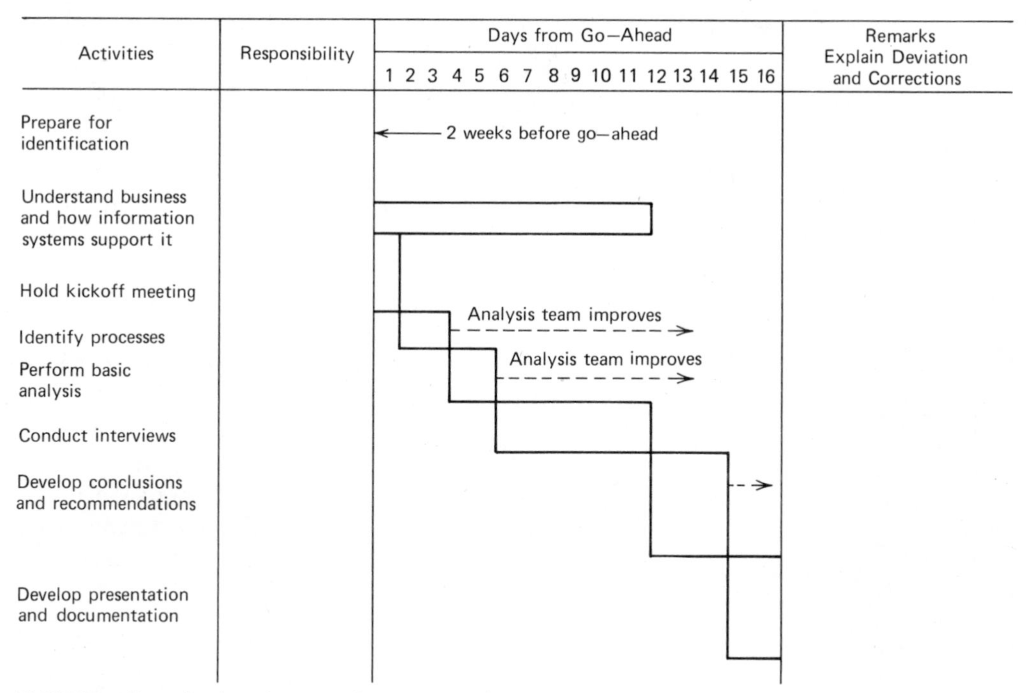

Activities
Responsibility
Days from Go-Ahead
1 2 3 4 5 6 7 8 9 10 11 12 13 14 15 16
Remarks
Explain Deviation
and Corrections
Prepare for identification
2 weeks before go-ahead
Understand business and how information systems support it
Hold kickoff meeting
Identify processes
Perform basic analysis
Conduct interviews
Develop conclusions and recommendations
Develop presentation and documentation
Analysis team improves
Analysis team improves

FIGURE 9 Sample identification phase action plan.

be accessible or used by the team, it gives the team direction in analyzing what is to be done and helps make the interviews more valuable.

A preliminary review is also advisable to ascertain whether the information systems management is capable of effectively planning, controlling, and operating the information systems function in the current environment. During this review, consideration should be given to identifying the capabilities that are lacking or that would impede the information systems organization in advancing the business toward a data base/data communication environment. A checklist is provided in Appendix A to assist in this review.

Study Control Room and Administrative Support

Certain administrative activities should be completed before the identification phase begins. Support personnel should be selected and given an orientation emphasizing their importance in helping to package and merchandise the output of the study. The type of administrative support required is:

Secretarial (letters, interview scheduling, text to be typed)
Art (visuals)
Reproduction (document production)

A conference room (study team control room) should be assigned for the team's use during the entire identification phase. This room should be of adequate size and comfort to allow the team to analyze the needed data, review the study's direction and status, plan and conduct interviews, validate findings, and prepare recommendations. Preferably, this conference room should not contain a phone, should be securable, and should contain at least one locked cabinet for the team's documentation. If possible, the conference room should be in a relatively central

Key Data to be Supplied to the Study Team		
Item	Business	Information Systems
Planning		
Objectives	X	X
Cycle	X	X
Responsibilities	X	X
Procedures and format	X	X
Update mechanism	X	X
Tracking	X	X
Project development		
Initiation	X	X
Priority setting	X	X
Justification/funding	X	X
Reporting/tracking	X	X
User involvement	X	X
Problem resolution	X	X
Organization	X	X
New or expected changes	X	X
Prime responsibilities	X	X
Size (number of people, cost, etc.)	X	X
Major related studies	X	X
Products (2–5 Yr.)		
Products/markets	X	
Products/competition	X	
Product trends	X	
Dependencies	X	
Life cycle	X	
Finance		
Profit trends and objectives	X	
Data required at executive level	X	X
Cost center: profit and loss reporting		X
Budgets	X	X
Information systems profile		X

location, not far from the study sponsor and easily reached by the interviewees. The room should have sufficient wall space for displaying charts and should be stocked with the ample supplies for the team's use—flip charts, markers, pencils, etc.

UNDERSTANDING THE BUSINESS

Today's organizations are bigger and more complex than ever before and must cope with technological advances and with the development of more complex products. Organizational philosophies and practices must keep pace with these changes. Therefore, during the identification phase the study team must develop a thorough understanding of the business and how information systems support the business. This understanding must be developed from the perspective of a general manager, and the team must continually keep the general manager's information needs in mind.

Management cannot be viewed as a single entity with a single set of activities. A typical three-level management hierarchy is shown in Figure 10. Each level is involved in planning, control, and operations activities. The general management level is at the top; the middle, or functional, level includes such managers as the manufacturing executive and marketing execu-

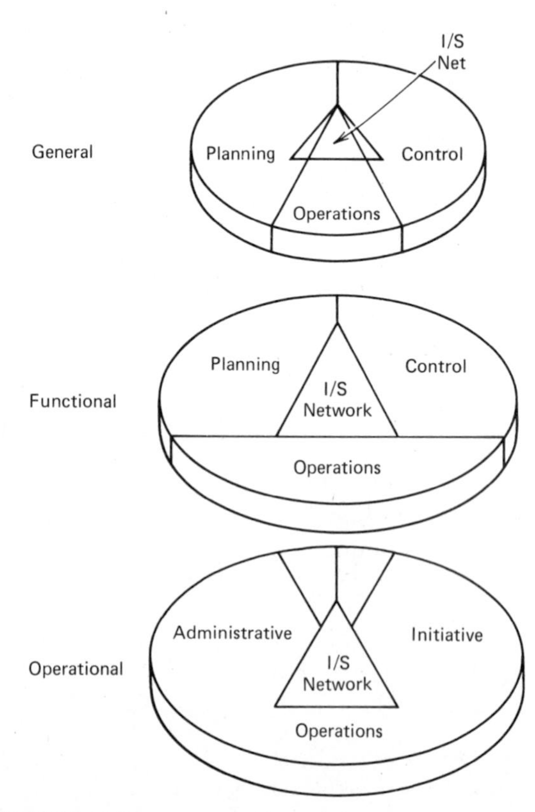

FIGURE 10 Management hierarchy.

tive; at the base is the operating level concerned with daily productive activities, for example, product sales manager or manager of a particular production line. At the bottom level, the time spent by sales managers to manage the processing of orders or perform administrative acts must be minimized to allow them more time for the actual selling.

The information needs to support each management level are essentially the same, the differences most being in type of information, level of detail, and required response time. From the general management point of view, the information system must tie the three levels together with the goal of allowing an increased proportion of time to be spent on activities such as planning and selling rather than on administrative operations.

A broad view of the organization is required to define an information system for the general manager. Simply stated, the role of a general manager is to maximize outputs with minimum resources. Depending on business scope, he is helped toward this goal by two classes of functional management: process management and resource management. These two classes are illustrated in Figure 11.

Process management functions, represented by the vertical bars, are for development, manufacturing, marketing, services, and finance. These processes are representative and will vary from organization to organization.

Resource management functions, represented by horizontal bars, are for cash, personnel, materials, and facilities. These resources will be the same regardless of organization, since they are basic. The primary role of the general manager is to allocate these resources among the various processes.

Today's large businesses have complex product mixes, subject to shifting markets in fast-

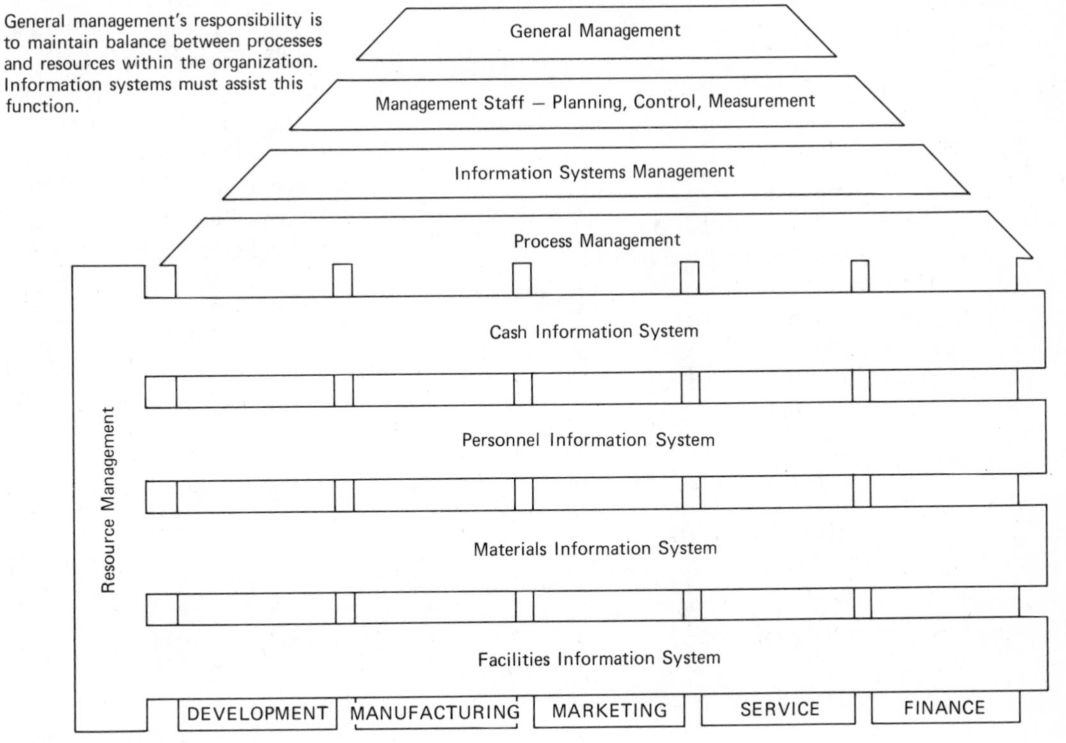

FIGURE 11 Process and resource management.

paced marketplaces. Faced with many varying decisions, the general manager must be able to assess the effects of both external conditions and internal plans as they relate to business objectives and operating performance. He must be able, for example, to recognize and assess alternative actions when his marketing force is selling more items than his plant is able to produce. The general manager's information system must be capable of supporting the management responsibility in a timely and accurate fashion.

The conversion of present operational and functional information systems into structures suitable for the general management concept is not a short-term undertaking. The progression toward such systems should be broken down into a logical sequence of activities. The identification phase—specifically, understanding the business and how information systems support it—is the first in this series of activities leading toward the implementation of an information systems network to support the general manager.

Business Processes

Through the 1950s and early 1960s the computer was primarily used as a tool for help in solving operational problems. It was often applied within a single functional group—finance, manufacturing, personnel, etc. In large organizations, separate computer systems were often found at multiple locations within a single function. The management of computer applications was autonomous.

Development of operational systems was generally uncoordinated and undisciplined. As a result, files were usually fractionalized. Some companies took steps to centralize key files within functional areas to provide functional management with a more consolidated and consistent information base.

Management has generally viewed the operation of its business, not as a series of interrelated processes, but rather as a collection of functional units with a set of responsibilities defined to carry out the required activities of the business. Consequently, when viewed in connection with processes, many functional organizations are found performing activities in several different process areas. Therefore, to achieve data independency in a business (that is, isolation from the impact of organizational change), information systems are defined to support processes.

Business processes are the essential decisions and activities required to manage and administer the resources and operations of a business. Business processes provide:

- A comprehensive understanding of how the business accomplishes its overall mission and objectives
- A tool for grouping organizational responsibilities and separating them from essential activities and decisions (the basis for defining systems independent of organization)
- A basis for understanding information needs and current data processing support
- A basis for separating management control processes from operational processes
- A basis for defining planned systems support, determining its scope, designing it to be modular, and setting priorities for it
- A basis for defining key data requirements

A business usually has between four to ten major processes. Each major process has a number of subprocesses, and activities for six industry types (distribution, education, health, manufacturing, process, and public sector) are included in Appendix B.

Because of the importance of business processes, they should be defined and classified in such a way as to:

- Obtain consistent level of detail
- Separate management control processes from operational control
- Reduce or eliminate confusion between a defined process and a particular organizational entity

Although no formula exists for defining the processes, the following approaches have been used successfully to aid in identifying and defining the processes.

1. Processes are defined in relation to resources:

- Resource management crosses functional (organizational) lines horizontally.
- Resource management is a key basis of management control activity.
- Resources go beyond traditional cash, facilities, personnel, and material to include products and services.
- A key resource (product or service) is typically selected when a business unit has common product areas (private industry) or common missions (public sector). Processes are defined in association with the overall supply/demand relationship of the key resource. Other resources are considered to support the key resource.
- Multiple resources may be chosen when significant information systems support is essential or of relative importance to several resource areas. For example, in a university environment, the students, curriculum, facilities, staff, and finances may all be considered key resources.
- Resource life cycle is used to logically order all important processes and help assure their identification.

  Life cycle stages may be:

  —Requirements, acquisition, stewardship, retirement

  —Feasibility, implementation, maintenance, disposition

  —Planning, installation, measurement and control, plan revision

2. Processes are defined in relation to business/information flow analysis:

- The total of business is considered.
- Diagrams of information that flow through the entire business unit are used to identify the major activities and decision points.
- The function or mission of the business is relatively constant (airlines, insurance).

3. Processes are defined in relation to common missions, services, activities or objectives:

- The level of the business unit encompasses multiple or unrelated product or service areas.
- Commonality of major processes exists.
- State or local governments or corporate conglomerate-type environments are studied.
- Logical flow may be difficult.

The study team should develop a brief description of the processes identified so that no question will arise during the study and no misinterpretation be made by those not involved in defining the processes as to the exact purpose and meaning of each. An example of the level of detail desired to describe the processes is shown in Appendix C.

Relationship of Organization to Business Processes

When defining the business processes, the team should ignore the existing organization structure except for referring to organization charts to check that the process definitions take in all the necessary activities currently being performed. Once the processes that describe what is necessary for the operation and management of a business have been defined, the organizational entities that perform the processes are identified. To present an overview of this relationship, a matrix is used. Other matrices, including systems and data bases, are later combined and related to it.

Figure 12 is an example of basic organization-to-process matrix. The processes and subprocesses of the business are identified horizon-

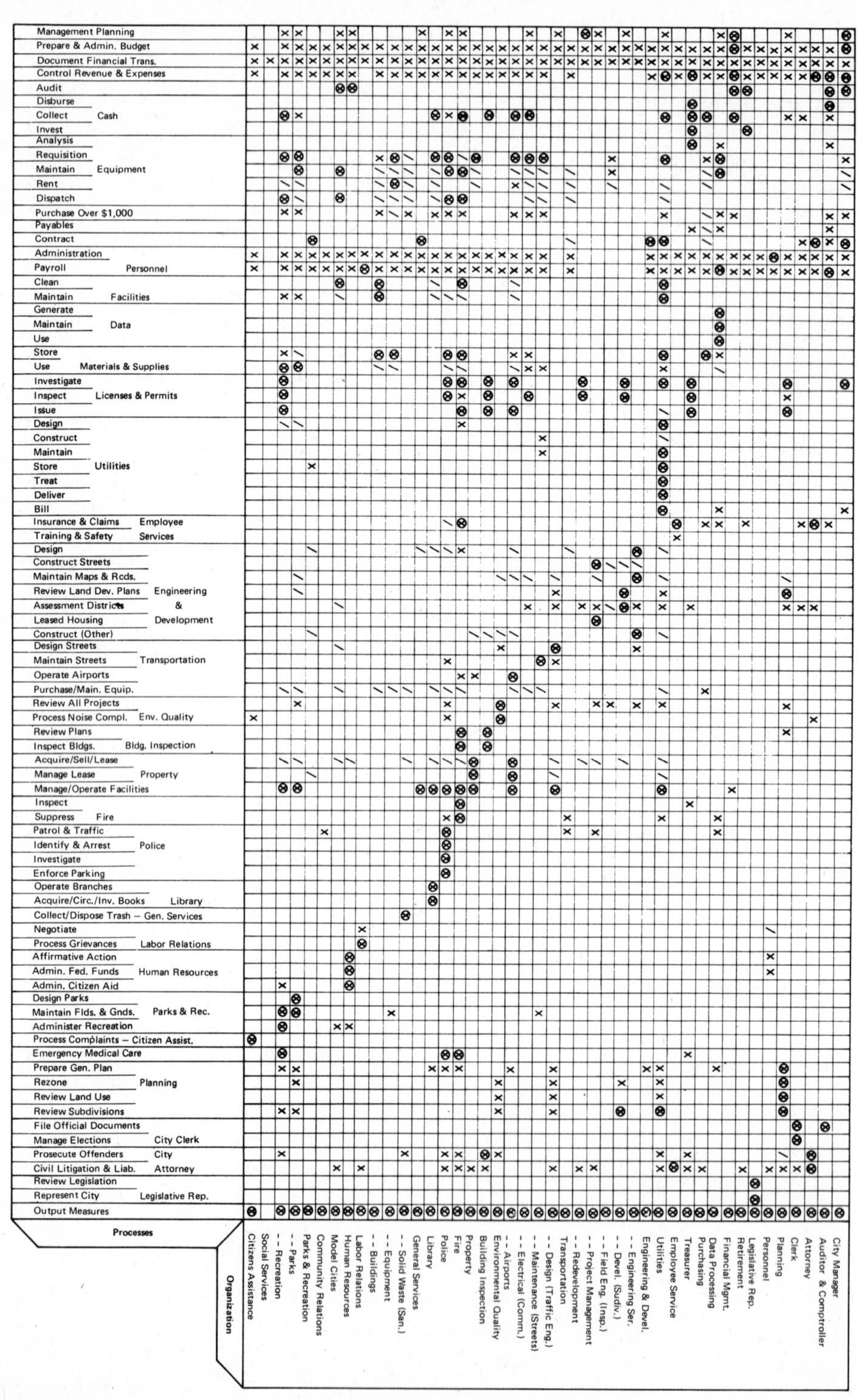

FIGURE 12 Organization-to-processes matrix (city government).

tally. The organizational units are identified vertically. The number of organizational units varies depending upon the size and complexity of the organization. Since the identification phase is meant to provide a broad overview of the business, or the portion of the business being studied, not every management entity is identified. Using an organization chart, the executive positions are identified as well as the functions reporting to these executives: the president or chief executive officer, and usually two levels below that position.

To keep this organization identification manageable, common or similar organizations are best represented as one organizational unit; for example, 100 sales offices may be shown as a single unit. Where feasible, plants and laboratories should also be grouped into units. One-unit representation is also generally appropriate when organizations of different scopes are doing the same job. For example, a financial planning organization may have three departments, each focusing on separate divisions. The mission of all three is still financial planning, and they can be represented by one organizational unit.

Once the processes and organization have been related, the study team completes the matrix by indicating the degree of involvement of each organizational unit to the processes. For example, in Figure 12 a code is used to show three levels of involvement:

Ⓧ Major responsibility and decision maker
X Major involvement in the process
/ Some involvement in the process

Such indicators do not describe the actual responsibilities of each of the organizational units but serve only as a guide to assigned responsibility for the involvement in a process or portion of a process. This level of indication is sufficient for the identification phase. Some businesses have found this matrix to be valuable after the study as an index to a management system manual, in which they develop actual responsibility and activity statements for each of the organization/process intersects.

The organization-to-process matrix is initially constructed by the study team using the members' knowledge and perspective of the business. It helps them identify key individuals to be interviewed and determine the questions to be asked the individuals responsible for the processes. It also helps in analyzing needed information systems support of the business. To authenticate the matrix, each person interviewed by the team should be asked to confirm or correct the portion of the matrix showing his responsibility or involvement. Details are presented under "Executive Interviews."

The organization-to-processes matrix provides the first quadrant of a four-quadrant matrix relating organizations, processes, information systems, and data bases required by the business. In Figure 13 the overall relationship is shown, and the functions within each of the quadrants are described. Development of the remaining quadrants is described below.

Current and Planned Data Processing Support

An important step in the successful development of a network of data base/data communications (DB/DC) systems is to understand current and planned data processing systems and applications—whom they support, how they are integrated, and the kinds of data (files) that support them. To gather this information, the study team should review the information systems plan, if one exists. If it does not, one of the study team members should work with someone in the information systems area who understands the current and planned information systems and can relate them to the business philosophy and problems. In reviewing the data processing support, the intent is to determine how it relates to the business processes, as shown in the sample matrix in Figure 14. However, since most sys-

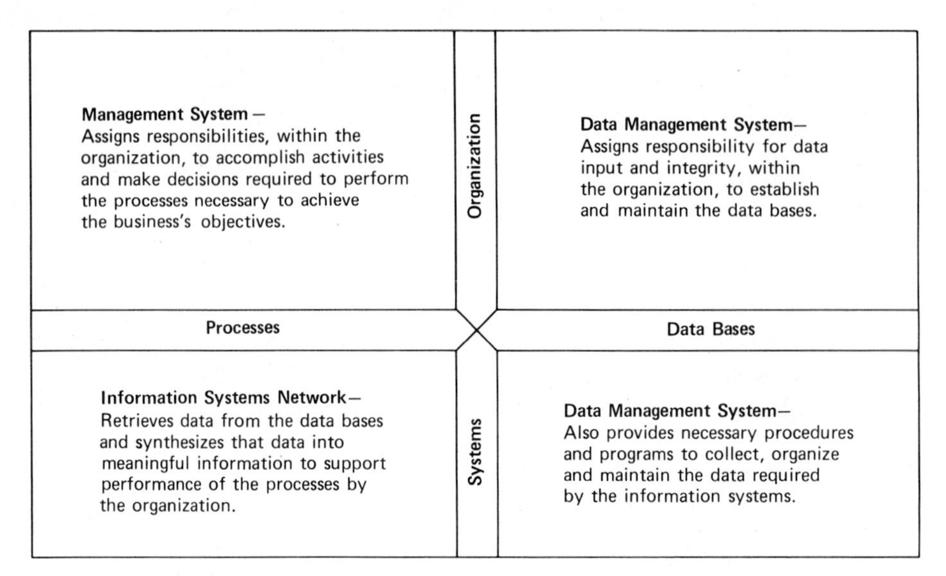

FIGURE 13 Interrelationship of management system to data management system and information systems.

tems and applications were developed by and for individual organizational units, establishment of a direct relationship may be difficult. The study team may find it advantageous to initially develop a matrix of data processing applications to functions as shown in Figure 15.

While the study team is collecting the information on data processing applications necessary to construct the applications-to-functions matrix, it should be identifying the files in use, their content, and the responsibility for the information contained in each.

Having collected and compiled the current data processing applications information into the format shown in Figure 15, the team is in a position to group similar applications vertically on the systems-to-processes matrix. By using the applications-to-functions matrix and referring to the organization-to-processes intersects, the team can identify the processes or portions of the processes currently being supported. A letter code is used to indicate the status of an application:

C Currently supporting a process
P Planned to support a process
C/P System currently in place; another system planned to enhance or replace it

The information compiled in Figure 16 permits the study team to fill in the remaining two quadrants of the four-segment matrix to complete a pictorial overview of the relationship of organization to processes and their relationship to current data processing support (see Figures 16 and 17). Just as similar subprocesses were grouped and identified, similar applications are grouped and identified in the systems segment, and the similar files can be grouped and identified on the data classes portion of the matrix.

These charts and matrices provide an overview of the current and presently planned data processing support of the business; however, they cannot indicate the needed degree of support and the value of such support to each of the processes. Such information should be obtained from the executive interviews regarding problems with current data processing support and additional needs for information. The study team should use these interviews to help determine the areas where emphasis should be applied in developing and enhancing appropriate information systems.

Processes	Vehicle Disposal					Vehicle Rent & Lease					Vehicle Acquisition					Marketing							Finance					Management						
Systems	Tax & Transfer	Collection	Market Analysis	Where to Sell	When to Sell	Contract Termination	Maintenance	Issue Contract	Credit Check	Reservation	Acceptance & Registration	Purchase Order Issuance	Quantity	When to Buy	Equipment Specifications	Competitive Analysis	Account Service	Account Selling	Advertising/Promotion	Franchise Sales/Service	Develop Mkt'g Strategy/Force	Marketing Research	Acquisition of Funds	General Accounting	Accts Receivable	Payroll	Accounts Payable	Asset Tracking	Asset Util. & Control	Personnel Devel. & Util.	Monitor & Control	Budgeting	Short Range Planning	Long Range Planning
Accounts Payable																								C			C				C			
Annual Volume Rebate																	C	C													C			
Air Care				C	C	C	C	C	C	C																					C			
Credit Cards									C																									
Centrex Phones																															C	C		
Data Entry Stats																															C			
Deposits																															C			
Documentation Library																															C			
Daily Reporting																								C							P			
Floating Fleet										C/P																		C	C		C			
General File Support																																		
General Ledger – Car Leasing				C	C								P	P									P	C				C	C	C	C	C	P	
General Ledger				P	P								C	C									C	C				C	C	C	C	C	P	
Input Processing (Car R/A's)						P																									C			
Truck Entry (Truck R/A's)						P																									C			
Leasing Commission								C																							C			
Licensee Lease																													C		C			
Licensee Receivables																													C		C			
Market Penetration			P																												C			
Reservations										C																								
National Discount Bulletin									C																									
Pool Control																												P	C		C			
Open Receivables		C							P								C	C					C		C			C	C		P	P		
Car Leasing – Receivables		C							P								C	C					C		C			C	C		P			
Payroll																										C				C	C			
R/A Control																															C			
Rental Properties																															C			
System Fees																				C											C			
Systems and Programming																															C			
Tabs																								C										
Taxes																								C										
Car Lease – Vehicles	C	C		C	P	P	C	C	C	C	C	C	C	C	C		C	C			C		C	C				C	C		C			
Vehicle Master – Corporate	C			C	C								C	C	C								C	C				C	C		P			
Vehicle Purchase Order				C	C						C	C	C	C									C	C					C		C			

FIGURE 14 Systems-to-processes matrix (transportation).

Key Financial and Statistical Data

In preparing for the identification phase, the team leader and the data processing representative gather planning, financial, and other statistical data for the study team's initial orientation. The objective of that activity is to help the team understand how resources are committed by examining investments, inventory, personnel, equipment, etc. If possible, an attempt should be

Major DP Applications \ Functions	Human resources	Store development	Advertising	Sales	Merchandising	Finance & Accounting	Warehousing & Transportation
Recon. & merchandise improvements					X		X
Inventory management & Reporting (items & $)			X	X	X	X	X
Retail accounting				X			
Products				X	X		
Accounts payable						X	
Accounts receivable						X	
FOCIS		X	X	X	X	X	X
Payroll						X	
Slash		X					
Billing						X	X

FIGURE 15 Data processing applications-to-functions matrix.

made to identify any major shift in the way resources are to be spent over the next 24 months. Although a detailed analysis cannot be accomplished during the identification phase, enough information can be evaluated to give the study team some confidence that it can identify the best area of opportunity for recommended improvements. Normally, knowledge of how resources are spent will help the team gain that confidence, since resource expenditures generally reflect profit opportunities, major concerns, and directional changes.

The study team may find that additional financial and statistical information is needed to complement that initially presented by the team leader and data processing representative. This area is the most sensitive for the team's questioning and analysis. The team must carefully weigh what answers are needed in each case to confirm level and scope, and what questions should be asked. Members should be prepared to justify these questions and explain how the team will use the information. The answers should be directed toward the objectives of the study in developing the preliminary information systems network and in identifying the first subsystems to be developed.

An approach to gathering this data is to identify the basic resources and associate those resources with major organization elements. Such an analysis can help quantify an existing management perception and allow the team to base its recommendations on fact instead of intuition. As the need for resource management comes to the surface, the need for information systems to support that management on a broad, business-wide basis becomes apparent.

Another possible approach is to prepare a matrix identifying the resources and expenses in terms of business processes or organization. Figure 18 is an example of such a matrix.

Comparing the results of the organization-to-processes matrix with the resources-to-functions matrix will point out new opportunity areas for data processing support. It is not unusual to find that 70% of the data processing applications are supporting 20% of the business process, and these are usually overhead processes rather than the key processes required for better management of the business.

The association of expenses to processes is not an easy task and may be impractical in the time alloted for identification. However, if the study team determines during the identification phase that these relationships need to be defined, some of its members can begin gathering the data necessary to perform the task during the follow-on-phase. This criterion of practicality is applicable for all data gathering during identification. Time is critical in this phase, and if the team members get too involved in trying to restructure data, they may neglect the basic objective of identification—to define the level and scope that offer the best opportunity for a follow-on effort and to develop a preliminary action plan to conduct that effort. Although this planning guide identifies a spectrum of activities, in the final analysis, the whole sense of priorities and re-

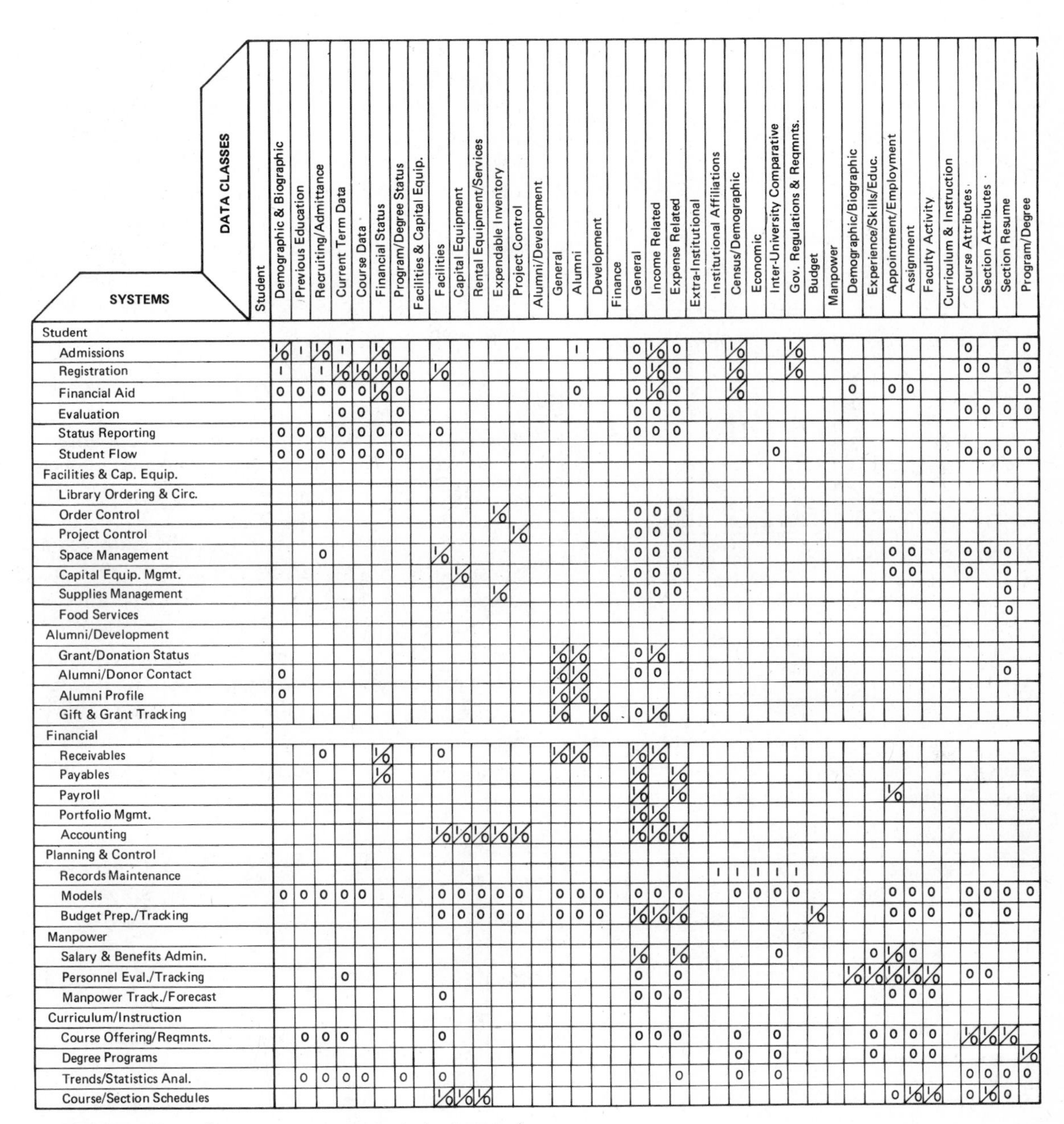

SYSTEMS \ DATA CLASSES	Student	Demographic & Biographic	Previous Education	Recruiting/Admittance	Current Term Data	Course Data	Financial Status	Program/Degree Status	Facilities & Capital Equip.	Facilities	Capital Equipment	Rental Equipment/Services	Expendable Inventory	Project Control	Alumni/Development	General	Alumni	Development	Finance	General	Income Related	Expense Related	Extra-Institutional	Institutional Affiliations	Census/Demographic	Economic	Inter-University Comparative	Gov. Regulations & Reqmnts.	Budget	Manpower	Demographic/Biographic	Experience/Skills/Educ.	Appointment/Employment	Assignment	Faculty Activity	Curriculum & Instruction	Course Attributes	Section Attributes	Section Resume	Program/Degree
Student																																								
Admissions		I/O	I	I/O	I		I/O										I			O	I/O	O			I/O			I/O									O			O
Registration		I		I	I/O	I/O	I/O	I/O		I/O										O	I/O	O			I/O			I/O									O	O		O
Financial Aid		O	O	O	O	O	I/O	O									O			O	I/O	O			I/O						O		O	O						O
Evaluation					O	O		O												O	O	O															O	O	O	O
Status Reporting		O	O	O	O	O	O	O		O										O	O	O																		
Student Flow		O	O	O	O	O	O	O																			O										O	O	O	O
Facilities & Cap. Equip.																																								
Library Ordering & Circ.																																								
Order Control													I/O							O	O	O																		
Project Control														I/O						O	O	O																		
Space Management				O						I/O										O	O	O											O	O			O	O	O	
Capital Equip. Mgmt.											I/O									O	O	O											O	O			O		O	
Supplies Management													I/O							O	O	O																	O	
Food Services																																							O	
Alumni/Development																																								
Grant/Donation Status																I/O	I/O			O	I/O																			
Alumni/Donor Contact		O														I/O	I/O			O	O																		O	
Alumni Profile		O														I/O	I/O																							
Gift & Grant Tracking																I/O		I/O		O	I/O																			
Financial																																								
Receivables				O			I/O			O						I/O	I/O			I/O	I/O																			
Payables							I/O													I/O		I/O																		
Payroll																				I/O		I/O											I/O							
Portfolio Mgmt.																				I/O	I/O																			
Accounting										I/O	I/O	I/O	I/O	I/O						I/O	I/O	I/O																		
Planning & Control																																								
Records Maintenance																								I	I	I	I	I												
Models		O	O	O	O	O				O	O	O	O	O		O	O	O		O	O	O			O	O	O	O					O	O	O		O	O	O	O
Budget Prep./Tracking										O	O	O	O	O		O	O	O		I/O	I/O	I/O							I/O				O	O	O		O		O	
Manpower																																								
Salary & Benefits Admin.																				I/O		I/O					O					O	I/O	O						
Personnel Eval./Tracking					O															O		O									I/O	I/O	I/O	I/O	I/O		O	O		
Manpower Track./Forecast										O										O	O	O											O	O	O					
Curriculum/Instruction																																								
Course Offering/Reqmnts.			O	O	O					O										O	O	O			O		O					O	O	O	O		I/O	I/O	I/O	
Degree Programs																									O		O					O		O	O					I/O
Trends/Statistics Anal.			O	O	O	O		O		O												O			O		O										O	O	O	O
Course/Section Schedules										I/O	I/O	I/O																					O	I/O	I/O		O	I/O	O	

FIGURE 16 Data classes-to-systems matrix (university).

sponsibilities—what is most important and what must be done—rests on the judgment of the study team leader and team members.

The activities described so far concern the gathering and analysis of resource allocation data currently published and available to the study team and their own perception of such allocation. Where data cannot be found, the executive interviews can provide the necessary information.

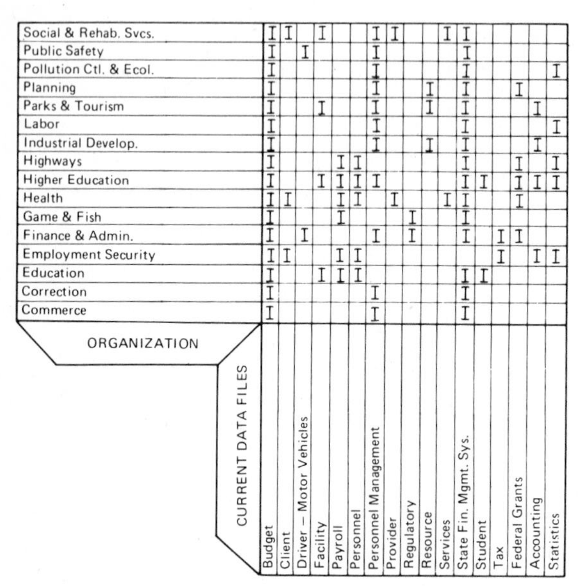

FIGURE 17 Organization-to-current data files matrix (state government).

Executive Interviews

Interviewing and interview analysis consume more time during the identification phase than any other activity because the executive interviews are the primary source of information for determining the business problems and management's need for overcoming the problems. Therefore, to determine how long this phase will take, the study team must establish the number of executive interviews that will be needed.

One of the recommended procedures for determining the number of interviews is the development of the organization-to-processes matrix. Two basic questions should be answered in this approach.

- What organizations are involved in each of the processes?
- Within the organization, who must be interviewed to determine the problems, objectives, and requirements?

For a given business process, the study team should determine what organizational unit is involved in the decision making and operational control of the process. The key word is decision making. The degree of involvement or responsibility can vary anywhere from providing data to final management and operational responsibility for a process. Generally, assignment of final responsibility to one specific organization is difficult. More often, several organizations reach a

Resources / Organization	Assets		Expense		People		Information Systems Expense	
	$ (000,000)	%	$ (000,000)	%	NO.	%	$ (000)	%
Senior management			3	6	15.2			
Buying	42	45	8	17	341.7	8	226.8	30
Selling			13	28	2759.1	66	8.4	1
Advertising and promotion			4	9	103.3	3	6.4	1
Financial	19	21	3	6	332.4	8	516.5	68
Personnel			4	9	51.6	1	.7	
Operations	32	34	12	25	582.3	14	2.1	
Totals	93	100	47	100	4185.6	100	760.9	100

FIGURE 18 Resources-to-organization matrix.

decision together, each organization concurring or dissenting and inputting its own point of view (for example, sales, manufacturing, and production planning developing a final production plan).

Once the organizations involved in a process are identified on the organization-to-processes matrix, the team must determine who must be interviewed within each organization. A rule-of-thumb in using organization charts to fix responsibilities is to look two levels below the president or chief operating officer. At that level, responsibilities will be fairly discrete.

When the team has drawn up its list of executives to be interviewed, the team leader prepares a letter for distribution to the interviewees. This letter stresses dates, times, and general questions that will be considered during the interview.

The interviews should be kept to a planned format and should be helpful to both the interviewees and the team. Since the study team is not interested in specific, detailed information requirements during this phase, the questions should be fairly general, designed simply to help the team identify the executive's information needs. The team can then recommend improvements in information processing that will aid in decision making as well as in operation of the business. The following questions have been used in many studies and are offered as guidelines.

- Briefly, what is your area of responsibility?
- What are your objectives within your area of assigned responsibility?
- How are you measured? How do you measure your subordinates?
- What major problem have you encountered within the last year that has made your job overly difficult or taken an inordinate amount of your time?

—What has prevented your solving it?

—What is needed to solve it?

—What would be the value to your area and the business if it were solved?

(*Note:* This series of questions is repeated for each major problem mentioned by the executive.)

- How satisfied are you with your current data processing support?

—Accuracy

—Timeliness

—Need for additional manual effort to present meaningful information

- What major changes do you envision within the next two to five years that will have a major impact on your area of responsibility?
- What other or additional information do you need to aid you in managing your area? What would its value be?
- What do you expect to result from this study effort—what does it mean to you and to the business?

An explanation of some of the questions may be beneficial. First, the team should learn from the executive, in his own words, what his responsibilities are, since conclusions drawn from the organization chart or job description may not be correct. Second, the objectives he has personally established for himself in performing his function may have a large bearing on his information needs. Third, since a major influence on business success is how effectively resources are allocated and used, the team should be concerned about methods of measurement.

When asking about problems, the team should make certain that the executive does not present only his last problem and overlook more serious ones that may have occurred months before. Throughout the questioning, the team should be asssessing how the executive is spending his time. The higher he is in the organization, the more time he should be spending in planning and control rather than in operations.

The information satisfaction and needs questions should emphasize the additional costs that may be incurred from inaccurate and untimely information, and the value of any additional information desired. Because major information systems development cycles can take from two to five years to complete, it is important whether plans are underway for centralization, decentralization, acquisitions, organizational changes, new products, etc., that would impact recommendations for information systems development.

The final question regarding the executive's expectation from the study may be of value to the study team in ascertaining how well management will accept and implement the study recommendations. The answers may also be illuminating in a business that has been involved in studies in the past that have not resulted in any measurable change.

Several guidelines are helpful in conducting the executive interviews. First, the interviews are best conducted in the study team's control room. This approach gets the executive away from his office and minimizes interruptions. Since the interview will normally take about two hours, only two interviews should be scheduled per day. Remember that the team is interested in gaining information from the interviewee; therefore, the session is primarily one of listening on the part of the team and further questioning for clarification and justification of needs. No more than four team members should attend an interview. The team leader should conduct the interview and ask the questions, helped by another team member. One of the team members should be designated as the primary note taker, although others on the team may also take notes.

At the outset of the interview the study team leader should describe the purpose of the interview, briefly review what the study team has done to date (pointing out some of the charts on the wall that have been developed), and state that notes will be taken during the interview, which will be summarized and returned to the interviewee for his additions, corrections, and clarification. The revised and approved summary should be promptly returned to the study team for use in drawing conclusions and making recommendations.

The business process definitions and the relationship of the organization to the processes as identified by the team should be described to the interviewee. This information can be given at the beginning of the interview, when the team's work to that point is reviewed. The executive should be asked to check the processes and make certain that the relationships of his organization(s) to the processes are correct. He should also be given a copy of the definitions and the matrix to take back to his office for review. The reviewed copy should be returned along with the approved interview summary.

In summary, the purpose of the interview is to validate what has been accomplished to date and to gain input that can be obtained only from the executive. The study team should concentrate on obtaining significant information rather than simply collecting interesting stories.

Business- and Information-Related Problems

As the interviews proceed and the problems are identified, the information should be compiled so that the team can readily draw meaningful conclusions. Since the aim of the study is to develop recommendations for information that will better support the processes and, in turn, the organizations responsible for the processes, the association of problems with processes is advisable. A chart similar to the one shown in Figure 19 may be valuable to the team in compiling the problems, needs, and value statements.

In documenting the problem and need statements, the following criteria govern the inclusion of a statement:

- It should be significant
- It should be at a low enough level of detail to deal with.

Organization	Problem Example	Impact	Need/ Recommendation	Value/ Benefit
Business affairs	The annual financial report requires weeks to compile from the manually maintained general ledger.	The production of this report requires attention of highly paid accountants when financial data for this report could be produced by a properly designed financial system.	Design a general ledger system that would automatically provide the financial reports to the university.	Financial reporting can be done more easily, quickly, and with less clerical effort.
Provost	EPA personnel reporting requires significant manual effort. There is no historical data in a current computer file. Although some EPA personnel data is available from computer files, it is difficult to assimilate for use in reporting. Considerable effort is required in collecting this data from various sources and manually preparing reports.	High-level staff members are occupied with routine clerical matters rather than policy interpretation, planning, and other functions more in line with the responsibilities of their office. There is difficulty in meeting reporting deadlines and concern about the accuracy of the information reported.	Establish a comprehensive personnel system and a complete personnel data base. The system should allow proper interfaces with the financial, student, and course data bases.	Reduction of the clerical effort of professional personnel in retrieving needed data for institutional study planning and reporting.
Financial aid Provost Personnel Student affairs research	Changes in reporting formats require continual modification of report programs. Examples: • Work study and NDSL reports • EEO and affirmative action • Ethnic enrollment reports	Consumes programmer effort in maintenance, thus reducing time for new applications development.	Recommended personnel, financial, and student data bases should provide some flexibility in meeting changing requirements. Some type of generalized inquiry program should be used to assist in this problem.	Reduction of maintenance and one-shot programming effort, thereby freeing more resources for development of new computer systems.
Business office	Monthly historical budget information cannot be provided to departments and offices using current manual methods.	Departmental management must manually maintain this information.	The financial system should incorporate this capability.	Provision of a better fiscal tool for departmental manager. Reduction of need for clerical effort in departments.

FIGURE 19 Sample problems, needs, and value statements chart.

- It should not have other problems embedded in it.
- It should have a cause and effect relationship explicitly stated or implied.

After the information is compiled, the chart will usually show several similar or related problems expressed in different terms. These problems should be combined to form a composite and provide a method of weighting to determine relative importance. The team is now in a position to start developing its formal conclusions and recommendations.

CONCLUSIONS AND RECOMMENDATIONS

The data gathered and information solicited through interviews with the key executives has now been structured to assist the study team to understand the business from general management's viewpoint. This data includes information pertaining to:

1. Business environment

- External—economics, customers, technology, competition, government, suppliers
- Internal—corporate policies, practices, constraints

2. Business plans

- Goals, objectives, strategies
- Resources, schedules
- Finances
- Measurements and controls

3. Planning processes
4. Organization

- Positions, names
- Number of people
- Responsibilities, objectives
- Key decision makers

5. Products and markets
6. Geographic distribution
7. Financial statistics
8. Industry position and industry trends
9. Major problem areas
10. Major studies, task forces, committees in last two years.

In addition to accumulating information to assist in understanding the business, the team has acquired information about data processing and how it is currently supporting the business. By supplementing this information through the executive interviews, the study team has data pertaining to:

1. Information systems plans

- Goals, objectives, strategies
- Resources, schedules
- Finances

2. Organization charts

- Positions, charts
- Number of people

3. Planning processes

- Systems, justification requirements
- Standards, guidelines

4. Geographic distribution

- Equipment, terminals
- Communications
- Facilities

5. Software environment
6. System and application profiles
7. Funding processes
8. Major studies, task forces, committees in last two years
9. Major problem areas

Having completed the executive interviews and gathered the above information, the study team is ready to develop conclusions and make recommendations as to how information systems can better assist in improving the management decision making and operation of the business.

Before drawing final, specific conclusions, the team should:

1. Complete the process definitions and the organization-to-process matrix. Make necessary revisions to include interview findings and add clarification.

2. Categorize the problems obtained from the interviews for meaningful analysis. Possible categories include:

- Environment
- Strategic planning
- Operational planning
- Measurement and control
- Productivity
- Products and services
- Information systems support

(*Note:* A problem may be placed in more than one category.)

3. Abstract the key findings—problems, opportunities, and needs.

4. List the key decision and activity areas, current data processing support, and the value of that support.

5. Complete the current data processing systems-to-process matrix.

6. Complete the data classes-to-systems and organization-to-data files matrices.

7. Display the general, functional level of information flow.

8. Abstract all value statements related to problems and improved information systems support.

The results of these analyses should be displayed on charts and graphs that are meaningful to management. The team can then complete its activities in relation to the study objectives and develop specific recommendations relative to:

- Identifying a network of information systems
- Identifying key problem areas related to business processes
- Selecting the first, most needed subsystems
- Reviewing information systems management capabilities
- Identifying key activities required by management to define the first subsystems

Network of Information Systems

After the study team has satisfied its basic objectives of understanding the business and how data processing supports it, the next objective is to identify a network of information systems that will more effectively provide general and operational management with information to meet management control requirements. The network of information systems to be described at the end of the identification phase should be a logical set of systems and data bases related to the decision and activity areas of the processes and subprocesses.

The network should provide a visual depiction of the strategic long-range objective of the information systems plan. The network is the logical relationship of information systems (and their external interfaces) deemed necessary to support the information requirements within the defined level and scope of the organization studied.

Normally, an information systems network consists of multiple operational-level information systems and one management-oriented information system (centered around planning, control, and measurement processes). In the light of the information gathered by the study team, a review of the basic principles of BSP can be helpful in identifying the network:

- Systems design to meet the objectives, level, and scope of the business being studied
- Systems to support business processes, not independent organizations only
- Planning from the top down. The study has considered the external environment, the business objectives, and the information needs of executives
- Implementation from the bottom up. The systems must be able to relate the existing data and applications to those needed to support the business processes and the organizations responsible and involved with the processes.

One generalized method of identifying the major systems of an information systems network involves relating the supply, demand, requirements, and administrative aspects of a business to the network. This approach provides a rationale for considering a logical grouping of processes and subprocesses within the network. Most businesses have basic supply and demand areas, even though the labels may vary. For example, in a university the community and the students who want an education can be considered the demand, which is supplied through finances and the university's teaching staff. Using these five major areas (demand, requirements, supply, administration and management) a generic information systems network can be drawn up, including interfaces to customers, suppliers, and the marketplace/environment. This generic network is shown in Figure 20.

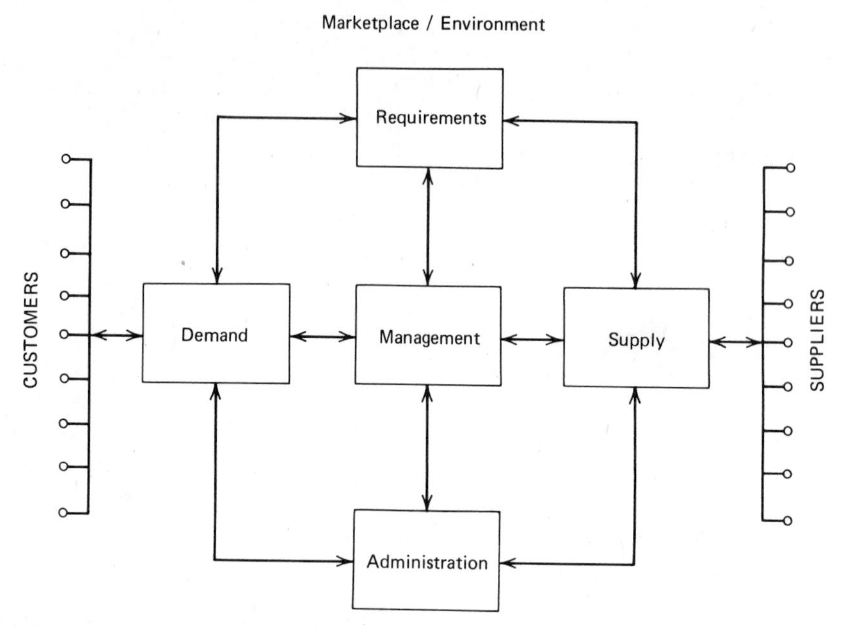

FIGURE 20 Generic information systems network.

- The *Demand* segment is concerned with the marketability of a product or service and relates directly to a set of customers.
- The *Supply* segment is concerned with providing the product or service as efficiently as possible. Its outside relationship is normally with a set of suppliers or vendors.
- The *Requirements* segment determines and defines the products or services of the business with an external interface to the market place and other environmental factors.
- The *Administrative* segment, in itself, adds nothing to the product or service but is necessary for overall accountability of the business. This segment has its own interfaces, including unions, employees, tax requirements, and other reporting requirements.
- The fifth segment, *Management*, ties together the other operational-level segments. Included in this segment are the planning, control, and measurement aspects of the business.

Figure 21 shows the information systems network with its major systems areas that was designed for a major portion of IBM. The demand segment is expanded to the distribution, business operations, and field systems areas, which interface to customers through local branch offices. Requirements are represented by product planning and development. The supply segment includes process design and release, manufacturing processes, and manufacturing control, which interface to suppliers and vendors through multiple manufacturing facilities. The planning, control, and measurement segment completes the network, making it a true business information system.

A similar network for state government is depicted in Figure 22, which shows an entire statewide system. Each of the agencies and department could be portrayed similarly.

Once the network has been identified, the study team can consider the major problems and information needs associated with each in prepa-

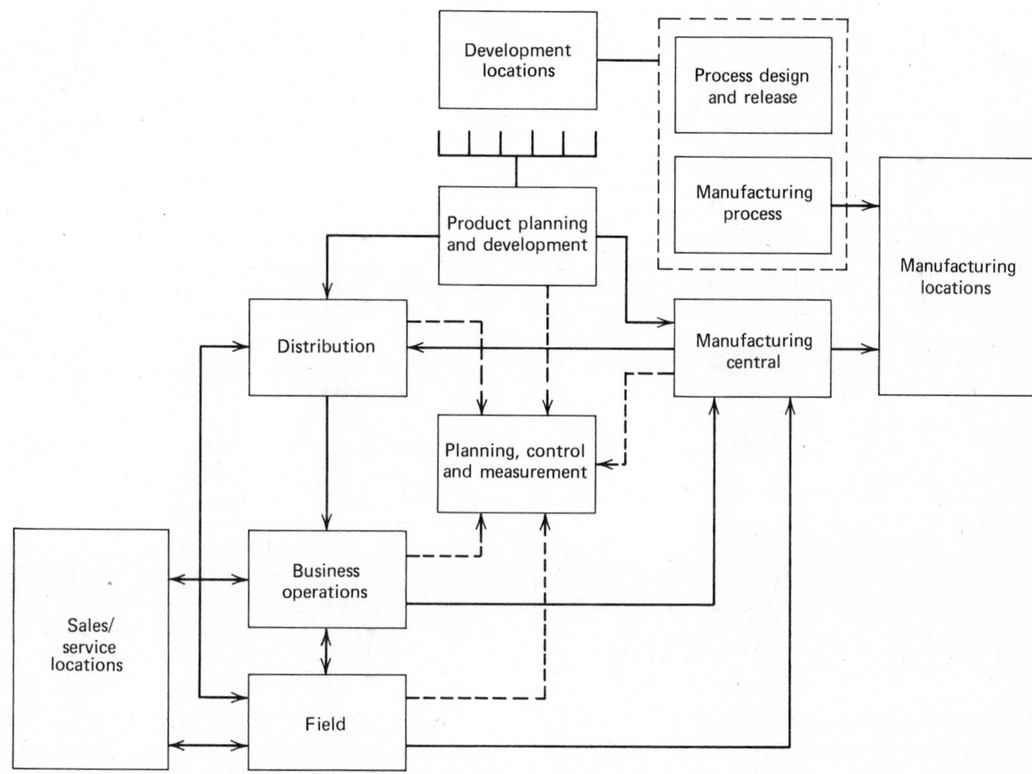

FIGURE 21 IBM's information systems network.

ration for selecting the first subsystems to be addressed.

Key Problem Areas Related to Processes

Through data gathering and executive interviewing, the study team has collected many problem statements. These statements should be grouped into meaningful problem areas that are directly related to the business processes and subprocesses.

In reviewing the list of problems, most of them will generally be found to be directly related to information or the lack of it. Since the principal purpose of the identification phase is to propose ways that information can improve the management and operation of the business, the problem statements can logically be grouped in terms of information deficiencies. These deficiencies usually fall into five basic classifications:

1. Accuracy. Inaccurate information is being furnished to users.
2. Timeliness. Users experience excessive delay in obtaining information.
3. Format. Information is furnished to users in a form that is too detailed, is not in appropriate units, or is difficult to use.
4. Availability. Information may have been collected but is not readily accessible to users.
5. Lack of information. The information may exist, but due to problems in communication its existence may not be known by those that need it.

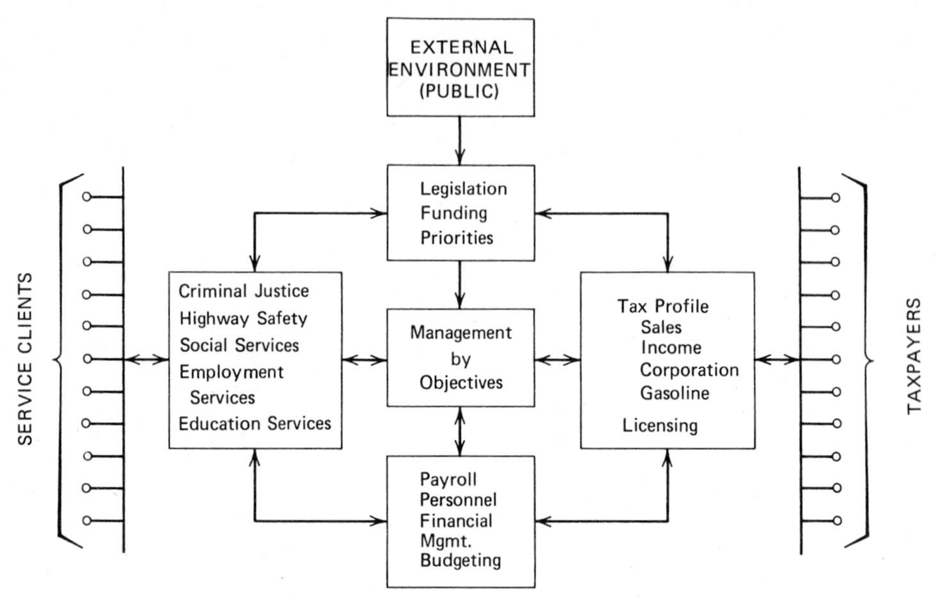

FIGURE 22 State government information systems network.

By classifying the problems and then relating these classifications to the business processes, the team should be in a better position to identify the key problem areas of the business. One problem statement may fall into more than one classification. For example, a common statement of executives about the information they receive is: "I get the information I need, but it is inaccurate and too late to be of much value." This problem falls into more than one information deficiency classification and may also be associated with more than one process in which the executive is involved. The statement is very meaningful, since it identifies the information needed, and the team need only propose ways for overcoming the deficiencies in accuracy and timeliness.

A matrix of processes versus information deficiencies may help in assigning priorities to the problem areas. Although this matrix is quantitative, showing the number of times a problem is reported, it gives no qualitative evaluation of the seriousness of the problem and its impact on the conduct of the business. The matrix does, however, show the number of organizational units affected by the problem.

The study team should remember, while analyzing the interviews to identify key problems, that relatively few of those interviewed have overall business responsibility. With the exception of the president and chief executive officer, most of the executives interviewed have only functional responsibility, and their problems are, for the most part, those that have a direct bearing on the performance of their function. Although these problems may be serious in relation to an individual function, they may not represent a significant problem to the overall operation and management of the business. Therefore, the team should make a special effort to identify and weight the problems in relationship to the total business.

Having completed this step, the team can move on to identify the subsystems areas that will offer the greatest promise in aiding management and developing an information systems network.

First and Most Needed Subsystems

Some of the questions to be answered in determining the first subsystems are:

- Will the subsystem provide a significant near-term savings and a substantial long-term return on investment?
- Whom will it impact, and how many people will be involved?
- Will it lay the groundwork for an initial data base structure for the network?

A method of determining logical priorities is to group the major criteria into four categories:

1. Return on investment

- Implementation cost (estimated)
- Dollar return (area of potential)
- Cost/benefit ratio

2. Impact

- Number of organizations and people affected
- Qualitative effect
- Effect on accomplishing overall objectives

3. Success

- Degree of business acceptance
- Probability of implementation
- Length of implementation
- Risk
- Resources available

4. Demand

- Value of existing systems
- Relationship with other systems
- Political overtones
- Need

The major processes and supporting subprocesses can be analyzed as potential first-subsystem candidates and ranked on a scale of 1 to 10 for each of the four categories above. Some type of pictorial representation can then be drawn to emphasize the most needed subsystems. Figure 23 is an example of this analysis prepared by a state government.

Because of resource limitations, the definition of more than the very highest priority subsystems may be impossible initially. However, the same method can be used in making later selections. After the definition of the first subsystems has been completed, the priorities of the remaining subsystems should be reassessed. For example, after the first four subsystems have been defined, the subsystem initially listed fifth may not be at the top of the list on the second round because the business's requirements and problems have changed in the interim.

When selecting the first subsystem, the study team should be alert to a potential problem. The most needed subsystem may prove to be impractical to implement until another subsystem(s) is available to furnish necessary data. This problem may not be clearly perceived until the follow-on phases, but should be identified as soon as possible. If it is encountered, management should be informed that the study team recognizes the most desirable subsystem, but that technical reasons and prerequisite demands make a lower priority choice advisable.

Information Systems Management

An important consideration during the identification phase is a review of the existing information systems organization and its capabilities. In most cases the acceptance of recommendations from this phase will force a change of direction for the organization from developing independent applications for individual units to developing integrated, business-wide DB/DC systems. Therefore, the study team should identify the requirements for change and possible expansion of the organization before attempting to develop such systems and perhaps failing if the essential data processing organization is not available.

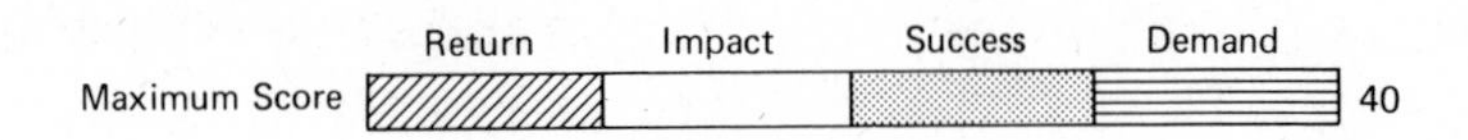

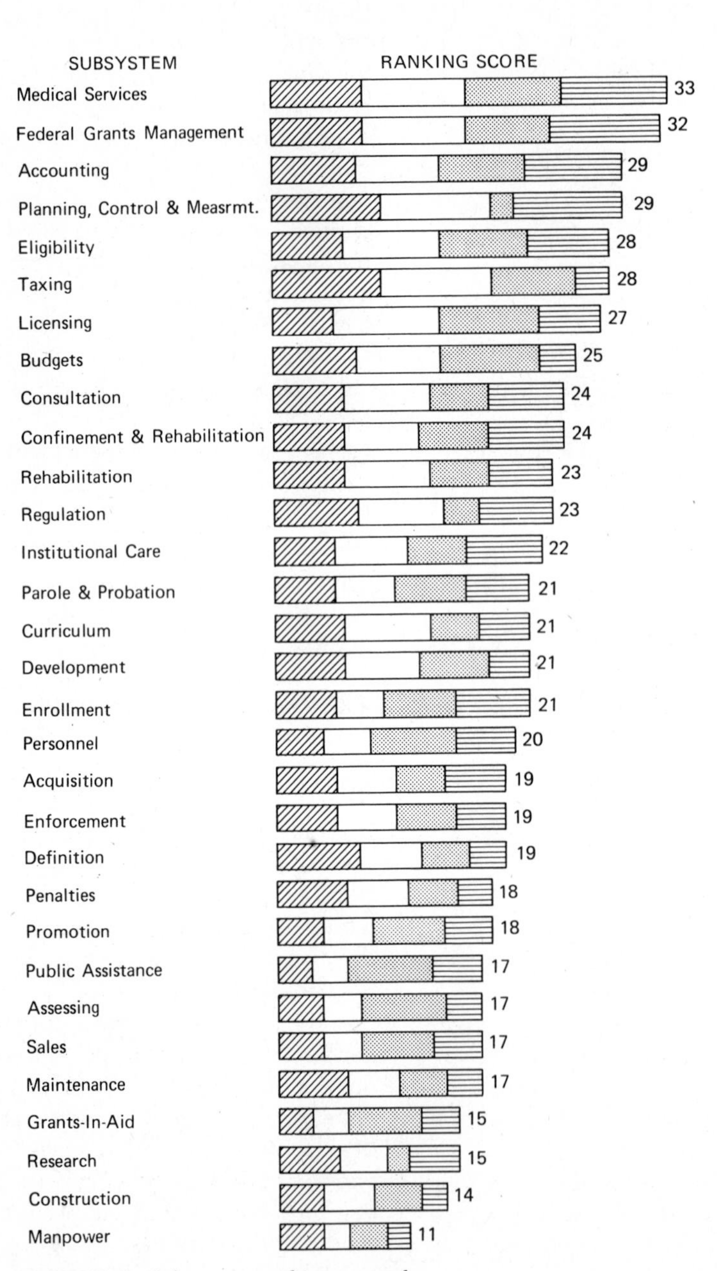

FIGURE 23 Subsystem ranking example.

The purpose of information systems management is to plan and control the resources available to support the development and operation of information systems. The management system helps ensure the continuous planning, control, measurement, and operation of the information system function. Specific policies, procedures, and relationships that effectively support the overall business objectives are needed before the defined logical information system network can be realized.

During the preparation stage for the identification phase, the information systems representative is expected to compile and present information regarding the data processing operation to the study team. This presentation is preferably made before the study to help shorten the study and allow adequate time to prepare and review the information thoroughly. The checklist in Appendix A should help the team assess the ability of the information systems organization to perform under the current environment and determine any shortcomings that should be addressed before continuing toward a DB/DC environment. An extensive review and assessment of each of these items is not expected during this phase; however, obvious deficiencies should be identified and plans made to correct them. The series of IBM publications on installation management may also help in this effort. These documents cover such topics as managing systems development, organizing the data processing (DP) activity, DP cost allocations, DP quality assurance, and policies and procedures.

Following this activity, the team is ready to take the steps necessary to proceed to the definition phase.

Preliminary Action Plan for Definition Phase

The intent of the action plan is to identify the key decisions and activities required to help management provide the proper direction for transition to the definition phase. Specifically, the objectives of the definition phase should be to:

- Define the first subsystems to be implemented.
- Identify the subsystems within the major systems of the information systems network.
- Develop a long-range strategic plan for the development of other information subsystems in concert with the business objectives.
- Identify proper information systems management capabilities to help ensure effective implementation of the network.

In addition to these definition phase objectives, the action plan should address the level and scope, tasks, expected output, network definition, and any necessary executive and personnel education.

The action plan should answer the following questions:

1. What will happen?

- Detailed definition of the first subsystems
- General description of follow-on systems composing the network
- Requirements for information systems management to implement the system

2. Who will participate?

- Team leader—businessman from the major process area initially addressed
- Team members and a definition of their role.

(Note: At least one team member should be held over from the identification phase to provide continuity.)

- Interviews with executives, managers, and key personnel within the areas involved with the designated process area

3. What will be delivered?

- Study report

- Long-range plan, including:
  - —In-depth definition of first subsystems
  - —Implementation schedule for first subsystems
  - —Resources required for first subsystems
  - —Cost and benefit analysis for first subsystems

4. How will it be controlled?

- PERT charts
- Periodic management reviews
- Work plan
- Resource plan
- Status reporting plan
- Briefings and education plan

The completion of these activities enables the identification phase study team to proceed to their final activities—preparation of the study report and presentation of the study results to executive management.

IDENTIFICATION PHASE REPORT AND EXECUTIVE PRESENTATION

Outline of Report

During the initial steps of preparing for the identification phase and developing an action plan for conducting the phase, it is recommended that a preliminary outline, essentially a table of contents, be written for the study report. The intent of preparing a basic outline at that early stage is to force the study team members to think of the audience to whom the output of the study is to be presented and also to consider the objectives of the study so that they fulfill their obligations. The study team leader is also better able to assign responsibility for the initial preparation of individual sections of the report while the study is progressing. The members of the team can each pay particular attention to the information needed to make their individual sections more meaningful. This preliminary work should also speed the wrap-up of the identification phase. Since the report is to be a consensus of the entire team, each of the sections should be reviewed by each of the members so that the final document reflects all their comments.

The final report may be structured in many ways, depending upon precedence and methods of presentation within the business conducting the study.

The most significant findings, conclusions, and recommendations should be summarized in the first few pages of the report for the use of top management. Supporting details should be included later in the appendices for other members of the organization and team members that will participate in future phases of systems implementation.

Presentation Media

The principal factors to consider in determining how the report should be presented is the type and size of the audience and the accepted ways of making such a presentation in that business environment. Consultation with the executive sponsor early in the phase to obtain his advice on this subject can be most helpful in establishing the proper direction.

If the presentation is made to a small group, easel flip charts are adequate and popular. If the audience numbers more than a dozen, viewgraphs or slides should be considered. If slides are to be used, adequate time should be planned for their preparation, whether in-house or by an outside vendor.

Content of the Report

The primary writing responsibility for each of the sections of the report should be assigned at the beginning of the identification phase. As the phase progresses, changes, additions, and deletions can be made to the preliminary outline. By the completion of the executive interviews, agreement should have been reached on the final table of contents, and the individual study team members should have available to them most of the information needed to complete their assigned sections. The study team leader usually assumes the responsibility for writing the background and overview because much of this information comes directly from the orientation information presented to the team at the beginning. Each of the other sections may be assigned to the individual who was delegated responsibility for that area during the study.

The whole team should concur on the conclusions, the recommendations, and an action plan. These areas should be reviewed with the executive sponsor before the team drafts the final report. Controversial areas or areas of high impact on the business may be reviewed with the executive involved to determine how best to present the recommendations.

Executive Presentation

The final draft of the report contains all of the material from which the executive presentation can be developed. The principal aims of the executive presentation are to inform management of the study's findings, make recommendations, and secure approval of the action plan.

The presentation should be brief, preferably no longer than 1½ hours. It should be logical and factual and should end with a recommendation for the next step. For example, the presentation may take the following form:

1. Introduction
   a. Background
   b. Study objectives
      (1) Understand the business
      (2) Understand information systems
      (3) Identify areas for improvement
      (4) Recommend action plan
   c. Study approach
      (1) Study business (level and scope)
      (2) Define processes
      (3) Conduct interviews
      (4) Relate organization to processes
      (5) Relate systems to processes
2. Major problems identified
3. Major conclusions
4. Recommendations for needed subsystems
5. Action plan for definition phase
   a. Specific activities
   b. Responsibilities
   c. Schedule dates

DEFINITION PHASE ACTIVITIES

The definition phase should not be initiated without top management's full understanding and approval of the recommendations developed during the identification phase. The objective of the definition phase is to develop a long-range information systems plan that includes the design, development, and implementation of a network of information systems based upon data base/data communication concepts. The information systems plan should be a single, broad, long-range plan to guide future information systems implementation efforts. The plan should be based on the business's own specific requirements so that it accurately reflects the needs of the organization. It should include sufficient detail to guide all levels of management concerning what needs doing, how, when, and by whom in the organization.

Such a long-range information system plan, based on the real needs of the business and prepared with full management involvement, should help to assure that the business's information systems resources will be effectively used. In addition it should:

- Provide a basis for communication between general, functional, operational management, and systems professionals.
- Provide management with statements of value associated with recommended information subsystems and related to business problems and opportunities.
- Accelerate the rate at which information systems can be designed and implemented.
- Minimize the effort associated with the design and development of an advanced information system and maximize the potential benefits that a new system can produce.
- Provide a basis for measuring the implementation progress.
- Provide disciplines for organizing and controlling the plan that can be readily understood and communicated throughout the organization.

This section contains the methodology and activities recommended for the definition phase. Factors such as the size of the business, scope and expected duration of the study, and level of data processing sophistication will determine how the methodology should be applied. Six major activities should be performed to accomplish the objectives of the definition phase:

1. Preparation
2. Interviews
3. Definition of information systems network
4. Definition of first subsystems
5. Definition of information systems development controls
6. Preparation of documentation

PREPARATION FOR DEFINITION PHASE

Before the formal start of the definition phase, certain activities should take place to ensure team continuity, maximum productivity, and a successful end product:

- Study team selection and orientation
- Education of the team
- Action plan for the definition phase

Study Team Selection and Orientation

For continuity, at least one and preferably more of the identification phase team members should be selected for the definition phase study team. The team leader is a particularly desirable holdover. The study team should include:

- A business-oriented member with a good knowledge of the area to be studied
- An information systems analyst with experience in systems design
- A member with technical experience in communications and data management
- A technical writer

Before the beginning of the definition phase, the team leader should conduct a team orientation. The team members should be introduced, and a brief rundown given on each individual's background. Since some of the team members may be unfamiliar with BSP, the BSP principles and objectives should be summarized.

The results of the identification phase and the objectives of the definition phase are then presented. The team leader should also cover the following items:

- Administrative details
- Action plan
- Project control
- Preparation for definition phase
- Team assignments

A similar orientation should be held for the executives and key personnel of the areas to be studied in the definition phase. This meeting should include an explanation of the type of information and data that the study team members will be requesting when they are studying the area.

Team Education

A review of the BSP identification phase report and the areas to be studied can help to determine whether team members need special education. If any specific expertise is required, appropriate classes can be scheduled on an individual basis.

If the team leader selected for the definition phase was not a member of the identification phase study team, he or she should attend classes on BSP, DB/DC, or other courses that are appropriate.

Action Plan for Definition Phase

Study scope, complexity of the business, number of people involved, and expected duration of the study are considerations for determining the specific content of the definition phase action plan. The plan should be in writing so that the study team can look at it, understand it, feel they are part of it, and participate in its creation and modification.

The content of the action plan is more important than its form; it should reflect the understanding between members of the team as well as between the team and management. Since the plan represents the best thinking at a particular time, it should be a living and evolving tool, designed for easy modification. Control should be established to assure that everyone knows about the latest changes.

The action plan is important because it specifies what is to be done, by whom, and when. It is the chief means of measuring planned versus actual performance. Generally, an action plan contains the following modules:

- Study announcement plan
- Manpower plan
- Work plan
- Financial plan
- Study review plan

Study Announcement Plan

The study team should carefully consider how best to inform management personnel about the study, emphasizing that it has been sanctioned by executive management. The announcement should be developed by the team leader and sent to all levels of the organization impacted by the study.

The announcement should include a list of people to be interviewed and tentative questions to alert the interviewees. It should identify the areas of information that will be covered during the interview, particularly any specific areas that require data gathering beforehand. The intent is to speed the interview by having as much of the data as possible immediately available to the interviewee. Announcement letters may be supplemented by a group orientation session.

Manpower Plan

The manpower plan states the number of team members required for the definition phase and how they should be phased into the study. Critical skills and eligible sources should be identified well before the study formally begins.

Personnel selection criteria should be defined precisely and clearly to help assure that only qualified people are assigned to the study team. The full-time team should be small, five to ten people, with appropriate liaison to the functional area.

Work Plan

The work plan has two main parts: (1) allocation of work, and (2) a method of measurement to evaluate acceptability of the work. The work plan formally identifies, defines, and sequences the major tasks and subtasks of the program.

One approach to allocating work and establishing a method of measurement is described below:

1. *Allocating work*. The two main tools for defining, allocating, and scheduling work are the Gantt chart and the activity network chart.

The Gantt chart defines the activities to be done, who will do them, and when. A sample is shown in Figure 24. The last column on the right is used to insert deviation-from-plan notations. The triangles indicate milestones or checkpoints. As significant changes occur, each team member should update the portion of the chart that depicts his responsibility area. In this way, the team leader and all team members can get a quich overview of the current study status. The team leader should follow up on deviation and major comment items immediately to assure that he thoroughly understands them and that he is in agreement with the planned remedial action.

The activity network chart should be a customized, detailed depiction of study activities showing interdependencies between tasks and subtasks. One popular version is a critical path diagram. This diagram helps the user comprehend critical time-sequence relationships of large and complex projects, pinpoint major critical nodes during the study, and see the effects of activity slippages on the entire project. Given this insight, the study team leader can consider shifting or increasing resources for critical activities before a major problem occurs. Another value of the critical path diagram is its use to mechanize the process of updating plan changes and progress. The technique can be used to examine the scheduling effects of redistributing manpower and other resources. For further information, refer to the IBM publication *Introduction to Minipert* (GH20-0852).

2. *Establishing measurement method*. The work plan should spell out measurable objectives. Two effective ways of ensuring measurability are milestone reviews and specific job assignments.

Milestones are generally established at the conclusion of any major task and subtask where a major segment of work is complete and a decision is required. Usually, the completed output from one task interfaces into others. One of the chief points of evaluation is to determine whether the work will fit with the segment it is impacting.

The team leader should write out the details of all job assignments so that each team member understands what is expected. The written assignments stress with whom and when each team member should cross-check information. Clarity in job assignment is essential to the success of the study.

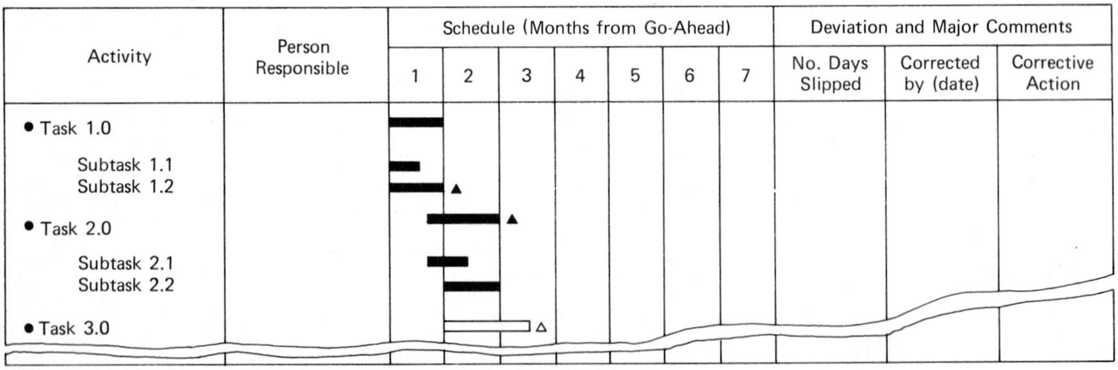

Activity	Person Responsible	Schedule (Months from Go-Ahead)							Deviation and Major Comments		
		1	2	3	4	5	6	7	No. Days Slipped	Corrected by (date)	Corrective Action
• Task 1.0											
Subtask 1.1											
Subtask 1.2											
• Task 2.0											
Subtask 2.1											
Subtask 2.2											
• Task 3.0											

FIGURE 24 Sample Gantt chart.

Financial Plan

The two main elements in a financial plan are budgeting and reporting. The budget should be expressed in terms of resources required for the study broken out by month and translated into dollars. Primary expenses generally are manpower and travel. Reporting of actual expenditures should be reviewed at regular intervals by the team leader.

Study Review Plan

A plan should be scheduled and organized by the study team leader for major, comprehensive reviews at selected study milestones. The major objectives of the reviews are to:

- Communicate program status to executive management and members of the study team in order to maintain visibility of the program and help assure that management's objectives are being met.
- Discuss actual and anticipated delays and decide how to get back on schedule.
- Assess the quality of the work performed and take remedial action as appropriate.
- Determine readiness to begin subsequent activities.
- Solicit recommendations for improvements in the study action plan or any other aspect of the study.
- Discuss and resolve any major problems on the part of executive management or members of the study team.
- Reconfirm responsibilities and schedules for major actions to be evaluated in the next milestone review.

INTERVIEWS

The study team should confirm the business processes and organizational data gathered during the identification phase before conducting user interviews. The organization-to-process matrix (Figure 12) will help identify user interview requirements. This matrix will undergo considerable expansion as more information is gathered during the user interviews.

Interviews with key user management should be scheduled so that users of the first subsystems to be implemented are interviewed first. This approach helps ensure the earliest possible start of the first subsystems definition. Information should be gathered to:

- Confirm the team's understanding of the business structure, organizational interrelationships, and plans for the future.
- Develop an understanding of the functional business objectives and how they relate to overall business objectives.
- Understand projections for growth of the function.
- Determine the key decisions made by the function.
- Understand major problem areas or control exposures.
- Determine the degree of user satisfaction with current information systems support, the support that is required for the future, and value of this projected support to the user.

Key aspects of the user interviews are the level of management interviewed and the depth of detail sought. Interviews in the functional area should be with management personnel designated by top-level management of that function. Top-level managers need to know the purpose of the interview and approximately what information is required so that they can suggest who should be interviewed. The study team is interested in interviewing the most knowledgeable member of user management.

NETWORK DEFINITION

A principal activity during the definition phase is to define the major systems of the information systems network, how they interrelate, and how the data flows between and among these systems.

The groundwork for defining the network was laid in the identification phase (see "Network of Information Systems"). The purpose in the definition phase is to further define the network on the basis of a more precise understanding of the objectives of the business. The network is developed by charting business processes, data classes, and systems groups, and showing clusters of systems that use certain data types to support processes. The information systems network shown in Figure 25 indicates a natural grouping of systems and flow of data between them. This information can then be used to define a network of data bases required to support the information systems network.

In addition to defining the interrelationships of major systems, the definition phase must address:

- Modifications that must be made to current systems (primarily to indicate how existing systems are to evolve into the network)
- Specific user problems that will be solved as a result of the network approach
- Integration of the network approach with business objectives to help ensure that the long-range information systems plan is based on sound logic
- Alternate approaches that were investigated and rejected, and the reasons for rejection
- Technical environment of the network, including statements about the control system environment and use of supervisory programs (such as data base/data communications systems control programs and degree of standardization)

Given the level of detail specified at this point, it is not expected that actual programming of any segment of the network can begin. The normal general and detailed design steps must be followed for each subsystem before the development effort can begin. As the network is being further defined, the first subsystems, which were recommended and approved in the identification phase presentation, will be defined in considerably more detail.

FIRST-SUBSYSTEM DEFINITION

One of the main activities in the definition phase is to define a generalized business information structure, in accord with the information systems network, that can address the specific business needs. It should describe in general terms the subprocesses and data bases that should compose the first subsystems and how these subprocesses and data bases relate to each other.

This process is an in-depth, top-down definition of the processes and users involved in the scope of the first subsystems. The method for defining the first subsystems is the same as that described above for the network definition except that it considers all levels of management in the process rather than only selected levels. In addition, projections of resource requirements, including equipment, should be developed for the first subsystems.

Assuming that the above criteria are considered and the first subsystems are confirmed, definitions of the first subsystems should be made with the users in the areas involved. This is a general definition and not a detailed design or specification. The definition can be depicted in several ways, such as the flowchart method (Figure 26) or the HIPO, a design aid and documentation technique. Documentation describing the HIPO method is available through local IBM branch offices.

When the definition has been formulated, the data base should be defined to consolidate the general data requirements for the first subsystems.

A data base can be defined as a nonredundant collection of interrelated data items processable by one or more applications:

- *Nonredundant* means that individual data el-

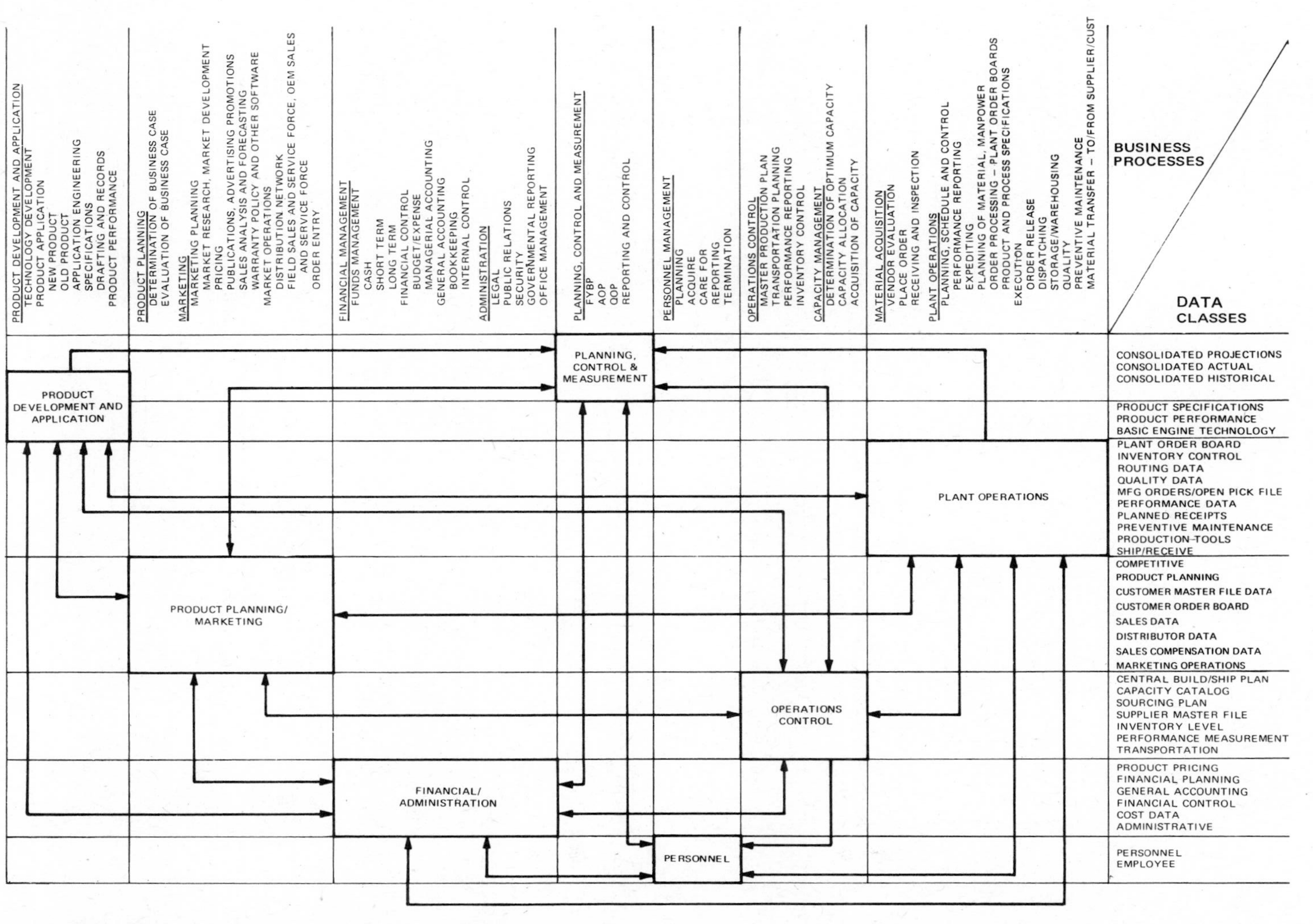

FIGURE 25 Information systems network (manufacturing).

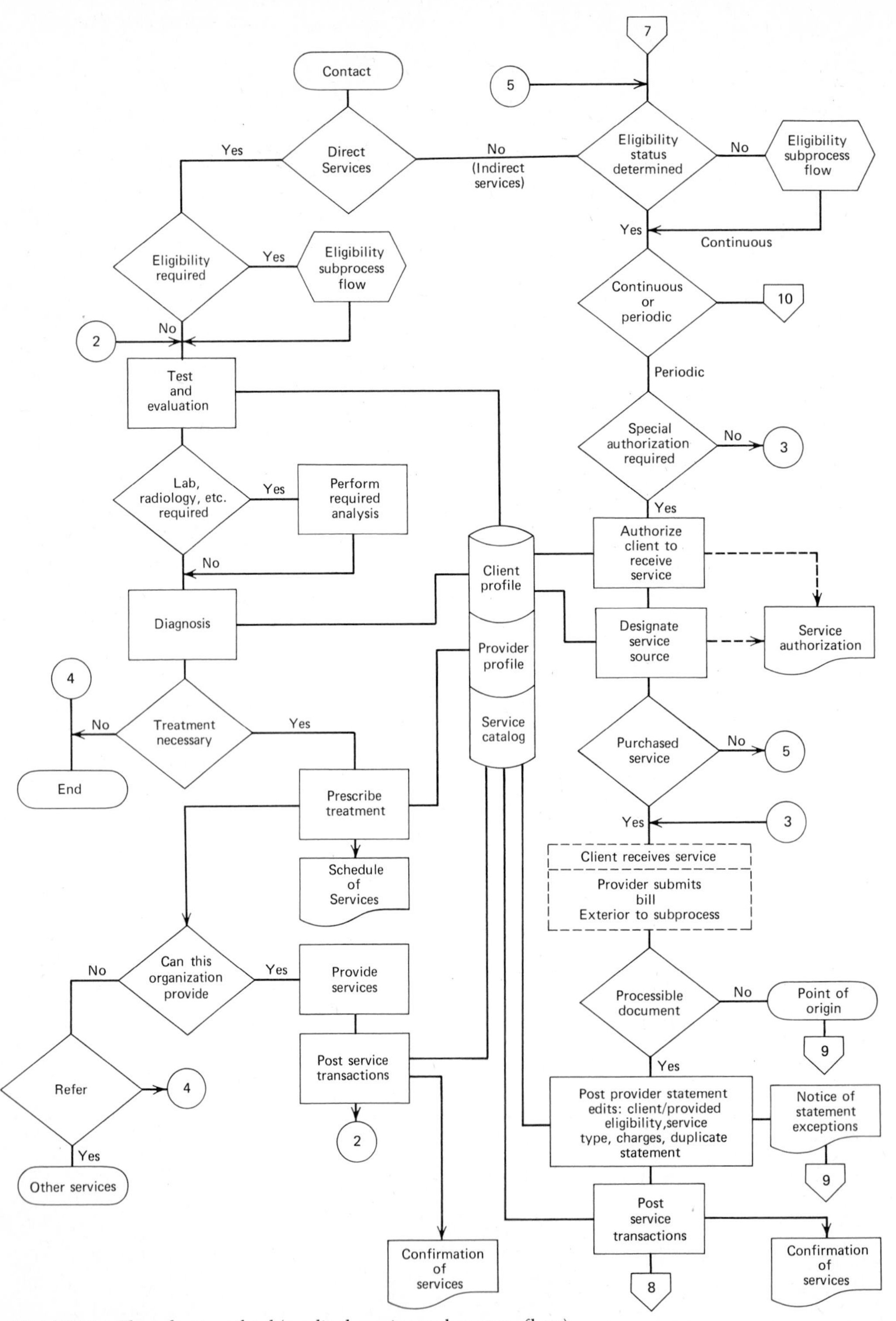

FIGURE 26 Flowchart method (medical services subprocess flow).

ements appear only once (or at least less frequently than in normal file organizations) in the data base.

- *Interrelated* means that the files are constructed with an ordered and planned relation ship that allows data elements to be tied together, even though they may not necessarily be in the same physical record.
- *Processable by one or more applications* means simply that data is shared and used by several different subsystems.

Development of a data base has some obvious benefits. By consolidating files, the user can obtain better control of data and reduce storage space and processing time. Equally important are the resultant data synchronization and timeliness. Use of a single information source makes processing more accurate because all subsystems refer to the same data.

A data base system helps overcome some of the complexities of data management by managing data centrally. It can provide additional data relationships while minimizing storage redundancy. Figure 27 illustrates how a data base system can support subsystem needs independently and concurrently.

The goal in data base definition should be to produce a practical, workable system that makes key information available, rather than trying to include every conceivable type of data. Therefore, the definition process should include an evaluation of each piece of data and a decision as to whether it should be included in the data base.

A comparison of the data base environment with the traditional approach to systems development and maintenance reveals the advantages of the data base concept. In the traditional approach, a system is usually designed, programmed, tested, and then implemented as a total entity. Its advantages cannot be realized by the end user until the entire system is completed. The time involved can cause frustrations, since business requirements cannot be kept frozen long enough to avoid changes, delays, and false starts. Also, when data or logic changes are required, considerable testing may be necessary to determine how the change affects other programs or system functions.

By contrast, the data base approach allows a gradual transition from existing systems to online, transaction-driven systems. The data base is created gradually, and a few transactions are implemented at a time. Current data base techniques facilitate this type of data base and system implementation. By gradually implementing transactions, user department signoff can be obtained more easily, and the user can enjoy the benefits earlier. This approach also helps to overcome the problem of not being able to freeze business requirements and technological advances during the development cycle. Changes that must be made along the way impact individual transactions or become new transactions or become new transactions themselves. Changing data needs can be accommodated without affecting programs that do not use that specific area or segment of the data base. Thus, the data base environment is a better way to accomodate change, deliver benefits to the user, and control development costs.

NETWORK MANAGEMENT SYSTEM

During the identification phase a cursory review was made of the existence of certain information systems management functions and activities that are required to successfully implement an information systems network; these functions and activities are described here. If deficiencies are found within the current organization structure, recommendations for overcoming these deficiencies should be incorporated in the action plan. Often, portions of the required management system can be implemented during the definition phase. The information systems functions or activities to be addressed are:

- Planning process
- Data management

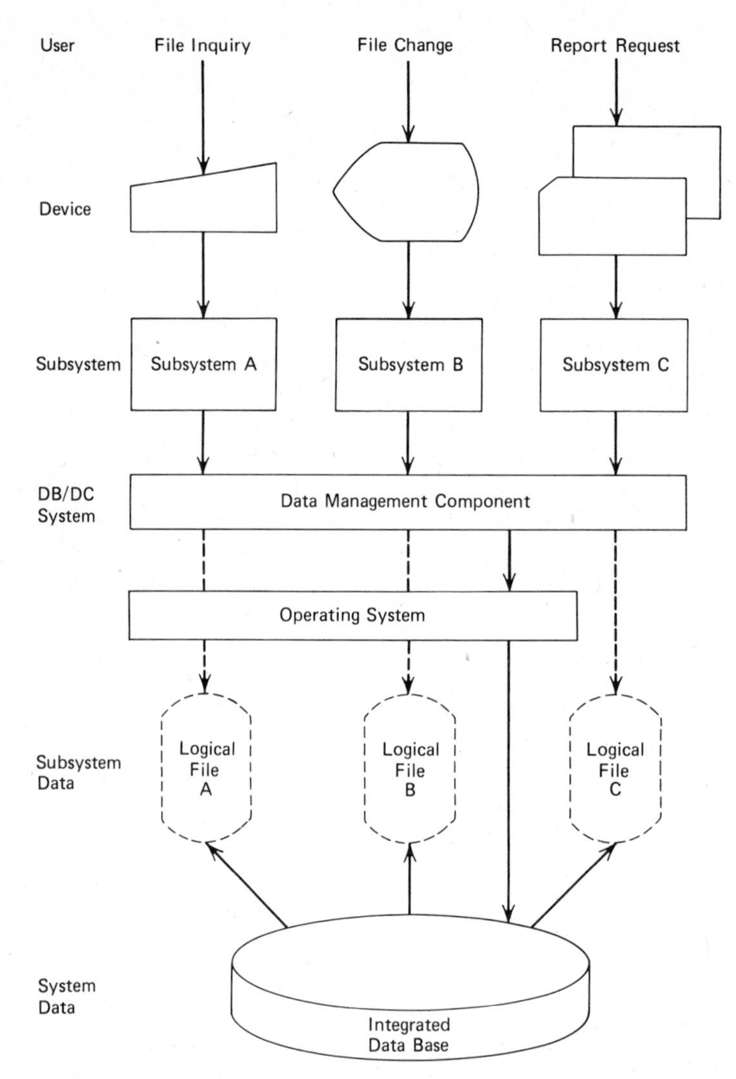

FIGURE 27 Data base system.

- Management practices
- Management system strategies and actions
- Control and measurement of network implementation

Planning Process

The business environment is constantly undergoing change, and some of these changes will require revision of certain parts of the information systems plan. Guidelines must be established for the format, content, and approval of these revisions. It should be clearly defined what organization within the business has responsibility for the planning process.

A plan is only as viable and effective as the planning process and the degree of user management involvement and commitment. Two elements should be considered for an effective planning process, the plan, and plan assessment. Typical problems associated with planning should also be taken into consideration.

The Plan. An effective information systems planning process must provide the basis for management directions, measurements, and control of resources. The process requires:

- A formal information systems plan. (The content of the plan is described at the end of this chapter.)
- A formal mechanism for preparing, integrating, and maintaining the plan at all levels
- User participation in preparing, assessing, reviewing, and concurring with the plan

The key objectives of the plan are to:

- Obtain top management commitment
- Identify high payback areas
- Allocate resources and measure performance
- Provide better decision-making capabilities
- Provide better business visibility
- Enhance user communication and cooperation
- Optimize the use of technology
- Display problems, risks, and opportunities

Plan Assessment. Equally important to development and maintenance of an efficient plan is an assessment of the plan against business goals and objectives. The assessment should be done at an appropriate organizational level to ensure follow-through and resolution of issues.

Plans are assessed for year-to-year changes, plan-to-plan changes, logical distribution of resources, interrelationships of projects, consistency of objectives and strategies with other plans, interlocation impacts and transfer of costs, and division of personnel, equipment, and dollars between development and operations.

Typical Planning Problems. Problems that must be recognized in the planning process may include:

- Incomplete definition of the information systems expense (communications, process control, user education, etc.)
- Strategic plans that do not realistically reflect the resource capabilities necessary to accomplish the plan
- Assignment of responsibility for determining the ultimate value of the systems (should not be data processing personnel)
- Overoptimism of the planning people themselves
- Lack of user involvement
- Lack of an established mechanism for resolving issues between operating units and between staff and line personnel
- Lack of a total architectural approach for the long term
- Inattention to the political side of planning
- Lack of an appropriate time horizon for the plan
- No development of tactical or operational plans from the long-range plan
- No identification of long-range products, markets, organizations, governmental constraints, etc.
- No projections concerning inflation, productivity, technology, learning curves, etc.

Data Management

Data procedures and standards are required to ensure timely, reconcilable, and well defined data. Although these data management requirements always existed, their effectiveness is vital to the information systems network.

Proper data management requires three major activities to be performed: data base administration, data base structuring, and establishment of systems standards.

Data Base Administration. To handle the administration aspects of data, the business must establish organizational responsibilities, procedures, and standards:

- A function or strategy that addresses data management throughout the organization

- Technical responsibility for the data base, including:
  —Logical and physical definition and creation
  —Monitoring for performance and security
  —Recovery and restart
  —Reorganization
  —Space allocation
- Nontechnical responsibility (user interface) for the data base, including:
  —Content determination
  —Input source
  —Integrity requirements
  —Requirements forecasting
  —Security requirements
  —Update frequency
  —User requirements and satisfaction
- A data dictionary/directory system that identifies key elements, primary and secondary users, source responsibility, location, etc.
- Procedures and practices defined to address such factors as data accuracy, failsafe, security, data base efficiency, and user liaison

Data Base Structure. Effective structuring of the data base requires that the following activities be performed:

- Determine what common data bases should be developed and establish a priority schedule for their development.
- Design data bases to support business processes.
- Restructure data to take advantage of technological changes and respond to changing requirements, etc., with minimum impact on the user/subsystem.
- Forecast data needs (such as volume growth in transactions or product line diversification) to assure that capacity and performance can be maintained.
- Communicate, evaluate, justify, and assign priorities to changes in data requirements.

Systems Standards. A function is essential to establish the rules, practices, guidelines, and procedures for developing the systems within the network:

The standards function should consider the following factors.

- Standards and conventions to govern data integrity:
  —Assignment of responsible source of data
  —Reasonableness checks of requested information
  —Data modification traceability
- Standards and conventions to govern data security:
  —Identification of sensitive data
  —Authorization for use or modification of data
  —Program audit and testing procedures
  —Operating recovery and physical security procedures
- Coding and classification standards and procedures for change nomination, concurrence, implementation, and documentation
- Delegated responsibility to help ensure appropriate systems control program solution and compliance with software strategy and to test the viability of the strategy
- Standards program that ties into the current information systems plan
- Monitoring for compliance to standards

Management Practices

An activity is required in information systems management to identify the procedures, standards, and responsibility assignments to be included in the management practices, and the appropriate level in the business at which management practices originate and are enforced. The practices should be created to help all organizations understand their role in information systems design, development, and operation.

This function should be at an equal or higher level in the organization than the level for which the network is being defined. It should consolidate and publish management practices, which should be continually reviewed and updated to reflect the changing environment. Each practice should include the effective date, areas of application, methods of monitoring compliance, and reason for the practice.

Management System Strategies and Actions

The objective of this activity is to define the strategies and actions required to achieve an effective information systems management system.

The strategies should:

- Specify necessary changes to the planning process.
- Specify the control and measurement processes needed to ensure implementation of systems in accordance with the plan.
- Specify a data management function that will allow data to be communicated efficiently between systems and subsystems.

Control and Measurement of Network Implementation

Since the network approach to systems design is evolutionary, typically extending over one year, effective control and measurement of the process are important. Six activities are used for these purposes: funding, accounting, phase reviews, auditing, facilities control, and key project status reporting.

Funding. Effective control over the funding and allocation of information systems resources is important and requires:

- Criteria for the release and distribution of funds
- Progressive funding for development projects
- Funding of development on the basis of tangible benefits, and measurement of performance against objectives
- User involvement in the funding commitments
- Setting of project priorities

Accounting. Proper accounting for total data processing resources (personnel, equipment, facilities, and funds) requires:

- Identification of resources expended on information systems broken out by such items as equipment, telecommunications, manpower, and travel.
- Monitoring of actual expenses against planned figures on a periodic basis for both development and operations
- Identification of expense transfers between locations and functions
- Resource accounting by functional users to identify expenses in support of functional information requirements
- Accurate measurement of expenses against plans

Phase Reviews. Phase reviews are helpful for development project evaluation and control. The goals of phase reviews include:

- Timely evaluation of a project's progress
- Assurance that agreed-upon objectives are being achieved
- Assessment of technical feasibility of objectives
- Milestone-oriented development projects
- Project reviews conducted on the basis of documentation at the conclusion of each major phase
- Resolution of problems

Auditing. Audit procedures are needed for management examination of development and operational projects to help ensure effective and efficient use of information systems resources. In setting up auditing procedures, the following requirements should be considered:

- Audit teams should be made up of personnel with greater experience and technical range than those assigned to the projects.
- Audits should be conducted by personnel other than those responsible for development, implementation, and operation of the projects.
- Audit findings should be reviewed by the appropriate levels of management to ensure resolution.
- Audit reporting should provide for periodic status meetings, immediate action reports, and a formal report at the conclusion of the audit.

Audits of development projects cover:

- Project plans and controls
- Financial and internal controls
- Mission status and goals
- User relationships
- Interorganizational requirements
- Security of data
- Fallback capability
- Hardware planning
- Morale of personnel

Audits of operational projects cover:

- Data processing operations
- Programming support
- Data integrity, control, and management
- Total resource utilization
- User relationships and requirements
- Security of data
- Physical working conditions
- Training of personnel

Facilities Control. Adequate control of the development or acquisition of facilities to support information systems requires:

- A function or activity responsible for information systems facilities planning
- Sequencing of facilities development or acquisition that is related to information systems development requirements
- Facility costs included in the funding and accounting for major systems
- Operational procedures for facilities such as security and emergency procedures

Key Project Reporting. Successful implementation of an information systems network requires continuous executive involvement. A brief quarterly status report (QSR) helps keep management informed of the progress of key projects.

The QSR should contain:

- A brief description of the project
- Schedule (start date; end date)
- Headcount (planned; actual)
- Costs (planned; actual)
- General appraisal (problems; control exposures; audit results)

DEFINITION PHASE DOCUMENTATION

Although the definition phase documentation is not prepared until the end of the study, its content and purpose should be considered at the beginning of this phase to determine the data to be gathered. The documentation can be prepared in sections, which can either stand alone or be combined into a comprehensive report. The recommended format includes:

- Executive summary
- Definition phase report
- Proposed action plan
- Information systems plan

Executive Summary

The summary is the most critical part of the study report. It should provide management with the key information it needs to make long-range decisions. Therefore, this section should be written from management's viewpoint. To help executives find the results of the study quickly in capsule form, the executive summary should present as briefly as possible, the major ideas contained in the balance of the report. Although it is usually prepared after the other sections of the document are completed, it is placed first to aid the reader. The executive summary may also be published as a separate volume for selected executives to avoid disclosure of major or sensitive recommendations to the general reader. The essential contents of the summary are outlined below.

1. Major findings, conclusions and recommendations. These are divided into information systems network and information systems management subsections.
   a. The information systems network recommendations should provide a schematic of the network and the systems in it. Additionally, the network recommendations should include:
      - A master schedule showing a prioritized approach to the implementation of the major subsystems in the network and an approximation of the total resources requirements to implement the network
      - A description of the purpose and value of each system
      - A specification of the tasks required to design, develop and implement the first subsystems along with a statement of resource requirements and value justification
   b. The information systems management recommendations should include:
      - A description of each function being recommended, including the purpose and value
      - A master schedule showing when the management functions will be phased into full operational existence
2. Major benefits and costs. Appendix D shows a format that can be used to display major benefits and cost information. The level of detail in this area depends on the depth of study for the network definition.

An outline of an executive summary follows.

Introduction

- *Purpose:* Describe the document's primary objective; that is, to define an information systems network that will furnish management with the information it needs to make its key business decisions and optimize its information systems resource expenditures. Point out how executives, users, and data processing can each use the plan.
- *History:* Summarize the major actions leading to the study.
- *Scope:* Define the study limits and emphasis.
- *Structure:* Discuss the major parts of the report and their function.
- *Acknowledgements:* Thank "prime movers" and reference appendix on participants.

Methodology (May be included in Introduction)

- List and discuss the major steps in conducting the study. Use simple one- or two-sentence statements explaining how it was done.

Environment (May be included in Introduction)

- Describe the overall structure of the business unit being studied.

- Describe the general and specific climate of the business.
  - *General:* Describe growth, demand for information, effectiveness of current systems, and obvious needs.
  - *Specific:* Reinforce with a few worst-case examples of shortcomings of existing systems and procedures.
- Describe the need for an information systems plan to identify, secure, organize, and communicate the required information.

Summary

MAJOR FINDINGS AND CONCLUSIONS

- Describe the team's findings in two major areas: current information systems and management practices.
  - Write a narrative describing the existing information systems and how they are managed. Analyze the major shortcomings.
  - Support the narrative with an illustration of the current systems and information flow.
- Discuss the problems created by the current structure and procedures. Typical problems are:
  - No single responsibility for establishing information systems requirements
  - Redundancies and gaps
  - Inconsistent, untimely, and incomplete data
- Reinforce each problem with a major example or two. Where possible, quantify the cost of these problems.

MAJOR RECOMMENDATIONS

- State what the team wants management to do and how the recommended actions will solve the problems. Distinguish between information systems network and management recommendations.
- Tell a narrative story describing the proposed network and management system. Describe each information system and function; explain why they are needed and how they solve the problems.
- Be more explicit on the first subsystems. Explain the purpose, functions, and users satisfied. This description may be in the form of a table or illustration.
- Support the narrative with the following illustrations and tables:
  - Proposed network and information flow
  - Master schedule divided into network and management, with manpower resources called out
- Typically, information systems management recommendations will pivot around:
  - Need for a director to create and maintain a plan and monitor planned versus actual performance
  - Need for centralization, with the accent on common systems to serve functions that are common across many organizations
  - Need for common coding as the base for more efficient and effective data sharing and systems growth.

BENEFITS/COST ANALYSIS

- Project gross benefits and costs for the overall network over the plan period. Emphasize first subsystems benefits and costs.
- Accent that more meaningful projections will come out of individual subsystems design work.
- Accent tangible and intangible values.

Definition Phase Report

The report should show that a thorough evaluation was performed and that the business, organization, philosophy constraints, and plans are understood. All major statements about shortcomings, as well as the value statements for justification and benefits, should be checked out and substantiated.

A sample of the contents and format for the report follows.

Introduction

- Similar to the executive summary. Add paragraphs describing how the purpose and structure of this part differs from the executive summary.
- Same option to include methodology and environment in introduction depending on their importance.

Methodology

- Similar to the executive summary but more detailed

Environment

- Similar to the executive summary but more detailed

Findings, Conclusions, and Recommendations

FINDINGS

- Discuss the existing systems and assess their strengths and weaknesses from the following standpoints.

Network Definition and Scope

- Define the area of the business studied and tell why it was selected.
- Present subsystems descriptions and information flows. Give subsystems name, description, functions, location, users, and major data bases.

Business Processes and Data Requirements

- Define the business processes and how they relate to one another.
- Discuss where the data originates for the subsystems supporting those processes.

System Flow and Data Interrelationships

- Show the relationships and dependencies between the various systems studied.

CONCLUSIONS

- State the basic conclusions derived from the findings described above.
- Use a chart similar to the example shown in Appendix E to identify the benefits of the systems in the network.

RECOMMENDATIONS

- Discuss the recommended actions. Start with an overview that describes the proposed network. Define the objectives and functions of each system. Emphasize the first subsystems and justify their selection.
- Summarize the major recommenda-

tions. They will generally pivot around:

- — Need for centralization. Accent the need for common systems by identifying information needs that are common to functions.
- — Need for common coding. Point out how this is basic to sharing data and promoting more effective and economical use of the information systems resource.

- Specific recommendations should be made in the subsections below.

Shared Data

- Group the business processes and users that require similar information-systems support.
- Identify the major systems required to serve these processes and users.
  - — Cross-reference the organization/processes matrix for a number of organizational functions involved in each process.
  - — Cross-reference the systems/processes matrix for the current and planned systems supporting each process, and those with minimal or no support. Use the organization/data files matrix for data to support the recommendations.

Common Systems

- Define and group the logical areas of the business that have common application requirements and can be supported by a common subsystem or group of subsystems.
- Develop a table showing which users require which common data (payroll, personnel, product design, financial, etc.).

Network Concept

- Define how the major systems can be related to form a network. Emphasize interrelationships and interfaces, and show how the work of one logically relates to the work of others.
  - — Refer to user satisfaction comments.
  - — Refer to analysis of processes and users that control the data.
  - — Define each system in terms of data base content (type of information, not data element level), business process supported, applications supported, key user impact areas, and value.
- Define the modifications required in existing applications (define evolution).
- Specify the user problems solved.
- Show how this strategy supports the business plan.
- Discuss the required control systems and supervisory programs.

First Subsystem Definition

SUBSYSTEM DESCRIPTION

- Discuss why this subsystem was selected as first.
- Present the subsystem description, define its uses and users, and discuss its value.
- Define the basic processes.
- Describe the present efforts in this area.
- Relate the processes and data processing systems to users.
- Define the resources currently committed to the subsystem.

PROBLEMS

- Discuss the major shortcomings of current procedures, emphasizing:
  - Management and organizational interfaces
  - Common data bases
  - Data interfaces

RECOMMENDATIONS

- Restate the system objectives.
- Define the steps required to achieve them, including the phases to follow.

Proposed Configuration

- Define configuration.
- Point out the wide range of users to be satisfied and the interactive nature of the system.

Common Software/Data Management

- Emphasize the need for common software and data management techniques and how they make it easier to fit future systems into the network.

Architecture

- Explain how the systems will evolve. Discuss the data base, data processing support and characteristics, and minimum hardware configuration. For data processing, cover:
  - Common software
  - Communications (terminal configuration)
  - Organizational independence

EVOLUTION

- Show the transition from existing to proposed systems, accenting how it will be accomplished without jeopardizing the current system.
- Present the major steps involved and the estimated timetable.
- Discuss the control and management system to be used to help ensure that actual events follow the plan.

Information Systems Management (Refer to Appendix I)

FINDINGS AND CONCLUSIONS

- Discuss and evaluate the effectiveness of existing management practices as they apply to:
  - Information systems objectives
  - Management practices, with emphasis on a formalized information systems plan and definition of functions responsible for its effective implementation and maintenance
  - The planning and control process
  - Information systems architecture
  - Data management
  - Design implementation
  - System justification and funding

RECOMMENDATIONS

- State what should be done in architecture, data management, and planning, control and measurement.
  - Emphasize the need for an information systems plan and the identification of someone to implement and maintain it.

  - Emphasize the need for a data administrator.
  - Define the organizational implications.
- Explain the value of each recommendation—why it should be done.

Benefits/Cost Analysis

- Present reasons why management should commit itself to the plan, emphasizing the first subsystems. Describe the potential benefits and cost for the rest of the network.
- Insofar as practical, put a dollar value on each improvement identified in the recommendations. Add the intangible values:
  - Savings from eliminating redundancies (support, personnel, and programs)
  - Savings through cost avoidance (extrapolate the cost of the present system)
  - Savings from eliminating errors
  - Value of using professional manpower better
  - Value of using equipment and facilities better
  - Value of rechanneling people to develop new applications, instead of maintaining old ones
  - Value of receiving accurate, timely, and consistent data
  - Value of having enough comparative information to make knowledgeable rather than intuitive decisions

Proposed Action Plan

The action plan is a schedule outlining the objectives, strategies, actions, schedule dates, and responsibility for achievement of the study recommendations. Strategies are the first-level courses of action necessary to accomplish the information systems plan objectives. They include:

- Administrative activities necessary to communicate the results of the study
- Introduction of information management practices
  - Establishment of the necessary organizations and relationships
  - Necessary changes to the planning process
  - Control and measurement needed to help ensure implementation of systems according to the plan.
  - Practices to help reduce information system efforts to manageable projects
  - Data management functions to allow data to be communicated efficiently between systems and subsystems
- Key project initiation
  - Concentration on the prime area of need for system support
  - Establishment of the physical facilities necessary to support the plan
- Effective communication through a course of education to achieve a continuing level of understanding
- Emphasis on the philosophy of top-down planning and bottom-up implementation at all levels of the business

Figure 28 shows a recommended format for the action plan, containing sample entries.

Information Systems Plan

Although the information systems plan may be included in the overall documentation it should be constructed so that it can stand alone.

An information systems plan has two basic elements:

1. A report on Business Systems Planning's findings, conclusions, and recommendations
2. A plan that defines the actions and resources required by a business to

- Develop an information systems network

Objectives	Strategies	Actions	Schedule	Responsibility
1. Administrative	Provide vehicle for communicating study results	1. Letter stating results of study 2. Publication of BSP documentation 3. Presentations	4-10-75 4-17-75 by 5-18-75	Director systems BSP team BSP team
2. Introduce information management practices	Establish the necessary system development organizations and charters	1. Assign system liaison personnel. 2. Align systems development groups to support business processes.	Prior to 1-1-76 1-1-76	Executive Director Systems Executive Director Systems
	Develop necessary policies and procedures for information management practices	1. Planning/Replan techniques for introducing I/S strategies with business planning strategies into the planning cycle. 2. Control and measurement-introduce project and management concepts of —funding —accounting —phase reviews —auditing	1-1-76 1-1-76	Executive Director Systems, I/S planning, staff-liaison Executive Director Systems, I/S planning, systems managers
3. Assure I/S plan reflects business needs and management viewpoint	Continuing update on current plan and status	1. Newsletters 2. Presentations to management 3. Conferences 4. Periodical technical update sessions with system development and technical personnel.	Ongoing	Executive Director Systems, I/S Planning

FIGURE 28 Sample action plan.

- Design, develop, and implement the subsystems in the network with emphasis on the first subsystems
- Define the management system needed to help ensure success of the total effort

Although the plan leads to system design, development, and implementation, it does not in itself present that level of detail. The plan defines the work that has to be done; the more detailed effort in the next phase will provide direction for doing the work.

The plan is the roadmap towards implementation of the information systems network and should be updated at regular intervals to reflect the current status. Management should determine what part of the organization has the responsibility of maintaining the plan and communicating its current status to the appropriate areas.

A recommendation on the contents and format of the I/S Plan follows.

Introduction

- Discuss the purpose of the plan:
  - Establish the goals, objectives, strategies, actions, and responsibilities required to establish a uniform approach to information systems design and development.
  - Provide a vehicle to communicate these plans throughout the business.

  - Improve information handling and management.
  - Establish guidelines to assess the plan's effectiveness.
  - Provide an orderly, economical way of developing the systems needed to meet the increased needs of the business.
  - Ensure that the plan does support overall business planning.
  - Provide a data source to analyze data processing resource utilization.
- Discuss the five major parts of the plan:
  - Section 1 Purpose
  - Section 2 Environment
  - Section 3 Resource Commitment
  - Section 4 Major Project Summaries and Schedules
  - Section 5 Action Synopsis

Environment

- Establish the need for the plan. Point out the demands and problems that underscore the need for improved data processing utilization.

Resource Commitment

- State the objectives of the section: to present the information systems budget, describe how it is committed, and track historical data.
- Summarize the five-year expenditures trend (19____ to 19____) along the following lines:
  - Ways in which current dollars and manpower are being spent to support the business
  - Information systems expense growth
  - Expenditure changes year-to-year
  - Manpower distribution by major function and project
  - Manpower distribution by skills
  - Equipment profile

Major Project Summaries and Schedules

- Write a summary for each major project, including:
  - Purpose and description
  - Schedule (with milestones identified)
  - Resource requirements
  - Definition of testing function
  - Programming standards and education requirements
  - Dependencies
  - Contingency plan

Action Synopsis

OVERALL

- Discuss the necessary activities
- Define the objectives, goals, strategies, actions, and responsibilities for each of these activities.

INFORMATION SYSTEMS NETWORK

- Discuss each proposed system from the same standpoints (objectives, goals, etc.). Typical objectives may revolve around functional systems, unified systems, data management, and standards development.

INFORMATION SYSTEMS MANAGEMENT

- Discuss management recommendations from the standpoint of architecture, data management, and planning, control, and measurement.

- Recommend the creation of a management function to direct plan development, implementation, and measurement.

Appendices

A Glossary of Terms
B Project Schedules by Phases
C Steps Involved in Project Design and Implementation Phases

APPENDIX A: INFORMATION SYSTEMS MANAGEMENT REVIEW CHECKLIST

- Project management
  - Planning
  - Specifications
  - Schedule and resources tracking and control
  - Change control
  - Phase reviews
  - Status reporting
- Data administration
  - Data dictionary/directory
  - Security
  - Control
  - Coding structures
  - User interface
  - Standards
- Information systems planning
  - Maintenance
  - Management
- Budgeting
  - Development
  - Tracking
- Organization
  - Structure
  - Responsibilities
- Financial justification
  - Procedures
  - Application of procedures
- Operations
  - Data entry
  - Job staffing
  - Scheduling
  - Library
  - Machine room
  - Performance against objectives and schedules
- Productivity improvement programs
- Application systems
  - Identification
  - Selection
  - Definitions
  - Development
- User interface
- Executive steering or policy committee
  - Membership
  - Responsibilities
- Standards
  - Definition
  - Control
  - Evaluation
  - Revision
- Training
  - Current level
  - Education programs
- Quality assurance
  - Procedures
  - Organizational responsibility
  - Effectiveness
- Applications systems evaluation and audit
  - Performance evaluation
  - Justification follow-up
  - User satisfaction
- Program and application system maintenance
  - Change control
  - Implementation procedures
- Information systems architecture
  - Definition

- Control
- Evaluation
- Revision

- Hardware/software/data communications
  - Planning
  - Selection
  - Measurement
  - Evaluation
  - Disposal
- Systems engineering
  - Technical support
  - Systems programming
  - Problem definition and solution
- Administration
  - Usage accounting and billing
  - Clerical support
  - Supply management
  - Space management

- Security

APPENDIX B: EXAMPLES OF PROCESSES AND SUBPROCESSES

DISTRIBUTION INDUSTRY

- Finance
  - Cash management
  - General ledger
  - Accounts payable
  - Payroll and labor distribution
  - Cost accounting
  - Sales audit
  - Route accounting
  - Credit management
  - Accounts receivable
- Administration
  - Personnel
  - Financial planning and budgets
  - Warranty
  - Stockholder relations
  - Security
- Physical distribution
  - Customer order entry
  - Credit authorization
  - Billing
  - Inventory control
  - Inventory forecasting and allocation
  - Shipping
  - Order allocation
  - Physical distribution planning
- Receiving
  - Warehouse operations
  - Vehicle scheduling
  - Labor scheduling
  - Vehicle maintenance
  - Finished goods purchase order entry
- Marketing and merchandising
  - Sales analysis
  - Brand management
  - Sales management
  - New-item tracking
  - Sales forecasting
  - Advertising planning/evaluation
  - Promotion planning/evaluation
  - New-product introduction
  - Market research
- Branch/store operations
  - Stock replenishment
  - Receiving and marking
  - Credit authorization
  - Financial reporting
  - Labor scheduling
  - Dealer services or direct store deliveries
  - Shelf space allocation
  - Pricing
  - Facilities control
  - Point-of-sale activity
- Production
  - Production planning
    - Manufacturing facilities control
    - Material requirements planning
    - Inventory management (raw material, WIP, packaging)
    - Raw material/component purchasing

Capacity planning and scheduling
Process/design/industrial engineering
Production operations control
Production process monitoring and control
Packaging/filling monitoring and control
Plant floor data collection
Plant equipment maintenance/cost analysis
Labor monitor and control
Product structure/routings maintenance

- Research and development
  - Scientific calculations
  - Information retrieval
  - Engineering administration
  - Instrument monitoring and control
  - Prototype process monitoring and control

EDUCATION INDUSTRY

- Alumni development
  - Resource acquisition
  - Resource utilization monitoring
  - Alumni programs and services
  - Alumni tracking
- Curriculum/instruction
  - Assessment of needs and resources
  - Design of curriculum
  - Course instruction
  - Curriculum and course evaluation
  - Scheduling (classes)
- External communication and relations
  - Reports (publicity)
  - Negotiation
  - Public service
  - Extra-university affiliation
- Facilities
  - Acquisition/disposition
  - Construction/alteration
  - Maintenance/custodial care
  - Allocation/scheduling
  - Rental property management
  - Security
- Financial
  - Collection of receivables
  - Stewardship of funds
  - Disbursement of funds
  - Management of funds
  - Financial services
  - Protection against financial liability
  - Procurement of support
  - Performance of inquiry
  - Publication of results
- Students
  - Promotion/recruiting
  - Admissions
  - Registrations
  - Student activities/life
  - Advising/evaluation of students
  - Student status
- Goods and services
  - Assessment of needs
  - Acquisition
  - Inventory of expendables
  - Inventory of nonexpendables
  - Distribution
- Institutional planning
  - Goals development
  - Strategy planning development and integration
  - Tactical planning
  - Control, measurement, and update
  - Budgeting of resources
- Personnel
  - Recruiting
  - Hiring/termination
  - Career development and evaluation
  - Salary and benefit administration
  - Employee relations
  - Compliance with government regulations
  - Assignment of teaching responsibility
  - Transfer status
- Research
  - Identification of areas of interest
  - Procurement of support
  - Performance of inquiry
  - Publication of results

APPENDIX C: EDUCATION PROCESS DEFINITIONS—UNIVERSITY EXAMPLE

Alumni/development. The process involved with obtaining and managing external and alternate internal sources of support for the university (excluding normally acquired funds for tuition, rent, etc.).

- *Resource acquisition.* The process of identifying, locating or creating alternate sources of support for the university and gaining the commitments for this support.
- *Monitoring resource utilization.* The process of managing the use of donated resources to comply with the conditions of the gift.
- *Alumni programs/services.* The development and administration of programs and services directed to former students of the university, specifically university graduates.
- *Alumni tracking.* The process of maintaining contact with former students of the university.

Curriculum/instruction. The development and presentation of academic programs and courses of instruction.

- *Assessment of needs and resources.* The determination of course and program needs of students and the resources available to support these needs.
- *Design of curriculum.* The definition of the structure, sequence, and requirements of academic courses to best meet the objectives of the academic department and the university.
- *Course instruction.* The presentation of the course information to students.
- *Curriculum and course evaluation.* The process of judging the adequacy of the current course/curriculum in meeting the objectives of the department and the university.
- *Scheduling.* The determination of class times so as to minimize conflicts of faculty, students, and other related classes.

External communication and relations. The processes involved with the relationship of the university to the community at large.

- *Speeches, reports, and publicity.* The process involved with maintaining and improving the image of the university in the community at large.
- *Negotiation.* The process of relating on business issues with the general community and public, private, and governmental agencies.
- *Public service.* The process of developing and administering gratuitous services given to the general community: legal and medical aides, community use of university facilities, student volunteer services, etc.
- *Extra-university affiliation:*
  Institutional affiliations. Programs, agreements, or understandings with institutions outside the university to provide access to facilities of other institutions, student clinical placement, interinstitutional referral of staff or clients, etc.
  Individual affiliations. Formal or informal agreements or understandings with individual members of the faculty and staff allowing the member to spend some part of his working time (daily, weekly, or yearly) on activities related to his profession, but outside of the university and compensated for by other than the university. Such activities include but are not limited to consulting and clinical practice.

Facilities and capital equipment. The process involved with the acquisition, maintenance and disposition of all major nonfiscal assets of the university.

- *Acquisition/disposition.* The processes involved with the procurement of all facilities and capital equipment and the eventual disposition of that property. Included are deci-

sions to acquire, evaluation for the purposes of selection, selection of vendors, legal transfer of ownership, and evaluation for the purpose of disposition.
- *Construction/alteration.* The processes involved with the development or modification of facilities including development of a design, selection of vendor services, compliance with legal requirements, and physical construction or alteration of the facility.
- *Maintenance/custodial care.* The upkeep of the physical condition of the plant.
- *Allocation and scheduling.* The processes involved with making facilities and capital equipment available to the university in an equitable and efficient manner (high utilization).
- *Rental property management.* The overseeing of rental properties.
- *Security (physical and financial).* The process of providing both physical and financial protection against loss or damage of facilities and capital equipment.

Financial. The processes involved with the management of the fiscal resources of the university.

- *Collection of receivables.* The administration of the university's accounts receivable, including alumni pledges.
- *Stewardship of funds.* The processes involved with the fiduciary aspect of asset management (e.g., legal, audit, monitoring, and compliance with government regulations).
- *Disbursement of funds.* The distribution of money in accordance with the university budget, including accounts payable.
- *Management of funds.* The long- and short-term investment of funds, including management of endowments.
- *Financial services.* Supplementary financial processes such as negotiating student loan conditions with banks and cashing checks.
- *Protection against financial liabilities.* The provision of financial protection (insurance) to cover suits against the university

Goods and services. The processes involved with the acquisition, maintenance, and disposition of the nonmajor real assets of the university.

- *Assessment of needs.* The determination of the type and quantity of goods and services required to support the activities of the university.
- *Acquisition.* The procurement of required goods and services. Included are evaluation of alternative products and services, evaluation and selection of vendors, purchase contract negotiation and signoff, and physical acquisition of the goods.
- *Inventory of expendables.* The process of maintaining and managing the supply of on-hand expendable goods.
- *Distribution.* The process of making goods available to the entire university in an equitable and efficient manner.

Institutional planning. The processes directed to the orderly establishment and achievement of the goals of the university.

- *Goal development.* The process in which the leaders of the institution decide the end results to be achieved.
- *Strategic planning—development and integration.* The determination of ways of achieving the university's goals. This process is likely to be interactive, involving central administration and the individual schools and colleges so as to develop compatible, collaborative plans and programs.
- *Tactical planning—integration and assessment.* The process of identifying short-term alternatives and setting priorities. The result is the allocation of resources across the university in an integrated manner.
- *Tactical planning—control, measurement, and update.* The monitoring of objective attainment and resource allocation through

budget status reports and milestone reviews. The budget status is measured against the tactical plan, and where deviations occur, either greater controls are instituted or the tactical plan is changed. Tactical planning results in a statement of the anticipated financial position of the university, not only for dollars, but also for staff and facilities for a definite period of time.

- *Budgeting of resources.* The documentation of the tactical plan. Preparation of the budget includes receiving requests, prioritizing needs, designating income and expenditure support levels to support the various departments, and obtaining approval.

Personnel. Those processes that have primary focus on the employees (faculty and staff) of the university.

- *Recruiting.* The process of attracting quality people to the staff of the university. Included are general recruiting efforts geared specifically to the identification, interviewing, and evaluation of staff candidates.
- *Hiring/terminating.* The processes of employing accepted staff candidates and also terminating this employment. Hiring includes determining or negotiating the conditions of employment, authorizing the employment, and entering the individual into the university system.
- *Career development and evaluation.* The establishment of policies and guidelines concerning career advancement, determination of tenure, evaluation of candidates for advancement, and promotion of staff members.
- *Salary and benefits administration.* The establishment of salary and benefit guidelines and policies, and the determination or negotiating of salary levels and benefits.
- *Employee relations.* The process of dealing with university personnel, either collectively (unions) or individually, on any matters concerning employee job satisfaction and general well-being.
- *Compliance with government regulations.* The process of ensuring that university systems and programs conform to and satisfy government requirements.
- *Transfer status.* The process of administering any nonpromotionary change in job responsibility or general nonpromotionary change in employment status (leave of absence, interdepartmental transfer, sabbatical).
- *Assignment of teaching responsibilities.* The allocation and scheduling of faculty to teach courses.

Research. The process of investigation or experimentation aimed at the discovery and interpretation of facts, revision of accepted theories or laws in the light of new facts, or practical application of new or revised theories or laws.

- *Identifying areas of interest.* The determination of the type of research that the university can physically support, that faculty talent can conduct, and that can reasonably be expected to attract funding.
- *Support procurement.* The process of getting a commitment for support of an identified area of research.
- *Performance of inquiry.* The process of conducting the research effort.
- *Publication of results.* The process of distributing the results of scholarly investigation.

Student. Those processes that have primary focus on the student (with the exception of the course instruction process).

- *Recruiting.* The process of attracting applicants to the university.
- *Admissions.* The evaluation of the qualifications of student applicants, and the formal acceptance that confers the rights and privileges of a student.
- *Registration.* The enrollment of admitted stu-

dents into university programs. Included are course enrollment, fee assessment, and dorm assignment.

- *Student activities and student life.* The provision of nonacademic services and programs to students. Included are extracurricular activities, placement, student government, health services, intramural programs, student services, etc.
- *Advising and evaluation of student.* The process of measuring and recording academic progress of students and providing guidance and counseling. Included are grading, advising, and tracking for purpose of recommendation.
- *Student status.* The process of determining changes of the relationship of a student to the university. Included are decisions involving graduation, leaving the university, returning to the university, and leaves of absence, but not transferring into the university, which is accomplished through the admission process.

APPENDIX D: SAMPLE SUMMARY OF BENEFITS AND COSTS

The costs of obtaining the benefits consists primarily of the costs of the information systems department and computing equipment at remote sites. To illustrate the total cost of the five-year plan accurately, remote site equipment costs have been consolidated with information systems department projected costs shown in Table K-1.

As shown in the table, the projected costs for 1975 and 1976 will not significantly change (1976 is a 4.1% increase over 1975). During those years, the cost of upgrading the technical staff will be offset by a reduction in equipment costs.

The projected costs for 1977 through 1979 will increase at an average annual rate of 7.7% for the period.

TABLE K-1
Information Systems Cost (000 Dollars)

	Actual		Projected				
Element	1973	1974	1975	1976	1977	1978	1979
Costs allocated to the action plan	—	—	1275	1350	1500	1575	1725
Other department costs	—	—	1950	2175	2250	2475	2700
Equipment costs	—	—	2250	2175	2400	2550	2700
Total	5025	5250	5475	5700	6150	6600	7125
Percent Increase		4.5	4.3	4.1	7.9	7.3	7.9
Gross Benefits	—	—	—	1500	4500	5700	7500

APPENDIX E: EXAMPLES OF INFORMATION SYSTEMS NETWORK BENEFITS

<table>
<tr><th></th><th></th><th colspan="2">Benefits</th></tr>
<tr><th>System</th><th>Subsystem/Applications</th><th>Tangibles</th><th>Intangibles</th></tr>
<tr><td>Advanced Administrative System</td><td>• Customer master records
• Accounts receivable
—Cash
—Collections
• Order entry
• Scheduling
• Commissions
• Inventory/lease billing
• CAI</td><td>• Administrative costs
—Administrative personnel
—I/S equip. and personnel
11% reduction in expense base
2% reduction in expense growth rate
• Improved asset management
—Less idle inventory
—Faster delivery to customers
• Better cash management
• Better collections
• Education/training cost reduction
• Reduction in lost revenue from improper billings</td><td>• Customer service
—Decreased errors
—Earlier delivery (work order entry cycle to one day)
—Faster response to customer inquiries
• Business flexibility
—Admin. process independent of organization
—Better response to marketing changes
• Better control, earlier measurement
—More accurate, timely, complete demand information
• Industry leadership</td></tr>
<tr><td>Common Manufacturing Information System</td><td>• Operations planning
• Manufacturing activity
• Release and control
• Product process description
• Procurement
• Warehousing
• Final assembly
• Logistics control</td><td>• Plant operations
—60+ % manpower reduction, clerical activities impacted
—Reduced overtime & equipment
• Inventory
—Inventory reductions
—1% purchased, 5% other
—10% reduction in inventory losses
—Faster processing of requisitions
• Procurement
—Improved lot prices and schedules
—Faster response to requirements
—Better supplier control
—Improved cash management</td><td>• Corporate philosophy
—Decentralized management in an integrated manufacturing complex
• Employee morale and employment practices
• Improved customer service
—Reduced errors
—Reduced lead time
• Better capacity information in support of general management demand/supply decisions
• Industry leadership</td></tr>
</table>

System	Subsystem/Applications	Benefits	
		Tangibles	Intangibles
		• Information systems —Displacement of existing I/S —Reduced growth rate • Manufacturing —Reduced cost of new plant startup ($3 million) —Reduced cost of product transfers ($1 million) —Effective workload balancing —Improved asset management through reduced lead time	
Accounting Reporting System	• Cash and sales • Store payroll accounts payable • Store reporting	• Clerical reductions (20%) • Improved cash management —Cash receipts —Accounts payable • Displacement of existing information systems • Communications costs (5–10%)	• Elimination of duplicate files • Improved management control —Shorter closing cycle —Improved analysis of budget performance • Strategic base for future I/S needs
Manufacturing Control System	• Bill of materials • Descriptive data • Authorized releases • Purchasing and inventory control • Gross requirements • Net requirements • Burden	• Reduction in clerical staff • Reduction in manufacturing personnel • Reduction in inventories (2%) • Reduced scrap and rework • I/S cost avoidance ($3 million/yr.) • Management productivity	• Management philosophy —Centrally managed integrated manufacturing complex • Improved customer service —Reduced lead time • Industry leadership • Improved long range capacity planning
Distribution	• Carrier records • Transit scheduling and control • Field warehouse management • Carrier payments	• Reduced transportation costs • Reduced delivery cycle • Reduced "pipeline" inventories • Reduced clerical costs • Improved loss/damage recoveries	• Better carrier relations • Long-range transportation and site planning • Improved customer service
Product Design and Release	• Technology planning • Design automation • Cost estimating	• Reduced clerical and professional costs • Shortened product development cycle • Only feasible way to handle many advanced technologies	• Improved product decisions • Competitive product advantage • Improved cost control and performance management
Personnel	• Personnel records • Skills inventory • Manpower planning	• Reduced clerical costs • Displacement of I/S • Improved utilization of manpower	• Improved employee morale • Corporate personnel policies

System	Subsystem/Applications	Benefits	
		Tangibles	Intangibles
		—Right man for the job —Reduce recruiting costs —Reduce training costs —Reduce turnover	• Better response to governmental requests (e.g., Fair Employment) • Reduced line management administrative efforts
Engineering Planning Information Coordination	• Product planning • Project planning • Input-update • Management inquiry	• Clerical savings (up 50%) • Professional management Productivity (15–20% improvement) • Displaced I/S expenses	• Management philosophy —Detailed business planning —Decentralized management • Organizational flexibility • Improved business planning • Improved resource utilization
Field Engineering/ Management Information System	• Personnel • Measurements • Computer-aided enrollment • Cross-training-utilization • Operating plan summary • Planning • Machine analysis data bank • Suggestions	• Reduced clerical and I/S expenses • Improved resource usage —Field —Education • Reduced data storage and communication costs	• Improved customer service • Improved education planning • Improved employee morale • Strategic base for future I/S needs

APPENDIX F: EXAMPLES OF INFORMATION SYSTEMS MANAGEMENT VALUE

Information Systems Management	Major Findings and Conclusions	Major Recommendations	Major Benefits
Information Systems Planning	The business has not adequately implemented a planning process as a basis for measuring and controlling information systems resources and objectives.	Implement an information systems planning process.	1. Highlights problems, risks, and opportunities for action. 2. Provides a basis for measuring actual versus planned performance and expenditure.

Information Systems Management	Major Findings and Conclusions	Major Recommendations	Major Benefits
			3. Acts as a vehicle for planning, allocation, and/or procurement of resources. 4. Facilitates business-wide information systems strategy development. 5. Provides discipline for commitments horizontally through the business as well as vertically to all levels of the business.
Control and Measurement	1. No visible development-oriented audits are being performed. 2. No phase reviews are conducted at the conclusion of the development phase. 3. Budgeting and accounting procedures do not provide a vehicle for identification of the total amount of resources expended on information systems.	1. Conduct audits or projects in development. 2. Specify and document development phase review procedures to be followed at the conclusion of the development phase of projects. 3. Develop systems and procedures that will identify total information systems resource expenditures and tie them into the budget and accounting systems as a continuous process.	1. Development audits provide the assurance that the developmental policies are being implemented, and that the information system expenditures, objectives, and so on, are within the guidelines of the information systems plan. 2. A phase review at the completion of the development phase (prior to implementation and conversion) will enable determination whether a completed system or subsystem meets objectives before granting approval to implement. 3. An accounting system identifying total information systems expenditures would: A. Enable analyses required to establish procedures for equitable distribution of costs to users. B. Provide a measure of the effectiveness of a resource application, a basis to justify expense levels, and a vehicle for analyzing the impact of budget modification.

Information Systems Management	Major Findings and Conclusions	Major Recommendations	Major Benefits
			C. Enable the periodic measurement of planned versus actual expenditures.
Data Management	1. Insufficient resources are being applied to assure that data bases are in keeping with the total business requirements. 2. Users are not sufficiently involved for the entire duration of the development of a system. 3. There is an inadequate perspective of the broad aspects of data management, such as arbitration, ownership of data bases, and control of such factors as security, validity, and timeliness. 4. There has been inadequate development of data management tools, such as a data dictionary/directory system.	1. Develop plans for the management of data to include the total organizational, technical, and administrative control of data. Tools such as a data/dictionary/directory system must be expanded to meet the growing needs of the business.	1. Provides the ability to obtain timely, reliable data. 2. Enables movement toward a consolidated data base necessary for the integrated information systems objective. 3. Maximizes use of resources by improving data interchangeability. 4. Reduces redundant development and maintenance of data bases. 5. Facilitates the use of shared data bases.
Information Systems Architecture	1. Existing information systems strategies are not sufficiently defined to provide direction for systems development to meet cross-function information requirements. 2. Responsibilities for developing, managing, and controlling information systems activities and data are not definitive enough to be effective. 3. Budgets for information systems are controlled by DP Dept., not allocated to individual functions based upon their use.	1. Define and implement those strategies required to provide uniform criteria for the design and development of integrated information systems. 2. Commensurate with the authority of the level of management, assign definitive organizational responsibilities for the planning, control, and measurement of information systems. 3. Allocate budgets and funds for information systems activities to the functional units,	1. Defines business-wide objectives and strategies on both a short- and long-term basis for future development. 2. Identifies development and implementation efforts. 3. Provides a basis for approving interim systems proposals. 4. Provides a consolidated, business-wide approach to support the integrated information needs of management functions. 5. Provides the ability to translate objectives and strategies into specific,

Information Systems Management	Major Findings and Conclusions	Major Recommendations	Major Benefits
		while preserving centralized control.	measurable actions and to assign responsibilities to organizational elements. 6. Enables advancement toward centralized data base information systems, while maintaining the authority and accountability of management. 7. Clarifies missions by assigning responsibility and bounds of innovative freedom. 8. Provides a point of accountability. 9. Controls and better defines the role and value of information systems to the missions of the various organizational elements.

READING QUESTIONS

1. Define BSP. What functions does it perform?
2. What is accomplished through BSP that was missing with the use of prior techniques?
3. Would BSP be useful only for a company beginning to use the computer? Explain.
4. Develop a schematic drawing that shows all the documents prepared in a BSP project and their relationship.
5. Write a one-sentence description of the purpose of each document in the schematic prepared in Question 4.
6. How does BSP relate to SOP—what are the advantages and disadvantages of one compared with the other?
7. How should a BSP project be conducted? Who is on the team and what are the responsibilities of each team member?
8. Explain how the information from each document produced in BSP will be used by the system developer.
9. Compare Figure 14 with the Information Matrix in BISAD (Figure 12 of Section II). What are the similarities and differences?
10. How would BSP be used in conjunction with PSL/PSA? Relate specific outputs of BSP to specific inputs to PSA/PSA.
11. BSP requires a massive effort. How would the BSP team be organized—who has responsibility for what?
12. How would BSP relate to PLEXSYS? Compare specific outputs of BSP to specific inputs of PLEXSYS.

13. How would BSP relate to SREM? Compare specific outputs of BSP with specific inputs of SREM.

14. How would BSP relate to structured analysis? Compare specific outputs of BSP with specific inputs of SADT.

PSL/PSA: A Computer-Aided Technique for Structured Documentation and Analysis of Information Processing Systems

Daniel Teichroew
Ernest A. Hershey, III

I. INTRODUCTION

Organizations now depend on computer-based information processing systems for many of the tasks involving data (recording, storing, retrieving, processing, etc.). Such systems are man-made, the process consists of a number of activities: perceiving a need for a system, determining what it should do for the organization, designing it, constructing and assembling the components, and finally testing the system prior to installing it. The process requires a great deal of effort, usually over a considerable period of time.

Throughout the life of a system it exists in several different "forms." Initially, the system exists as a concept or a proposal at a very high level of abstraction. At the point where it becomes operational it exists as a collection of rules and executable object programs in a particular computing environment. This environment consists of hardware and hard software such as the operating system, plus other components such as procedures which are carried out manually. In between, the system exists in various intermediary forms.

The process by which the initial concept evolves into an operational system consists of a number of activities each of which makes the concept more concrete. Each activity takes the results of some of the previous activities and produces new results so that the progression eventually results in an operational system. Most of the activities are data processing activities, in that they use data and information to produce other data and information. Each activity can be regarded as receiving specifications or require-

Source: Daniel Teichroew and Ernest A. Hershey, III, "PSL/PSA: A Computer-Aided Technique for Structured Documentation and Analysis of Information Processing Systems," *IEEE Transactions on Software Engineering*, Vol. SE-3, No. 1, January 1977, pp. 41–48.

ments from preceding activities and producing data which are regarded as specifications or requirements by one or more succeeding activities.

Since many individuals may be involved in the system development process over considerable periods of time and these or other individuals have to maintain the system once it is operating, it is necessary to record descriptions of the system as it evolves. This is usually referred to as "documentation."

In practice, the emphasis in documentation is on describing the system in the final form so that it can be maintained. Ideally, however, each activity should be documented so that the results it produces become the specification for succeeding activities. This does not happen in practice because the communications from one activity to succeeding activities is accomplished either by having the same person carrying out the activities, by oral communication among individuals in a project, or by notes which are discarded after their initial use.

This results in projects which proceed without any real possibility for management and review and control. The systems are not ready when promised, do not perform the function the users expected, and cost more than budgeted.

Most organizations, therefore, mandate that the system development process be divided into phases and that certain documentation be produced by the end of each phase so that progress can be monitored and corrections made when necessary. These attempts, however, leave much to be desired and most organizations are attempting to improve the methods by which they manage their system development [20], [6].

This paper is concerned with one approach to improving systems development. The approach is based on three premises. The first is that more effort and attention should be devoted to the front end of the process where a proposed system is being described from the user's point of view [2], [14], [3]. The second premise is that the computer should be used in the development process since systems development involves large amounts of information processing. The third premise is that a computer-aided approach to systems development must start with "documentation."

This paper describes a computer-aided technique for documentation which consists of the following:

1. The results of each of the activities in the system development process are recorded in computer processible form as they are produced.
2. A computerized data base is used to maintain all the basic data about the system.
3. The computer is used to produce hard copy documentation when required.

The part of the technique which is now operational is known as PSL/PSA. Section II is devoted to a brief description of system development as a framework in which to compare manual and computer-aided documentation methods. The Problem Statement Language (PSL) is described in Section III. The reports which can be produced by the Problem Statement Analyzer (PSA) are described in Section IV. The status of the system, results of experience to date, and planned developments are outlined in Section V.

II. LOGICAL SYSTEMS DESIGN

The computer-aided documentation system described in Sections III and IV of this paper is designed to play an integral role during the initial stages in the system development process. A generalized model of the whole system development process is given in Section II-A. The final result of the initial stages is a document which here will be called the System Definition Report. The desired contents of this document are discussed in Section II-B. The activities required to produce this document manually are described in Section II-C and the changes possible through the use of computer-aided methods are outlined in Section II-D.

A. A Model of the System Development Process

The basic steps in the life cycle of information systems (initiation, analysis, design, construction, test, installation, operation, and termination) appeared in the earliest applications of computers to organizational problems (see for example, [17], [1], [4], and [7]. The need for more formal and comprehensive procedures for carrying out the life cycle was recognized; early examples are the IBM SOP publications [5], the Philips ARDI method [8], and the SDC method [23]. In the last few years, a large number of books and papers on this subject have been published [11], [19].

Examination of these and many other publications indicate that there is no general agreement on what phases the development process should be divided into, what documentation should be produced at each phase, what it should contain, or what form it should be presented in. Each organization develops its own methods and standards.

In this section a generalized system development process will be described as it might be conducted in an organization which as a Systems Department responsible for developing, operating, and maintaining computer based information processing systems. The System Department belongs to some higher unit in the organization and itself has some subunits, each with certain functions (see for example, [24]. The System Department has a system development standard procedure which includes a project management system and documentation standards.

A request for a new system is initiated by some unit in the organization of the system may be proposed by the System Department. An initial document is prepared which contains information about why a new system is needed and outlines its major functions. This document is reviewed and, if approved, a senior analyst is assigned to prepare a more detailed document. The analyst collects data by interviewing users and studying the present system. He then produces a report describing his proposal system and showing how it will satisfy the requirements. The report will also contain the implementation plan, benefit/cost analysis, and his recommendations. The report is reviewed by the various organizational units involved. If it passes this review it is then included with other requests for the resources of the System Department and given a priority. Up to this point the investment in the proposed system is relatively small.

At some point a project team is formed, a project leader and team members are assigned, and given authority to proceed with the development of the system. A steering group may also be formed. The project is assigned a schedule in accordance with the project management system and given a budget. The schedule will include one or more target dates. The final target date will be the date the system (or its first part if it is being done in parts) is to be operational. There may also be additional target dates such as beginning of system test, beginning of programming, etc.

B. Logical System Design Documentation

In this paper, it is assumed that the system development procedure requires that the proposed system be reviewed before a major investment is made in system construction. There will therefore be another target date at which the "logical" design of the proposed system is reviewed. On the basis of this review the decision may be to proceed with the physical design and construction, to revise the proposed system, or to terminate the project.

The review is usually based on a document prepared by the project team. Sometimes it may consist of more than one separate document; for example, in the systems development methodology used by the U.S. Department of Defense [21] for non-weapons systems, development of the life cycle is divided into phases. Two documents are produced at the end of the Definition subphase of the Development phase: a Functional Description, and a Data Requirements Document.

Examination of these and many documentation requirements show that a Systems Definition Report contains five major types of information:

1. A description of the organization and where the proposed system will fit; showing how the proposed system will improve the functioning of the organization or otherwise meet the needs which lead to the project.
2. A description of the operation of the proposed system in sufficient detail to allow the users to verify that it will in fact accomplish its objectives, and to serve as the specification for the design and construction of the proposed system if the project continuation is authorized.
3. A description of its proposed system implementation in sufficient detail to estimate the time and cost required.
4. The implementation plan in sufficient detail to estimate the cost of the proposed system and the time it will be available.
5. A benefit/cost analysis and recommendations.

In addition, the report usually also contains other miscellaneous information such as glossaries, etc.

C. Current Logical System Design Process

During the initial stages of the project the efforts of the team are directed towards producing the Systems Definition Report. Since the major item this report contains is the description of the proposed system from the user or logical point of view, the activities required to produce the report are called the logical system design process. The project team will start with the information already available and then perform a set of activities. These may be grouped into five major categories.

1. *Data collection.* Information about the information flow in the present system, user desires for new information, potential new system organization, etc., is collected and recorded.
2. *Analysis.* The data that have been collected are summarized and analyzed. Errors, omissions, and ambiguities are identified and corrected. Redundancies are identified. The results are prepared for review by appropriate groups.
3. *Logical Design.* Functions to be performed by the system are selected. Alternatives for a new system or modification of the present system are developed and examined. The "new" system is described.
4. *Evaluation.* The benefits and costs of the proposed system are determined to a suitable level of accuracy. The operational and functional feasibility of the system are examined and evaluated.
5. *Improvements.* Usually as a result of the evaluation a number of deficiencies in the proposed system will be discovered. Alternatives for improvement are identified and evaluated until further possible improvements are not judged to be worth additional effort. If major changes are made, the evaluation step may be repeated; further data collection and analysis may also be necessary.

In practice the type of activities outlined above may not be clearly distinguished and may be carried out in parallel or iteratively with increasing level of detail. Throughout the process, however it is carried out, results are recorded and documented.

It is widely accepted that documentation is a weak link in system development in general and in logical system design in particular. The representation in the documentation that is produced with present manual methods is limited to:

1. Text in a natural language.
2. Lists, tables, arrays, cross-references.
3. Graphical representation, figures, flowcharts.

Analysis of two reports showed the following number of pages for each type of information.

Form	Report A	Report B
Text	90	117
Lists and tables	207	165
Charts and figures	28	54
Total	335	336

The systems being documented are very complex and these methods of representation are not capable of adequately describing all the necessary aspects of a system for all those who must, or should, use the documentation. Consequently, documentation is

1. Ambiguous: natural languages are not precise enough to describe systems and different readers may interpret a sentence in different ways.
2. Inconsistent: since systems are large the documentation is large and it is very difficult to ensure that the documentation is consistent.
3. Incomplete: there is usually not a sufficient amount of time to devote to documentation and with a large complex system it is difficult to determine what information is missing.

The deficiencies of manual documentation are compounded by the fact that systems are continually changing and it is very difficult to keep the documentation up-to-date.

Recently there have been attempts to improve manual documentation by developing more formal methodologies [16], [12], [13], [22], [15], [25]. These methods, even though they are designed to be used manually, have a formal language or representation scheme that is designed to alleviate the difficulties listed above. To make the documentation more useful for human beings, many of these methods use a graphical language.

D. Computer-Aided Logical System Design Process

In computer-aided logical system design the objective, as in the manual process, is to produce the System Definition Report and the process followed is essentially similar to that described above. The computer-aided design system has the following capabilities:

1. Capability to describe information systems, whether manual or computerized, whether existing or proposed, regardless of application area.
2. Ability to record such description in a computerized data base.
3. Ability to incrementally add to, modify, or delete from the description in the data base.
4. Ability to produce "hard copy" documentation for use by the analyst or the other users.

The capability to describe systems in computer processible form results from the use of the system description language called PSL. The ability to record such description in a data base, incrementally modify it, and on demand perform analysis and produce reports comes from the software package called the Problem Statement Analyzer (PSA). The Analyzer is controlled by a Command Language which is described in detail in [9] (Fig. 1).

The Problem Statement Language is outlined in Section III and described in detail in [10]. The use of PSL/PSA in computer-aided logical system design is described in detail in [18].

The use of PSL/PSA does not depend on any particular structure of the system development process or any standards on the format and

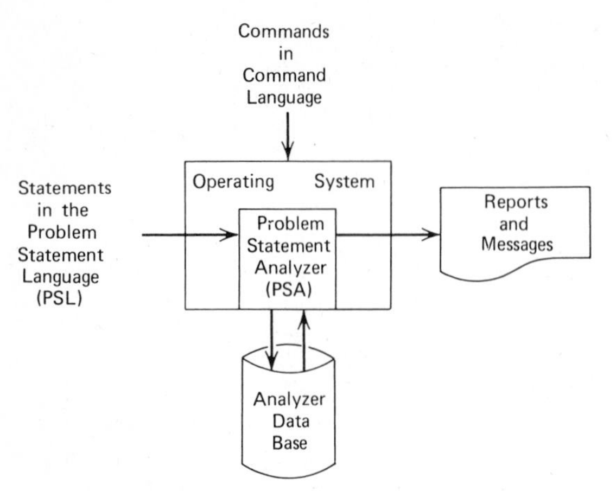

FIGURE 1 The Problem Statement Analyzer.

content of hard copy documentation. It is therefore fully compatible with current procedures in most organizations that are developing and maintaining systems. Using this system, the data collected or developed during all five of the activities are recorded in machine-readable form and entered into the computer as it is collected. A data base is built during the process. These data can be analyzed by computer programs and intermediate documentation prepared on request. The Systems Definition Report then includes a large amount of material produced automatically from the data base.

The activities in logical system design are modified when PSL/PSA is used as follows:

1. Data collection: since most of the data must be obtained through personal contact, interviews will still be required. The data collected are recorded in machine-readable form. The intermediate outputs of PSA also provide convenient checklists for deciding what additional information is needed and for recording it for input.
2. Analysis: a number of different kinds of analysis can be performed on demand by PSA, and therefore need no longer be done manually.
3. Design: design is essentially a creative process and cannot be automated. However, PSA can make more data available to the designer and allow him to manipulate it more extensively. The results of his decisions are also entered into the data base.
4. Evaluation: PSA provides some rudimentary facilities for computing volume or work measures from the data in the problem statement.
5. Improvements: identification of areas for possible improvements is also a creative task; however, PSA output, particularly from the evaluation phase, may be useful to the analyst.

The System Definition Report will contain the same material as that described since the documentation must serve the same purpose. Furthermore, the same general format and representation is desirable.

1. Narrative information is necessary for human readability. This is stored as part of the data but is not analyzed by the computer program. However, the fact that it is displayed next to, or in conjunction with, the final description improves the ability of the analyst to detect discrepancies and inconsistencies.
2. Lists, tables, arrays, matrices. These representations are prepared from the data base. They are up-to-date and can be more easily rearranged in any desired order.
3. Diagrams and charts. The information from the data base can be represented in various graphical forms to display the relationships between objects.

III. PSL, A PROBLEM STATEMENT LANGUAGE

PSL is a language for describing systems. Since it is intended to be used to describe "proposed" systems it was called a Problem Statement Language because the description of a proposed system can be considered a "problem" to be solved by the system designers and implementors.

PSL is intended to be used in situations in which analysts now describe systems. The descriptions of systems produced using PSL are used for the same purpose as that produced manually. PSL may be used both in batch and interactive environments, and therefore only "basic" information about the system need to be stated in PSL. All "derived" information can be produced in hard copy form as required.

The model on which PSL is based is described in Section III-A. A general description of the system and semantics of PSL is then given in Section III-B to illustrate the broad scope of system aspects that can be described using PSL. The detailed syntax of PSL is given in [10].

A. Model of Information Systems

The Problem Statement Language is based first on a model of a general system, and secondly on the specialization of the model to a particular class of systems, namely information systems.

The model of a general system is relatively simple. It merely states that a system consists of things which are called OBJECTS. These objects may have PROPERTIES and each of these PROPERTIES may have PROPERTY VALUES. The objects may be connected or interrelated in various ways. These connections are called RELATIONSHIPS.

The general model is specialized for an information system by allowing the use of only a limited number of predefined objects, properties, and relationships.

B. An Overview of the Problem Statement Language Syntax and Semantics

The objective of PSL is to be able to express in syntactically analyzable form as much of the information which commonly appears in System Definition Reports as possible.

System Descriptions may be divided into eight major aspects:

1. System Input/Output Flow
2. System Structure
3. Data Structure
4. Data Derivation
5. System Size and Volume
6. System Dynamics
7. System Properties
8. Project Management

PSL contains a number of types of objects and relationships which permit these different aspects to be described.

The System Input/Output Flow aspect of the system deals with the interaction between the target system and its environment.

System Structure is concerned with the hierarchies among objects in a system. Structures may also be introduced to facilitate a particular design approach such as "top down." All information may initially be grouped together and called by one name at the highest level, and then successively subdivided. System structures can represent high-level hierarchies which may not actually exist in the system, as well as those that do.

The *Data Structure* aspect of system description includes all the relationships which exist among data used and/or manipulated by the system as seen by the "users" of the system.

The *Data Derivation* aspect of the system description specifies which data objects are involved in particular PROCESSES in the system. It is concerned with what information is used, updated, and/or derived, how this is done, and by which processes.

Data Derivation relationships are internal in the system, while System Input/Output Flow relationships describe the system boundaries. As with other PSL facilities System Input/Output Flow need not be used. A system can be considered as having no boundary.

The *System Size and Volume* aspect is concerned with the size of the system and those factors which influence the volume of processing which will be required.

The *System Dynamics* aspect of system description presents the manner in which the target system "behaves" over time.

All objects (of a particular type) used to describe the target system have characteristics which distinguish them from other objects of the same type. Therefore, the PROPERTIES of particular objects in the system must be described. The PROPERTIES themselves are objects and given unique names.

The *Project Management* aspect requires that, in addition to the description of the target system being designed, documentation of the project designing (or documenting) the target system be given. This involves identification of people involved and their responsibilities, schedules, etc.

IV. REPORTS

As information about a particular system is obtained, it is expressed in PSL and entered into a data base using the Problem Statement Analyzer. At any time standard outputs or reports may be produced on request. The various reports can be classified on the basis of the purposes which they serve.

1. *Data Base Modification Reports:* These constitute a record of changes that have been made, together with diagnostics and warnings. They constitute a record of changes for error correction and recovery.
2. *Reference Reports:* These present the information in the data base in various formats. For example, the Name List Report presents all the objects in the data base with their type and date of last change. The Formatted Problem Statement Report shows all properties and relationships for a particular object (Fig. 2). The Dictionary Report gives only data dictionary type information.
3. *Summary Reports:* These present collections of information in summary from, or gathered from several different relationships. For example, the Data Base Summary Report provides project management information by showing the totals of various types of objects and how much has been said about them. The Structure Report shows complete or partial hierarchies. The Extended Picture Report shows the data flows in a graphical form.
4. *Analysis Reports:* These provide various types of analysis of the information in the data base. For example, the Contents Comparison Report analyzes similarity of Inputs and Outputs. The Data Process Interaction Report (Fig. 3) can be used to detect gaps in the information flow, or unused data objects. The Process Chain Report shows the dynamic behavior of the system (Fig. 4).

After the requirements have been completed, the final documentation required by the organization can be produced semiautomatically to a presented format, e.g., the format required for the Functional Description and Data Requirements in [21].

V. CONCLUDING REMARKS

The current status of PSL/PSA is described briefly in Section V-A. The benefits that should accrue to users of PSL/PSA are discussed in Section V-B. The information on benefits actually obtained by users is given in Section V-C. Planned extensions are outlined in Section V-D. Some conclusions reached as a result of the developments to date are given in Section V-E.

A. Current Status

The PSL/PSA system described in this paper is operational on most larger computing environments which support interactive use, including IBM 370 series (OS/VS/TSO/CMS), Univac 1100 series (EXEC-8), CDC 6000/7000 series

```
Parameters: DB=-EXBDB NAME=hourly-employee-processing NOINDEX NOPUNCHED-NAMES
  PRINT EMPTY NOPUNCH SMARG=5 NMARG=20 AMARG=10 BMARG=25 RNMARG=70
  CMARG=1 HMARG=60 NODESIGNATE SEVERAL-PER-LINE DEFINE COMMENT NONEW-
  PAGE NONEW-LINE NOALL-STATEMENTS COMPLEMENTARY-STATEMENTS LINE-NUMBERS
  PRINTEOF DLC-COMMENT

 1 PROCESS                                         hourly-employee-processing;
 2      /* DATE OF LAST CHANGE - JUN 26, 1976, 13:56:44 */
 3      DESCRIPTION;
 4       this process performs those actions needed to interpret
 5       time cards to produce a pay statement for each hourly
 6       employee.;
 7      KEYWORDS:               independent;
 8      ATTRIBUTES ARE:
 9       complexity-level
10                   high:
11      GENERATES:              pay-statement, error-listing,
12                              hourly-employee-report;
13      RECEIVES:               time-card;
14      SUBPARTS ARE:           hourly-paycheck-validation, hourly-emp-update,
15                              h-report-entry-generation,
16                              hourly-paycheck-production;
17      PART OF:                payroll-processing;
18      DERIVES:                pay-statement
19        USING:                time-card, hourly-employee-record;
20      DERIVES:                hourly-employee-report
21        USING:                time-card, hourly-employee-record;
22      DERIVES:                error-listing
23        USING:                time-card, hourly-employee-record;
24      PROCEDURE;
25        1. compute gross pay from time card data.
26        2. compute tax from gross pay.
27        3. subtract tax from gross pay to obtain net pay.
28        4. update hourly employee record accordingly.
29        5. update department record accordingly.
30        6. generate paycheck.
31       note: if status code specifies that the employee did not work
32         this week, no processing will be done for this employee.;
33      HAPPENS:
34          number-of-payments TIMES-PER pay-period;
35      TRIGGERED BY:           hourly-emp-processing-event;
36      TERMINATION-CAUSES:
37                              new-employee-processing-event;
38      SECURITY IS:            company-only;
39
40 EOF EOF EOF EOF EOF
```

FIGURE 2 Example of a FORMATTED PROBLEM STATEMENT for one PROCESS.

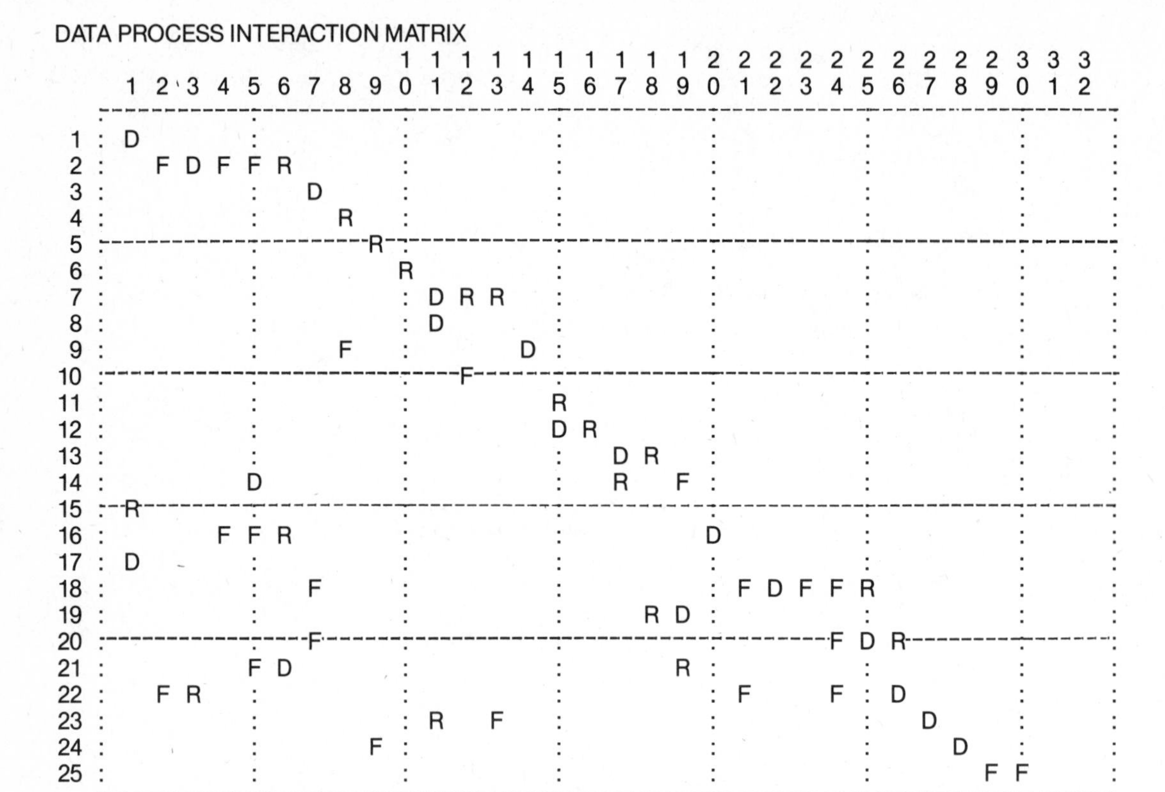

DATA PROCESS INTERACTION MATRIX

	1	2	3	4	5	6	7	8	9	10	11	12	13	14	15	16	17	18	19	20	21	22	23	24	25	26	27	28	29	30	31	32
1	D																															
2		F	D	F	F	R																										
3							D																									
4								R																								
5									R																							
6										R																						
7											D	R	R																			
8											D																					
9								F						D																		
10												F																				
11															R																	
12															D	R																
13																	D	R														
14					D												R		F													
15	R																															
16				F	F	R														D												
17	D																															
18							F														F	D	F	F	R							
19																		R	D													
20							F																	F	D	R						
21					F	D													R													
22		F	R																		F			F		D						
23											R		F														D					
24									F																			D				
25																													F	F		

FIGURE 3 Example of part of a Data Process Interaction Report.

(SCOPE, TSS), Honeywell 600/6000 series (MULTICS, GCOS), AMDAHL 470/VS (MTS), and PDP-10 (TOPS 10). Portability is achieved at a relatively high level; almost all of the system is written in ANSI Fortran.

PSL/PSA is currently being used by a number of organizations including AT&T Long Lines. Chase Manhattan Bank, Mobil Oil, British Railways, Petroleos Mexicanos, TRW Inc., the U.S. Air Force and others for documenting systems. It is also being used by academic institutions for education and research.

B. Benefit/Cost Analysis of Computer-Aided Documentation

The major benefits claimed for computer-aided documentation are that the "quality" of the documentation is improved and that the cost of design, implementation, and maintenance will be reduced. The "quality" of the documentation, measured in terms of preciseness, consistency, and completeness is increased because the analysts must be more precise, the software performs checking, and the output reports can be reviewed for remaining ambiguities, inconsistencies, and omissions. While completeness can never be fully guaranteed, one important feature of the computer-aided method is that all the documentation that "exists" is the data base, and therefore the gaps and omissions are more obvious. Consequently, the organization knows what data it has, and does not have to depend on individuals who may not be available when a specific item of data about a system is needed. Any analysis performed and reports produced are up-to-date as of the time it is performed. The coordination among analysts is greatly simplified since

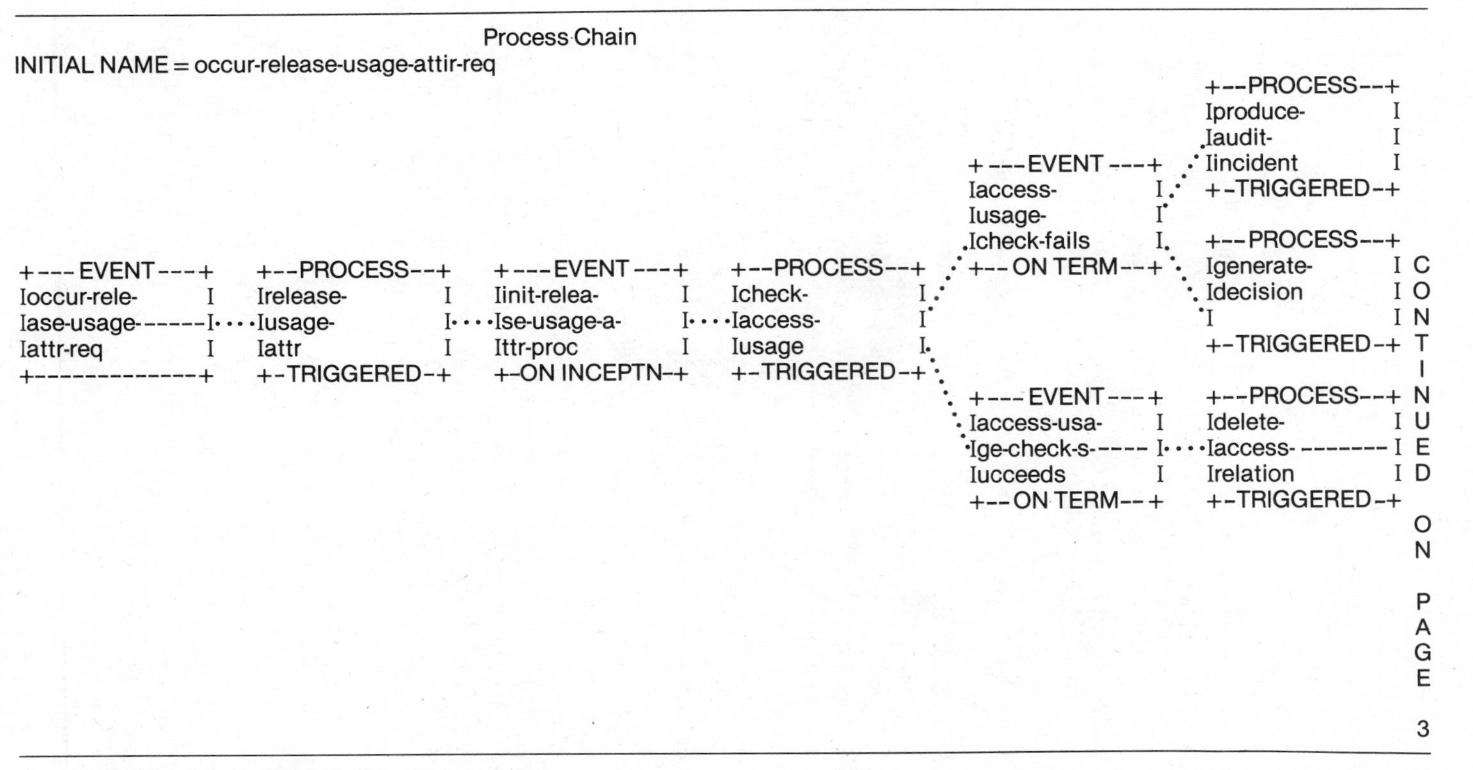

FIGURE 4 Example of a Process Chain Report.

each can work in his own area and still have the system specifications be consistent.

Development will take less time and cost less because errors, which usually are not discovered until programming or testing, have been minimized. It is recognized that one reason for the high cost of systems development is the fact that errors, inconsistencies, and omissions in specifications are frequently not detected until later stages of development: in design, programming, systems tests, or even operation. The use of PSL/PSA during the specification stage reduces the number of errors which will have to be corrected later. Maintenance costs are considerably reduced because the effect of a proposed change can easily be isolated, thereby reducing the probability that one correction will cause other errors.

The cost of using a computer-aided method during logical system design must be compared with the cost of performing the operations manually. In practice the cost of the various analyst functions of interviewing, recording, analyzing, etc., are not recorded separately. However, it can be argued that direct cost of documenting specifications for a proposed system using PSL/PSA should be approximately equal to the cost of producing the documentation manually. The cost of typing manual documentation is roughly equal to the cost of entering PSL statements into the computer. The computer cost of using PSA should not be more than the cost of analyst time in carrying out the analyses manually. (Computer costs, however, are much more visible than analysts costs.) Even though the total cost of logical system design is not reduced by using computer-aided methods, the elapsed time should be reduced because the computer can perform clerical tasks in a shorter time than analysts require.

C. Benefits/Costs Evaluation in Practice

Ideally the adoption of a new methodology such as that represented by PSL/PSA should be based on quantitative evaluation of the benefits and costs. In practice this is seldom possible; PSL/PSA is no exception.

Very little quantitative information about the experience in using PSL/PSA, especially concerning manpower requirements and system development costs, is available. One reason for this lack of data is that the project has been concerned with developing the methodology and has not felt it necessary or worthwhile to invest resources in carrying out controlled experiments which would attempt to quantify the benefits. Furthermore, commercial and government organizations which have investigated PSL/PSA have, in some cases, started to use it without a formal evaluation; in other cases, they have started with an evaluation project. However, once the evaluation project is completed and the decision is made to use the PSL/PSA, there is little time or motivation to document the reasons in detail.

Organizations carrying out evaluations normally do not have the comparable data for present methods available and so far none have felt it necessary to run controlled experiments with both methods being used in parallel. Even when evaluations are made, the results have not been made available to the project, because the organizations regard the data as proprietary.

The evidence that the PSL/PSA is worthwhile is that almost without exception the organizations which have seriously considered using it have decided to adopt it either with or without an evaluation. Furthermore, practically all organizations which started to use PSL/PSA are continuing their use (the exceptions have been caused by factors other than PSL/PSA itself) and in organizations which have adopted it, usage has increased.

D. Planned Developments

PSL as a system description language was intended to be "complete" in that the logical view of a proposed information system could be described, i.e., all the information necessary for functional requirements and specifications could be stated. On the other hand, the language should not be so complicated that it would be difficult for analysts to use. Also, deliberately omitted from the language was any ability to provide procedural "code" so that analysts would be encouraged to concentrate on the requirements rather than on low-level flow charts. It is clear, however, that PSL must be extended to include more precise statements about logical and procedural information.

Probably the most important improvement in PSA is to make it easier to use. This includes providing more effective and simple data entry and modification commands and providing more help to the users. A second major consideration is performance. As the data base grows in size and the number of users increases, performance becomes more important. Performance is very heavily influenced by factors in the computing environment which are outside the control of PSA development. Nevertheless, there are improvements that can be made.

PSL/PSA is clearly only one step in using computer-aided methods in developing, operating, and maintaining information processing systems. The results achieved to date support the premise that the same general approach can successfully be applied to the rest of the system life cycle and that the data base concept can be used to document the results of the other activities in the system life cycle. The resulting data bases can be the basis for development of methodology, generalized systems, education, and research.

E. Conclusions

The conclusions reached from the development of PSL/PSA to date and from the effort in having it used operationally may be grouped into five major categories.

1. The determination and documentation of requirements and functional specifications can be improved by making use of the computer for recording and analyzing the collected data and statements about the proposed system.
2. Computer-aided documentation is itself a system of which the software is only a part. If the system is to be used, adequate attention must be given to the whole methodology, including: user documentation, logistics and mechanics of use, training, methodological support, and management encouragement.
3. The basic structure of PSL and PSA is correct. A system description language should be of the relational type in which a description consists of identifying and naming objects and relationships among them. The software system should be data-base oriented, i.e., the data entry and modification procedures should be separated from the output report and analysis facilities.
4. The approach followed in the ISDOS project has succeeded in bringing PSL/PSA into operational use. The same approach can be applied to the rest of the system life cycle. A particularly important part of this approach is to concentrate first on the documentation and then on the methodology.
5. The decision to use a computer-aided documentation method is only partly influenced by the capabilities of the system. Much more important are factors relating to the organization itself and system development procedures. Therefore, even though computer-aided documentation is operational in some organizations, that does not mean that all organizations are ready to immediately adopt it as part of their system life cycle methodology.

REFERENCES

1. T. Aiken, "Initiating an electronics program," in *Proc. 7th Annu. Meeting. Systems and Procedures Assoc.*, 1954.
2. B. W. Boehm, "Software and its impact: A quantitative assessment," *Datamation*, pp. 48–59, May 1973.

3. ____, "Some steps toward formal and automated aids to software requirements analysis and design," *Inform. Process.*, pp. 192–197, 1974.
4. R. G. Canning, *Electronic Data Processing for Business and Industry,* New York: Wiley, 1956.
5. T. B. Glans, B. Grad, D. Holstein, W. E. Meyers, and R. N. Schmidt, *Management Systems.* New York: Holt, Rinehart, and Winston, 1968, 340 pp. (Based on IBM's study Organization Plan, 1961.)
6. J. Goldberg, Ed., "The high cost of software," in the proceedings of a symposium held in Monterey, CA, Sept. 17–19, 1973, sponsored by the U.S. Air Force Office of Naval Research. Menlo Park CA: Stanford Research Institute, 1973.
7. R. H. Gregory and R. L. Van Horn, *Automatic Data Processing Systems.* Belmont, CA: Wadsworth Publishing Co., 1960.
8. W. Hartman, H. Matthes, and A. Proeme, *Management Information systems Handbook,* (ARDI). New York: McGraw-Hill, 1968.
9. E. A. Hershey and M. Bastarache, "PSA—Command Descriptions," ISDOS Working Paper no. 91, 1975.
10. E. A. Hershey, E. W. Winters, D. L. Berg, A. F. Dickey, and B. L. Kahn, *Problem Statement Language—Language Reference Manual,* ISDOS Working Paper no. 68, 1975.
11. G. F. Hice, W. S. Turner, and L. F. Cashwell, *System Development Methodology.* Amsterdam, The Netherlands: North-Holland Publishing Co., 1974, 370 pp.
12. IBM Corporation, Data Processing Division, White Plains, NY, "HIPO—A design aid and documentation technique," Order no. GC-20-1851, 1974.
13. M. N. Jones, "HIPO for developing specifications," *Datamation,* p. 112–125, Mar. 1976.
14. G. H. Larsen, "Software: Man in the middle," *Datamation,* pp. 61–66, Nov. 1973.
15. G. J. Meyers, *Reliable Software Through Composite Design,* New York: Mason Charter Publishers, Inc., 1975.
16. D. T. Ross and K. E. Schoman, Jr., "Structured analysis for requirements definition" in *Proc. 2nd Int. Conf. Software Eng.*, San Francisco, CA, Oct. 13–15, 1976.
17. H. W. Schrimpf and C. W. Compton, "The first business feasibility study in the computer field," *Computers and Automation* Jan. 1969.
18. D. Teichroew and M. Bastarache, *PSL User's Manual,* ISDOS Working Paper no. 98, 1975.
19. TRW Systems Group, *Software Development and Configuration Management Manual,* TRW-55-73-07, Dec. 1973.
20. U.S. Air Force, "Support of Air Force automatic data processing requirements through the 1980's," Electronics Systems Division, L. G. Hanscom Field, Rep. SADPR-85, June 1974.
21. U.S. Department of Defense, *Automated Data Systems Documentation Standards Manual,* Manual 4120.17M, Dec. 1972.
22. J. D. Warnier and B. Flanagan, *Entrainment de la Construction des Programs D'Informatique,* vol. I and II. Paris: Editions d'Organization, 1972.
23. N. E. Willworth, Ed., *System Programming Management,* System Development Corporation, TM 1578/000/00, Mar. 13, 1964.
24. F. G. Withington, *The Organization of the Data Processing Function,* Wiley Business Data Processing Library, 1972.
25. E. Yourdon and L. Constantine, *Structured Design.* New York: Yourdon, Inc., 1975.

READING QUESTIONS

1. What is PSL?
2. What is PSA?
3. What information is prepared for a systems definition report? (Provide a one-sentence description of each function covered in the report.)
4. The logical system design consists of what activities? (Provide a one-sentence description of each activity.)
5. How can the computer aid in logical system design?
6. What functions are performed by PSL/PSA?
7. How does PSL/PSA affect the system definition report?
8. Give a one-sentence description of each of the elements of PSL.
9. Give a one-sentence description of each of the outputs of PSA.
10. List the advantages and disadvantages of PSL/PSA, compared with second generation system development techniques.
11. What enhancements are planned for PSL/PSA?

The PSL/PSA Approach to Computer-Aided Analysis and Documentation

Daniel Teichroew
Ernest A. Hershey, III
Y. Yamamoto

PROBLEMS ADDRESSED

The manual method for defining the requirements of a proposed information processing system requires a number of different activities. Analysts collect data about the present system and about requirements for the new system. This data is then collated, analyzed, and summarized into a set of specifications for a new system. These specifications are usually included with an implementation plan and cost/benefit analysis in a document variously called a feasibility report, a system definition report, a system specification report, or a functional requirements report. PSL/PSA is a tool for describing systems, recording the descriptions in machine-processable form, and storing them in a data base (see Figure 1). PSL/PSA supplements manual systems analysis procedures by providing the capability to use a computer for some of the clerical activities.

Source: Daniel Teichroew, Ernest A. Hershey, III, and Y. Yamamoto, "The PSL/PSA Approach to Computer-Aided Analysis and Documentation," copyright © Daniel Teichroew, 1979.

With the PSL/PSA approach, data is expressed in a formal language called the Problem Statement Language (PSL). This name was chosen because the language is used to express a problem to be solved by the system implementers. PSL statements can be entered into a computerized data base incrementally (i.e., as it is obtained) or in batches, as desired. As it is entered, the Problem Statement Analyzer (PSA) checks the cor-

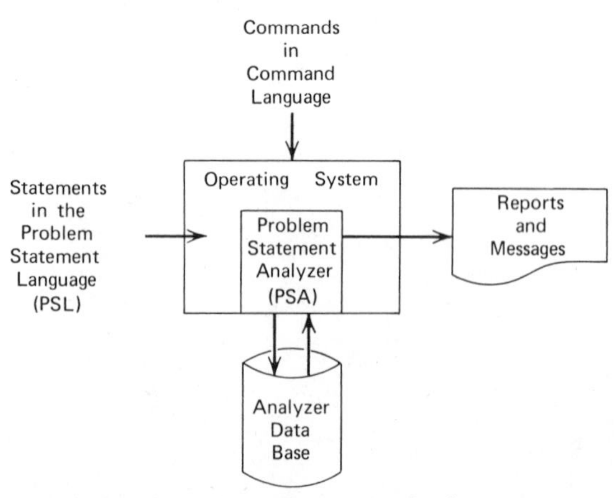

FIGURE 1 The Problem Statement Analyzer.

rectness and consistency of the new data with the data already in the data base.

PSA can produce reports of all or any part of the data base upon request. Note that the data shown on a given report may have been entered by different analysts at different times. Note also that during the production of a report, PSA performs numerous checks and analyses and generates appropriate warnings and diagnostics.

When a project is completed, PSL/PSA can be used to generate copies of the various project reports. These documents can be assembled into the format of the final specification that is required by the organization.

PROBLEM STATEMENT LANGUAGE (PSL)

The requirements for a target system should be expressed in an unambiguous machine-processable form that can accurately and completely express all relevant requirements for the logical design of the system. PSL is designed with a precise syntax and semantics in order to accomplish this. Target system subsystems and components, properties, and relationships among the components can be established.

In PSL, components are called *objects*. The concepts of objects, names of objects, types of objects, and relationships among objects are illustrated in the following example:

> A system called payroll processing takes employee information which comes from departments and employees and produces outputs which go to the departments and employees. The system also maintains payroll master information.

This paragraph represents a typical text description of a required system. The analyst receiving the description would probably produce a chart of the format like the one shown in Figure 2.

The first step in using PSL is to identify the objects in the system being described. This can be done by underlining them in the text description:

> A <u>system</u> called <u>payroll processing</u> takes <u>employee information</u> which comes from <u>departments</u> and <u>employees</u> and produces <u>outputs</u> which go to the <u>departments</u> and <u>employees</u>. The <u>system</u> also maintains <u>payroll master information</u>.

Each of the defined objects has a unique name and each of these objects is described in a different context: *employee information* represents information passing from *departments and employees* to *payroll processing; payroll master information* represents information maintained by *payroll processing*, and so forth. Actually, each of these objects represents a different type of class of object (e.g., the type of object suggested by *employee information* is an INPUT, and *payroll master information* suggests a SET). Each of the objects defined in the narrative de-

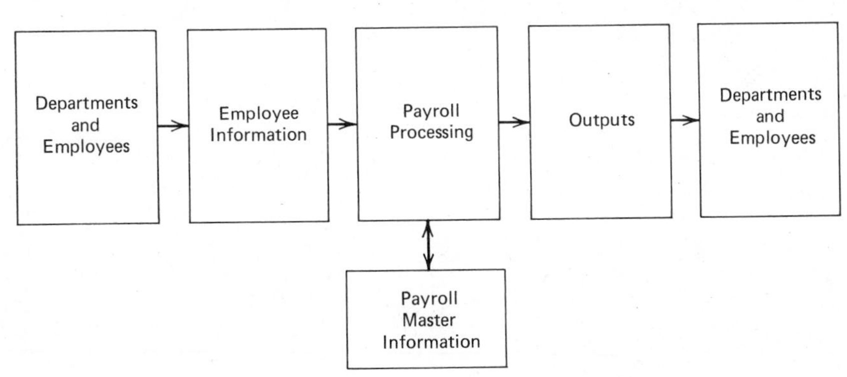

FIGURE 2 System flowchart for the payroll processing example.

scription can be given a corresponding PSL name and object type:

Narrative	PSL Name	PSL Object Type
Payroll processing	Payroll-processing	PROCESS
Employee information	Employee-information	INPUT
Departments and employees	Departments-and-employees	INTERFACE
Outputs	Paysystem-outputs	OUTPUT
Payroll master information	Payroll-master-information	SET

(Note that PSL does not allow blanks in the names of objects; hyphens are normally used to connect names consisting of more than one word.) In addition, an effort is made to assign meaningful PSL names such as paysystem-outputs instead of outputs.

The assignment of types of objects formalizes what the analysts must achieve; with this type of information, the analyst would probably have drawn the kind of system chart shown in Figure 3 rather than the one shown in Figure 2.

The next step in using PSL is to identify the relationships among the objects that have been identified. The relationships implied in the text description are:

> A system called payroll processing takes employee information which comes from departments and employees and produces outputs which go to the departments and employees. The system also maintains payroll master information.

These relationships are expressed in the following terminology:

Text Relationship	PSL Relationship
Takes	RECEIVES
Comes	GENERATED BY
Produces	GENERATES
Go	RECEIVED BY
Maintains	UPDATES

Thus, the description of the system using PSL would be:

Object	Relationship	Object
Payroll-processing	RECEIVES	Employee-information
Employee-information	GENERATED BY	Departments-and-employees

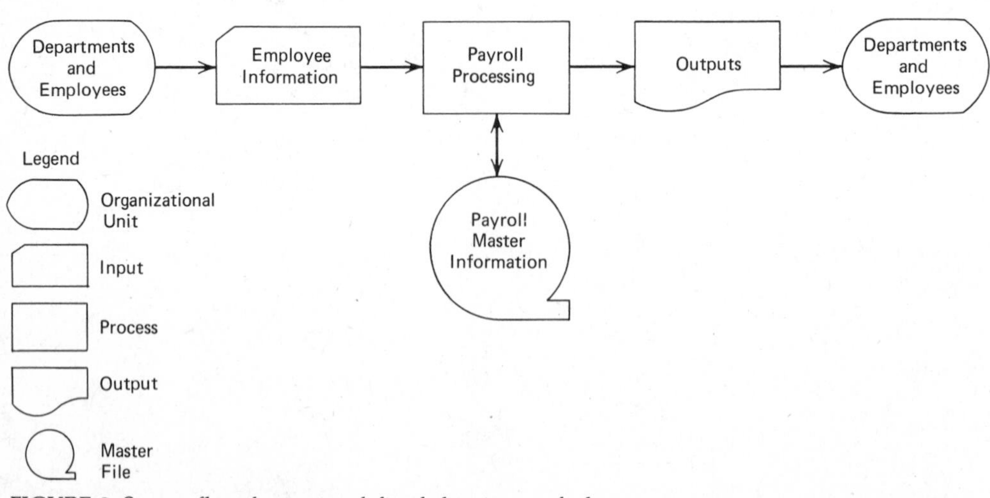

FIGURE 3 System flowchart using defined charting symbols.

Payroll-processing	GENERATES	Paysystem-outputs
Paysystem-outputs	RECEIVED BY	Departments-and-employees
Payroll-processing	UPDATES	Payroll-master-information

And the complete PSL problem statement for the example would be:

```
SET              Payroll-master information;
INTERFACE        Departments-and-employees
  GENERATES*       Employee-information;
  RECEIVES*        Paysystem-outputs;
PROCESS          Payroll-processing
  UPDATES*         Payroll-master-informa-
                     tion;
  RECEIVES*        Employee information;
  GENERATES*       Paysystem-outputs;
```

Note that there are many ways that this information could be stated that would be equivalent for the Analyzer.

Once these statements have been entered into the Analyzer Data Base, the Analyzer can be used to generate a number of standard outputs. Figure 4 shows the Formatted Problem Statement that would be generated from this example. This report contains all information stored about selected objects in the data base. Note that this report can be generated for some or all of the objects that have been defined.

Note that the Formatted Problem Statement presents all implied relationships as well as explicitly defined ones. This is why only five relationships (marked with asterisks) are shown in the previous example, although 10 are shown in the Formatted Problem Statement. This example also illustrates five types of objects (i.e., INPUT, OUTPUT, INTERFACE, SET, and PROCESS). Note that the language allows 22 types of objects and 36 relationships, which are sufficient to describe information processing systems.

The available relationships can be split into eight major groups based on the *aspect* of the system they describe. The eight major system aspects are:

- System input/output flow
- System structure
- Data structure
- Data derivation
- System size and volume
- System dynamics
- Communication and analysis aids
- Project management

Each is defined below. Note that specifying information about each of these aspects involves one or more object types and relationships.

System Input/Output Flow. This aspect of a system deals with the interaction between the target system and its environment. This involves describing the objects (INPUTS) supplied by the environment (INTERFACES) to the target system, the objects (OUTPUTS) produced by the target system and accepted by the environment, and the responsibility of the environment for information (SETS) within the system. Transfers of data within the system are not considered because there is no interaction with the environment.

System Structure. This aspect is concerned with the structures inherent in most types of systems. Structures may be introduced to facilitate a particular design approach (e.g., top-down); system structures that actually exist in the system can also be defined.

Data Structure. This aspect represents the relationships among data used and/or manipulated by the system as seen by the users of the system. Data structures also exist in the way data is grouped in collections of information such as documents. The description of data structures

```
      PSA Version A4.0R2                                 JUN 29, 1976  11:28:44     Page      1
                                University of Michigan - MTS

                                 Formatted Problem Statement

Parameters:    DB=-EXADB[1] FILE=-PSANAME  NOINDEX  NOPUNCHED-NAMES  PRINT' EMPTY  NOPUNCH  SMARG=5
  NMARG=20  AMARG=10  BMARG=25  RNMARG=70  CMARG=1  HMARG=60  NODESIGNATE  ONE-PER-LINE  DEFINE
  COMMENT  NONEW-PAGE  NONEW-LINE  NOALL-STATEMENTS  COMPLEMENTARY-STATEMENTS  LINE-NUMBERS
  PRINTEOF  DLC-COMMENT

  1 INPUT                                                        employee-information;
  2       /*   DATE OF LAST CHANGE - JUN 26, 1976, 15:27:33 */[2]
  3      GENERATED BY:  departments-and-employees;
  4      RECEIVED BY:   payroll-processing;
  5
  6 INTERFACE                                                    departments-and-employees;
  7       /*   DATE OF LAST CHANGE - JUN 26, 1976, 11:11:57 */
  8      GENERATES:     employee-information;
  9      RECEIVES:      paysystem-outputs;
 10
 11 OUTPUT                                                       paysystem-outputs;
 12       /*   DATE OF LAST CHANGE - JUN 26, 1976, 15:27:33 */
 13      GENERATED BY:  payroll-processing;
 14      RECEIVED BY:   departments-and-employees;
 15
 16 PROCESS                                                      payroll-processing;
 17       /*   DATE OF LAST CHANGE - JUN 26, 1976, 11:11:57 */
 18      GENERATES:     paysystem-outputs;
 19      RECEIVES:      employee-information;
 20      UPDATES:       payroll-master-information;
 21
 22 SET                                                          payroll-master-information;
 23       /*   DATE OF LAST CHANGE - JUN 26, 1976, 11:11:57 */
 24      UPDATED BY:    payroll-processing;
 25
 26 EOF EOF EOF EOF
```

[1]The basic data from which this report was produced is stored in a data base named EXADB.

[2]The contents of this data base were last changed on June 26, 1976.

FIGURE 4 Formatted problem statement.

also involves specification of relationships among logical collections of data and the data associated with such relationships.

Data Derivation. This aspect of system description specifies the way in which data is manipulated or derived by the system. It specifies what information is used, updated, and/or derived and by which processes. Note that data derivation can deal with data operations at any level.

System Size and Volume. This aspect involves the size of the system and the factors that influence the volume of the required processing. All parameters and time intervals are named as objects.

System Dynamics. This aspect of system description presents the manner in which the target system behaves over time. This aspect examines what causes a particular event to occur and what happens when this event occurs. The importance lies in identifying when things happen and under what conditions.

Communication and Analysis Aids. Communication and analysis aids are available to describe characteristics that distinguish specific objects from other objects of the same type. The properties of particular objects in the system must be described. Generally, properties involve any description particular to a given object. In addition to the formal relationships, any information that is needed to describe an object (and that cannot be specified by using one or more relationships) can be specified in a narrative or text description called a *comment entry*. Comment entries are not named (as objects are); they apply to only one named object. Several types of comment entries can be defined depending on the type of object to which they pertain.

Project Management. This aspect is concerned with the documentation of the project in terms of the persons involved and their responsibilities.

THE PROBLEM STATEMENT ANALYZER (PSA)

The Problem Statement Analyzer is designed to operate in an interactive environment, using the facilities of the host operating system (although it can be used in batch mode). It is relatively independent of the computing environment for ease of transportability.

PSA's subsystems are the Command Language Interface, the Data Base Update Facility, the Report Generator Facility, and a Library Facility. The Command Language Interface module interprets user commands and causes the execution of the appropriate module to handle that command. The command processing modules fall into two categories:

- Data base update modules
- Report generation modules

The command processing modules interface with the data base through a data base management system (DBMS) that is part of the Library Facility. The Library Facility also performs such peripheral functions as data base initialization and dump and restore.

Report Output

Standard output and reports can be produced or requested in any time. The format of the reports is described in Figure 5.

Several classes of reports are generated as a by-product of specific PSL/PSA functions. The Data Base Modification Reports, for example, are generated whenever the data base is modified. These constitute a record of changes that have been made or attempted, together with associated diagnostics and warnings. These reports can be used by systems analysts for error correc-

Version of Analyzer producing the report[1] Date and Time produced[1] Page Number[1]

Computing Environment on which Analyzer was used[1]

Name of the Report

Parameters in effect in the production and printing of the report

Line numbers[2] Body of the report[3,4]

—
—
—
—
—
—
—
—

*	*
*	*
*	*
*	*
*	*
*	*

End of report

[1]This information is supplied by the operating system and appears at the top of each page.
[2]Optional.
[3]PSL reserved words are printed in upper case, and the system description is printed in lower case.
[4]The body of the report can contain information not initially expressed in PSL; this information is enclosed in /* and */.

FIGURE 5 Standard format for analyzer reports.

tion and can also be filed as a chronological record of changes to the data base.

Other reports, which do not affect the contents of the data base, can be produced upon request. Production of these reports is usually a two-step process. In the first step, the part of the system which is of interest, as specified by options in the command language, is determined by producing a list of object names (see Figure 6). In the second step, the specified report is produced for the objects in the list. Many of the reports have a variety of format options, including lists, matrices, and other graphic forms. Many reports also contain options regarding the amount of information displayed. This facility can be used in generating reports from a terminal, such that the analyst can determine whether the report is valid before having it printed.

Available reports can be classified based on its purpose. The complete reference report is the Formatted Problem Statement as illustrated in Figures 4 and 7. Note that Figure 7 shows a variety of relationships that are applicable to one process.

Another class of reports provides directory and dictionary information. There are also structure reports, which present object type hierarchies, either as a list of nodes or graphically.

Data structure can be represented in various formats showing, for example, all the data elements of an object (e.g., inputs) with or without the intervening data structure. The data structure can also be analyzed to show the number of elements in common among any combination of data objects (see Figure 8).

Some reports show the flow of data from one process to another. Since the data objects can be defined individually, these reports can be used to detect gaps in the flow, unused data objects, and the like. The flow can be represented in matrix form, or graphically as in the Extended Picture report (see Figure 9).

The dynamics of the system in terms of time intervals, events, and conditions are represented

```
     PSA Version A4.0R2                          JUN 26, 1976   15:43:52      Page        1
                              University of Michigan - MTS

                                    Name Selection

Parameters:    DB=Unknown   PRINT   PUNCH=Unknown   EMPTY   NOINPUT
  SELECTION='(ELE : GR) & (ATTR=data-standard : ATTR=type)'   ORDER=type,data-standard

   1     city                                  ELEMENT
   2     employee-name                         GROUP
   3     job-title                             ELEMENT
   4     state                                 ELEMENT
   5     street                                ELEMENT
   6     supervisor                            ELEMENT
   7     department                            ELEMENT
   8     employee-identification-number        ELEMENT
   9     job-number                            ELEMENT
  10     number-of-deductions                  ELEMENT
  11     pay-grade-code                        ELEMENT
  12     salary                                ELEMENT
  13     total-hours                           ELEMENT
  14     birthdate                             ELEMENT
  15     current-date                          ELEMENT
  16     employment-date                       ELEMENT
  17     pay-date                              ELEMENT
  18     termination-date                      ELEMENT
```

The Name Selection Report is an ordered selection of a list of names that satisfy given criteria. In this case, selected objects are those that are either ELEMENTS or GROUPS and have value for either ATTRIBUTES 'data-standard' or 'type.' The selected objects are sequenced first by the value of 'type' and secondly by the value of 'data-standard.'

FIGURE 6 Name selection report.

```
        PSA Version A4.0R2                              JUN 26, 1976  14:04:34      Page      1
                                   University of Michigan - MTS

                                    Formatted Problem Statement

Parameters:   DB=-EXBDB  NAME=hourly-employee-processing  NOINDEX  NOPUNCHED-NAMES  PRINT  EMPTY
  NOPUNCH  SMARG=5  NMARG=20  AMARG=10  BMARG=25  RNMARG=70  CMARG=1  HMARG=60  NODESIGNATE
  SEVERAL-PER-LINE  DEFINE  COMMENT  NONEW-PAGE  NONEW-LINE  NOALL-STATEMENTS
  COMPLEMENTARY-STATEMENTS  LINE-NUMBERS  PRINTEOF  DLC-COMMENT

  1 PROCESS                                                    hourly-employee-processing;
  2      /*   DATE OF LAST CHANGE - JUN 26, 1976, 13:56:44 */
  3     DESCRIPTION;
  4          this process performs those actions needed to interpret
  5          time cards to produce a pay statement for each hourly
  6          employee.;
  7     KEYWORDS:        independent;
  8     ATTRIBUTES ARE:
  9          complexity-level
 10                         high;
 11     GENERATES:       pay-statement, error-listing,
 12                      hourly-employee-report;
 13     RECEIVES:        time-card;
 14     SUBPARTS ARE:    hourly-paycheck-validation, hourly-emp-update,
 15                      h-report-entry-generation,
 16                      hourly-paycheck-production;
 17     PART OF:         payroll-processing;
 18     DERIVES:         raw-statement
 19       USING:         time-card, hourly-employee-record;
 20     DERIVES:         hourly-employee-report
 21       USING:         time-card, hourly-employee-record;
 22     DERIVES:         error-listing
 23       USING:         time-card, hourly-employee-record;
 24     PROCEDURE:
 25          1. compute gross pay from time card data.
 26          2. compute tax from gross pay.
 27          3. subtract tax from gross pay to obtain net pay.
 28          4. update hourly employee record accordingly.
 29          5. update department record accordingly.
 30          6. generate paycheck.
 31        note: if status code specifies that the employee did not work
 32            this week, no processing will be done for this employee.;
 33     HAPPENS:
 34          number-of-payments TIMES-PER pay period;
 35     TRIGGERED BY:  hourly-emp-processing-event;
 36     TERMINATION-CAUSES:
 37                      new-employee processing event;
 38     SECURITY IS:     company only;
 39
 40 EOF EOF EOF EOF EOF
```

[1] The command requested all information about an object called 'hourly-employee-processing

[2] Properties of the object.

[3] System Input/Output Flow

[4] Structure

[5] Data derivation

[6] Size and volume

[7] Dynamics

FIGURE 7 Formatted problem statement—description of a process.

```
      PSA Version A4.0R2                                        JUN 26, 1976  12:05:10     Page      1
                               University of Michigan - MTS

                                Contents Comparison Report

Parameters:   DB=Unknown  FILE=Unknown  BASIC-CONTENTS-MATRIX  CONTENTS-SIMILARITY-MATRIX
  CONTENTS-SIMILARITY-SUMMARY  PERCENTAGE-OVERLAP-MATRIX  PERCENTAGE-OVERLAP-SUMMARY  PERCENT=75
  NOEXPLANATION


  MATRIX ROW AND COLUMN DEFINITIONS


    Column Names                                                   Row Names

  1 surname                          ELEMENT                  1 error-listing                OUTPUT
  2 initial                          ELEMENT                  2 hired-employee-report        OUTPUT
  3 first-name                       ELEMENT                  3 hourly-employee-report       OUTPUT
  4 employee-identification-number   ELEMENT                  4 salaried-employee-report     OUTPUT
  5 social-security-number           ELEMENT                  5 terminated-employee report   OUTPUT
  6 error-code                       ELEMENT
  7 employment-status                ELEMENT
  8 employment-date                  ELEMENT
  9 department                       ELEMENT
 10 gross-pay                        ELEMENT
 11 status-code                      ELEMENT
 12 total-hours                      ELEMENT
 13 termination-date                 ELEMENT
 14 house-number                     ELEMENT
 15 street                           ELEMENT
 16 apartment-number                 ELEMENT
 17 city                             ELEMENT
 18 state                            ELEMENT
 19 zip-code                         ELEMENT
 20 pay-grade-code                   ELEMENT
 21 cumulative-gross-pay             ELEMENT
 22 cumulative-tax-deductions        ELEMENT
 23 cumulative-fica-deductions       ELEMENT
 24 cumulative-hours                 ELEMENT
 25 cumulative-state-deductions      ELEMENT
 26 cumulative-federal-deductions    ELEMENT
```

FIGURE 8 Contents comparison report (Part I of III).

```
      PSA Version A4.0R2                            JUN 26, 1976  12:05:10     Page     2
                                  University of Michigan - MTS

                                   Contents Comparison Report
  BASIC CONTENTS MATRIX

                                            11111111112222222
                                   12345678901234567890123456
                                  +---------+---------+------+
                                1 :******   :         :      :
                                2 :***** ** :         :      :
                                3 :****    ****       :      :
                                4 :****     ***       :      :
                                5 :***** ** :  **************:
                                  +---------+---------+------+
  CONTENTS SIMILARITY MATRIX


                                       1  2  3  4  5
                                     +--------------+
                                   1 I 6  5  4  4  5I
                                   2 I    7  4  4  7I
                                   3 I       8  7  4I
                                   4 I          7  4I
                                   5 I            21I
                                     +--------------+

  CONTENTS SIMILARITY SUMMARY
ROW     N A M E                     TYPE        RELATION     ROW     N A M E                    TYPE
----    -------                     ----        --------     ----    -------                    ----

  2 hired-employee-report           OUTPUT    IS A SUBSET OF   5 terminated-employee-report     OUTPUT
  4 salaried-employee-report        OUTPUT    IS A SUBSET OF   3 hourly-employee-report         OUTPUT

      This summary identifies OUTPUTS which are subsets of others or contain the same data elements as others.
```

FIGURE 8 Contents comparison report (Part II of III).

```
PSA Version A4.0R2                                   JUN 26, 1976  12:05:10        Page        5
                              University of Michigan - MTS

                               Contents Comparison Report

PERCENTAGE OVERLAP MATRIX

                                   1   2   3   4   5
                               +-------------------+
                             1 I   6  83  67  67  83I
                             2 I  71   7  57  57 100I
                             3 I  50  50   8  88  50I
                             4 I  57  57 100   7  57I
                             5 I  24  33  19  19  21I
                               +-------------------+
```

NAME ONE	TYPE	% OF ONE CNTD IN TWO	NAME TWO	TYPE	% OF TWO CNTD IN ONE
error-listing	OUTPUT	83	hired-employee-report	OUTPUT	N/A
error-listing	OUTPUT	83	terminated-employee-report	OUTPUT	N/A
hired-employee-report	OUTPUT	100	terminated-employee-report	OUTPUT	N/A
hourly-employee-report	OUTPUT	88	salaried-employee-report	OUTPUT	100

The Contents Comparison Report can be used to analyze the data contents of objects such as OUTPUTS. In this example the five OUTPUTS contain twenty-six different data elements.

FIGURE 8 Contents comparison report (Part III of III).

```
     PSA Version A4.0R2                                   JUN 26, 1976  11:41:39      Page      2
                                  University of Michigan - MTS

                                      EXTENDED PICTURE
INITIAL NAME = create-access-relation
                                                                          +--PROCESS--+       +--ELEMENT--+
                                                                          Igenerate-  I       Idecision-  I
                                                                        . Idecision   I.......Itype       I
                                       +--PROCESS--+     +--ELEMENT--+  . I           I       I           I
                                       Icheck-need-I     I           I .  +USES TO DRVI       +--DERIVED--+
                                       Ito-        I.......Ierror-code I
                                     . Iknow-usage I     I           I .  +--PROCESS--+       +--ENTITY---+
+--PROCESS--+     +--ELEMENT--+   .    +USES TO DRV+     +--DERIVED--+  . Iproduce-   I       Iincident-  I  C
Icreate-    I     Iusage-     I .                                       . Iaudit-     I.......IinformationI  O
Iaccess-    I.......Iattribute  I                                         Iincident   I       I           I  N
Irelation   I     I           I.                                        +USES TO DRV+       +--DERIVED--+  T
+-----------+     +--UPDATED--+ .                                                                             I
                                 .     +--PROCESS--+     +--ELEMENT--+                                        N
                                  .    Icheck-     I     I           I     NAME OCCURS                        U
                                   ...Iaccess-     I.......Ierror-code I     ELSEWHERE.                         E
                                       Iusage      I     I           I     SEE INDEX.                         D
                                       +USES TO DRV+     +--DERIVED--+
                                                                                                              O
                                                                                                              N

                                                                                                              P
                                                                                                              A
                                                                                                              G
                                                                                                              E

                                                                                                              3
```

This report begins with one object, in this case a PROCESS called 'create-access-relation' and follows all process-data process links that have been defined. The relation of an object to the one on its left is shown in the lower boundary of the object box. The paths are followed until a previously identified object is encountered, until no next object has been defined or until the maximum link parameter is exceeded.

FIGURE 9 The extended picture report.

```
   PSA Version A4.0R2                                JUN 26, 1976  11:43:09      Page      2
                                       University of Michigan - MTS

                                              Process Chain
INITIAL NAME = occur-release-usage-attr-req
                                                                                                   +--PROCESS--+
                                                                                                   Iproduce-   I
                                                                                                 . Iaudit-     I
                                                                                 +---EVENT---+   .  Iincident   I
                                                                                 Iaccess-    I .   +-TRIGGERED-+
                                                                                 Iusage-     I
                                                                               . Icheck-failsI .   +--PROCESS--+
+---EVENT---+       +--PROCESS--+        +---EVENT---+       +--PROCESS--+   .  +--ON TERM--+   .  Igenerate-  I  C
Ioccur-rele-I       Irelease-   I        Iinit-relea-I       Icheck-     I .                     . Idecision   I  O
Iase-usage--I.......Iusage-     I.......Ise-usage-a-I.......Iaccess-    I                        I           I  N
Iattr-req   I       Iattr       I        Ittr-proc   I       Iusage      I.                        +-TRIGGERED-+  T
+-----------+       +-TRIGGERED-+        +ON INCEPTN-+       +-TRIGGERED-+ .                                       I
                                                                           .    +---EVENT---+      +--PROCESS--+  N
                                                                            .   Iaccess-usa-I      Idelete-    I  U
                                                                             ...Ise-check-s-I......Iaccess-    I  E
                                                                                Iucceeds    I      Irelation   I  D
                                                                                +--ON TERM--+      +-TRIGGERED-+
                                                                                                                  O
                                                                                                                  N

                                                                                                                  P
                                                                                                                  A
                                                                                                                  G
                                                                                                                  E

                                                                                                                  3
```

```
This report is similar to the Extended Picture Report except that the path
followed here is the one defined by System Dynamics relationships.  Here the
EVENT 'occur-release-usage-attr-req' TRIGGERS the PROCESS 'release-usage-
attr' which when it starts CAUSES the EVENT 'init-release-usage-attr-proc', etc.
```

FIGURE 10 Process chain report.

by several reports. This information is presented in a table, or graphically as in the Process-Chain report (see Figure 10).

While PSL/PSA is not directly oriented toward project management, a few reports that are useful to project managers are available. There are also facilities that enable analysts to develop their own reports and include them in the standard report repertoire.

Cost Benefit/Analysis of Computer-Aided Documentation

The major benefits claimed for computer-aided documentation are improved quality and decreased design, implementation, and maintenance costs. The quality of the documentation, measured in terms of preciseness, consistency, and completeness is increased with PSL/PSA because the analysts must be more precise. In addition, the software conducts checking, and the output reports can be reviewed for remaining ambiguities, inconsistencies, and omissions. Although completeness can never be fully guaranteed, one important feature of the computer-aided method is that all documentation is stored in the data base; therefore, any gaps or omissions are obvious. Consequently, the organization knows what data it has and does not have to depend on individuals who may not be available when a specific item of data is needed. Any analysis performed and reports produced are current. In addition, coordination among analysts is greatly simplified, since analysts can work in their own areas and have consistent system specifications.

Development takes less time and costs less because errors that are not discovered until the programming or testing phase are minimized. Maintenance costs are considerably reduced because the effect of a proposed change can easily be seen, thereby reducing the probability that one correction will cause other errors.

The costs of various analyst functions (e.g., interviewing, recording, and analyzing) are not generally recorded. It can be argued, however, that the direct cost of documenting specifications using PSL/PSA should be approximately equal to the cost of producing the documentation manually. Similarly, the cost of typing manual documentation roughly equals the cost of entering PSL statements into the computer. PSA computer costs should not be more than the cost of carrying out the analyses manually (computer costs, however, are much more visible than analyst costs). Although the total cost of logical system design is not reduced by using computer-aided methods, the elapsed time should be shortened.

APPLICABILITY

PSL/PSA can be used regardless of the system development procedure, project management system, or documentation standards in use. It can be used in place of or to complement manual documentation methods such as HIPO [6] or SADT [7]. M. N. Jones [8] stated that documentation is difficult to update when frequent changes are made to a developing system. Because all PSL/PSA data is in an easily modified data base and documentation is "extracted" from this data base, frequent changes are not difficult for the PSL/PSA user. In addition, PSL/PSA accomplishes many of the functions of data dictionary systems. Note that this is not limited to data but includes all of the objects specified in the system.

CURRENT STATUS

PSL/PSA is fully operational and is currently used for documentation preparation by AT&T Long Lines, Boeing, Chase Manhattan Bank, British Railways, North American Rockwell, Royal Bank of Canada, IBM Corporation,

SPERRY UNIVAC, Fujitsu, Southern California Edison, and others. It is operational in most larger computing environments that support interactive use, including the IBM 370 Series (OS/VS/TSO/CMS), the SPERRY UNIVAC 1100 Series (EXEC-8), the CDC 6000/7000 Series (SCOPE, TSS) the Honeywell 600/6000 Series (MULTICS, GECOS), AMDAHL 470/VS (MTS) PDP-10 (TOPS 10), Siemans 4004 (BS2000), and Fujitsu M190 (OSI V/F4).

PSL/PSA has been developed and is maintained by the ISDOS Project at the University of Michigan. This project is supported in part by annual grants from participating organizations. These organizations may send representatives to various project meetings, receive reports on work in progress, and have the opportunity to use such software as PSA as it is developed. The funds received are used for additional development, as well as for the maintenance and support of the system.

As with any computer-aided information system, the costs include computer time, facilities for input and output, training and consulting for users, and software development. The cost of acquiring PSL/PSA from the ISDOS Project is nominal compared with developing the technology in-house. The ISDOS Project can also provide initial training and consulting help, although most organizations prefer to develop their own in-house capability for training and consulting.

PSL/PSA Implementation. Usually the first step in considering the use of PSL/PSA is a pilot project. This project should be relatively small but large enough to demonstrate the advantages of computer-aided documentation methods (e.g., one that requires two or three analysts for two or three months). A project with other problems (e.g., that is behind schedule) should not be selected. Responsibility for evaluation should be assigned to a person who is supportive of new methods. After the pilot project is completed, the organization is in a better position to decide on computing facilities, procedures for input/output, and provisions for training and consulting.

IMPACT ON THE ORGANIZATION

PSL/PSA, if used properly, should improve the productivity of systems analysts by replacing clerical tasks in manual documentation methods with computer-aided facilities. It is not intended to replace analysts or reduce the number required in an organization; nor is it intended to enable untrained or unqualified analysts to do systems analysis. It should permit qualified analysts, adequately trained in PSL/PSA and with the right computing facilities and management support, to improve the quality of their work.

It must be stressed that PSL/PSA is only a tool. The potential benefits described in this portfolio can be obtained only if the tool is used correctly. This requires close management supervision of the complete system development methodology used by the organization.

REFERENCES

1. Teichroew, Daniel and Sayani, H. "Automation of System Building," *Datamation*, (Aug. 1971), Vol. 17, No. 8.
2. Teichroew, Daniel. "A Survey of Language for Stating Requirements for Computer Based Information Systems." *Proceedings of the Fall Joint Computer Conference*, 1972.
3. National Cash Register Company, "Accurately Defined Systems." 1967.
   Lynch, H. J., "ADS: A Technique in System Documentation," Database, Vol. 1, No. 1 (Spring 1969), pp. 6–18.
4. Myers, D. H. *A Time Automated Technique for the Design of Information Systems*. New York: IBM Systems Research Institute, 1962.

"The Time Automated Grid System (TAG): Sales and Systems Guide." IBM Publication No. Y20-0358-1 (Approx. 1968. Reprinted in J. F. Kelly Computerized Management Information Systems MacMillan 1970, pp. 367–400.)

5. Teichroew, Daniel. "Improvements in the System Life Cycle." *Information Processing 74*. North-Holland Publishing Company, 1974.
6. *HIPO—A Design Aid and Documentation Technique*. Order No. GC20-1851, IBM Corporation, Data Processing Division, White Plains NY 10504.
7. Ross, D. T. and Schoman K. E., Jr. "Structured Analysis for Requirements Definition." *Proceedings of Second International Conference on Software Engineering*. San Francisco CA, October 13–15, 1976.
8. Jones, M. N. "HIPO for Developing Specifications." *Datamation*, (March 1976), pp. 112–125.
9. U.S. Department of Defense, *Automated Data Systems Documentation Standards Manual*, Manual 7935. 13 September 1977.
10. *U.S. Department of Commerce Guidelines for Documentation of Computer Programs and Automated Data Systems*. Federal Information Processing Standards Publication. FIPS PUB 38, February 15, 1976.

READING QUESTIONS

1. What are the steps in the PSL procedure for defining a system (developing a problem statement)?
2. Revise the problem statement to reflect the following changes:
   (a) The master file is expanded. It is now an employee information master, containing payroll and all other employee information.
   (b) Separate output for hourly and salary employees.
3. Revise the problem statement to reflect the following changes:
   (a) The system also produces information for labor costs, as an input to the general ledger system.
   (b) The system receives input on cost centers against which labor costs will be distributed.
4. Show the effect of the changes in Question 2 on all the PSL/PSA reports.
5. Show the effect of the changes in Question 3 on all the PSL/PSA reports.
6. Revise the problem statement and all affected PSL/PSA reports to reflect the following data element changes:
   (a) Add: sex code
   date of last pay increase
   (b) Delete: first name
   (c) Combine: house number and apartment number
7. Revise the problem statement and all affected PSL/PSA reports to reflect the following changes:
   (a) Calculate and report pay earned from date of employment (for use in determining insurance benefits)
   (b) This report is produced upon request by supervisor rather than on a regular basis.
8. Develop a table that compares PSL commands/descriptions with those of the five predecessor problem statement languages described in the survey paper on "Third Generation Development Techniques for Computer-Based Systems."

A Requirements Engineering Methodology for Real-Time Processing Requirements

Mack W. Alford

1. INTRODUCTION

The problems of correctly specifying the requirements for large software systems (particularly real-time weapon systems) have been highlighted by Royce [1], Boehm [2], APL [3], and the AIAA Software Management Conferences [4]. Department of Defense Directive 5000.29 [5] emphasizes the need for early software visibility, risk reduction through software requirements analysis prior to DSARC II (the second Defense System Acquisition Review Council review of a weapon system) and greater "front end" development. This problem has been under study the past three years in the Software Requirements Engineering Program (SREP), performed by TRW Defense and Space Systems Group* for the Ballistic Missile Defense Advanced Technology Center (BMDATC). SREP is one part of an overall BMDATC Software Development System described by Davis and Vick [6].

The objective of the research was to synthesize a methodology which addressed the technical and management aspects of generating software requirements. The technical aspects were to include the identification of the activities to be performed, the intermediate products (e.g., functional simulation of the processing), the form in which the requirements were to be specified, and any language or support software to be used to improve the requirements quality or speed up the requirements generation. The management aspects were to include techniques for scheduling and evaluating intermediate milestones so as to improve the visibility of the status of the requirements. The goal was to develop procedures and tools to lower the life cycle cost and schedule for developing software by making the requirements generation phase more manageable, enabling the generation of software specifications with fewer errors, and making the requirements modification more manageable and error free.

Source: Mack W. Alford, "A Requirements Engineering Methodology for Real-Time Processing Requirements," September 1976, *TRW*, Redondo Beach, Calif., pp. 1–28.

*Under Contract DASG60-75-C-0022.

The results of this research include the following:

- a survey of the state-of-the-art in Software Requirements Engineering [7],
- analysis of the types of software requirements problems [8],
- an approach to specifying testable functional and performance requirements [9, 10],
- the definition and implementation of the Requirements Statement Language, RSL [11], and the Requirements Engineering and Validation System, REVS [12], which embody that approach, and
- a set of steps for generating and evaluating the software requirements utilizing RSL and REVS, reported here.

The combination of the language, support software, and steps is called the Software Requirements Engineering Methodology, or SREM. SREM treats the phase of the software development which starts when the system requirements and responsibilities are first allocated to the Data Processing Subsystem. Its product, a specification of the processing to be performed, is independent of the architecture of the hardware and software which is to satisfy it. The processing requirements constrain all of the data processing which could be accomplished in software (even though it may ultimately be implemented in special purpose hardware, e.g., Fast Fourier Transforms). This provides the foundation of an informed selection of adequate data processing hardware, and a subsequent software design to meet the processing requirements.

The synthesis of the methodology was aided by frequent experimental application of preliminary results. Due to length limitations, this paper will present only an overview of the methodology concepts, steps, and experimental results. Section 2 discusses the problem characteristics identified in previous BMDATC research efforts. Section 3 describes the key concepts of the methodology. Section 4 describes the methodology steps, products, and techniques for their evaluation. Section 5 presents an example application of SREM to illustrate its products and benefits. Section 6 briefly describes the results of its experimental application, and Section 7 presents a summary of the lessons learned.

2. PROBLEM DEFINITION

The problems of generating software requirements have been a topic of continuing research sponsored by BMDATC since 1970. For example, between 1970 and 1973, the Terminal Defense Program (TDP) was sponsored in part to demonstrate that software requirements for large unmanned weapon systems could be written in a computer independent fashion. A set of software requirements was written for a prototype real-time BMD system, and a process design methodology was utilized to design, develop, and test software for a vector processor.

The Terminal Defense Program demonstrated the following principles:

1. A specification of the real-time software containing highly complex logic and algorithms can be written which state equations to be solved, not how to solve them. Although the resulting software was implemented on a vector processor, it was evaluated to be implementable on any large centralized processor, i.e., computer independent.
2. A Computer Independent Software Specification provided more design freedom in the design of the operating system and applications software than had previously been obtained in a specification of that detail.
3. The development of an analytical simulation of the requirements by implementing example algorithms for each type of processing aided greatly in the identification of specification errors before software development. Approximately 80 errors were reported in the software development stage from a specification of over 400 pages.

Areas identified as needing improvement included the following:

- traceability of the simulation to the requirements—without extreme discipline, they tended to drift away from each other,
- testability of the performance requirements—even when equations are provided, accuracies and response times for the processing must be constrained in testable terms, and
- traceability of requirements—traceability was generally weak from a specification paragraph back to an originating specification document paragraph.

It was with this background that the Software Requirements Engineering Program was initiated. During its first phase, the nature of the requirements generation problem was analyzed. The symptoms were obvious: in almost every software project which fails, the requirements are accused of being late, incomplete, over-constraining, and just plain wrong. The published literature (e.g., [2], [13], [14], and [15]) discussed these symptoms. For example, the CCIP-85 study [2] indicated that errors generated during the requirements phase are the most expensive to fix—they are discovered late, and usually require new code to be added, the fundamental design of the software to be modified, and a large amount of retesting performed. This implies that the generation of more error-free requirements has a high cost leverage, and that even small improvements would be worthwhile. Bell and Thayer [8] analyzed existing software development programs to determine the nature of the requirements errors and their frequencies. Such information is necessary to assure that the right issues are addressed by the methodology. Ten desirable properties of a software specification were then summarized [10]: completeness, consistency, correctness, testability, unambiguity, design freedom, traceability, communicability, modularity (or change-robustness), and automatability.

A survey was made of the state-of-the-art techniques and methodologies for requirements generation [6, 7]. Its primary conclusion was that no overall methodology existed which specifically addressed the generation of requirements for large, real-time software satisfying the above properties.

As a result of the foregoing analyses, three goals were then identified for a Software Requirements Engineering Methodology:

- a structured medium or language for the statement of requirements, addressing the properties of unambiguity, design freedom, testability, and modularity and communicability.
- an integrated set of computer aided tools to assure consistency, completeness, automatability, correctness, and
- a structured approach for developing the requirements in this language, and for validating them using the tools.

3. KEY CONCEPTS

The conventional way to describe software requirements is in terms of a hierarchy of functions (e.g., see MIL-STD-490 [16]). Figure 1 illustrates this approach, in which all processing is divided into functions (e.g., communications control, device scheduling, surveillance processing, engagement control), and each of these is further subdivided into subfunctions and sub-subfunctions. A fundamental difficulty of this approach is that the processing for one input message might be described by a number of the sub-subfunctions; and the input required to exercise a specific sub-subfunction may be difficult to construct. Such a specification is hard to test, since most requirements are stated at the sub-subfunction level or below. Also, it implies a specific design, subroutines to implement sub-sub-functions.

The first key concept of SREM is based on the observation that real-time software is tested

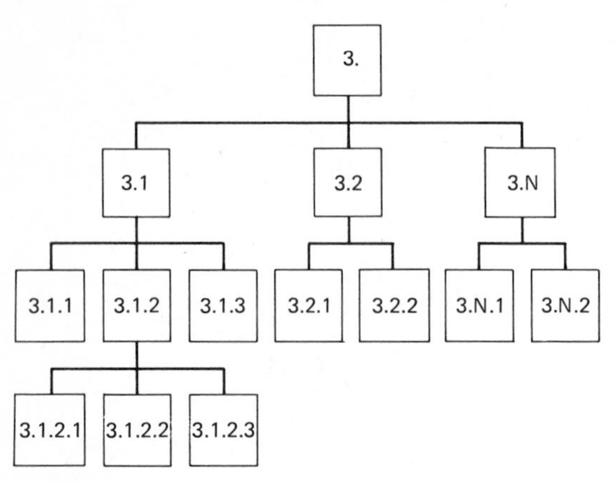

FIGURE 1 Decomposition by functional hierarchy.

by inputting an interface message and extracting the results of its processing—output messages and the contents of memory. To illustrate this, consider Figure 2. When MESSAGE 1 is input, three results are to occur: two messages are to be output by the processor, MESSAGE 2 and MESSAGE 3, and data resulting from processing the message is to be saved to process a subsequent message, MESSAGE 4. Thus, testable requirements must be specified in terms of data input and output.

The second key concept of SREM addresses these problems by defining the processing to be performed in terms of paths of processing. Figure 3 illustrates such a description of processing. When MESSAGE 1 is received, it will be processed by Steps A and B; then each of the processing Steps C (resulting in the output of MESSAGE 2), D (resulting in the storage of data), and E and F (resulting in the storage of data and the output of MESSAGE 3) take place in any order, indicated by the "AND" node joining the processing steps to B. Note that the sequence of processing steps ABC, ABD, and ABEF describe the processing to be performed in a design-free manner, i.e., no subroutine hierarchy is implied.

A third key concept addresses the statement of performance requirements: a test is defined in terms of variables measured on the paths. The places on the paths where data is measured are called validation points, and are analogous to test points in electrical circuits where voltages and currents are to be measurable. The definition of performance in terms of a test serves two purposes: it assures testability, and it communicates the requirement unambiguously.

The above approach was applied to a number of simple examples, and appeared to be sufficient for the statement of requirements. The approach was then applied to the preliminary translation of the Terminal Defense Program (TDP) software specification. This was the most complex problem which had been addressed to date, and the complexity factor soon made its presence felt. The processing of one particular message was found to have a total of 137 distinct results depending on the previous history of processing. The specification of processing in terms of the stimulus-response combinations was found to be testable and unambiguous, but difficult to comprehend, and difficult to determine whether all combinations had been addressed.

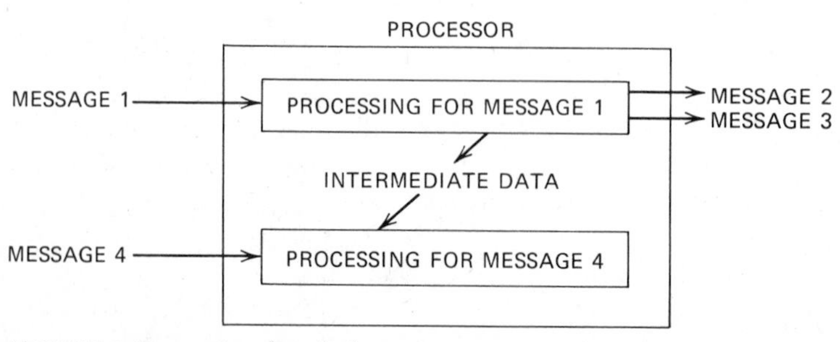

FIGURE 2 Processing description.

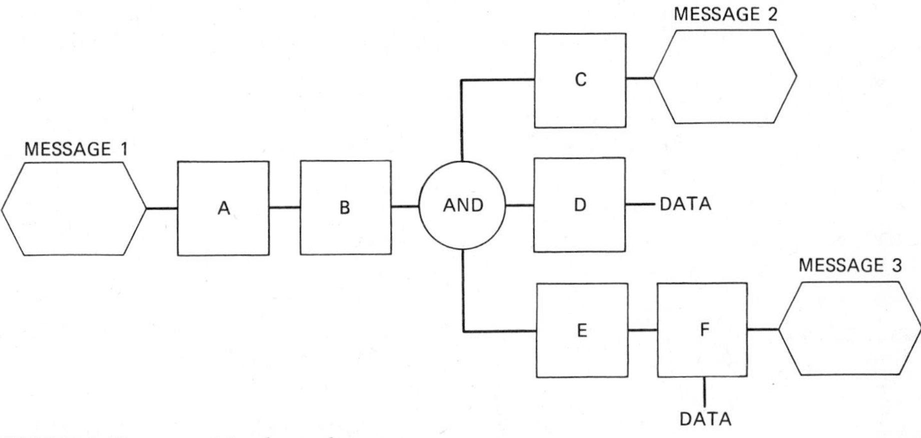

FIGURE 3 Decomposition by paths.

This demonstrated the need for a more compact notation.

To address this problem, the fourth key concept of SREM is that the paths processing a given type of stimulus should be integrated into a network, called a Requirements Network (R—Net for short). If a path execution is conditioned upon the contents of the message or the data base, this condition is summarized as a value of a selector variable; the path of processing is selected based on the value of this variable. An example R—Net is provided in Figure 4. Note that

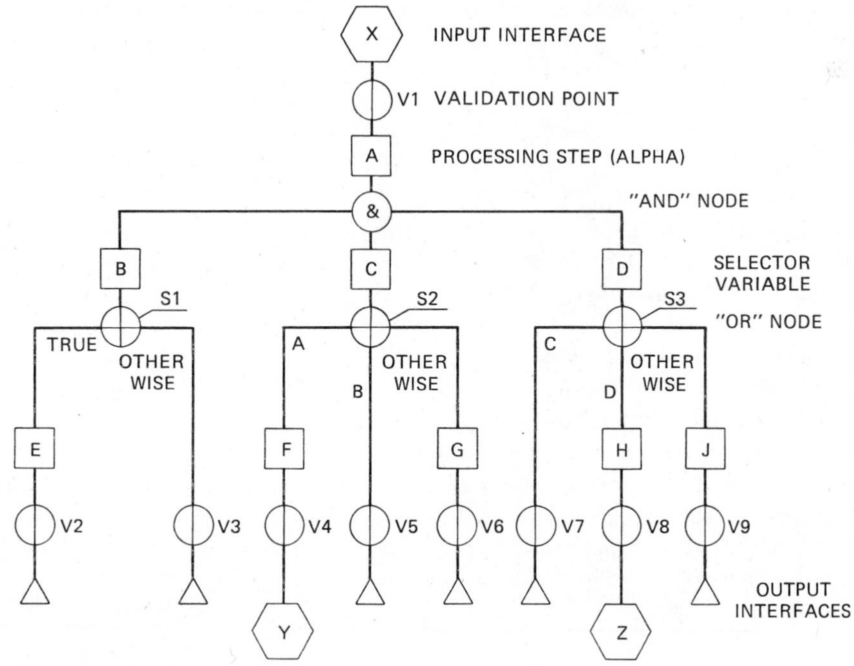

FIGURE 4 R—Net nomenclature.

18 combinations of results are summarized by combinations of eight paths and the values of three selector variables. The 137 distinct results discussed above were summarized by 21 paths.

Specification in this format maintains the advantages of testability and design freedom (all paths were in fact still explicit), while providing for visibility of the relationships between the paths.

These results were formalized using an extension of the Graph Model of Computation [17] developed at UCLA to describe the operation of software. The details of this extension are reported in [9] and [10].

With this approach, a number of other experiments were carried out to write preliminary processing requirements for a proposed Air Traffic Control System (which included some man-in-the-loop considerations), an Underseas Surveillance System, and a Medical Information System. Confidence in the approach was gained, and no further modifications to the approach were found to be necessary.

The fifth key concept of SREM is the use of a formal language (Requirements Statement Language, or RSL) for the statement of requirements based on the above concepts. A formal language is needed to reduce ambiguity and serve as input to the support software (automatability).

The sixth key concept is the use of automated tools to speed up and validate the requirements. These tools are integrated into a Requirements Engineering and Validation System (REVS) accepting RSL as input. These tools check the requirements for completeness and consistency, maintain traceability to originating requirements and simulations, and generate simulations to validate the correctness of the requirements. Modularity is enhanced by the maintenance of the requirements and their traceability in a centralized data base; a flexible facility to extract such information from the data base provides for the documentation of the requirements.

The capabilities of REVS and a preliminary version of RSL were defined. Both were refined by experimental application. The resulting RSL and REVS are described in [11] and [12].

The seventh key concept is that the methodology steps produce intermediate products which are evaluated for completeness. The production of intermediate products is necessary for planning and scheduling the steps. The ability to evaluate the products for their completion is necessary to assure that a step is in fact completed, and provide for management visibility and control. An overview of these steps is provided below.

4. METHODOLOGY STEPS AND THEIR PRODUCTS

The SRE Methodology steps address the sequence of activities and usage of RSL and REVS to generate and validate the requirements. It assumes that system functions and performances have been allocated to the data processor, and have been collected into a Data Processing Subsystem Performance Requirement, or DPSPR (see [1] for more details). Each step produces intermediate products which are evaluated for their completion. The description which follows is of necessity simplified to present the main ideas in limited space. An example is provided in the next section to illustrate the use of the methodology.

Figure 5 provides an overview of the steps of the methodology, indicating the products of each step, and criteria for evaluating for its completion. The first step translates and interprets the DPSPR into a requirements baseline written in RSL. This provides the mechanism for an early review of the DPSPR for adequacy and for the planning of the remainder of the requirements generation activities. The second step addresses the traceability of performance requirements on the processing paths back to the DPSPR. In the third step, the sensitivity of the path performances to the satisfaction of the DPSPR is determined, the test for the performance requirements is established, and the Process Performance Re-

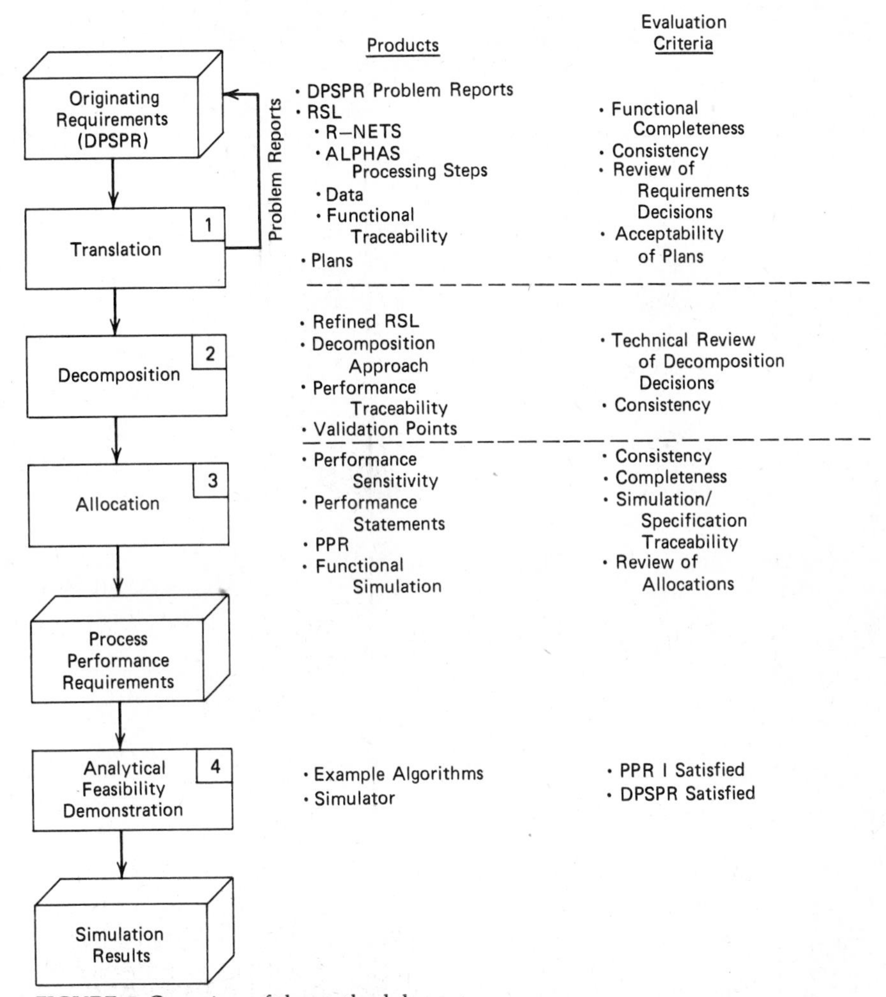

FIGURE 5 Overview of the methodology steps.

quirements (PPR) is published. In the fourth step, an example algorithm is selected for each processing step, and the analytical feasibility of the algorithmic requirements is demonstrated via simulation. Each of these steps is discussed below in more detail.

4.1 Step 1—Translation

The issues addressed in Step 1 are: the adequacy of the DPSPR for the generation of the processing requirements, the early baselining of the functional requirements, and the budgeting and scheduling of the requirements generation activities. The activities of this step include the summary of the DPSPR requirements paragraphs as RSL ORIGINATING REQUIREMENTS entered into the REVS data base; the generation of the R—Nets, data, and processing steps with traceability back to the DPSPR; the analysis of the consistency and completeness of those requirements; the generation of DPSPR Problem Reports for all problems identified; and the planning of the remainder of the requirements generation activities. The translation activity is complete when the source of every functional processing requirement and its associated performance requirement is either identified, or is covered by a DPSPR Problem Report, and every DPSPR requirement is mapped onto some processing.

Products and Their Evaluation. When this analysis is complete, a formal review of the DPSPR is held at which the times for the answers to the DPSPR Problem Reports are set, thereby allowing the scheduling of the requirements generation activities. This ensures early visibility of when the requirements products will be available.

The conduct of the review is assisted by the use of the data extraction and analysis capabilities of REVS, e.g., the functional completeness and consistency can be verified by REVS, and the traceability of the DPSPR statements to the R—Nets can be displayed. The review of the DPSPR is complete when each problem previously identified has been withdrawn, fixed, or a date set for its resolution.

The review of the plan for subsequent activities is complete when the budgets, schedules and technical approaches for defining performance requirements for the processing paths are consistent with the resources allocated to the requirements engineering phase.

When these conditions are satisfied, the translation step is complete and the resulting requirements can be placed under configuration control.

Whenever modifications to the DPSPR are generated, the activities of Step 1 are repeated, and the requirements and schedules are updated accordingly. If portions of the DPSPR are modified, the traceability features can be used to identify the affected portions of the baselined requirements in the REVS data base, thereby allowing a quick response for the processing of such changes.

4.2 Step 2—Decomposition

The decomposition step addresses two issues: the incorporation of the processing to satisfy the subsystem performance constraints into the processing requirements, and the preliminary definition of the performance requirements. The step is finished when all requirements on the data processor identified in the interface specification are accounted for in the processing description, and the form of the performance requirements and the engineering trade studies to obtain the performance bounds are identified.

The design and development of the other subsystems of the overall system frequently place restrictions and additional requirements on the processing to be performed. For example, it is in the subsystem design phase that the detailed restrictions for scheduling radar pulses or communication messages are determined. Such information is usually recorded in the interface specifications as interface constraints. Such constraints are expressed in RSL and are entered into the REVS data base as additional ORIGINATING REQUIREMENTS, and are traced to the appropriate paths of the modified R—Nets. This highlights the implications of a TBD (To Be Determined) in the originating requirements. If additional paths are added to an R—Net, some portions of Step 1 may need to be repeated, e.g., the modification of the requirements activities budgets and schedules.

The preliminary definition of the performance requirements involves three activities: the identification of the form of the requirements, specification of the variables of the software to be measured (i.e., the data to be collected at validation points), and the recording of the decisions made to relate these accuracy and timing requirements back to the satisfaction of the DPSPR performance requirements. These engineering decisions (their assumptions, the alternatives considered, and the rationale for the final selection) are necessary for understanding the requirements traceability; they are vital to an orderly assessment of the impact of the requirements changes.

In order to allocate the DPSPR performance values to the values of accuracy and timing for the paths, a set of performance models linking the path performances to the DPSPR performance may be necessary. If so, the form of these models is identified, and the data and resources necessary to accomplish their development are estimated.

Products and Their Evaluation. The products of this step include a set of decisions which describes the manner and rationale for the allocation of performance, the traceability of the path performances to the DPSPR performances via the decisions; the revised R—Nets, including the validation points and the data to be recorded, and models linking the path performance to the DPSPR performance.

The decisions constituting the allocation approach should be evaluated by a formal design review for invalid assumptions, inadvertent overconstraints, etc. This review is assisted by the capability of REVS to extract the traceability of the engineering decisions to the DPSPR, the paths, and other decisions. The adequacy of the projected performance models should also be reviewed to assure that all measures of performance have been accounted for. This should be a technical review conducted much like the Preliminary Design Review for a software design. The key to having such a design review is the recording of the traceability, the decomposition decisions, and the performance studies. Because the R—Nets may have been refined, a set of completeness and consistency checks should again be done using the automated analyzers in REVS. In addition, the consistency of the test for the requirement and the data available at the validation points should be determined. If requirements are being developed in phases, this review is repeated for each phase.

Finally, the plan for the third step is compared to the performance allocation approach to assure that the work can be accomplished with the allocated resources; if not, the modifications to the plan may be necessary.

4.3 Step 3—Allocation

During the third step, the sensitivities of the path performances to the DPSPR performances are determined. This is necessary to establish the tradeoff between timing and accuracies of the different paths which are to satisfy the requirements and to select an allocation which is not overly restrictive in any particular dimension. It is also desirable to use the least constraining form of the performance requirements (e.g., the volume of an ellipsoid might be constrained rather than the length of any particular axis).

For complicated systems like weapons systems, it is necessary to develop a functional simulator of the process which simulates the operation of the processing at a message-by-message level. This is used to check the expected behavior of the system, and it is frequently the only way that meaningful data processing loading information can be determined from the definition of the system load (i.e., the system load may be in terms of the number of objects which it must deal with, while the data processing load is in terms of the number of messages to be handled). The simulator is generated using the REVS simulation generation facilities, thereby assuring the traceability of the simulation and the requirements.

When the performance for each processing path has been established, the requirement and its test are written in RSL. The procedure which tests the requirement uses the data at the validation points. The results of the engineering trade studies are recorded to maintain traceability. The data extraction facilities of REVS are then used in the preparation of the software requirements specification This eliminates the introduction of additional errors in the publication activity.

Products and Their Evaluation. The products of this step include the identification of the sensitivity of the DPSPR performance to the path performances; the software requirements specification, including the performance statements and their tests; and a functional simulator representing the processing.

The sensitivity analyses and the establishment of the performance values for the paths are subject to design review. The performance statements and tests must be consistent, and the final version of the R—Nets, the data, and processing test descriptions and their relationships must satisfy the properties of consistency and completeness. When all of these measures have been satisfied, the PPR can be given to a process designer for the design of software which implements the process on a specific selection of processing hardware.

4.4 Step 4—Analytical Feasibility Demonstration

In Section 2, we noted that analytical simulation of requirements by implementing example algorithms for each type of processing greatly aids the requirements validation for highly complex systems. Such simulation is sometimes desirable to demonstrate that the critical processing requirements can, in fact, be met. For example, it may not be entirely obvious that a specific tracking accuracy can be achieved within a specific number of radar pulse returns. If so, analytical feasibility should be demonstrated before attempting the design of an algorithm for the real-time software. In addition, it provides for a direct check that algorithms which meet the PPR will, in fact, meet the originating DPSPR requirements.

This approach also aids in the communication of the requirements. Meseke, evaluating his experiences with the Safeguard software requirements, reported:

> *. . . experience suggests that it is probably best to state the performance requirement and then provide a recommended technique to be used at the designer's option. [13]*

The activities involved are those of building any large simulator: algorithm packages are established with requirements on input, output, and processing. Algorithms are selected for each processing step; if algorithms are not available which satisfy the requirements, they must be developed. The requirements engineer can then use the REVS facilities to combine algorithms into an analytic simulator. This simulator can be driven with realistic interface data produced by a simulation of the environment. If REVS is used to generate the simulation, the traceability of the simulator to the specification is guaranteed.

Products and Their Evaluation. The product of this step is a simulator which embodies sample algorithms and which demonstrates that the critical processing can be performed (although not necessarily on any specific machine within the specified response times).

The measure of whether this has been accomplished is simply the test for meeting the DPSPR and the tests in the PPR for the processing accuracy. Note that the feasibility of accomplishing the desired processing within the required time responses for a specific processor cannot be determined without a preliminary software design. Thus this step demonstrates computational feasibility, not real-time feasibility.

4.5 Discussion

The above description identifies the sequence of steps in the development and validation of the software requirements. There is no implication that all requirements must be platooned through the steps in unison: rather, it is possible to develop a portion of the requirements through Step 4 before the requirements for others have finished Step 2. All of this scheduling and sequencing information is the proper subject of the plan for the requirements development. There is a considerable body of evidence and agreement that a software development should be specified in terms of a Software Development Plan, with milestones for groups of capabilities. The discussion above indicates that the same type of discipline can and should be applied to the development of the requirements as well.

The identification of the reviewable products for the requirements generation activities is the key to the scheduling and control of those activities. Whenever changes to the DPSPR are forwarded to the SRE activity, a modified version of Step 1 is accomplished. An estimate of the technical, cost, and schedule impact of implementing such a change is tied to the paths of processing. Visibility of the generation of requirements is then achievable.

5. EXAMPLE

In order to illustrate the steps of the methodology, their products and the kinds of errors identified by those steps, consider the "patient-monitoring" problem used by Stevens, Myers, and Constantine [18]. Although simply stated, it contains many of the elements of actual real-time systems, and illustrates the need for a definitive statement of the requirements before software design is initiated.

The problem is the following:

> *(1) A patient monitoring program is required for a hospital. (2) Each patient is monitored by an analog device which measures factors such as pulse, temperature, blood-pressure, and skin resistance. (3) The program reads these factors on a periodic basis (specified for each patient) and stores these factors in a data base. (4) For each patient, safe ranges for each factor are specified (e.g., patient X's valid temperature range is 98 to 99.5 degrees Fahrenheit). (5) If a factor falls outside of the patient's safe range, or if an analog device fails, the nurse's station is notified. [18, p. 135, sentence numbers added]*

In a real-life case, the problem statement would contain more detail. However, it is sufficient to initiate the development of the requirements.

5.1 Step 1—Translation

First, each sentence is written as an ORIGINATING_REQUIREMENT in RSL. Sentence 1 imposes no specific requirement. Sentence 5 identifies two output messages, i.e., factor out of range, and device failure notification. Sentences 2 and 5 identify two input messages: device data, and a message indicating a device failure of some sort (this latter is implied, not explicit). Sentences 2 and 3 identify required contents of the processing data base, i.e., a factor history, and a set of safe factor ranges for each patient. This data is partially described in RSL in Table 1. Note that patient data is associated with the patient; RSL describes this by defining an ENTITY_CLASS PATIENT, which has data ASSOCIATED with it. In this way, RSL describes data contents of the software without imposing a data base design. Note that the measurement and the file containing the history of the measurements have different names because they represent different data.

Sentences 3 and 5 identify processing to be performed. Four paths of processing are identified for device data to cover the cases of device failure, factors within safe ranges, and factors outside safe ranges. In addition, Sentence 3 identifies a scheduling function; these are represented in two R—Nets shown in Figure 6. Note that "STORE_FACTOR_DATA" is parallel to checking the factors of safe ranges.

Each of the processing steps on the R-Net is described in RSL as an ALPHA, which includes a definition of the INPUT and OUTPUT DATA, a description, and other attributes. When attempting to define the ALPHA EXAMINE_FACTORS, the input data are found to be the device

TABLE 1
Example RSL Data Descriptions

```
ORIGINATING_REQUIREMENT: SENTENCE_2.
  DESCRIPTION:     "DEFINES ANALOG DEVICE MEASUREMENTS".
  TRACES TO:       MESSAGE DEVICE_REPORT.

MESSAGE: DEVICE_REPORT.
  PASSED THROUGH: INPUT_INTERFACE FROM_DEVICE.
  MADE BY:         DATA DEVICE_NUMBER, DATA TYPE_MESSAGE,
                   DATA DEVICE_DATA.
  TRACED FROM:     SENTENCE_2.

DATA: DEVICE_DATA.
  INCLUDES:        DATA PULSE, DATA TEMPERATURE,
                   DATA BLOOD_PRESSURE,
                   DATA SKIN_RESISTANCE.

ENTITY_CLASS: PATIENT.
  ASSOCIATES:      DATA PATIENT_NUMBER,
                   DATA SAFE_FACTOR_RANGE FILE
                   FACTOR_HISTORY.

DATA: SAFE_FACTOR RANGE.
  INCLUDES:        DATA LOW_PRESSURE, DATA HI_PRESSURE,
                   DATA LOW_TEMPERATURE, DATA HI_TEMPERATURE,
                   DATA LOW_SKIN_RESISTANCE,
                   DATA HI_SKIN_RESISTANCE.
  TRACED FROM:     SENTENCE_4.

FILE: FACTOR_HISTORY.
  CONTAINS:        DATA MEASUREMENT_TIME, DATA HPULSE,
                   DATA HTEMPERATURE, DATA HBLOOD_PRESSURE,
                   DATA HSKIN_RESISTANCE.
  TRACED FROM:     SENTENCE_3.
```

data and the safe ranges for the patient's factors; the output data consists of the variable RANGE which is used to determine which branch of the R—Net is next executed. In trying to describe the processing, it becomes clear that some correlation is necessary between patient number and device number; therefore PATIENT_DEVICE_NUMBER was added to the patient description. The resulting RSL is contained in Table 2.

The RSL is analyzed for completeness and consistency. Two things are identified with the assistance of REVS: the safe factor ranges data associated with the patient are never created, and the patient factor history is created but not used. The creation of the safe factor ranges requires the definition of another interface FROM_NURSES_STATION with a message to set the safe factors, and another message is created to retrieve the patient history data so that it is used.

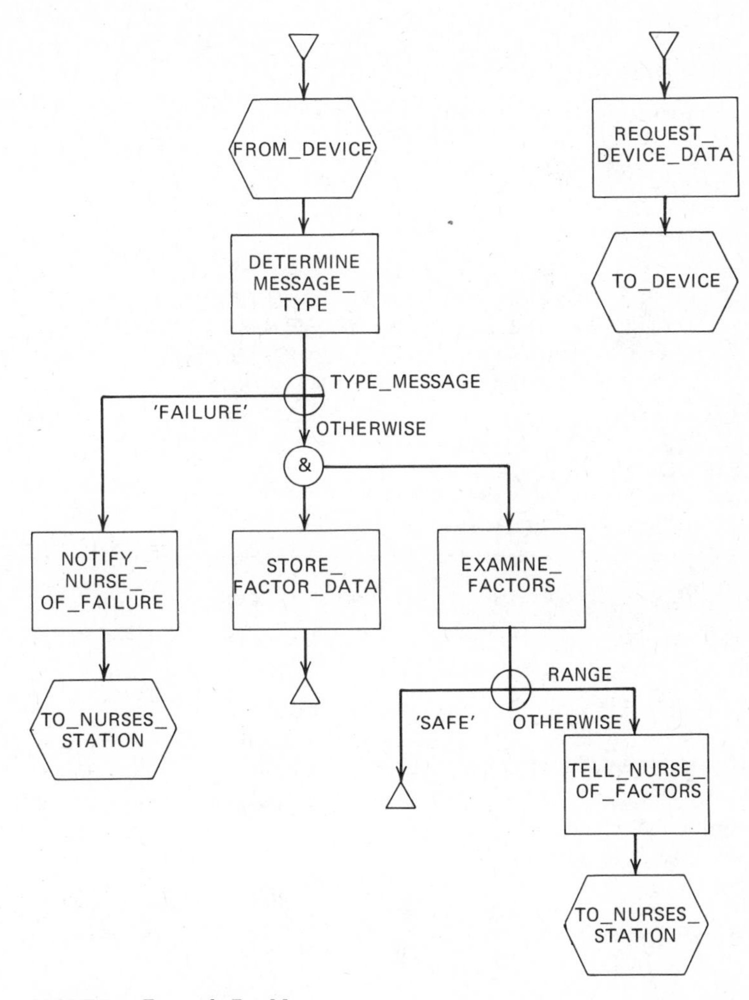

FIGURE 6 Example R—Net.

Both of these changes are traced to a decision documenting the assumption, and a DPSPR problem report is generated.

Further analysis reveals that the units of all of the variables at the interfaces are not specified, and the contents of the device error message are not even known. These are also documented in DPSPR problem reports.

Finally, the source of performance requirements for the processing paths is addressed. It is clear that some sort of time-response requirements should be imposed upon the processing which notifies the nurse of a device failure or a factor out of range. Accuracies for the processing should also be specified. Also, no limits are provided on the number of devices or the total required message arrival rates. No information is provided in the problem statement from which this information can be derived. Without this information, the performance requirements (response times and accuracy requirements) for the paths will TRACE TO no ORIGINATING_REQUIREMENTS. Thus the lack of this information is identified in a DPSPR problem report.

TABLE 2
Additional RSL for Example

```
ALPHA: EXAMINE_FACTORS.

    INPUTS:   DATA DEVICE_DATA, DATA SAFE_FACTOR_RANGE.

    OUTPUTS: RANGE.

    DESCRIPTION: "THIS PROCESSING STEP FIRST RETRIEVES THE SAFE FACTOR DATA ASSO-
                 CIATED WITH THE DEVICE, COMPARES THE DEVICE DATA TO THE SAFE FAC-
                 TOR RANGES FOR THE PATIENT BEING MONITORED, AND DETERMINES
                 WHETHER THE FACTORS ARE IN BOUNDS (RANGE_SAFE) OR OUT OF
                 BOUNDS".

DATA: PATIENT_DEVICE_NUMBER.

    INCLUDED IN: DATA SAFE_FACTOR_RANGE.
```

The products of Step 1 are the RSL and the DPSPR problem reports. Note that the problem reports were not prompted by some superior skill or experience in analyzing requirements, but by attempts to "fill in the blanks" provided by the Requirements Statement Language. Thus RSL provides not only a language in which to record the requirements, but prompts the Requirements Engineer to ask the right questions.

5.2 Step 2—Decomposition

To illustrate the impact of the detailed design of the monitoring device and communication hardware on the processing requirements, consider the following questions:

- What is the exact manner in which the analog device can fail and inform the data processor of the failure? Additional processing may be required to identify a failure, e.g., measurements unchanged for three messages.
- What time delays are experienced from the query of the device until a message is sent to the data processor? Is a failure mode that it simply not respond? This may require a "time-out" feature.
- Is communication between the devices and the processor accomplished by time sharing one message net, or by one line per device? If time shared, additional scheduling constraints may be introduced.
- Does calibration of the analog devices require data processing support? If so, what are the calibration procedures and what processing is required?

The answers to each of the above questions could require a refinement of the RSL data, ALPHAs, and R—Nets.

To illustrate the performance decomposition, consider the response time requirements to notify the nurse of a device failure. If a response time constraint is not derivable from the originating requirements, it will be determined by a decision not traceable directly to the originating specification (e.g., within 10 seconds of reading the device). The fact that this was an "arbitrary" decision will be recorded as a requirements decision, and appear in the performance traceability.

Consider the scheduling of the patient monitoring. The problem statement identified that each patient should be monitored at the specified rate. The decomposition step would attempt to quantify the requirement by addressing what should be measured (e.g., the patient monitoring times, and the rate requested for each patient), the form of the requirement (e.g., time of measurement plus reciprocal of rate), and allowed tolerances (e.g., plus or minus N seconds). This will result in a set of validation points and associated data to be measured.

5.3 Step 3—Allocation

In Step 3, the values of the tolerances are selected, and a test is written in terms of the data at validation points. Thus, the tolerance for meeting the scheduling of the monitoring is established (e.g., plus or minus 60 seconds) and is recorded as a specific test. The maximum load is defined, and a PPR can be published.

The utility of a functional simulator can be seen for even this simple example. A simulator would project the message rates into and out of the processor, the length of the patient factor files as a function of time (or nursing shift), and the message rate to the nurses station. Thus issues related to the dynamic behavior of a data processor satisfying the requirements and the nurse/machine interface can be surfaced. Note that these are functions of the requirements, not a specific implementation.

5.4 Step 4—Analytical Feasibility Demonstration

Because of the simplicity of this example, no analytical simulators appear to be necessary for this problem. For more complicated extensions of the problem, e.g., having the computer control injections based upon the patient's monitored factors, a "non-real-time" demonstration of feasibility might well be in order.

5.5 Summary

Even though this example is simple, it illustrates many of the problems of real-time software requirements: the initial problem statement does not specify all of the functions to be performed nor all of the interface data, and it lacks testable performance requirements. Software designed from these requirements will reflect these inadequacies. SREM systematically surfaces these problems early, and substantially aids the development of sound processing requirements.

6. EXPERIMENTATION

The goal of the Software Requirements Engineering Program was the synthesis of a methodology which could be applied to large, complex problems, not just simple problems like that of Section 5. To that end, experimental applications of the methodology were continuous. Some of the early development experiments were discussed in Section 3, others are discussed below.

6.1 Real-Time Test Control

An early independent application of the preliminary concepts provided confidence in the correctness of the approach. When the software requirements specification for a real-time Test Control Program needed extensive modification, it was rewritten in terms of stimulus/response relationships, with all conditions for execution of different paths identified.

This version of the specification was found to be an improvement in several dimensions. Because of its design-free nature, it did not need modification when several modifications to the software design occurred. The cost of maintaining the requirements during the next two years was less than half of the cost before the rewrite for the same level of change traffic. Response time to respond to requirements changes was similarly reduced. Traceability of the requirements to the software test plan was simple and direct due to the input/output nature of the requirements. Although qualitative in nature, these evaluations indicated that the projected benefits of a specification written in terms of paths were achievable.

6.2 Track Loop Experiment

In the Track Loop Experiment, the methodology was applied to the generation of the processing requirements for the tracking portion of the Terminal Defense Program software specification. The purpose of this experiment was to demonstrate that the contents of the software requirements could be written in RSL, to evaluate the products of the methodology steps, and to evaluate the steps of the methodology previously defined. The products experiment included an example of the DPSPR contents used to generate the requirements (about 20 pages), and a prototype software specification written in RSL (80 pages).

In the first phase of the experiment, an originating specification was written containing the previously specified DPSPR contents. This DPSPR was heavily reviewed before publication. In the second phase, the processing functional and performance requirements were developed in RSL following the steps of the methodology; REVS was not yet available, so the RSL was keypunched on cards and periodically listed.

At the end of Step 1, a review was held of the DPSPR contents. A surprising number of ambiguities (2-4 per page) were surfaced as a result of the Step 1 activities (e.g., did the elevation constraint apply to altitude or to the angle between horizontal and the radar line-of-sight to a target?). A number of "holes" in the specification of the performance were identified by attempting to trace the effect of the performance of a path of processing back to DPSPR required performances. An early conclusion of the experiment was that the Step 1 DPSPR review was explicit and thorough.

During Steps 2 and 3, the performance requirements approach was formulated, and a number of the requirements were written in RSL. The documentation of the design decisions and their use to describe the traceability of DPSPR and PPR performance requirements was met by enthusiasm; the communication of the requirements, as well as the understanding of the traceability, was enhanced. The specification of the performance requirements in terms of a test flushed out further ambiguities. As one of the participants explained, "If I could have written the requirements in English, I would have been done in three days; but you can't wave your hands in RSL."

When completed, the RSL was formatted into a PPR and evaluated. The general conclusion was that the SREM steps did produce an acceptable PPR, that RSL was sufficient to define the requirements, and that these requirements were more design free, modular, traceable, and testable than a "standard" specification.

The Methodology steps, RSL, and the projected REVS capabilities were reviewed against the experimental results, and all were modified where a weakness was detected. For example, the statement of design free performance requirements in RSL was strengthened, and the capabilities of REVS to extract requirements data for presentation in a specification were upgraded.

During the next phase of the experiment, the requirements were written in the upgraded RSL, and a message-by-message functional simulator was written using RSL and executed using the REVS Simulation Generation and Execution facilities. Several inconsistencies were discovered by REVS—some simple, some quite subtle. The resulting requirements were demonstrably free of many types of common errors (i.e., inconsistent units, data used before initialized, required processing which resulted in data not used, etc.). The resulting requirements were used as a "test case" to validate the REVS capabilities. These requirements also form the basis for the Methodology User's Manual, a "textbook" on software requirements generation.

6.3 Summary

The development of the methodology for generating software requirements required not only analysis and synthesis, but the experimental application of the concepts to realistic situations. The experiments proved necessary to define a set of steps which would generate software requirements for realistic cases.

7. REFERENCES

1. W. W. Royce, "Software Requirements Analysis, Sizing, and Costing," in *Practical Strategies for Developing Large Software Systems*, Edited by E. Horowitz, Addison-Wesley, 1975.
2. B. W. Boehm, "Software and Its Impact: A Quantitative Assessment." *Datamation*, Volume 19, No. 3, (May 1973), pp. 48–59.
3. DOD Weapon Systems Software Management Study, Johns Hopkins University, Applied Physics Laboratory, Report SR95-3, June 1975.
4. *Proceedings of Software Management Conference*, First Series, 1976, AIAA.
5. DOD Directive 5000.29, Management of Computer Resources in Major Defense Systems, April 26, 1976.
6. C. Davis and C. Vick, "The Software Development System," *Proceedings of the Second National Conference on Software Engineering*, October 1976.
7. I. F. Burns, et al., "Current Software Requirements Engineering Technology," TRW Systems Group, Huntsville, Ala., August 1974.
8. T. E. Bell and T. A. Thayer, "Software Requirements: Are They Really a Problem?," TRW Software Series No. TRW-SS-76-04, July 1976.
9. M. W. Alford and I. F. Burns, "An Approach to Stating Real-Time Processing Requirements," Presented at the Conference on Petri Nets and Related Methods, M.I.T., Boston, Mass., 1–3 July 1975.
10. M. W. Alford and I. F. Burns, "R—Nets: A Graph Model for Real-Time Software Requirements," Presented at MRI Conference on Software Engineering, New York City, New York, April 1976.
11. T. E. Bell and D. C. Bixler, "A Flow-Oriented Requirements Statement Language," TRW Software Series No. TRW-SS-76-02, April 1976.
12. T. E. Bell, D. C. Bixler, and M. E. Dyer, "An Extendable Approach to Computer-Aided Software Requirements Engineering," TRW Software Series No. 76-05, July 1976.
13. D. W. Meseke, "Safeguard Data Processing System: The Data Processing System Performance Requirements in Retrospect," *Bell System Technical Journal*, Special Supplement, 1975.
14. F. J. Buckley, "Software Testing—A Report From the Field," *IEEE Symposium on Computer Software Reliability*, New York City, New York, 30 April to 2 May 1973.
15. J. D. McGonagle, "A Study of a Software Development Project," James P. Anderson and Co., September 1971.
16. Department of Defense, "Military Standard Specification Practices," MIL-STD-490, October 1968.
17. V. C. Cerf, "Multiprocessors, Semaphores, and a Graph Model of Computation," Report No. UCLA-ENG-7223, Department of Computer Science, University of California, Los Angeles, April 1972.
18. W. P. Stevens, G. F. Myers, and L. C. Constantine, "Structured Design," *IBM Systems Journal* 13, No. 2, 1974, pp. 115–139.

READING QUESTIONS

1. What is SREM?
2. Explain each of the seven key concepts in the SRE methodology.
3. Give a one-sentence description of each of the four steps in the SREM procedure.

4. Explain the translation step in SREM.
5. Explain the decomposition step in SREM.
6. Explain the allocation step in SREM.
7. Explain the analytical feasibility demonstration step in SREM.

8a. Review the patient monitoring example at the end of the Stevens, Myers, and Constantine paper. Compare the structured design approach to computerizing that problem with the one described in the SREM paper. How would these two tools (structured design and SREM) be used—as complimentary tools or as alternative tools? Provide sound support for your conclusions.

8b. What are the advantages of SREM compared with structured design? What are its disadvantages?

9. Describe the results of the track loop experiment with SREM.

An Extendable Approach to Computer-Aided Software Requirements Engineering

Thomas E. Bell
David C. Bixler
Margaret E. Dyer

1. INTRODUCTION

The development of data processing systems has too often resulted in cost overruns, schedule slippages, or failures to produce a system which satisfies the original requirements [1]. As Boehm stated, "Software (as opposed to computer hardware, displays, architecture, etc.) is 'the tall pole in the tent'—the major source of difficult future problems and operational performance penalties" [2]. His Air Force study showed that the annual expenditures by the Air Force and NASA for software are twice the expenditures for hardware. This indicates the tremendous impact that problems in developing software can have on the budgets of the services or on the cost of any large software-based system that is being developed.

A number of studies have identified the symptoms and suspected root causes of problems in software development. Among these are the McGonagle [3] study for the Air Force, studies by Bartlett [4], studies on software problems by Thayer [5, 6], and a study by Bell and Thayer [7]. In addition, studies by the Mitre Corporation [8] and the Applied Physics Laboratory of Johns Hopkins University [9] concentrated on how software is managed. Foremost among the problems in software development identified in these studies is the generally undisciplined approach which is usually taken.

The Ballistic Missile Defense Advanced Technology Center (BMDATC) is sponsoring an integrated software development research program [10] to improve the techniques for developing correct, reliable BMD software. Reflecting the critical importance of requirements in the development process, the Software Requirements Engineering Program has been undertaken as a part of this program by TRW Defense and Space Systems* to improve the quality of requirements specifications. The product of this program (SREM, the Software Requirements Engineering Methodology) includes techniques and procedures for requirements decomposition, and for

Source: Thomas E. Bell, David C. Bixler, and Margaret E. Dyer, "An Extendable Approach to Computer-Aided Software Requirements Engineering," July 1976, *TRW*, Redondo Beach, Calif., pp. 1–48.

managing the requirements development process. In addition, SREM includes the Requirements Statement Language (RSL), a machine processable language for stating requirements, and the Requirements Engineering and Validation System (REVS), an integrated set of tools to support the development of requirements in RSL. SREM was designed to bring the computer's aid to the requirements engineering phase of software development in order to reduce the number and severity of problems encountered there. The purpose of this paper is to describe the type of automated system SREM needed to be, the characteristics of SREM that we chose to meet these needs, and how we have checked our work to be sure that both human creativity and computer-imposed discipline are achieved.

2. SREM NEEDS FOR AUTOMATION

Developing software requirements is generally a difficult intellectual job, and the job begins to look nearly impossible when it involves large data processing systems like those in a Ballistic Missile Defense (BMD) System. The requirements document for the current BMD System (the System Technology Program, Site Defense Project) contains 8248 requirement and support paragraphs in a 2500 page specification. Manually checking each paragraph against all others is an enormous task, and generating them initially is even harder; automated techniques are clearly needed. However, these automated techniques should not force the engineer to spend large amounts of mental energy dealing with simulations and control languages; he should be able to state the requirements in a reasonable, natural language and then have the automated techniques keep track of the changes, ensure consistency, and report the interactive effects of his statements.

In order to allow the requirements engineer to use his creativity, the system should be natural and flexible; it should allow him to express requirements in terms of concepts which are familiar to him. If severe restrictions are built into the system, such as highly constrained syntax rules or a paucity of concepts, the engineer would spend more time trying to 'work the system' and less time working the requirements development problem. In addition, the technologies involved in a BMD system are so extensive and advance so rapidly that no predefined set of concepts could ever be expected to satisfy the specific needs of all future projects. Therefore, such a system should be extensible at the concept level so that a particular project with an application requiring a new concept may add it to the system.

A computer-aided system should enforce some measure of discipline on the creativity of the engineer so that the development process always moves in the direction of reduced ambiguity and increased consistency. For example, the computer could perform static checking of the requirements to illuminate inconsistencies such as conflicting names, improper sequences of processing steps, and conflicting uses of items of information which must be present in the system. With a flow-orientation, the computer could additionally check the dynamic consistency of the system through the use of a simulation.

The Software Requirements Engineering Program has produced a computer-aided system for the development of requirements which fulfills the needs described above. The following four sections of this paper provide an overview of that system, followed by descriptions of the Requirements Statement Language, the Abstract System Semantic Model (the central repository for requirements), and the tools that provide automated aid for the engineer.

3. OVERVIEW

The Requirements Engineering and Validation System (REVS) consists of three major segments:

- A translator for the Requirements Statement Language (RSL)

- A centralized data base, the Abstract System Semantic Model (ASSM)
- A set of automated tools for processing the information in the ASSM.

A diagram of the system is shown in Figure 1.

Central to REVS is the ASSM, a relational data base similar in concept to the system used in the ISDOS Problem Statement Language/Problem Statement Analyzer (PSL/PSA) system [11,12]. However, our need for extensibility and configuration management, as well as the flow approach needed for simulation, have necessitated many differences from the concepts used in PSL/PSA, and therefore differences in data base design.

The design of the ASSM provides a decoupling between the input language, RSL, and the analysis tools. This decoupling permits extending RSL without having to consider issues such as controlling the tools, interfacing with the host operating system, and doing other things which would compromise the naturalness needed in RSL. The decoupling has also permitted us to exercise great freedom in the design of RSL; we were free to develop the most natural way of expressing requirements without making concessions to control languages or problems of configuration management.

RSL is designed to be a means for stating requirements naturally while still being rigorous enough for machine interpretation. We pursued this goal, in part, by orienting the design around the specification of flow graphs of required processing steps. These flow graphs are expressed in RSL in terms of "structures," which are the products of a mapping of the two-dimensional graph (e.g., Figure 2) onto a one-dimensional stream suitable for computer input (e.g., Figure 3). Structures are built from primitive flow specification blocks much as the control flow of a computer program is built from control specification primitives. The types of structure primitives available in RSL are fixed in order to provide discipline by precluding the formation of structures whose meaning may be unclear. The specification of the processing steps themselves, and of other information related to these processing steps (e.g., the data items which they use), is done in a much more flexible manner. In fact, the concepts which may be expressed in this nonprocedural segment of RSL are not necessarily fixed. The language is extensible at the concept level in order to respond to situation-specific needs and

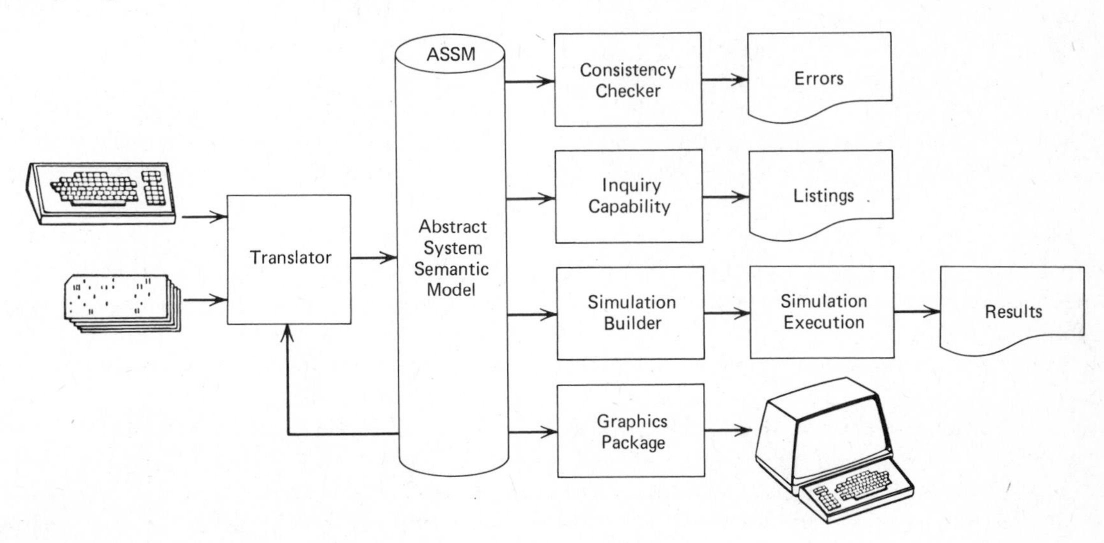

FIGURE 1 Information flows in REVS.

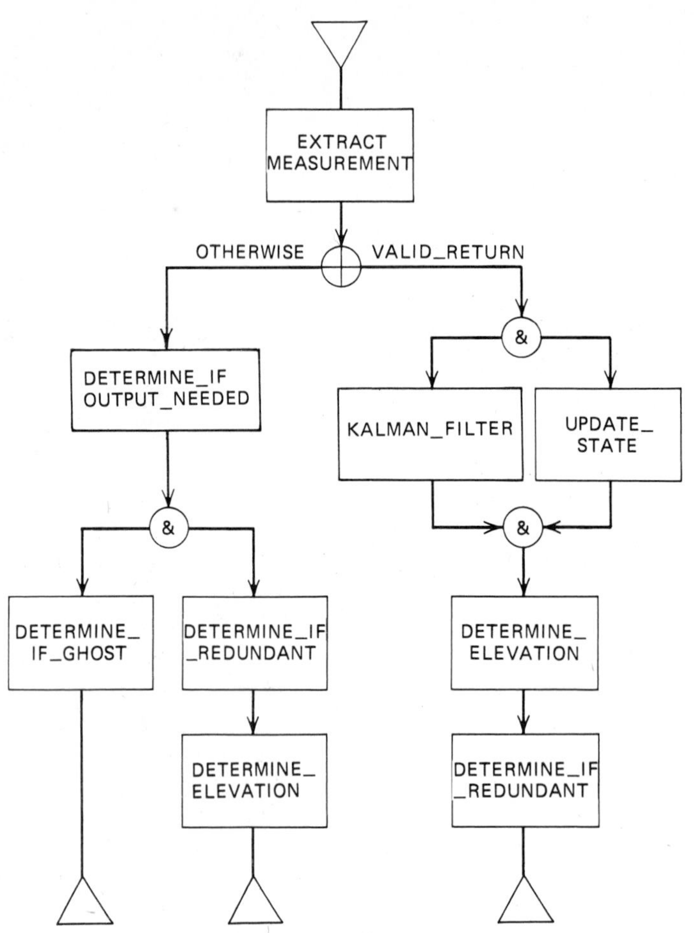

FIGURE 2 Flow graph of a sample R-Net.

new, unanticipated needs for stating requirements. The structures and the non-procedural statements of RSL are input to REVS through a translator which analyzes them to ensure individual correctness. The meaning of the statements is then abstracted and entered into the ASSM; no executable code is generated, only entries in the data base that can be used by the tools.

The tools designers are freed from the syntax of RSL; they work with the abstracted information in the ASSM. Thus, a syntax change in RSL does not usually compromise any of the tools; tools may be added or modified as the requirements engineering methodology (SREM) evolves, or as the application of REVS changes. The tools merely access the ASSM and in no way are dependent on RSL syntax.

The information available in the ASSM will support a wide variety of analysis tools. We have implemented a baseline set of widely applicable tools which perform analyses primarily related to flow properties of the information in the specification. Our analysis of requirements problems [7] indicated that these capabilities are very important in a methodology like SREM [13] for generating consistent, correct requirements and enforcing the desired discipline on the require-

```
R_NET: PROCESS_RADAR_RETURN.
        STRUCTURE:
                EXTRACT_MEASUREMENT
                DO (STATUS = VALID_RETURN)
                        DO UPDATE_STATE AND KALMAN_FILTER END
                        DETERMINE_ELEVATION
                        DETERMINE_IF_REDUNDANT
                        TERMINATE
                OTHERWISE
                        DETERMINE_IF_OUTPUT_NEEDED
                        DO DETERMINE_IF_REDUNDANT,
                            DETERMINE_ELEVATION,
                            TERMINATE
                        AND DETERMINE_IF_GHOST,
                            TERMINATE
                        END
                END
        END.
```

FIGURE 3 Sample R-Net in RSL.

ments generation process. Among these tools are an interactive graphics package to aid in the specification of the flow paths, static consistency checkers which primarily check for consistency in the use of information throughout the system, and an automated simulation generator and execution package which aids in the study of dynamic interactions of the various requirements. Situation-specific reports and analyses which a particular user may need in order to augment the information given by the baseline tools are generated through the use of a generalized extraction and reporting system. This system is independent of the extensions to RSL so that new concepts added to the language may be included in queries to the data base.

A unifying concept throughout the SREM, including RSL and REVS, is the specification of requirements for software in terms of flows through the system [14]. The use of these flows helps to bring about a disciplined approach to the development of software requirements, much as structured programming has done for the implementation of software.

4. RSL

Current software requirements documents are typically written in natural English and provide information about relatively isolated portions of the data processing system. Some of these pieces of information are very nebulous, and some are in such great detail that they represent an implementation—an implementation that may have been chosen without consideration of the remainder of the data processing system. The unevenness alone is annoying, and the probability is high that some parts of the system will be left without any written requirements. The largest deficiency of such documents, however, is that they often fail to provide critical information about how the pieces of the system will fit together. The implementor is then left without adequate information, and the requirements engineer cannot be certain the various pieces are consistent; often they are not, and no discipline is applied to make them consistent.

Our approach involves writing requirements in RSL [15], an artificial language. RSL provides information on how pieces of the system will fit together through use of the flow approach to defining the requirements. With this approach, the connectivity information is central to the development, and consistency may be easily checked, often through automated means. Another benefit of using an artificial language is the ability to define precisely the meaning of concepts in the language. With a natural language such as English, each person has a different idea about the connotation of a particular phrase and these differing ideas lead to ambiguities in interpretation of the specification. When an artificial language is used, the precise meaning of each concept may be fixed and documented. This leads to unambiguous interpretation of specifications using this language. Finally, an artificial language enables the designer to constrain the semantics (and thus the requirements statements) to a single, appropriate level of detail. Therefore, requirements in RSL can be (and are) limited to statements of true requirements that are testable in the software and are not over-constraining.

RSL is the primary means that a requirements engineer uses to communicate with REVS, so its characteristics can easily spell the success or failure of SREM. Therefore, we devoted considerable effort to RSL design and provide some of the important details of that design below.

4.1 Flows

The most obvious way to state system requirements for software is to describe the operations that each software module shall perform; these requirements differ little from program specifications. This type of "requirements specification" contains little information about the truly required sequence of processing, or even about communication between design modules. Stating the requirements (even if they are at the correct level) without indicating the required sequence invites problems in BMD systems and most other process control systems. The basic approach therefore must be statement of required operations as flows through the system. This orientation is facilitated by the stimulus-response nature of process control systems; each flow originates with a stimulus and continues to the final response. Specifying requirements in this fashion makes explicit the sequences of processing required. It also provides for direct testability of the requirements; the software may be tested to see if it provides the responses which were specified when given the associated stimulus.

Flows through the system are specified in RSL as Requirements Nets, also called R-NETs. In addition to enabling conditions, information requirements, and other material in R-NETs, they have flow structures consisting of nodes, which specify processing operations, and the arcs which connect them. The basic nodes include ALPHAs, which are the specifications of functional processing steps, and SUBNETs, which are specifications of processing flows at a lower level in the hierarchy. In essence, the SUBNET is an ALPHA which has been expanded to include internal details of the processing. All of the basic nodes are single-entry, single-exit. In addition to the simple sequential flow which may be represented by connecting this type of node, more complex flow situations are expressible in RSL by the use of structured nodes which fan-in and fan-out to specify different processing paths; the structured nodes are the AND, OR and FOR EACH.

The AND node is the first of three structured nodes. The meaning of an AND structure, which contains several paths, is that these paths are mutually order-independent. The processes on parallel paths may be executed in any order, or even in parallel. The fan-in at the end of the AND structure is a synchronization point; all of the parallel paths must be completed before any of the processes following the rejoin are performed.

The second structured node, the OR node, also has several paths. A condition is attached to each path at an OR-type fan-out. The one path with a true condition is processed; the others are ignored. Each OR node must have an OTHERWISE path that is processed if none of the conditions is true. In case more than one condition is true, the first path (as indicated by either implicit or explicit ordering) is processed.

The final structure type is the FOR EACH. This structure contains only one path; this path is processed once for each element of a set of data-processing system entities that is currently present. For example, a requirement might state that proximity of interceptors be determined FOR EACH reentry object. The FOR EACH, therefore, specifies that a particular process or set of processes is to be interacted upon without sequential implications.

The syntax of the structures in R-NETs makes the basic nodes and the several types of fan-out nodes explicit. The fan-in nodes and all of the arcs are visible only by implication. An example of such a structure is the one shown in Figure 2 and the syntax to represent it is shown in Figure 3.

The syntax of structures in R-NETs is similar to the syntax of many structured programming languages and is designed to achieve the same effect; that is, enforcing a discipline on the user. Through the use of a fixed set of flow primitives, flow structures which are ambiguous or unclear are precluded from appearing on R-NETs. This helps the requirements engineer himself see where he is vague or ambiguous, aids in the communication of requirements between requirements engineers, and permits a highly automated analysis of the requirements by the tools.

4.2 Extensions

RSL is an extensible language in order to permit the inclusion of new concepts that may be needed for future requirements. Typically, BMD software must perform guidance (largely an analytical problem), must allocate resources (a heuristic problem), and must interact with all other parts of the BMD system (a coordination problem). The processing demands of such systems are so large that the state-of-the-art (in hardware, software, algorithm concepts, etc.) is regularly stretched by the design of the data processing system, and the art advances quite rapidly. Combined with these difficulties for providing a language to state data processing requirements is our inability to know about future potential developments in other parts of the BMD system (radar, kill mechanism, etc.) which may require special interfaces or processing techniques. Any language with fixed concepts would quickly become inappropriate for stating such requirements because it would lack at least one needed concept. Therefore, RSL needed to be extensible.

To facilitate the implementation of extensibility at the concept level and to provide a clear framework for the concepts, the underlying architecture of RSL has been kept very simple. There are four primitives, outlined below, which form the foundation of the language.

a. *Elements*. Elements in RSL correspond roughly to nouns in English. The element types are simple standard prototypes which are used to describe the properties possessed by each element of the type. Some examples of standard element types in RSL are ALPHA (the class of functional processing steps), DATA (the class of conceptual pieces of data necessary in the system), and R-NET (the class of processing flow specifications).
b. *Relationships*. The relation (or relationship) in RSL may be compared with an English verb. More properly, it corresponds to the mathematical definition of a binary relation, a statement of an association of some type between two elements. The RSL relation is noncommutative; it has a subject element and an object which are distinct. However, there exists a complementary relationship for each specified relationship which is the converse of that specified relationship. DATA being IN-

PUT TO an ALPHA is one of the relationships in RSL; the complementary relationship says that the ALPHA INPUTS the DATA.

c. *Attributes.* Attributes are modifiers of elements somewhat in the manner of adjectives in English; they formalize important properties of the elements. Each attribute has associated with it a set of values which may be mnemonic names, numbers, or text strings. Each particular element may have only one of these values for any attribute. An attribute may pertain to all element types or may be restricted to be used with only a certain set of types. An example of an attribute is INITIAL_VALUE which is applicable to elements of type DATA. It has values which specify what the initial value for the data item must be in the implemented software and for simulations.

d. *Structures.* The final RSL primitive is the structure, the RSL representation of the flow model mentioned earlier. It is a mapping of a two-dimensional graph structure into a one-dimensional stream of computer imput. It models the flows through the functional processing steps (ALPHAs) or the flows between places where accuracy or timing requirements are stated (VALIDATION_POINTS).

All concepts in RSL are expressed in terms of these four primitives. The structures are not extensible; this preserves the necessary discipline for flow descriptions. On the other hand, new types of elements, relationships, and attributes may be added to the language at will to express new concepts. In fact, all concepts (both new and old) are defined as extensions.

As an example of using the extension capability, a requirement often involves information entering the data processing system from outside hardware (radars, operator consoles, etc.). The port into the data processing system is embodied in a type of element called an INPUT_INTERFACE. The complex of information which it handles is called a MESSAGE; the INPUT_INTERFACE then PASSES the MESSAGE into the data processing system. The definitions of these two types of elements and their connecting relationship in RSL are as follows:

```
DEFINE ELEMENT_TYPE:  INPUT_INTERFACE
                      (*A port between the data processing system and the rest of the system which
                      accepts data from another part of the system*).

DEFINE ELEMENT_TYPE:  MESSAGE
                      (*An aggregation of DATA and FILES that PASS through an interface as a logical
                      unit*).

DEFINE RELATIONSHIP:  PASSES
                      (*An INPUT_INTERFACE 'PASSES' a logical aggregation of data called a MES-
                      SAGE from the outside system into the data processing system*).

   COMPLEMENTARY RELATIONSHIP:  PASSED ("BY").

   SUBJECT:                     INPUT_INTERFACE.
   OBJECT:                      MESSAGE.
```

After these definitions are processed by the RSL translator and entered into the ASSM, they are available for use by any requirements engineer working with the data base. Of course, to make these particular definitions useful, additional relationships must be defined; of particular importance is the relationship which associates particular items of DATA with the MESSAGE.

With an extension mechanism this powerful, it would be a tempting proposition to provide only this mechanism so that the requirements engineer would have a means to provide his own concepts. This, however, would demand that the engineer design his own language for requirements statement as well as engineering the requirements themselves. We have resisted this temptation and have provided a core set of concepts that appears to be needed with great regularity, and have also provided the mechanism for extending the concepts as needed in particular problems. The examples of extensions given above are a part of this core set along with 19 other element types, 20 relationships, and 20 attributes.

4.3 Core Concepts

The examples below show the use of the core concepts to define requirements. They include the definitions of two required processing steps (ALPHAs), of a required item of information (DATA) which is to be OUTPUT from one of the ALPHAs, and of a statement from another document which necessitates the inclusion of several items in the requirements (ORIGINATING_REQUIREMENT).

4.4 The Translator

The RSL translator is a component of REVS; its purpose is to analyze the RSL statements which are input to it and to make entries in the ASSM corresponding to the meaning of the statements. It does this by extracting the RSL primitives (elements, relationships, attributes, and structures) which exist in the input statements, and by mapping them to constructs in the ASSM. The translator also processes modifications and deletions from the data base which are commanded by RSL statements specifying

```
ALPHA: EXTRACT_MEASUREMENT.
   INPUTS: CORRELATED_RETURN.
   OUTPUTS: VALID_RETURN, MEASUREMENT.
   DESCRIPTION: "DOES RANGE SELECTION PER CISS REFERENCE 2-7".
   ENTERED BY: "M. RICHTER".

ALPHA: DETERMINE_IF_REDUNDANT.
   INPUTS: CORRELATED_RETURN.
   OUTPUTS: REDUNDANT_IMAGE.
   DESCRIPTION: "THE IMAGE OF THE RADAR RETURN IS ANALYZED TO DETERMINE IF IT IS REDUN-
            DANT WITH ANOTHER IMAGE".
   ENTERED BY: "F. BURNS".

DATA: MEASUREMENT.
   INCLUDES: RANGE_MARK_TIME, AMPLITUDE, RANGE_VARIANCE, RD_VARIANCE,
            R_AND_RD_CORRELATION.
   DESCRIPTION: "THIS IS THE ESSENCE OF THE INFORMATION IN THE RETURN".
   ENTERED BY: "F. BURNS".

ORIGINATING REQUIREMENT:
            DPSPR_3_2_2_A_FUNCTIONAL.
   DESCRIPTION: "ACTION: SEND RADAR ORDER INFORMATION: RADAR ORDER. IMAGE (REDUNDANT)".
   TRACES TO: ALPHA COMMAND_PULSES
              ALPHA DETERMINE_IF_REDUNDANT
              MESSAGE RADAR_ORDER_MESSAGE
              DATA REDUNDANT_IMAGE
              ENTITY IMAGE.
   ENTERED BY: "T.E. BELL".
```

changes to already-existing instances in the ASSM. This is done in a manner similar to that used in processing the additions to the ASSM; that is, the primitives are extracted and the referenced construct is modified or deleted. For all types of input processing, the translator references the ASSM to do simple consistency checks on the input. This prevents the occurrence of disastrous errors such as the introduction of an element with the same name as a previously-existing element or an instance of a relationship which is tied to an illegal type of element. Besides providing a measure of protection for the data base, this type of checking catches, at an early stage, some of the simple types of inconsistencies that are often found in requirements specifications.

In addition to processing requirements statements given in terms of the core set of extensions, the translator accepts further extensions and enters them into the ASSM. Obviously, the extensions must be treated with even more care than the requirements statements since the deletion of, say, a previously legal element type may invalidate a large segment of the requirements. On the other hand, adding new concepts in response to each individual's desires is a pernicious practice. For these reasons, a lock mechanism has been built into the translator to enable it to reject any extensions while locked. This allows the management of a project to control the use of the extensions in their project and to enforce a disciplined use of the power of RSL.

The translator has been implemented using a compiler writing system [16]; this adds additional flexibility to the language. Changes to the syntax or the primitives of the language, which would be unthinkably expensive with a hand-built translator, may be accomplished with relatively small changes to the inputs of the compiler writing system. Thus, for example, if evolutions in the Software Requirements Engineering Methodology introduce a type of flow structure which does not currently exist in the language, the structure portion of RSL may be changed through the use of the compiler writing system. Of course, the impact of changes of this magnitude on the ASSM and tools are significant. Therefore, the use of this final level of flexibility is tightly controlled.

5. THE ABSTRACT SYSTEM SEMANTIC MODEL

The RSL statements that an engineer inputs to the Requirements Engineering and Validation System (REVS) are analyzed, and a representation of the information is put into a special data base. This data base is called the "Abstract System Semantic Model" (ASSM) because it maintains information about the required data processing system (RSL semantics) in an abstract, relational model. Each statement is checked for syntactic and elementary semantic correctness prior to being put into the ASSM. Therefore, a number of different tools can easily access the data without need to check for these types of correctness with each access.

In addition to maintaining information about the requirements, the ASSM maintains the concepts used to express the requirements. This means that all extensions, both the core concepts and the additions and modifications of specific projects, are in the ASSM. This allows the RSL translator to process extensions in a manner similar to that in which it processes the requirements statements. The modified concepts are then available for use as soon as they are entered.

The information in the ASSM is not simply a collection of text. Instead of this, it exists as a relational model of the information contained in the RSL statements. In this model, elements are represented by nodes (records) in the data base, and relationships are represented as connections between the nodes. Attributes and their values then consist of a node for the value and a connection to the element node with which the attribute is associated. The structures are expanded to the graph which they represent. This type of representation facilitates retrieval of information from the data base in terms of queries about the relationships between elements; complex combinations of relationships can be traced simply by following the proper connections in the ASSM. In order to support extensions of the language, another level of model exists. At this level prototypes for each type of element, relationship, or attribute may legally be represented. Adding concepts then translates to adding new prototypes to the model. Each instance of an element, relationship, or attribute is linked to its prototype. This facilitates concept-oriented retrieval operations such as "Find all elements of type DATA which are not INPUT to anything" as well as facilitating extensibility.

In addition to providing a means to enhance the efficiency of processing, the ASSM provides a central repository for all information about the system data processing requirements. In the environment of a large BMD system, this type of centralization is necessary since many individuals are continually adding, deleting, and changing information about requirements for the data processing system. The ASSM provides a means for all of them to work with the same base of current information; they can find out about the effects of their work on other requirements engineers, the characteristics of parts of the system that other people are defining, and the current status of their own work. The centralization allows both the requirements engineers and the analysis tools to work from a common baseline and enables implementation of management controls on changes to this baseline.

The function of the ASSM as a central repository for all information is crucial to both the extensibility and disciplined approach aspects of SREP. Extensibility of the language at the level of adding, deleting, or modifying concepts would be nearly impossible if those concepts were spread throughout the system. In addition, the residence of all information about both the software requirements and the concepts used for describing them in the ASSM makes it possible to take a modular and extendable approach to the tools. It also permits the imposition of configuration management controls in an efficient manner since blocking of modifications to the ASSM freezes the configuration. Finally, the analysis tools can easily scan through the data base to check for consistency, a task which would be extremely difficult without access to the description of the entire system at once.

6. AUTOMATED TOOLS

For a large software system, many people develop requirements for different segments of the system; using SREM, each person develops RSL descriptions of the requirements of his particular part of the system. REVS assists in this activity, and also provides mechanisms for imposing discipline and control on the requirements and the requirements engineering process. This is accomplished by the third segment of REVS, the automated tools. The requirements engineer uses the tools to identify those areas which need further resolution, to aid him in resolving problems, and to evaluate his inputs. At various milestones in the development, the tools are also used to evaluate the entire system. Typical requirements engineering efforts will have several iterations of this type, with the tools being used at each stage to show areas which need further work.

The baseline set of tools in REVS contains flow-oriented analysis and validation aids for verifying the completeness, consistency and correctness of the specified requirements. A generalized extraction and reporting system provides additional analysis and status information. While the flow-oriented analysis aids are not easily extensible, the flexible extraction system is a powerful extension capability adaptable both by the engineer to evaluate his requirements, and by management to maintain visibility and control.

6.1 Flow Orientation

The stating of requirements in RSL is based on identifying and relating other requirements information to specified functional flows of processing steps. In REVS, we have implemented tools to provide graphics input/output, to perform static analysis, and to create simulations based on this flow approach. These tools aid the engineer in developing the flows and in validating the consistency, completeness and correctness of the specified flow structures and their related requirements.

6.1.1 Interactive Graphics. The interactive R-NET generation tool provides graphics capabilities for users of REVS. Through this facility, the requirements engineer may input, modify, or display R-NETs. It also provides an alternative to the RSL translator for specification of the flow portion of the requirements. Using this tool, the user may even develop a graphic representation of an R-NET previously entered in RSL. These flow diagrams are inherently two-dimensional, so their display on a graphics system provides a more-easily understood representation than the (one-dimensional) language. Through the use of the ASSM, the user may work with either the graphic or RSL language representation of R-NETs; they are completely interchangeable.

The interactive R-NET generation facility possesses full editing capabilities. The user may input an R-NET "from scratch" or he may modify one previously entered. At the conclusion of the editing session, the new R-NETs replaces the old one (if any) in the ASSM. An alphanumeric keyboard and a trackball-driven cursor are the means by which the user communicates with the graphics system. He can select any of a series of functions from a menu of available functions. The editing functions provide means to position, connect, and delete nodes, to move them, to disconnect them from other nodes and to enter or change their associated names and commentary. Menu selection and screen positioning are done using the trackball; names and commentary are entered through the keyboard. The size of an R-NET is not limited by the screen; zoom-in, zoom-out, and scroll functions are provided.

6.1.2 Simulation. The most thorough automated test of the consistency, completeness, and validity of requirements is often performed with a simulation. For a large software system, the building of simulations must be automatic to preclude divergence of the requirements from the simulation and to allow rapid response and analysis of change.

The automatic simulation generation in REVS takes the ASSM representation of the requirements of a data processing system and generates from it simulations of the system. These simulations are discrete event, somewhat in the manner of GPSS [17]. They are driven by externally generated stimuli; the baseline system generates simulations to be driven by a System Environment and Threat Simulation (SETS) program [18] which models components of a ballistic missile defense system external to the data processing system, the threat, and system environment.

Two distinct types of simulations may be generated by REVS. The first is a simulation which uses functional models of the processing steps. These models may employ shortcuts to simulate the required processing, including the use of "artificial" data (those which are not required to appear in the ultimate real-time software). However, REVS recognizes such data only if they are declared as artificial. The discipline of requiring these declarations reduces the proliferation of data that might (or might not) be required in the ultimate software. This type of simulation serves as a means to validate the overall required flow of processing against higher level system requirements.

The other type of simulation uses analytic models, i.e., models that use algorithms similar to those which will appear in the software to perform complex computations. This type of simulation may be used to define a set of algorithms for the system which have the desired accuracy and stability. This does not establish feasibility of the set for any particular implementation; instead it provides an existence proof of an analytic solution to the problem. Both types of simulation are used to check dynamic system interactions, a type of analysis that is necessary for the dynamic, non-linear, closed-loop control problems occurring in ballistic missile defense systems.

The simulation generator transforms the ASSM representation of the requirements into simulation code in the programming language PASCAL [19]. The flow structure of each R-NET is used to develop a PASCAL procedure whose control flow implements that of the R-NET structure. Each processing step (ALPHA) on the R-NET becomes a call to a procedure consisting of the model or algorithm for the ALPHA. The models or algorithms are written in PASCAL. The data definitions and structure for the simulation are synthesized from the required data elements, their relationships, and their attributes in the ASSM.

By automatically generating simulations in this manner from the ASSM, we ensure that the simulations match and trace to the requirements. New simulations can be generated readily as requirements change, but discipline is enforced by precluding direct change to simulation code; all changes are made to the requirement statements themselves.

6.1.3 Static Analysis. Of course, many requirements inconsistencies do not require the dynamics and cost of a simulation for detection. Therefore, a group of tools is included in REVS to statically (without simulation) check for completeness and consistency in the requirements specification. These tools detect deficiencies in the flow of processing and data manipulation stated in the requirements. Three classes of static analysis tools are included in REVS for this type of detection.

The first class of these tools checks the structure of the R-NETs entered interactively for correctness prior to permanent entry in the ASSM; this includes such things as checking for one and only one start node, proper branching and rejoining of paths, and proper termination of all paths. These checks ensure complete interchangeability between the graphic representation and the RSL form of the R-NET. Additional checks are performed on R-NETs in the ASSM to ensure proper branching and rejoining of paths which include SUBNETs.

The second class of analyzers deals with data flow through the R-NETs. They operate using the R-NET structure in much the same manner that data flow analyzers for programming languages use the control flow of the program [20], but are complicated by the concurrency of path specifications allowed in RSL. These tools detect definite and potential errors in data use (such as local data being used before it is set, and data being set in more than one of a set of parallel paths). Other reports concerning the lifetime and use of data may be generated from the information gathered through these data flow analyzers.

The third class of analysis tools checks for proper hierarchy in the specification. This means that definitions must be specified for all SUBNETs used in R-NETs, that SUBNETs must not make reference to each other in a recursive manner, and that all ALPHAs and SUBNETs must appear on at least one R-NET. A similar analysis is performed on data hierarchies specified in the requirements.

6.2 Extensions

The interactive graphics, flow analysis aids and generalized query system provide the REVS user with a powerful set of automated tools. He may, however, want to have tools specialized to a particular application. The architecture of REVS and the flexibility of the extractor system facilitate such extensions at two levels—addition of completely new tools and creation of special reports by using the extractor.

6.2.1 Addition of Tools. REVS consists of layers of software surrounding the ASSM; the innermost layers contain the ASSM itself and a set of ASSM access routines, and the outermost layer is composed of the REVS executive. Between these reside the RSL translator and the automated tools which operate on data extracted from the ASSM as shown in Figure 4. By using the access routines to isolate the ASSM and by centralizing the control of REVS, the layered structure facilitates extensions to REVS and minimizes the impact on the software of extensions to RSL. Communicating in terms of RSL primitives, the ASSM access routines are unaffected by extensions to RSL and insulate the tools from the detailed organization of the data base. The ASSM can be physically restructured without impacting the tools.

The REVS executive invokes the automated tools based on command from the user. A new tool may be easily incorporated into REVS as an integral part by modifying the executive to recognize a new command and invoke the added tool.

6.2.2 Special Reports. The stucture of REVS easily accommodates extensions to the baseline tools. However, adding a new tool each time a requirements engineer needs a special report or analysis is costly and does not allow timely responsiveness to needs. REVS alleviates this problem by providing the requirements engineer an extensible tool (the extractor) to produce specialized reports. The user completely controls the scope of the analysis and content of his reports; he is not burdened with the details of specifying the format nor with looking at tabular forms to extract needed information.

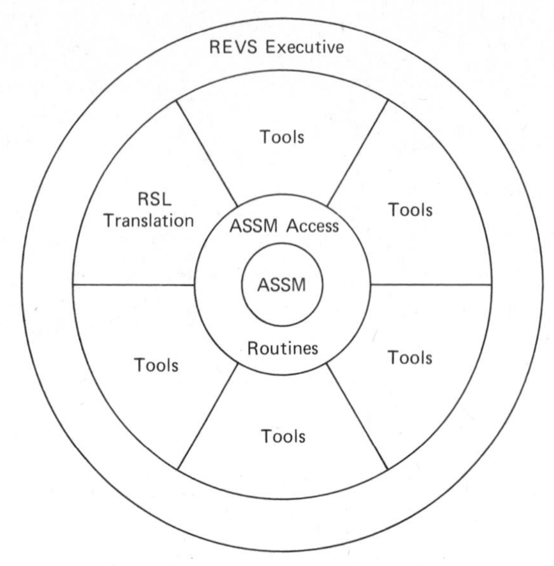

FIGURE 4 REVS layered design.

This flexibility of extraction from the data base is provided by the generalized extractor system. Using this system, the user can subset the elements in the ASSM based on some condition (or combination of conditions) and display the elements of the subset with any appended information he selects. Output is in a standardized form compatible with the RSL input form. Prepositions and additional punctuation are added so that a formal documentation of the requirements can be generated in an order standardized to the needs of a particular application. Being standardized RSL, outputs of the extractor can be easily correlated with the RSL inputs and the engineer has to deal with only one requirements format.

Information to be retrieved is identified in terms of RSL concepts. For example, if the user wants a report listing all DATA elements which are not INPUT to any ALPHA (processing step), he enters the following commands.

```
SET A = DATA THAT IS NOT INPUT.
LIST A.
```

By combining sets in various ways, he can detect the absence and presence of data, trace references on the structures, and analyze interrelationships established in the ASSM. In analyzing user requests and extracting information from the ASSM, the extractor system uses the definition of the concepts contained in the ASSM. Thus, as RSL is extended, the extensions and their use in the requirements are immediately available.

The extractor system can be used both for ad hoc inquiries and for routinely generated extractor special reports. Both the requirements engineer and management may pre-define reports and enter the requests to the extractor as needed. This allows the requirements engineer to establish a repertoire of situation-specific consistency and completeness checks, and to perform automatic regressive testing. Managers can periodically request reports on the status of the ASSM to impose standards and control the development of requirements. For example, RSL supports the traceability of software requirements to system requirements to permit rapid, economical and comprehensive change control. Using the generalized extractor system, a report can be produced at any time to show, for example, which requirements have no documented traceability and are therefore suspect.

7. CONCLUSIONS

The REVS software is now operational on the Texas Instruments Advanced Scientific Computer at the BMDATC Advanced Research Center in Huntsville, Alabama. The software executes in the manner described in this paper. This fact, though comforting, is hardly adequate for a justification of the effort to produce the system. If we were to conclude this paper with nothing more than the proud assurance that the system does not abnormally terminate, we would fall into a category of system builders that Barry Boehm has called "computer basket weavers."

Referring to people in this category, Boehm stated "A basket weaver has a very difficult job. He must plan his basket very carefully and he puts a lot of loving care into it; he builds it, studies it from various angles, discusses it with other basket weavers, and then goes off to build another basket. Very rarely though does he go out and sample users to find out whether they are interested in baskets with handles or with several compartments rather than one compartment, and the like. And, unless something changes considerably in computing, it will remain a kind of computer basket weaving." [21] Boehm is criticizing computer science researchers for not doing adequate requirements analyses and for not checking the degree to which the systems fulfill the requirements after their implementation. A system purporting to improve requirements engineering certainly has a special responsibility to identify its requirements and then to check the degree of fulfillment.

The central nature of RSL led us to concentrate on its requirements in our early development—and to document the requirements so that we could track our progress. The material in Section 2 of this paper is a summary of that document after we had expanded our requirements to include the software in REVS. Merely having some set of requirements, of course, does not prove anything; how well does the system actually satisfy the users' needs?

Some of the test cases used to evaluate requirement satisfaction using SREM are noted in Alford's paper [13]. Most of those, however, involved other parts of SREM than RSL and REVS. The cases noted below were particularly important in finding out whether users were interested in the "basket" that we have woven with RSL and REVS.

One of our first test cases involved restating into RSL a particularly involved part of the requirements for a medical information system. We hoped that the RSL statements would be clear enough that the user (a professional data processing specialist deeply involved in medical systems) would find graphical representations superfluous. Instead, we found that the graphics were really necessary for him to understand the flows through his system. When he used both RSL and graphics, he found five critical problems in the English requirements statements. One of these problems, for example, involved requiring a physician to check the same medical record five times when a single check would suffice. Implementing this system would have required changes in (and wastage of) physicians' time—if they had used the system at all! Clearly, our software system needs graphical output if it is to handle situations like this.

We were particulary interested in the medical system for this evaluation because its flows were far shorter than in a BMD system. Therefore, any indication of confusion in the absence of graphics would be magnified in the BMD situation. With the results described above, we were certain that graphics were required for our "basket" to satisfy the users' needs. Subsequently, we implemented the graphics capabilities needed in this, and subsequent, test cases.

A later test case involved having BMD data processing engineers develop the requirements for an actual BMD function (tracking). In this case the RSL translator was implemented to the point that we could input the RSL statements and have them checked for consistency. Our experience in this case indicated that some of the concepts in the core set of RSL were inappropriately conceived; they were unambiguous, but they did not supply the power required to state the requirements.

This BMD case resulted in revisions to RSL concepts that were incorporated into the baseline version of the language description. It also provided a true test of the extensibility built into RSL since we needed to incorporate the revisions into the translator as well as the language description. Fortunately, we found that the extensibility features were quite adequate, and we were able to revise the translator with trivial effort. Therefore, we found that we needed some features of our "basket" revised (the core concepts) but that other features were solid (the extensibility).

After the revisions to the core concepts, another BMD data processing engineer (one new to the project and not particularly sympathetic to many of its conclusions) started over on track-loop (the tracking function) using the revised RSL. He found a few further revisions that he desired, but they were quite minor. The "basket" seemed to be getting quite close to the users' needs; the translator even produced diagnostics at the right times.

One type of user is the person writing the requirements in RSL, but another is the person reading the resultant statements. The information needed by the reader must be in the RSL statements written by the requirements engineer or the system is, at least, severely crippled. RSL text is rather cryptic, and readers expressed concern about whether adequate information existed in its easily-written form. We performed another test case to evaluate whether our "basket" fulfilled the readers' information needs.

The RSL from the track-loop test case had given readers the impression that inadequate information existed there, so we used the actual RSL from that test case (an extract appears in Figure 5). Then we modified the RSL with phrases that substituted for the standard RSL element, relationship, and attribute names. Finally, we had the material typed in a conventional format and added the usual paragraph numbers. All of this added no new information; all the changes could easily have been done by a computer; in fact, we documented our algorithms to be sure that we exercised no discretion in making the changes.

We tested the sufficiency of the RSL that we had put into a familiar format (with lots of redundant material) by presenting it to readers unaware of its origin. Without exception they thought the text (an extract is shown in Figure 6) was from a real, normal specification. In fact, they sometimes needed to see the original RSL to be convinced that it was merely the same mate-

```
OUTPUT-INTERFACE: RADAR_ORDERS_BUFFER.
    ABBREVIATED BY: ROB.
    CONNECTS TO: RADAR.
    PASSES: RADAR_ORDER.
    ENTERED BY: "MIKE RICHTER".
MESSAGE: RADAR_ORDER.
    MADE_BY: RADAR_COMMAND.
DATA: STARTUP.
    INCLUDES: RA_ORDER_ID
              STARTUP_TIME.
DATA: SHUTDOWN.
  INCLUDES: RB_ORDER_ID.
DATA: TRANSMIT_RECEIVE.
    INCLUDES: RO_ORDER_ID
              RO_IMAGE_ID
              ALPHA_PHASE_TAPER
              BETA_PHASE_TAPER
              TRANSMIT_INFORMATION
              RECEIVE_INFORMATION
              NUMBER_OF_RANGE_GATES
              RANGE_GATE_INFORMATION.
ALPHA: INITIATE_STATE_VAL_DATA.
  ARTIFICIALITY: VALIDATION
  INPUT: HANDOVER_DATA, HANDOVER_TIME.
  OUTPUT: UPDATE_STATE_VALIDATION_DATA.
  DESCRIPTION: "THE REFERENCE ALGORITHM, KALMAN FILTER, IS INITIALIZED.
               HANDOVER_DATA IS COPIED INTO UPDATE_STATE_VALIDATION_DATA WITH
               PLACE_IN_TRACK_TIME SET TO HANDOVER_TIME".
ALPHA: INITIATE TRACK_ON_IMAGE.
  INPUT: HANDOVER_DATA.
  OUTPUT: HOIQ, STATE_DATA, IMAGE_ID.
  CREATES: IMAGE.
  SETS: IMAGE_IN_TRACK.
  DESCRIPTION: "A REQUEST FOR PULSES IS MADE BY ENTERING A FORMAL RECORD REQUEST
               INTO THE HOIQ WHICH FEEDS THE PULSE SENDING PROCEDURES".
```

FIGURE 5 RSL from BMD track loop.

rial reformatted. The needed information, at least at the paragraph level, is contained in RSL statements; perhaps an automated capability to produce more familiar-looking output should be included in future versions of REVS to make our "basket" as useful as possible.

Our evaluation of RSL and REVS is clearly not yet complete since users have not yet employed the final versions of the software. We will continue our evaluation of SREM utility to ensure that we have actually produced an extensible system that encourages disciplined thinking in engineering correct, complete, meaningful software requirements. Our evaluations to date

3.2.2.1 RADAR ORDERS BUFFER. There shall be an output interface from the data processing system called RADAR ORDERS BUFFER. The data processing system shall communicate through this interface with RADAR. Across this interface shall be passed RADAR ORDERS.

It is abbreviated by ROB. It was entered by MIKE RICHTER.

3.2.2.2 RADAR ORDER. When transmitted across an interface the software shall handle the message RADAR ORDER. This message is made up of RADAR COMMAND.

3.2.2.3 STARTUP. Information shall be maintained about STARTUP. This information shall include RA ORDER ID and STARTUP TIME.

3.2.2.4 SHUTDOWN. Information shall be maintained about SHUTDOWN. This information shall include RB ORDER ID.

3.2.2.5 TRANSMIT RECEIVE. Information shall be maintained about TRANSMIT RECEIVE. This information shall include:

RO ORDER ID
RO IMAGE ID
ALPHA PHASE TAPER
BETA PHASE TAPER
TRANSMIT INFORMATION
RECEIVE INFORMATION
NUMBER OF RANGE GATES
RANGE GATE INFORMATION

4.1.3.1 INITIATE STATE VAL DATA. Logical processing shall be done to INITIATE STATE VAL DATA. This shall have as input HANDOVER DATA and HANDOVER TIME. This shall have as output UPDATE STATE VALIDATION DATA.

NOTE: The reference algorithm, Kalman filter, is initialized. HANDOVER_DATA is copied into UPDATE_STATE_VALIDATION_DATA with PLACE_IN_TRACK_TIME set to HANDOVER_TIME.

In interpreting this requirement, note that the degree of artificiality in its statement is VALIDATION.

4.1.3.2 INITIATE TRACK ON IMAGE. Logical processing shall be done to INITIATE TRACK ON IMAGE. This shall have as input HANDOVER DATA. This shall have as output HOIQ, STATE DATA, and IMAGE ID. This logical processing shall, when appropriate, identify a new instance of IMAGE. This logical processing, when appropriate, shall identify the type of entity instance as being IMAGE IN TRACK.

NOTE: A request for pulses is made by entering a formal record into the HOIQ which feeds the pulse-sending procedures.

FIGURE 6 "Conventional" Format for Track Loop.

give every indication that RSL and REVS actually satisfy a critical need in furthering the development of large-scale software.

8. REFERENCES

1. Goldberg, J. (ed), "Proceedings of a Symposium on the High Cost of Software," Held at the Naval Postgraduate School, Monterey, California, September 17–19, 1973.
2. Boehm, B. W., "Software and Its Impact: A Quantitative Assessment," *Datamation*, Volume 19, Number 9 (May 1973), pp. 48–59.
3. McGonagle, J. D., "A Study of a Software Development Project," James P. Anderson & Co., September 1971.
4. Bartlett, J. C. et al., "Software Validation Study," Logicon, Inc., March 1973.

5. Thayer, T. A., "Understanding Software Through Empirical Reliability Analysis," *Proceedings of 1975 National Computer Conference*, pp. 335–341, June 1975.

6. Thayer, T. A. et al., "Software Reliability Study: Final Technical Report," Study Performed by TRW Defense and Space Systems Group for the Air Force Systems Command's Rome Air Development Center, Griffiss Air Base, New York, February 27, 1976.

7. Bell, T. E. and T. A. Thayer, "Software Requirements: Are They Really a Problem?," TRW Software Series, TRW-SS-76-04.

8. Asch, A., D. W. Kelliher, J. P. Locher III, and T. Connors, "DOD Weapon System Software Acquisition and Management Study, Volume I, MITRE Findings and Recommendations," MITRE Technical Report MTR-6908, The MITRE Corp., McLean, Va., May 1975.

9. Kossiakoff, A., T. P. Sleight, E. C. Prettyman, J. M. Park, and P. L. Hazan, "DOD Weapon Systems Software Management Study," APL/JHU SR 75-3, The Johns Hopkins University Applied Physics Laboratory, June 1975.

10. BMD Advanced Technology Center, "BMDATC Software Development System: Program Overview," Ballistic Missile Defense Advanced Technology Center, Huntsville, Alabama, July 1975.

11. Teichroew, D., E. A. Hershey and M. J. Bastarache, "An Introduction to PSL/PSA," ISDOS Working Paper No. 86, University of Michigan, March 1974.

12. Hershey, E. A., "A Data Base Management System for PSA Based on DBTG 71," ISDOS Working Paper No. 88. University of Michigan, September 1973.

13. Alford, M., "A Requirements Engineering Methodology for Real-Time Processing Requirements," Proceedings of the Second International Conference on Software Engineering, held in San Francisco, October 13–15, 1976. (Also published in the TRW Software Series as TRW-SS-76-07, dated October 1976.)

14. Alford, M. W., and I. F. Burns, "R-Nets: A Graph Model for Real-Time Software Requirements," *Proceedings of a Symposium on Computer Software Engineering, MRI Symposium Series,* Volume XXIV, Polytechnic Press, Brooklyn, N.Y.

15. Bell, T. E. and D. C. Bixler, "A Flow-Oriented Requirements Statement Language," *Proceedings of a Symposium on Computer Software Engineering, MRI Symposium Series,* Volume XXIV, Polytechnic Press, Brooklyn, N.Y., (Also in TRW Software Series, TRW-SS-76-02.)

16. Lecarme, O. and G. V. Bochmann, "A (Truly) Usable and Portable Translator Writing System," in: Rosenfeld, J. L. (ed.), *Information Processing 74,* North-Holland, Amsterdam, 1974.

17. Gordon, G., "A General Purpose Systems Simulation Program," *Proceedings 1961 Eastern Joint Computer Conference,* pp. 87–104.

18. F. J. Mullin, "Software Test Tools," *Proceedings of the TRW Symposium on Reliable, Cost Effective, Secure Software,* TRW Software Series Report TRW-SS-74-14, March 1974, pp. 6–47 - 6–48.

19. Jansen, K. and Wirth, N., "PASCAL: User Manual and Report," *Lecture Notes in Computer Science,* Volume 18, Springer-Verlag, Berlin, 1974.

20. Allen, F. E. and Cocke, J., "A Program Data Flow Analysis Procedure," *CACM,* Volume 19, Number 3 (March 1976), pp. 137–147.

21. Boehm, B. W., "Command/Control Requirements for Future Air Force Systems" in *Multi-access Computing: Modern Research and Requirements,* (pp. 17–29) Hayden Book Co., Inc., Rochelle Park, New Jersey, 1974.

READING QUESTIONS

1. What is RSL? How does it relate to REVS?
2. How do RSL and REVS relate to SREM?
3. Give a one-sentence description for each of functions shown in Figure 1.
4. Prepare a diagram that shows how RSL relates to SREM (refer to Alford's first paper). Explain each block in the diagram.
5. Review the test cases discussed in the conclusions portion of the paper; develop a table showing the RSL features and which were utilized in each test case.
6. What are the advantages and disadvantages of RSL?
7. In which phases of the system life cycle (defined in Couger's paper on "Evolution of System Development Techniques") would RSL/REVS be useful? Explain your reasoning.
8. Revise Figures 2 and 3 to include logic to compare the current measurement with the previous measurement, to determine the amount of difference.
9. Revise Figures 2 and 3 to include logic to store the results of the measurement in a transaction data base.

Software Requirements Engineering Methodology (SREM) at the Age of Two

Mack W. Alford

1.0 INTRODUCTION

The Software Requirements Engineering Methodology (SREM) was presented to the Software Engineering community two years ago at the Second International Software Engineering Conference [1]; the SREM support software, the Requirements Engineering and Validation System (REVS), was also presented then [2]. SREM was developed for the Ballistic Missile Defense Advanced Technology Center (BMDATC) to address the generation and validation of software requirements for Ballistic Missile Defense Weapons Systems—the motivation and environment for this research has been previously described [3]. At that time, REVS was operational only on the Texas Instruments Advanced Scientific Computer (TI ASC), and the methodology had been applied to one moderate sized "proof of principle" demonstration problem.

Source: Mack W. Alford, "Software Requirements Engineering Methodology (SREM) at the Age of Two," March 1978, *TRW*, Redondo Beach, Calif., pp. 1–48.

Since then, SREM has been successfully applied to both the generation and independent validation of software requirements for several systems. REVS has been transported to a number of other host computers, and its performance has been improved. The methodology has been successfully transferred to a number of other organizations, and applied to a wider class of problems. The purpose of this paper is to provide a status report of the SREM requirements development procedures, requirements language, support software, and transfer of this technology to other organizations, and to provide an overview of plans for extensions and improvements.

Section 2.0 contains a summary of the status of REVS installations and the transfer of the SREM technology to other organizations. Section 3.0 contains an overview of the results of several diverse SREM applications. Section 4.0 contains an overview of planned extensions to SREM and REVS. Section 5.0 contains some conclusions to our experiences to date.

The objectives of the research leading to SREM were to reduce the ambiguity and errors in software requirements, to make the software requirements development process more manageable, and to provide more automation in validating the software requirements. More details can be found in references [1], [2], and [3].

The SREM requirements development procedures identify the steps and objective completion criteria necessary to define software requirements using the Requirements Statement Language (RSL) and the Requirements Engineering and Validation System (REVS). SREM thus provides a road map of the sequence of activities necessary for the definition of software requirements and the manner in which REVS can be used to ensure that an activity is complete—thereby providing a high degree of management visibility of the requirements development process.

SREM is based on a Graph Model of Software Requirements [4] which is an extension of the Graph Model of Computation [5]. The basic concept underlying SREM is that design-free functional software requirements should specify the required processing in terms of all possible responses (and the conditions for each type of response) to each input message across each interface. Thus, functional requirements identify the required stimulus/response relationships, and autonomously generated outputs. These required actions of the software are expressible in terms of Requirements Networks (or R-Nets) of processing steps. Each processing step is defined in terms of input data, output data, and the associated transformation. Figure 1 presents an R-Net for a Hospital Patient Monitoring System [6] which accepts a measurement of the blood pressure, temperature, skin resistance, etc., for a patient, tests it for validity, records it, requests the next measurement, and tests the measurement against a pre-specified set of upper and lower limits. Note that five paths of processing are identified, which combine into three possible stimulus/response requirements—the paths to request the next measurement and record the current measurement occur regardless of whether the measurement violates the constraints.

The concepts of the Graph Model of Software Requirements are embodied in the design of the Requirements Statement Language (RSL), a machine-processible language for the unambiguous statement of software requirements. RSL is composed of elements (e.g., R—Nets, Processing Steps, Data Messages, Input Interfaces—the "nouns" of the language), their attributes (e.g., units of data, required response times of processing paths, descriptions—the "adjectives" of the language), their relationships (e.g., data is input to a processing step, a message passes an output interface—the "verbs" of the language), and structures (used to define the conditions and sequences of processing steps which comprise the required stimulus/response relationships to be satisfied by the software). Table 1 presents a subset of the requirements for the Patient Monitoring System expressed in RSL. In addition to nouns, verbs, and adjectives for stating requirements, RSL also contains elements, attributes, and relationships that express management concepts (e.g., traceability, completeness, authorship, and version). In all, RSL is composed of 21 types of elements, 21 types of attributes, 23 types of relationships, and three types of structures. It is the structures (R—Nets, Subnets, and Paths) and their formal mathematical foundations, and the stimulus-response orientation which distinguish RSL from the traditional techniques for stating software requirements (e.g., the PSL [7] approach, or standard DoD Military Specifications [8]).

REVS is a large software tool that handles a potentially large data base of requirements, therefore requiring a host computer with a large effective memory space and a moderately fast instruction rate.

REVS accepts RSL as input, translates it into an automated requirements data base, and provides a set of capabilities for analyzing and manipulating this data base. Specific capabilities include the following:

- Translation of an RSL expression of requirements into a central requirements data base.

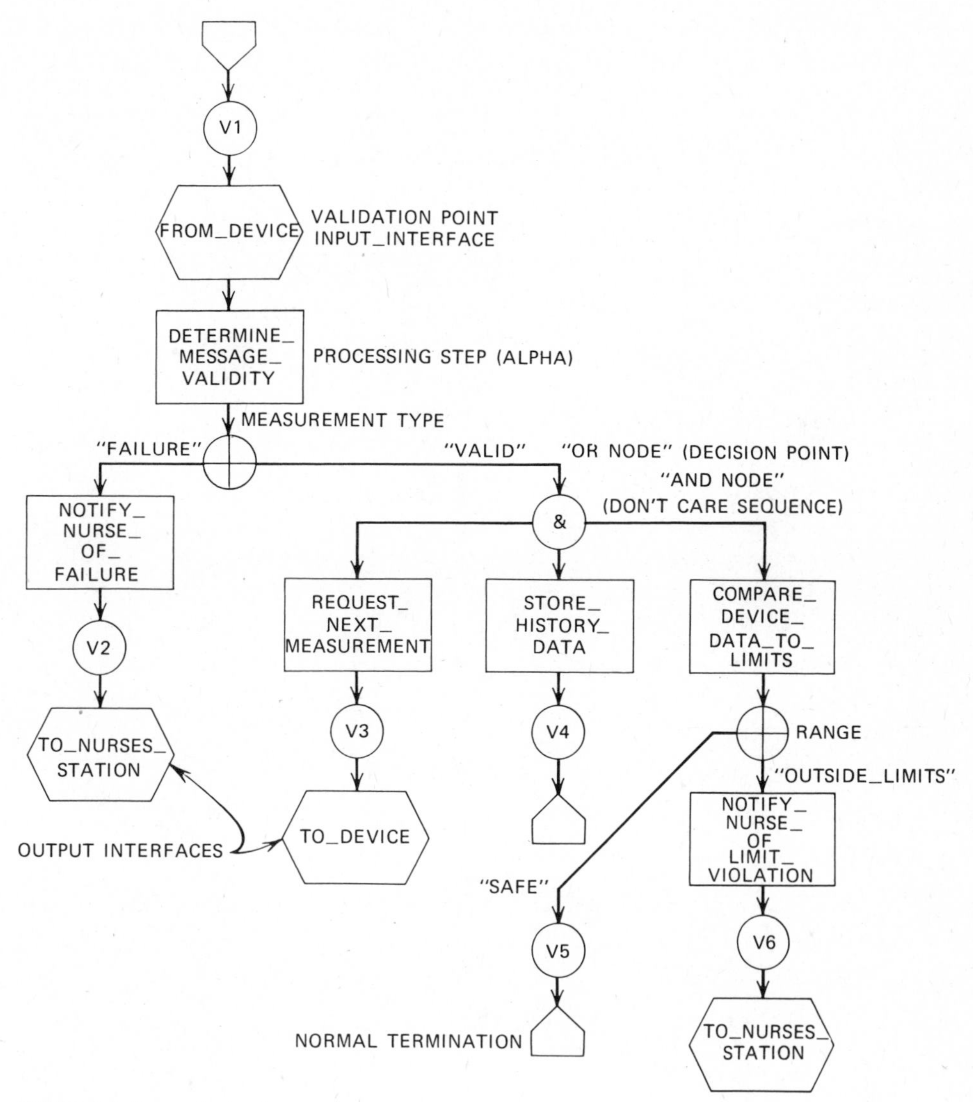

FIGURE 1 Example R—Net.

- Extraction, under user control, of information from the requirements data base for analysis and documentation.
- Identification, under user control, of subsets of the data base for automatic consistency, completeness, and traceability analyses.
- Automated checking of the requirements data base for specific properties of data flow consistency (made possible only because of the underlying formal foundations).
- Graphical representation of the requirements structures, both on-line and off-line.
- Automated generation and execution of stimulations directly traceable to the requirements definition.

The REVS program itself consists of over 40,000 executable PASCAL statements, making it the largest PASCAL program known to us. (In comparison, the PASCAL compiler consists of

TABLE 1
Example RSL Definitions

```
MESSAGE: DEVICE_REPORT.
  PASSED THROUGH: INPUT_INTERFACE FROM_DEVICE.
  MADE BY:    DATA DEVICE_NUMBER
              DATA TYPE_MESSAGE
              DATA DEVICE_DATA.
DATA: DEVICE_DATA.
  INCLUDES:   DATA PULSE
              DATA TEMPERATURE
              DATA BLOOD_PRESSURE
              DATA SKIN_RESISTANCE.
FILE: FACTOR_HISTORY.
  CONTAINS:   DATA MEASUREMENT_TIME
              DATA HPULSE
              DATA HTEMPERATURE
              DATA HBLOOD_PRESSURE
              DATA HSKIN_RESISTANCE.
  TRACED FROM: SENTENCE_2
               SENTENCE_3.
  ASSOCIATED WITH: ENTITY PATIENT.
ALPHA: STORE_HISTORY_DATA.
  INPUTS:     DEVICE_DATA.
  OUTPUTS:    FACTOR_HISTORY.
  DESCRIPTION: "THE DATA PROCESSOR SHALL RECORD EACH VALID MEASUREMENT FOR
                EACH PATIENT".
VALIDATION_PATH: MEASUREMENT_OUT_OF_LIMITS.
  PATH: VALIDATION POINT V1, VALIDATION POINT V6.
MAXIMUM TIME: 1.
  UNITS: SECONDS.
```

approximately 6,000 PASCAL statements.) An additional 10,000 FORTRAN statements perform data base management functions. The RSL Translator is produced from the Backus-Normal Form (BNF) definition of RSL using the Lecarme-Bochman Compiler Writing System from the University of Montreal, plus additional code for the definition of the semantic actions.

The SREM requirements development procedures identify in detail how RSL and REVS are used to generate and validate the processing requirements. The approach is to define functional requirements in terms of paths of processing, then to attach performance requirements to the paths.

First, all interfaces of the data processor are identified, together with the messages that cross interfaces and the message contents. Next, R-Nets are developed to specify stimulus/response relationships required of the software. This top-level information is then checked to assure that each output message is produced at least once. If this is not true, the processing definition is incomplete. The methodology continues by identifying data to be maintained by the software, the data flow between processing elements, and a series of consistency and completeness checks that identify errors and holes in the specification. After these functional requirements are identified and checked for static consistency, functional simulations are generated to verify the dynamic consistency of the requirements, and to derive required processing rates for interactive software systems. When all processing paths have been identified, performance requirements are derived and expressed with respect to these functional requirements. The functional requirements represent the "what" of the processing, while the performance requirements represent the "how well" attributes. Analytical level simulations can then be used to demonstrate the nonreal-time feasibility of the processing requirements, i.e., that algorithms exist which can meet the processing accuracy requirements—only a real-time software design will demonstrate the real-time feasibility of the processing requirements.

At the time of our presentations two years ago, SREM, RSL, and REVS had been applied to one moderately sized BMD problem to demonstrate capability and to illustrate the sequence of steps and associated outputs of the methodology. At that time, it was concluded that the research objectives had been achieved, but that SREM's applicability to other environments and utility in realistic software development environments had yet to be determined.

The remainder of this paper presents an overview of where we are today and where we are going in the future—in order to definitize and automate a comprehensive software requirements methodology.

2.0 WHERE WE ARE TODAY

The availability of SREM technology for use on software requirements development projects is dependent on two factors: the availability of people who are knowledgeable in the SREM concepts, techniques, and procedures; and the availability of REVS to support the use of SREM. The substantial progress which has been made in both of these areas in the past two years is discussed below.

2.1 TECHNOLOGY TRANSFER

Technology development is fruitless without the mechanism for its transfer to others working in the field. Mere existence of tools, or codification of experience, does not necessarily lead to a transfer of working knowledge which can be exercised by others in operational environments. The true test of a methodology is whether it can be absorbed and applied by others.

To successfully transfer SREM technology to others, three techniques have been used: transfer of documentation of the SREM procedures, language, and software capabilities; presentation of a "short course" in Requirements Engineering; and on-the-job training. Table 2 presents a list of the organizations which are applying these techniques. More details of the transfer are provided below.

Some individuals at the University of California at Berkeley, and at TRW, relied only on the published documentation and achieved mixed results in later applications. Requirements engineers from Johns Hopkins University/Applied Physics Laboratory (APL) and TRW have been provided with the basic documentation and with some on-the-job training consisting of assistance in SREM on a problem of their choice. The on-the-job training provided an intensive information transfer, and the opportunity to correct any misconceptions about the software requirements engineering process and misunderstandings of the language concepts. For example, requirements engineers with previous experience writing software requirements usually have an overwhelming urge to try to define a queueing scheme for buffering input messages in RSL;

Organization	Transfer Method: Documentation Only	Transfer Method: On-the-Job Training	Transfer Method: Short Course	Applications: Experimental Usage	Applications: Validation	Applications: Specification Generation
Applied Physics Laboratory		x		x	x	x
Hughes			x	x		
McDonnell Douglas Astronautics Corp.			x	x		x
RCA			x	x		x
Teledyne Brown Engineering			x	x		
TRW	x		x	x	x	x
University of California, Berkeley	x			x		x
U.S. Army Computer Systems Command			x	x		

TABLE 2 Technology Transfer

elimination of this design leads to statements of response time requirements which allows software designers to decide which buffering scheme meets such requirements.

In November 1977, a three-week course was held at McDonnell Douglas Astronautics Corporation (MDAC) in Huntington Beach, Cal., sponsored by BMDATC, to assist the technology transfer to a wider audience. Participants included requirements engineers from MDAC, Hughes, RCA, Teledyne Brown Engineering, and the U.S. Army Computer Systems Command. Course notes were provided to detail the methodology and to provide examples of inputs and outputs of each methodology step. All participants had the opportunity to define functional requirements for two example problems: a common example for the whole class, and then independent team projects. The course was evaluated by all as successful. Table 2 indicated partial results of this transfer: APL, MDAC, RCA, and TRW all have performed, or are performing, demonstration and/or validating software requirements using the technology.

Although continual on-the-job training is the ideal, the training course approach was found to be an effective and cost-effective mechanism for technology transfer. Several future courses are under consideration.

2.2 REVS AVAILABILITY

A critical element in the availability of SREM is the availability of REVS to support its application. Two years ago all REVS capabilities were operational, but REVS was available only on the TI ASC in Huntsville, Alabama, which had no remote access capability. This severely limited the scope of its application. Since then, REVS has been installed in several more locations on both TI ASC and CDC host computers with remote access capability. Table 3 summarizes the installation history of REVS; more details are provided below.

In early 1977, REVS was transported to the TI ASC at the Naval Research Laboratories (NRL) in Washington, D.C., for experimental use by APL. The NRL version of REVS is run in an off-line batch or on-line teletype access mode, and accessed via teletype for data extraction, correction, and consistency checking. When the

Location	Host	I/O Capabilities: On-line Graphics	I/O Capabilities: Remote Batch	I/O Capabilities: Teletype	I/O Capabilities: Off-line Graphics
Advanced Research Center, Huntsville, Alabama	CDC 7600	x	x		x
Naval Research Labs, Washington, D.C.	TI – ASC			x	x
MDAC, Huntington Beach, California	CDC 7700		x		x
TRW, Redondo Beach, California	CDC CYBER 74/174 TSS			x	x

TABLE 3 Current REVS Installations.

TI ASC in Huntsville was phased out in 1978, this became the only operational TI ASC version of REVS.

In July 1977, REVS was converted to a Huntsville CDC 7600, with access by on-line graphics and remote batch. This provided a substantial improvement in availability for demonstration projects. (As a by-product of this installation, the CDC 7600 was provided with a production level PASCAL Compiler with all defined PASCAL capabilities, and the ability to compile very large PASCAL programs with FORTAN subroutines.) This version of REVS was partially optimized for the CDC 7600 memory structure, resulting in program execution times compatible with the development of software requirements in an operational environment.

REVS was then transported to the CDC 7600 at MDAC in Huntington Beach California, for experimental use in writing specifications for Army BMD software. Due to the lack of on-line graphics hardware, only batch and remote batch capability is supported at that installation.

REVS also was transported to the CDC CYBER 74/174 Time Share System at TRW, Redondo Beach, Calif. This version of REVS is currently being optimized to reduce execution time and on-line memory requirements to a level compatible with operational usage requirements.

Thus, REVS has become available on various CDC computers. This has been instrumental in aiding technology transfer and capability assessment. It also has supported the generation and validation of software requirements in operational environments.

3.0 APPLICATIONS

SREM was an out-growth on research addressing the statement of software requirements for BMD systems which are fully automated (no man–machine interactions), and are real-time control and engagement oriented (engagement rules defined before software requirements definition is initiated). In addition, SREM addressed only the development of requirements for software to be executed on a single processor (no distributed processing). Although the mathematical foundations of SREM were believed to be applicable to other types of software requirements development efforts, no explicit claims were made about the efficacy of SREM for those requirements. Since the SREM approach was published, it has been applied on a variety of projects and environments. Table 4 presents characteristics of a number of these projects. Some of these projects are competitive in nature, thus sensitive details have been omitted. Results of these applications are discussed briefly below.

Projects A through E were performed at TRW under the direction of those responsible for the development of the methodology and tools. Project A involved the definition of top-level software requirements for air defense engagement control software during the conceptual design and phase of the project. This project involved the definition of requirements for processing to be distributed across more than five processors, and included the definition of man/machine interactions. Conclusions from this project included the following:

- SREM was applicable to the definition of testable requirements for the distributed processing network as a whole. The statement of such requirements enables meaningful tests to be performed after integration of the processors into the distributed network.
- The allocation of the network processing requirements to processing nodes was supported by REVS in an awkward fashion. Several approaches to ease this transition were identified but not implemented.
- The definition of the overall processing requirements provided the framework for identification of the man/machine requirements allocation issues (i.e., what role a man can plan in an engagement in terms of response

TABLE 4
SREM Applications

Project	Performer	Type System	Type Project	Characteristics			Estimated Software Size (Instructions)
				Real-Time	Distributed Processing	Man/Machine	
A	TRW	Air Defense	CQ PHASE RQMTS	X	X	X	200 K
B	TRW	Operating System	RQMTS RE-DEFINITION				20 K
C	TRW	Real-Time Info	RQMTS VALIDATION & RE-DEFINITION			X	200 K
D	TRW	Air Defense	DEMO	X	X		16 K
E	TRW	Command/Control	DEMO RQMTS RE-DEFINITION	X	X	X	—
F	TRW	Real-Time Info	CD PHASE RQMTS		X	X	150 K
G	TRW	Radar Control	CD DEFINE RQMTS	X			100 K
H	RCA	Test Software	CD DEFINE RQMTS				100 K
I	APL	Communications	DEMO/RQMTS VALIDATION	X		X	50 K
J	MDAC	BMD	DEMO/DEFINE RQMTS	X			200 K
K	UCB	Reactor Control	DEMO/DEFINE RQMTS	X			—
L	TRW	Relational DB	DEMO				N/A
M	TI	Relational DB	DEMO				N/A

times, judgment, capability, and the need to maintain positive control over the weapons).

Project B involved the definition of the requirements for an existing operating system for which software requirements were undocumented; some extensions to the operating system were contemplated, and, hence, requirements on the existing operating system capabilities had to be definitized in order to address requirements for the augmented system. Conclusions from this project included the following:

- Software requirements for an operating system can be written (contrary to popular opinion that operating systems are designed without requirements).
- A major known "bug" in their operating system (i.e., the removal of a file directory could result in making the files in that directory inaccessible to all users) was found to be a requirements problem, not a design problem.
- Requirements can be extracted from existing description and design documents; the SREM concepts for stating software requirements provide considerable aid in separating the requirements form the design.
- Methodology steps for definitizing requirements for existing software follow the same steps as those for generating software requirements, i.e., first identify the interfaces, the messages crossing the interfaces, their contents, the required stimulus/response relationships, etc.; thus, the SREM steps for using RSL and REVS were applicable to address the problem.

Project C involved the application of SREM to the redefinition of a software specification for an interactive (man/machine) near real-time information management system. A major conclusion from this project was that the redefinition of software requirements follows essentially the same steps as the generation of software requirements using SREM. The major problem in performing the validation was the identification of the required sequences of processing steps defined in the various processing "functions," and identification of the data flow between processing "functions"—just the problems that would have to be addressed to make the requirements testable.

Project D involved the redefinition and validation of requirements for existing air defense engagement software, in order to address the inclusion of requirements for new capabilities involving distributed processing. Again, the requirements were extracted from a combination of existing requirements and design documents, and sometimes from interpretation of the code itself. Conclusions from this project included the following:

- The SRE Methodology is, with little modification, applicable to requirements redefinition. The technique of first redefining the requirements, and then modifying the requirements to assess the impact on the existing software, yields significant advantages in comparison to "patching the code one more time."
- The simulation generation capability of REVS provides a rapid, cost-effective means for the generation of both functional and analytical simulations of required processing. In particular, the consistency and completeness checking capabilities of REVS speed up the process of debugging the simulation of the processing, and produce a simulation directly traceable to the requirements.

Project E involved the redefinition of top-level man/machine processing requirements for existing distributed software in order to provide an orderly means of defining augmentations, modification, and up-grade in performance requirements. This project is still on-going. Preliminary conclusions from this project include the following:

- Definition of the requirements for a network of distributed processors in terms of the stimulus/response relationships of the network provides an effective approach for separating requirements from design.
- The SRE Methodology is, with little modification, applicable to requirements redefinition and augmentation problems.
- When these requirements are established, performance improvements for the system can be allocated to require performance improvements of the individual nodes in an objective manner. In other words, application of SREM provides the tools and viewpoint to address performance improvement in a top-down manner.

Projects F and G involved the generation of requirements for software during the conceptual development phase at TRW. Project F addressed the definition of requirements for a real-time distributed information management system, while Project G addressed requirements for software to control a radar and process its data. Both of these projects started with a customer-furnished system concept, and were to generate preliminary software requirements, a preliminary design of the software, and a fixed-price bid on the software development in about six months. Similar conclusions were arrived at in both projects:

- It is difficult to develop complete and consistent, testable software requirements on that kind of schedule with any technique, even SREM. The software requirements development process is squeezed between the system design (selecting hardware, system operating rules) and the preliminary software design (to a sufficient detail to enable a fixed-price bid).

Traditionally, the specification problems are swept under a rug by writing a fuzzy software specification; SREM makes the requirements and the quality quite visible (e.g., it is entirely obvious from the automated checks of REVS whether or not all messages have been generated by the proposed functional processing, and whether or not the input/output data relationships are consistent).

- The use of SREM, particularly in the initial stages of drawing the first R-Nets, provides the communication mechanism for more meaningful discussions for clarifying the initial customer requirements. The necessity of identifying stimulus/response relationships in the R-Nets makes the top-level system logic meaningful to the user, identifies ambiguities in the initial requirements, and provides the benefits of structured walk-throughs in the very early requirements phase so the customer can identify misconceptions on the part of the requirements engineers. The customer expected perhaps a dozen questions of clarification on the requirements; to define R-Nets, he received many more questions and then commented on the R-Nets to identify misconceptions.

Project H was performed by RCA to successfully demonstrate the utility of SREM in defining requirements for test software. This comes close to the traditional batch-oriented data reduction type of software. SREM had the effect of making the requirements more clear and understandable than the traditional methods of writing requirements for such software.

Projects I, J, and K have just begun; no significant conclusions have yet been reached. They are, however, being carried out as demonstrations to provide information about the utility of SREM in operational environments.

Projects L and M took advantage of the user-extensibility features of REVS to define a new language for creating a relational data base. One significant feature of REVS is that RSL is user-extensible (i.e., new elements, attributes, and relationships can be defined by the user), and in the same (or subsequent) runs, information of this type can be translated by REVS into the automated data base, and then REVS can use these definitions to document and analyze these relationships. In Project L, Texas Instruments defined their own language to identify the type of information which might be stored to manage an on-line data base development of software. In Project M, TRW is developing an experimental on-line data base to aid in the rapid estimation of characteristics of data processing architectures to support advanced sensors. In both cases, REVS has been found to have the ability to define proposed data base contents in terms of elements, attributes, and relationships; define a language to implement it and use it experimentally, both for data input and retrieval, in order to explore the utility of the concepts. REVS thus provides a laboratory for experimenting with relational data base concepts due to the flexibility of its design.

3.1 EVALUATION

Only one of the above applications involved the definition of requirements for U.S. Army Ballistic Missile Defense software, the original target for SREM creation. The other projects involve SREM usage in roles not originally addressed (e.g., specification validation, specification redefinition, and augmentation), and having characteristics not originally addressed (e.g., distributed processing, man/machine interactions).

The specification validation role is a natural application for SREM, requiring very little modification of the steps, and no modification of RSL or REVS. It can be applied to a project with a preliminary software specification in order to methodically identify specification problem areas, or to a project already in software development in order to verify the completeness, consistency, and testability of the requirements. This role has been found to be a near-ideal type of project to demonstrate the utility of the SREM concepts to specific operational environments, to provide the vehicle for on-the-job training, and to provide useful results. It is one of the best mechanisms available to satisfy the new regulations (e.g., DoD Regulation 5000.29) requiring validation of software requirements before proceeding with engineering development of a large system.

The specification redefinition role is similar to the above role in that the requirements for a piece of software are defined first and then augmentations are defined with respect to the baseline specification. This approach provides significant advantages to this software augmentation process, i.e., it provides for a precise definition of the required augmentations, allows augmentations to be discussed in terms of requirements instead of specific design approaches, and provides testable requirements for the end product—the software.

Both SREM and REVS are applicable to the definition of distributed processing and man/machine interaction software requirements as they currently exist; improvements have, however, been identified. The application of SREM to these types of software provides a top-down viewpoint and testable requirements on the processing as a whole, in addition to detailed requirements directly traceable to these top-level requirements. It thus provides the framework for making the decisions for allocating the processing to the distributed nodes, for identifying communication requirements, and for allocating the total processing requirements between the software and the manual procedures. Experience with these problems has led to the identification of extensions and augmentations to RSL, REVS, and the SREM procedures. These will improve the ability to develop the detailed specifications from the top-level specifications which are directly traceable, and to provide automated support for the partitioning process.

4.0 PLANS FOR EXTENSIONS AND IMPROVEMENTS

The applications discussed in Section 3 provided experience in addressing a wider class of problems (i.e., distributed processing man/machine interactions, operating systems, on-line information systems) in various roles (i.e., specification generation, specification validation, specification redefinition and augmentation) and in various operational environments (e.g., conceptual development phase working with system and software designers vs. independent validation). Assessment of this experience has led to the identification of several types of extensions and improvements to SREM, RSL, and REVS. Specific areas of research and development currently underway include the following:

- REVS operational improvements.
- Distributed Processing Augmentations.
- Smoother transition to Process Design.
- Smoother transition from System Engineering.
- Smoother transition to Software Validation and Test Planning.
- Extensions to Business Data Processing problems.

Each of these is discussed in more detail below.

4.1 REVS OPERATIONAL IMPROVEMENTS

As used in an operational environment, it becomes cost-effective to transport and optimize REVS for specific installations, and to add user-oriented capabilities. TRW is optimizing the performance of REVS on the CDC CYBER 74/174 TSS installation by reducing run time for frequently used operations, and reducing memory size required to execute REVS. Although this work will address the operational characteristics of REVS on a particular installation, most benefits are to other installations. In addition, a number of transferrable features have been identified to allow the user more facility in manipulating the entire requirements data base (e.g., insert a "data entered" attribute for all data base elements quickly). Their incorporation into REVS has been planned. These improvements are of the kind expected when any utility software is placed into an operational environment (as opposed to a laboratory or research environment).

4.2 DISTRIBUTED PROCESSING AUGMENTATIONS

In the course of generating specifications for processing which ultimately would be distributed among several geographically distant nodes, extensions to RSL and REVS were identified which would automate the process of producing a specification for the nodes from the specification of the network, and define requirements for the communication links. In this way, the specification for the processing at a node could be automatically checked for consistency against the specification for the network processing and modifications to the node specification could be automatically incorporated in the network specification.

A methodology for performing the allocation of distributed processing to the processing nodes, and selecting the data processing hardware configurations and software design, has been a topic of separate research. The results of this research are planned to be incorporated into SREM to form an integrated methodology for the definition of engagement-oriented distributed processing systems.

4.3 PROCESS DESIGN INTERFACE

One of the lessons learned in the experience of using SREM in the Concept Definition (CD) phase is the necessity of quickly closing the loop between the software requirements and the preliminary design of the software. The process design of real-time software addresses the problem of defining the software tasks (the schedulable units of software), the global data base, and task scheduler which allows satisfaction of the response time requirements at the expected loads. In the CD phase environment, changes to the requirements occur with frightening rapidity and volume, and good mechanisms are required to keep track of these changes and assure that they have been translated into the preliminary process design.

Assessment of this experience has led to the identification of techniques for integrating the preliminary design concepts into the requirements data base, thereby assuring consistency of requirements and design. Plans for performing this augmentation are underway.

4.4 SYSTEMS ENGINEERING INTERFACE

A similar lesson learned from using SREM in the CD phase was the necessity of quickly responding to changes in the system design. Identification of different system hardware components, or different ways of using the components to achieve the system mission, leads to new or modified requirements on the data processing to be performed. During the CD phase, these system changes can occur rapidly, and require quick identification of the impact on data processing feasibility, as well as integration into the data processing requirements. Techniques for rapidly translating system design changes into software requirements changes are under development, and the integration of these techniques into SREM is in the planning stages.

4.5 SOFTWARE VALIDATION INTERFACE

It was identified early in the development of SREM that the technique of defining software requirements from the stimulus/response point of view would have a substantial impact on the software test planning cycle, both in terms of activities to be performed, and the techniques for planning and managing these activities. For example:

- During the software design phase, the software design can be validated to assure that all processing paths in the requirements exist in the software, and that only those paths exist (except for error detection and error handling).
- Invariants can be identified from the requirements which can be proven to be met by the software design and code.
- Software tests can be validated against simulation of the required processing before being applied to the real-time software, rather than debugging the real-time software and the test software concurrently.

The full implications of testing to a complete and consistent, testable software specification have not yet been identified, but it appears that the impact of SREM on the software validation process is substantive. Research is currently being planned to address this subject.

4.6 APPLICATION TO BUSINESS DATA PROCESSING

A significant difference between Business Data Processing (BDP) and BMD engagement software requirements development is that the systems analysis in BMD is performed by the BMD systems analysts, while in BDP problems, the systems analysis is performed by the software systems analysts. Thus, in BDP problems, the software requirements role shifts from one of translating systems requirements into software requirements, into the role of defining the system actions and then representing them as software requirements. This shift in roles has a substantial impact on methodology and tools necessary to support such requirements definition activities, in addition to the previously discussed attributes of man/machine interface requirements definition and identification of distributed processing requirements.

Research to adapt the concepts of SREM, and its extensions to the distributed processing, man/machine processing, and interface with systems engineering activities to the Business Data Processing environment, is currently underway.

5.0 CONCLUSIONS

There has not yet been time, since the availability of REVS, for a project to go from a conceptual design to a software specification written using the SREM technology, to a complete software design development, and test. However, from the current experiences of using SREM, in both demonstration and actual software requirements development activities, we can draw the following conclusions:

1. With the increasing availability of REVS on CDC computer, REVS is maturing to the level of capability necessary to support the definition of software requirements in operational environments.
2. SREM and REVS now have demonstrated utility in operational environments in defining and validating requirements for a wide class of software with the domain of demonstrated applicability increasing. Augmentations to REVS have been identified to improve the capabilities of SREM to deal with an expanded class of problems.
3. The SREM technology has been successfully transferred to others for specification generation, and specification validation activities.

Thus, it appears that SREM, even at the age of two, successfully addresses the problems of defining and validating software requirements.

6.0 REFERENCES

1. M. W. Alford, "A Requirements Engineering Methodology for Real-Time Processing Requirements," *IEEE Transactions on Software Engineering*, Vol. SE-3, No. 1, Jan. 1977, pp. 60–69.

2. T. E. Bell, D. C. Bixler, and M. E. Dyer, "An Extendable Approach to Computer-Aided Software Requirements Engineering," *IEEE Transactions on Software Engineering*, Vol. SE-3, No. 1, Jan. 1977, pp. 49–60.

3. C. G. Davis and C. R. Vick, "The Software Development System," *IEEE Transactions of Software Engineering*, Vol. SE-3, No. 1, Jan. 1977, pp. 69–84.

4. M. W. Alford and I. F. Burns, "An Approach to Stating Real-Time Processing Requirements," presented at Conf. on Petri Nets and Related Methods, Massachusetts Institute of Technology, Cambridge, MA, July 1–3, 1975.

5. V. C. Cerf, "Multi-Processors, Semaphores, and a Graph Model of Computation," Dept. of Computer Science, University of California, Los Angeles, CA, Report UCLA-ENG-7223, April 1972.

6. W. P. Stevens, G. F. Myers, and L. C. Constantine, "Structured Design," *IBM Systems Journal*, Vol. 13, No. 2, 1974, pp. 115–139.

7. D. Teichroew, E. Hershey, and M. Bastarache, "An Introduction to PSL/PSA," ISDOS Working Paper 86, Department of Industrial and Operations Engineering, University of Michigan, Ann Arbor, MI, March 1974.

8. Department of Defense, "Military Standard Specification Practices," Report MIL-STD-490, October 1968.

READING QUESTIONS

1. Is there any difference in the stated objective for SREM in this paper than in the first paper by Alford?
2. What enhancements have been made to SREM?
3. What factors are necessary for successful implementation of SREM?
4. What computing configuration is required to be able to use SREM?
5. Expand Table 4 to include other features of the applications of SREM, to enable a prospective user to better determine feasibility of its use.
6. What enhancements are planned for SREM?
7. Alford indicates that some revision is necessary for SREM to be utilized for business applications. What are those revisions? Couger has expressed it differently; explain the difference. (Refer back to the discussion on software engineering at the first of his paper on "Evolution of System Development Techniques.")
8. The figure on the R-Net for the Hospital Patient Monitoring System (Figure 1) is different than the one in the previous paper. What are those differences? What do you think are the reasons for these differences?
9. What is the graph model of software requirements? How does this technique relate to the original version of SREM?
10. Develop a schematic model of the interaction of REVS and RSL, with annotation to explain what is occurring in each block.
11. How does the present version of SREM differ from the previous version?
12. Develop a table comparing the features of PSL/PSA and SREM.

Plexsys: A System Development System

Benn R. Konsynski
Jay F. Nunamaker

With more than 10 years of experience in the development of computer aids in support of information system design and construction, several fundamental principles have emerged as important and useful in the specification of such support systems. The following list includes several principles that should be considered in the design and construction of software system development tools:

The system should serve as a workbench/workstation environment for the system development team. This means that a collection of integrated tools, procedures, transformations and models are available to the developer with no predefined procedural invocation schedule. The developer is free to invoke analysis and design processes according to his/her own procedures.

The system in effect should shorten the life cycle by facilitating a fast implementation of a prototype system. This not only reduces the heavy burden of high maintenance costs, but also improves the important activity of specification of user requirements.

The design process is not, at this time, fully subject to automation. This means that the system must be designed as a computer-aided support system with appropriate data base access, model management and inquiry facilities.

A key function of an analyst support tool is the continuous feedback process. The system must generate alternative views and projections of the current state of the requirements/design. This is necessary to amplify the analyst's cognitive processing.

In line with the above, the system must serve as a memory for the development process. A computer accessible memory of the requirements/design offers considerable improvement over past narrative and flowchart mechanisms of documentation.

The system must also serve as a dynamic documentation of both the requirements/design and of the design process itself. Design decisions are documented for future evaluation and requirements reconsideration in modification efforts.

The flexibility of the data base, the analysis procedures, and the design processes must offer support in the modification of existing or generated systems. The same analysis functions useful in design are available for evaluation of the impact of change.

ADDITIONAL PRINCIPLES IMPORTANT IN PLEXSYS

In addition to the concepts fundamental to the support of the software development process, several other "principles" were determined in the initial design of the PLEXSYS system. These include the following:

The specification language, data base of requirements and the interface command dialogue language should evolve as the development process progresses. No single language, specification form or complement of design transformations will sustain and be relevant to the entire design process. The interaction must change as the process proceeds from initial problem examination to the review of the target system implementation. Further, this should be evolutionary and not discrete, as different "phase" boundaries are perceived by different analysts. This is illustrated in Figure 1.

At every point in the process, multiple levels of logic are relevant. Thus "nested" languages are appropriate as well as multiple language forms (statement, graphic, tabular, inquiry, etc.).

Further, the multiple forms of specification (graphic, tabular, etc.) are necessary to accommodate the various cognitive styles of the users and analysts and prevent boredom. In addition, a variety of forms facilitates communication.

The nonprocedural language form will be translated to a machine language form, or to a compiler level language. Thus, we are expanding the scope of our past language translation efforts (see Figure 2).

The system should accommodate the consideration of several design alternatives. Past systems offered only one path to solution, due to the complexity of the management process. This no longer has to be the case.

The system should support requirements translation to a generic procedural form (in PLEXYS, the Procedural Primitives) rather than derivation of single target high level language (e.g., COBOL). In this way, the transportation of designs is facilitated.

The information system design process is a classical form of unstructured decision making. Therefore, the evolving concepts and foundations of Decision Support Systems (DSS) are quite relevant.

The system should facilitate the prototyping process to whatever degree that is adopted by the organization. This may require high level simulation or "quick and dirty" implementations that communicate the operation of the current design specification for evaluation.

It is clear that information systems of the future will consist of interrelated processes existing in a network of processors connected by intelligent communication facilities. The tools must not prematurely bind data and processes to processors. The distribution of data, processes and control functions must be accommodated according to performance, economic and security objectives.

PLEXSYS was designed as an analyst's and user's workbench to facilitate the development of information systems. The computer-aided approach allows many diverse analysis and design models to interact on an evolving database that is the target system requirements/design on the development "conveyor belt." The system serves

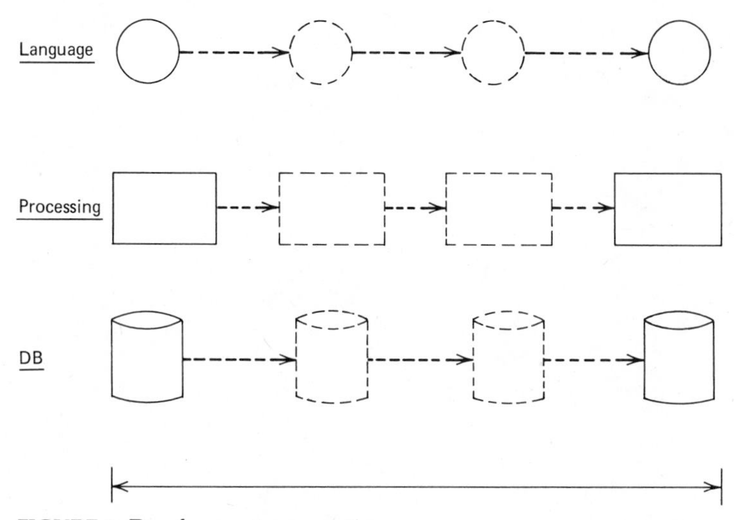

FIGURE 1 Development process time.

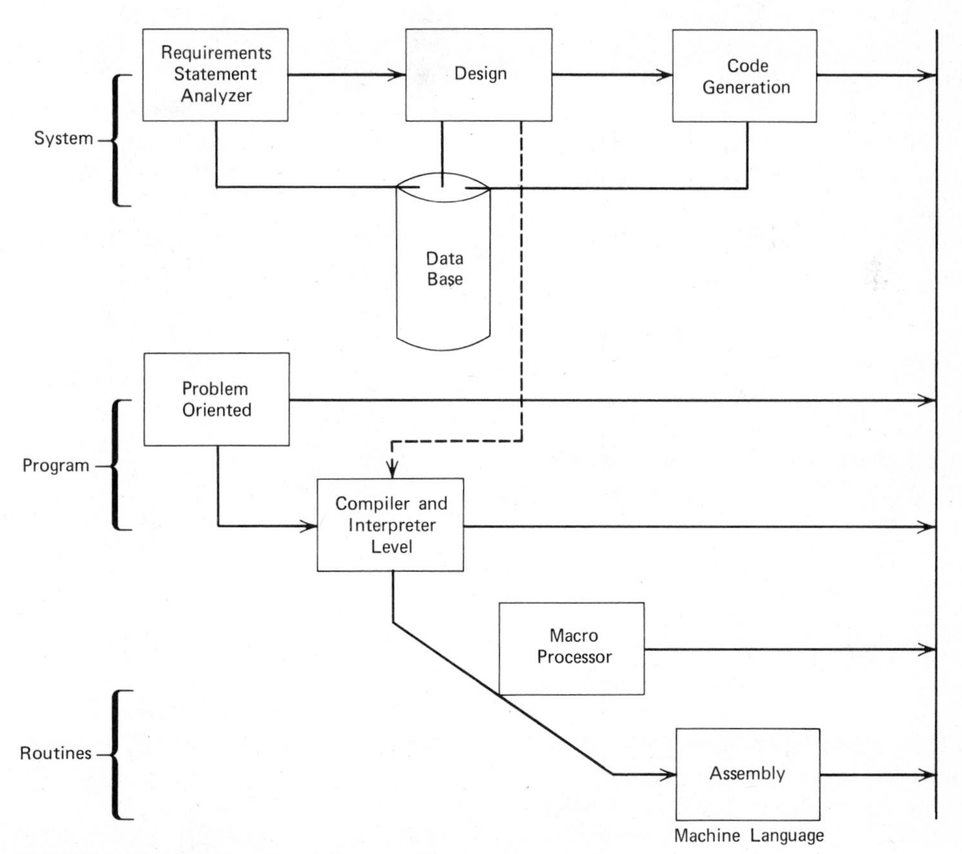

FIGURE 2 The development and scope of language processors.

to document the development process and provide continuous feedback on alternative views or projections of the evolving design. The above concepts and principles were taken into consideration in the design of PLEXSYS.

OVERVIEW OF PLEXSYS

The PLEXSYS methodology and system was designed and constructed to provide computer-aided assistance in the system development process. PLEXSYS was designed to facilitate the front end problem definition and strategic planning phases through all of the efforts necessary to actually produce executable code for a target hardware system.

Some of the tasks included in this process are creation of a business system plan, structure of the problem definition, specification of a set of requirements, analysis of requirements, logical and physical design, specification of file structuring, construction of the target system in a common language form (e.g., PASCAL or COBOL), and the review by the user for requirements modification.

PLEXSYS is an extension of the work of the PSL/PSA project, directed by Professor Daniel Teichroew at the University of Michigan. PLEXSYS differs from PLS/PSA in several important dimensions. PLEXSYS includes all of the features of PSL/PSA with several added dimensions. In one direction, it addresses the front end problem structuring required prior to the use of PSL/PSA, in this phase, PLEXSYS deals with the development of the business systems plan. The second major distinction is the consideration given to the problems dealing with detailed logical and physical design, leading to automatic code generation.

The following brief discussion focuses on the prototype version of PLEXSYS. The purpose of the prototype has been to demonstrate the feasibility of the approach. The number of languages and processors involved may at first appear cumbersome. It should be understood that in the integrated PLEXSYS life-cycle support system under development, the various features of languages and processors meld into a single "dynamic" language and analyzer.

Efforts have been made to develop facilities which permit non-programmers to define problems for computer solution through direct interaction with a computer. Despite these developments, defining, analyzing, designing and construction of information systems are primarily manual, ad hoc operations; there has been very limited application of analysis and design techniques as computer aids in the construction of software systems.

PLEXSYS consists of four interconnected parts or subsystems, structured to aid the participants in the system development process. An overview of PLEXSYS is presented graphically in Figure 3. The four phases of PLEXSYS overlap, as no clear phase boundaries are identifiable. As mentioned earlier, the intent is to support the continuum that is the system development process.

The four "phases" and the supportable activities are:

1. System Problem Structuring
   Strategic Planning
   High Level Feasibility Studies
   Problem Definition
   Determination of System Boundaries
   Coordination with Organizational Planning
2. Description and Analysis of Requirements
   Specification of Requirements
   Capture of Requirements
   Documentation of Requirements
   Logical Consistency and Completeness Analysis
3. Logical and Physical Design
   Logical Structure of Data and Databases
   Logical Structure of Processes, Modules and Programs
   Physical Structuring of Data and Processes
   Configuration of Hardware and Software

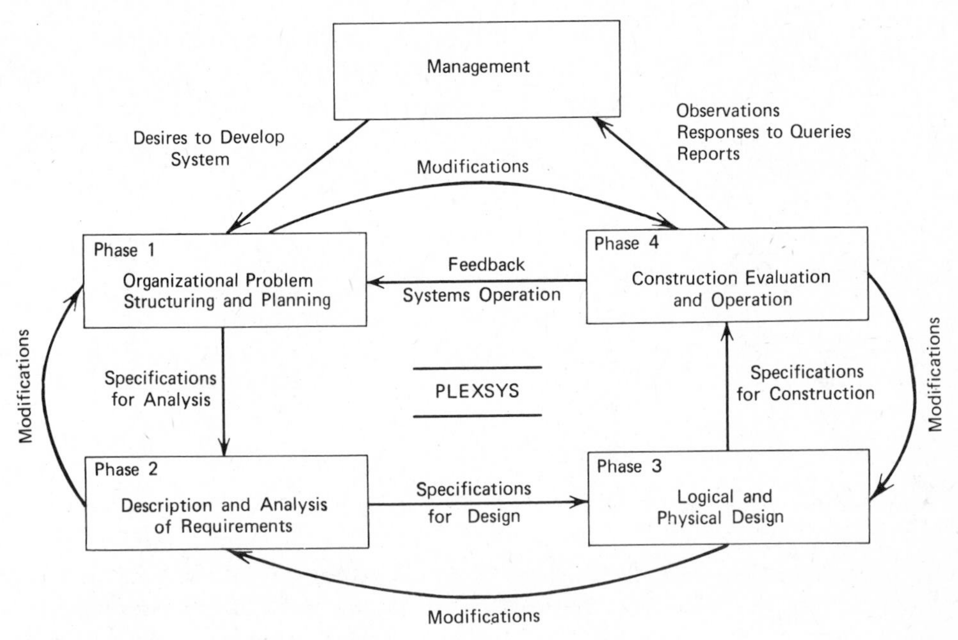

FIGURE 3 PLEXSYS: System development system.

Performance Estimation
Optimization of System Interconnection

4. Construction and Evaluation
   Code Generation
   Initialization of Databases

The following sections discuss each of the four phases in greater detail.

PHASE 1: PROBLEM STRUCTURING AND PLANNING

Phase 1, Problem Structuring, is concerned with aspects related to problem definition and business systems planning at a conceptual level. The essence of System Structuring is to address the aspects of problem boundary identification and to deal with the strategic planning process. Refer to Figure 4.

In System Structuring, one must continually ask: Is this really the problem? Am I unnecessarily constraining the problem definition based on the group's experience? We want to avoid constraining the problem at too early a point in the system development process. This front end phase involves the definition of objectives and goals of the organization or subunit. The resources of the organization that can be utilized in the system development process must be included in the organizational plan. The primary function of Phase 1 is to produce the system plan for the organization. In order to generate the system plan for the organization, the analyst must get everyone that is concerned with the system *actually* involved in the process of defining the system's objectives, goals, and so on. Everyone talks about how to do this; however, very few organizations have been provided with tools to assist with the structuring of the problem. Phase 1 is a collection of cognitive aids to assist in this process. In addition, the analyst must determine characteristics of the user, so we can match the system to the style of the user. The analyst must develop a user profile(s) and classify user types, and experience and education of the user in order

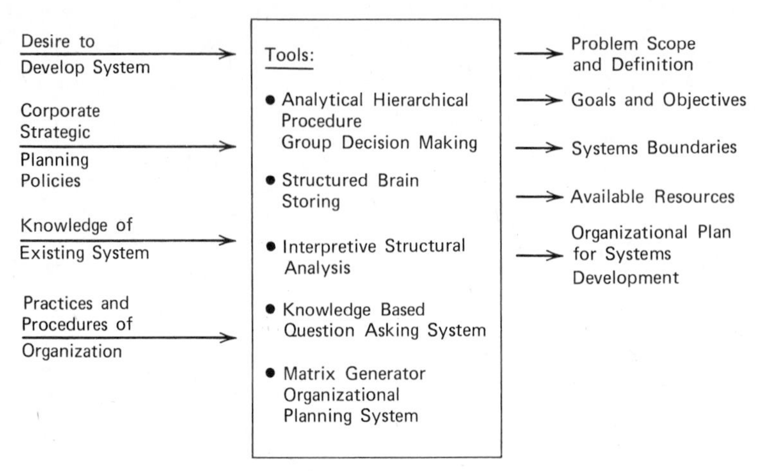

FIGURE 4 Phase 1: Organizational problem structuring and planning.

to customize the information system. Another analyst task is a profiling of the level of sophistication of the intended users.

The objective of Phase 1 is to determine the goals and objectives of the system. In order to accomplish this, one must obtain considerable information on the resources available for system development. The analyst must develop an accurate "picture" of the type of system desired or necessary. The analyst must document the interface with all other systems in the organization. The interfacing process is extremely critical when planning an integrated corporate data base. Frequently, insufficient time and energy are devoted to understanding and recognizing the constraints and restrictions that may be involved.

Most corporate systems today have hundreds and hundreds of subsystems that must be interfaced. PLEXSYS is a tool for organizing the information needed to develop an overall strategic plan. The interfaces, goals, and objectives must be considered from the beginning, for once we get to Phase 2 and Phase 3, it will be too late. Designs will already be constrained and it will be difficult to accommodate the design desired by the users. This means that one needs to know things like availability of data, whether it will be machine readable or whether it must be transcribed repeatedly. The scheduling requirements, inputs, outputs, and processes must be known, even at the front end conceptual level. Why? Because the analyst needs to understand the constraints, it's too late to discover them in the detailed statement of requirements. Certain constraints may become imbedded in the system and cannot be changed. Therefore, it is important to establish system boundaries and determine whether the boundaries are fixed or not. The analyst must test the boundaries and evaluate this definition against the statement of goals, objectives, and constraints. A check list of items that must be considered is generated. PLEXSYS provides this check list system for keeping track of information that may appear to be unrelated and/or extraneous when one is first getting started on a project. PLEXSYS Phase 1 provides the software for organizing information needed to develop the strategic plan. The PLEXSYS software triggers questions and provides reminder lists. In this way the user or analyst or manager will not overlook. Even very experienced people forget. Because they go through the same procedures frequently, they often overlook considerations that are quite important. The surgeon sometimes leaves a sponge in a patient's stomach, although everyone knows that you don't leave sponges in stomachs. PLEXSYS Phase 1 "keeps track of sponges." As each routine is completed, completion is indicated on the terminal until the system can respond affirmation to the question, "Have you removed all sponges?"

The check list feature can trigger us to continuously evaluate goals, objectives, and system boundaries. By going through a standard set of questions, we are often surprised by responses that we did not anticipate. Often we find we can relax one constraint or another. We find that the system constraints or boundaries are not as fixed as we assumed. This is very important, for with this new information, the system takes on different characteristics. We must spend considerable time on the front end problem, because situations and systems are often very different than they appear at first glance. Different perspectives must be presented and different views must be evaluated. PLEXSYS has the capability of presenting information from alternative perspectives and viewpoints. The PLEXSYS routines are designed to effectively assist the user in challenging system boundaries and constraints. This is, in effect, a way to restructure the problem. This may involve a reformulation of the problem in a different context so that issues may be made more clear. If users can obtain alternative views of a situation, then they can develop a different perspective regarding the problem.

PLEXSYS Phase 1 provides, in addition to the check list systems, the question generation system and the tests for systems boundaries. In addition, it provides a system called Structured Brainstorming System (SBS).

SBS is a methodology for capturing the information that goes on in a brainstorming session in a very structured way. It can be done at a terminal, with users in various locations around the country. SBS can be implemented on an electronic mail system or as a manual system.

Manually, the procedure is very simple. A set of related questions concerning issues to be addressed in the planning process is identified. The questions are placed, one per page, at the top of the page. The users respond to each question and each comment on a page. The sheet is passed to the next person, who then comments on the question and the previous comments. This process is continued for about one-half hour. After the exercise has been completed, the leader of the exercise has documentation or possibly a database of what happened in the structured brainstorming process.

Let's look at the type of questions that might be used in this exercise in the development of a strategic plan.

Question 1. What are the major long-term business problems facing Company Z?

Question 2. What are the major disadvantages of the present information system with respect to Co. Z's business?

Question 3. What are the major advantages of the present information system with respect to Co. Z's business?

Given a particular situation, the questions would be specific, and they would be related to the applications in question.

SBS provides a structured way to get started on the planning process. It also involves parallel processing, rather than sequential processing. The process is much more efficient than oral discussion, where only one person can talk at a time. Further, intimidation is reduced; one retains a degree of anonymity when writing comments on paper. It is clearly not as intimidating as challenging the boss in a discussion during a meeting. The prior documentation of comments is helpful in structuring one's own comments. In an oral discussion people are thinking of what they are going to say instead of listening to the speaker. They are concerned with the response they plan to give and do not listen. SBS, then, provides a mechanism for concentrating on the task. The set of responses can be grouped and categorized and then listed and reviewed so that they provide a starting point for further discussion.

The issues raised in SBS can then be structured and another technique called, ISSUE analysis, can be used to evaluate the full set of alternatives available to address the issues. The automation of these tools adds value to PLEXSYS utility in the planning process. PLEXSYS consists of a number of other techniques similar to the checklist systems, question generation system, structured brainstorming, and ISSUE analysis system.

PHASE 2: DESCRIPTION AND ANALYSIS OF REQUIREMENTS

A few of the statements of the PLEXSYS language are used in Phase 1 to assist in the structuring of the problem and in general design of the business systems plan. The majority of the problem description takes place in Phase 2. The PLEXSYS analyzer includes a computer-aided technique for structured documentation and analysis of information processing systems. It is a tool for describing system requirements, analyzing the descriptions, recording the descriptions in machine processable form, and storing them in a database. The PLEXSYS language and analyzer was designed as a "user friendly" to assist the user to a significant degree in comparison with earlier Requirements Definition Languages. The inputs and outputs of Phase 2 are described in Figure 5.

The PLEXSYS language was developed as a single comprehensive language that can be used at the very high level as well as the procedure definition level. The objective of PLEXSYS is to translate a very high level descriptive statement of a system into executable code for a target hardware/software configuration. Creation of a system using PLEXSYS includes the generation of a nonprocedural statement of requirements, selection of file and data structures based on data utilization, determination of the interfaces with a data management system, and generation of program code for the target system in COBOL or Pascal.

It should be remembered that the PLEXSYS system is concerned with supporting the decision making process at all levels of systems development. The capability of the PLEXSYS language ranges from high level conceptual definitions down to the "nitty gritty" decision level required to produce executable code.

The PLEXSYS language supports descriptions at five levels. A user operates at whichever

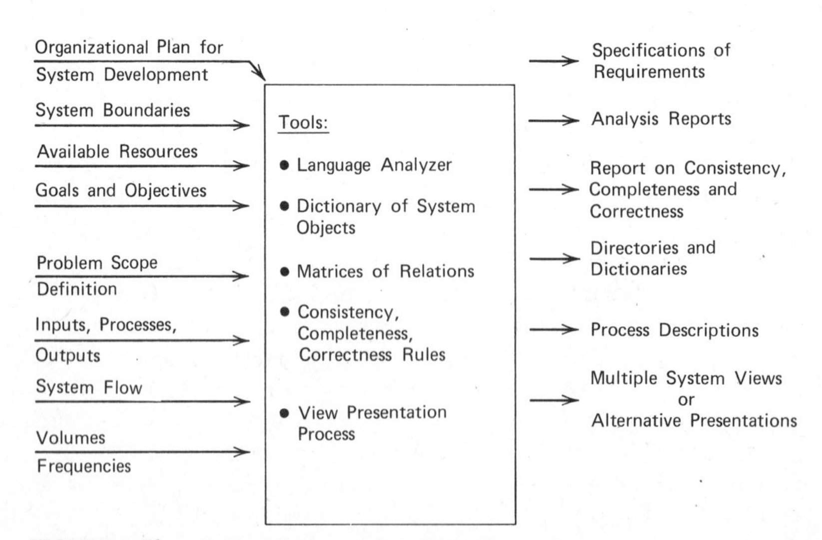

FIGURE 5 Phase 2: Description and analysis of requirements.

level is of interest at a particular time and place in the life cycle. The PLEXSYS language is not made up of distinct sublanguages, however. The PLEXSYS language can be roughly divided into the following levels.

Level 1
- System Flow
- System Activities

Level 2
- Data Definition
- Process Definition and Structures
- Procedure Definitions

Level 3
- Logical Structure
- Process Organization
- Data

Level 4
- Physical Environment
- Hardware Description
- Software Description

Level 5
- Code Generation
- Procedural Primitives

Form of Problem Description

The user of PLEXSYS has a choice of form of expression in specification of the problem description. The user can select from the following types of representation:

1. Statement
2. Graphical
3. Tabular

The user selects the mode of representation with which he/she is most comfortable. The usefulness of PLEXSYS is enhanced considerably through the effective use of a graphical display. The color graphics terminal offers the obvious advantage of being able to display output system flow descriptions and logical designs in a more useful and appealing form. However, the most important advantage appears to be related to the entry of information. The graphical approach, as illustrated in Figure 6, describes how the user is prompted from one stage to the next in the problem definition process. Figure 6*a* illustrates the prompt commands used to describe the objects and relations in the PLEXSYS system. Figure 6*b* illustrates the prompt commands used to run various analysis reports.

Figure 6*c* illustrates the full screen of the graphics terminal with the objects and relationships of the PLEXSYS language shown on the top of the screen. The list of reports available to be run is shown to the right of the screen. A typical system graph built by a PLEXSYS user of objects with associated relationships is shown in the center of the display scope.

The activities involved in Phase 2 begin once the conceptualization of the system has been completed and the business systems plan has been prepared. The requirements to be derived in Phase 2 include the definition of the required inputs and outputs and the computations necessary to produce them. We need some facility for expressing these requirements.

The PLEXSYS system was developed with the following characteristics in mind:

- The language must be capable of describing all requirements: conceptual; system flow; data structures; hardware environment; software environment and detailed procedural levels.
- The language must be extremely user friendly such that management will interact with the system.
- The structure of the language must allow for evaluation.
- The language must permit ease of modification of the requirements contained/stored in the various data bases.
- The requirements delineated using the lan-

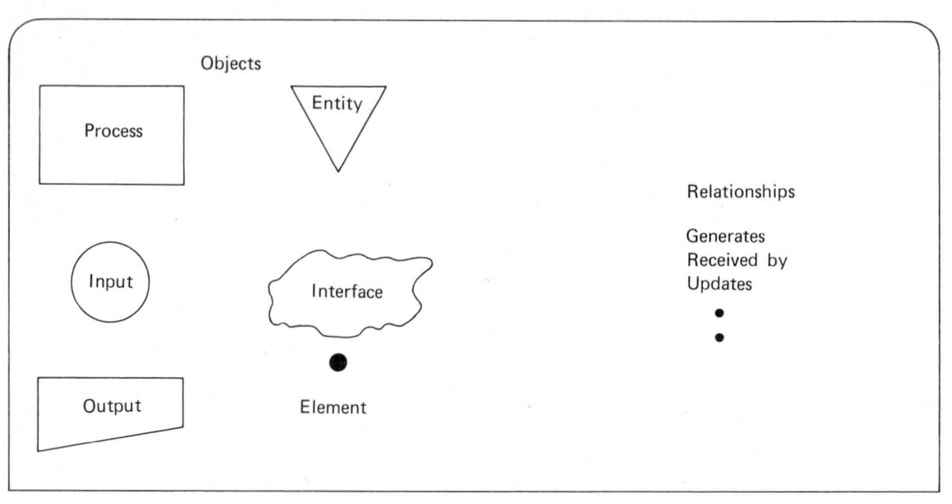

FIGURE *6a* Objects and relations.

guage must be compatible with the use of the language in the design and code generation phases.

- The language must facilitate portability.
- The language must facilitate documentation.

The PLEXSYS language was derived from the best features of other systems. However, it can best be described as a significant extension of PSL in terms of scope. Two important principles remain with respect to language structure:

- The language is as nonprocedural as possible where nonproceduralness is required and it is, at the same time, as procedural as necessary to produce code.
- The language supports hierarchical structuring in both process and data definition.

The first of these is especially important. By stating "what is to be done" rather than "how it is to be done" we prevent the premature bindings that hinder the portability of the design and the consideration of alternative designs. Secondly, PLEXSYS releases the user from many bookkeeping aspects of programming. Therefore, there is less chance for error because the user need concentrate on only those aspects relevant to the problem at hand.

The problem definition phase of system design and construction is necessarily an iterative procedure. The user cannot be involved until he or she has something to "grasp" such as feedback from the analyst in the form of feedback to initial system requirements proposal. Thus, the problem definition process is handled in an iterative manner facilitated by the many reports available through the PLEXSYS analyzer.

The interactive procedure of problem statement development is also enhanced by features which facilitate a top-down approach to the requirements definition. The basic language structure allows for hierarchical structuring.

A Procedure Definition Facility (PDF) was included for facilitating definition of logic at all levels of design. The PDF portion of PLEXSYS is based on the constructs determined necessary for structured programming top-down definitions.

Hierarchical structuring facilitates problem definition and assures a degree of correctness. Using a top-down structured approach, correctness can be established at the higher levels and maintained at each succeeding level.

A simple top-down description of a portion of a small example will illustrate the use of the language in defining the system requirements.

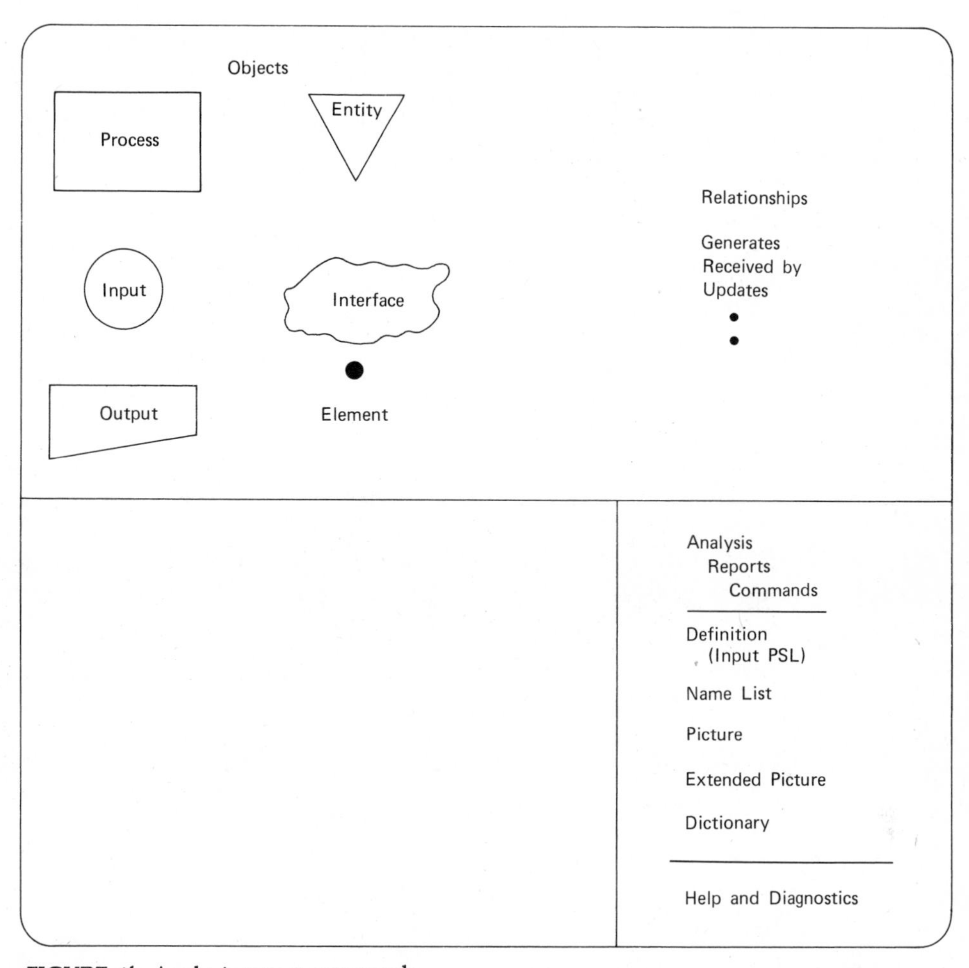

FIGURE *6b* Analysis report commands.

The structure defined is illustrated in Figure 7. Network structures as well as tree structures are permitted. The structure is determined through analysis of the relationships described in the defined process (see Figure 8).

To appreciate the benefit of a requirements statement tool, one must remember that the problem statement definition is an iterative procedure and each component is defined as the user builds the problem statement through the use of feedback from the analyzer.

A partial logical data structure, and representative PLEXSYS statements in deriving this structure and defining other aspects of the system are illustrated in Figure 9.

PLEXSYS Analysis

The process of defining the system requirements is handled in an iterative fashion by use of a PLEXSYS analyzer. The problem definers periodically process the problem statement or interactively update a problem statement through use of PLEXSYS. Thus, PLEXSYS serves as a tool in the problem statement process by continually

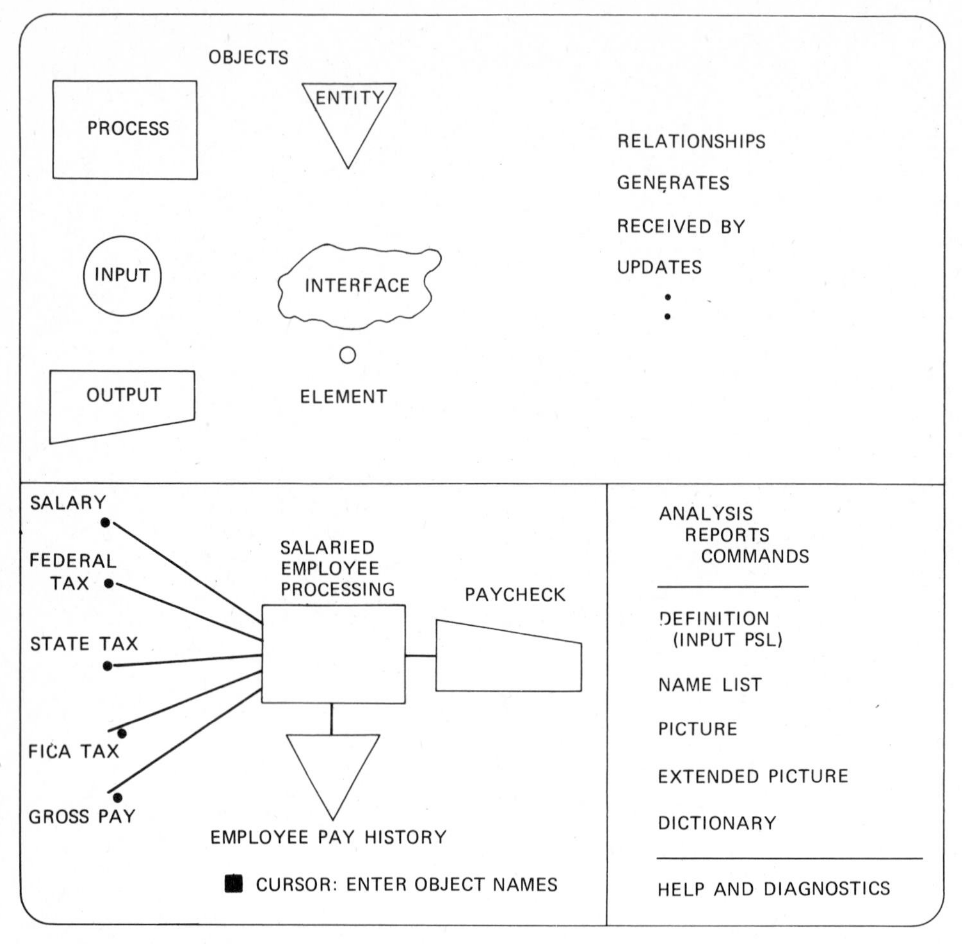

FIGURE 6c Screen layout for PLEXSYS.

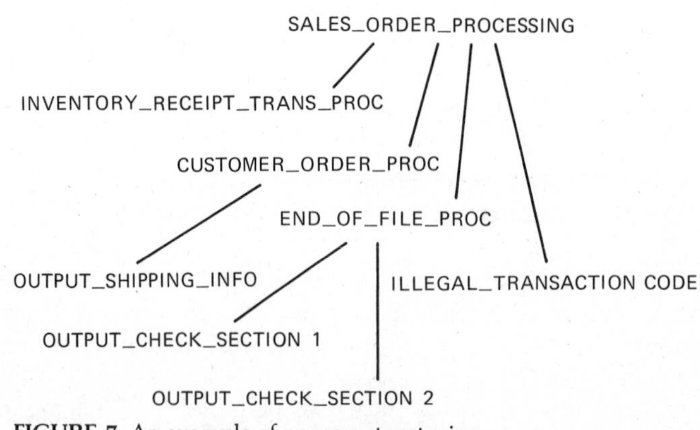

FIGURE 7 An example of process structuring.

```
PROBLEM DEFINER B. R. KONSYNSKI; MAILBOX IS BPA-505;
RESPONSIBLE FOR SALES_ORDER_PROCESSING;
DEFINE BPA-505 MAILBOX;

PROCESS SALES_ORDER_PROCESSING;
  PROCEDURE:
    IF INVENTORY_RECEIPT_TRANSACTION THEN INVENTORY_RECEIPT_TRANS_PROC
    ELSE IF CUSTOMER_ORDER_TRANSACTION THEN CUSTOMER_ORDER_PROC
    ELSE IF END_OF FILE_TRANSACTION THEN END_OF_FILE_PROC
    ELSE ILLEGAL_TRANSACTION CODE;
  USES SALES_ORDER_PROC_HEADERS;
  RECEIVES SALES_ORDER_PROC_INPUT;
  SUBPARTS ARE INVENTORY_RECEIPT_TRANS_PROC, CUSTOMER_ORDER_PROC,
   END_OF_FILE PROC, ILLEGAL_TRANSACTION_CODE;

PROCESS INVENTORY_RECEIPT_TRANS_PROC:
  USES NUMBER_RECEIVED TO UPDATE NUMBER_ON_HAND;
  UPDATES NUMBER_OR_INVENTORY_RECEIPT_TRANS;
  USES NUMBER_RECEIVED TO UPDATE TOTAL_NUMBER_RECEIVED;
  GENERATES INVENTORY_RECEIPT_TRANS_OUTPUT;
  PROCEDURE:
    ADD NUMBER_RECEIVED TO NUMBER_ON_HAND
    ADD 1 TO NUMBER_OF_INVENTORY_RECEIPT_TRANS
    ADD NUMBER_RECEIVED TO TOTAL_NUMBER_RECEIVED;

PROCESS CUSTOMER_ORDER_PROC;
  PROCEDURE:
    ADD NUMBER_ORDERED TO TOTAL_NUMBER_ORDERED
    ADD 1 TO NUMBER_OF_CUSTOMER_ORDERS
    IF NUMBER_ORDERED_GREATER_EQUAL_NOH THEN
      NUMBER_TO_SHIP = NUMBER ORDERED ELSE
      BLOCK NUMBER_TO_SHIP = NUMBER_ON_HAND
      BACK_ORDER = NUMBER_ORDERED - NUMBER_ON_HAND
      OUTPUT BACK_ORDER  TOTAL_BACK_ORDER = TOTAL_BACK_ORDER + BACK_ORDER
      END BLOCK
    IF NUMBER_TO_SHIP_GT_ZERO THEN BLOCK TOTAL TO SHIP =
      TOTAL_TO_SHIP + NUMBER_TO_SHIP NUMBER_ON_HAND =
      NUMBER_ON_HAND - NUMBER_TO_SHIP
      OUTPUT_SHIPPING_INFO END BLOCK;

PROCESS END_OF_FILE_PROC;
  PROCEDURE:
    CHECK_NOH = NUMBER ON HAND - (TOTAL_NUMBER_RECEIVED - TOTAL_TO_SHIP)
    CHECK QUANT_BACK ORDERED = TOTAL_BACK ORDER + (TOTAL_TO_SHIP -
     TOTAL_BACK_ORDER)
    IF CHECKS_NOT_ZERO THEN OUTPUT_CHECK_SECTION 1
    ELSE OUTPUT_CHECK_SECTION_2;

PROCESS OUTPUT_CHECK_SECTION 1;
  GENERATES CHECK_SECTION_1_OUTPUT;
```

FIGURE 8 Process structuring in PLEXSYS. (Continued on page 412)

```
PROCESS OUTPUT_CHECK_SECTION 2;
  GENERATES CHECK_SECTION_2_OUTPUT;
PROCESS OUTPUT_SHIPPING_INFO;
  GENERATES SHIPPING_INFORMATION_OUTPUT;
PROCESS OUTPUT_BACK_ORDER;
  GENERATES BACK_ORDER_OUTPUT;
PROCESS ILLEGAL_TRANSACTION_CODE;
  GENERATES ILLEGAL_TRANSACTION_MESSAGE;
```

FIGURE 8 Continued.

providing feedback to the problem definer of many aspects of the system.

The functions of PLEXSYS-Phase 2 include:

- Accepts and analyzes PLEXSYS language input.
- Maintains a data base containing an up-to-date problem statement.
- Provides output of the present system requirements in PLEXSYS language.
- Allows several problem definers to state requirements of different parts of the system while maintaining a coherent system requirements definition.
- Produces reports on demand which allow alternative views of the status of the requirements statement.
- Maintains directories and dictionaries that aid problem statement.
- Analyzes the problem statement for completeness and consistency.

The reports produced by PLEXSYS can be used by project management as well as the problem definer. Reports are produced under defined conditions or on demand and each type of report can provide detailed or summary information. Further facilities are provided for the generation of special reports according to user specifications. In addition to aiding in the problem definer's work, the reports aid him or her in communication with the user and thus speeds the problem statement process. After the system requirements are fully defined, the PLEXSYS outputs and data base serve as detailed documentation of the system.

The reports produced by PLEXSYS at various phases in the system's development life cycle are from among the following classes:

- Graphical presentation of static relationships.
- Graphical presentation of system flows.
- Tabular representation of relations.
- Tabular representation of similarities.
- Dictionaries and directories.
- Statistical presentation of completeness, frequency of occurrence, etc.
- Evaluation of consistency and completeness.
- User friendly help commands and presentation control commands.
- Data base manipulation such as update, delete, input, report.
- Data base query processing.

Following the process of analyzing the reports and cleaning up the set of requirements, the data base of requirements is passed to the next phase, which consists of the logical and physical analysis and design. Experience has shown that numerous iterations are required to generate a complete, consistent, and correct problem statement.

PHASE 3: LOGICAL AND PHYSICAL DESIGN

The Logical and Physical Design phase is the application of transformations that translate the now complete and consistent, logical description

```
INPUT SALES_ORDER_PROC_INPUT;
  CONSISTS OF TRANSACTION_CODE, TRANS_PARM;
  RECEIVED BY SALES_ORDER_PROCESSING;
ELEMENT TRANSACTION_CODE;
  VALUE 0 THRU 9;
ELEMENT TRANS_PARM;
  SYNONYMS ARE NUMBER_ORDERED, NUMBER_RECEIVED;
  VALUE 0 THRU 9999;

DESIGNATE QUANTITY_RECEIVED AS SYNONYM FOR TRANS_PARM;
ENTITY SALES_ORDER_PROC_SUBHEADERS;
  CONSISTS OF HEADER_NOH, RECEIVED_HEADER, SHIPPED_HEADER, CHECK_NOH_HEADER,
    CHECK_ORDER_HEADER;
SET SALES_ORDER_PROC HEADERS;
  CONSISTS OF SALES_ORDER_HEADER, SALES_ORDER_PROC_SUBHEADERS;
ENTITY SALES_ORDER_HEADER;
  CONSISTS OF SECTION_1_HEADER, SECTION_2_HEADER;

OUTPUT CHECK_SECTION_1_OUTPUTS
  CONSISTS OF SECTION_1_HEADER, HEADER_NOH, NUMBER_ON_HAND, RECEIVED_HEADER,
  TOTAL_NUMBER_RECEIVED, SHIPPED_HEADER, TOTAL_TO_SHIP, CHECK_NOH_HEADER,
  CHECK_NOH, CHECK_ORDER_HEADER, CHECK_QUANT_BACK_ORDERED;
  ELEMENT NUMBER_ON_HAND;
    VALUE 0 THRU 9999;
  ELEMENT TOTAL_TO_SHIP;
    VALUE 0 THRU 9999;
  ELEMENT SECTION_1_HEADER;
    VALUE 'CONTROL CHECK REPORT: SECTION 1;'
```

FIGURE 9 Data structuring in PLEXSYS.

of system requirements into formal specifications of program modules and data organizations (file specifications and data base schema). The system interactively supports the determination, presentation, and evaluation of design alternatives. The key to the logical design is the effective organization of processes into program modularizations and data reorganization to the hierarchical and network oriented data organizations. Refer to Figure 10.

The Logical design begins with a general analysis of information and process communication and control flow. A major objective of this analysis is the determination of process and data structuring that satisfies several design criteria (e.g., minimize data transport and information distribution). This analysis results in effective program data and module specifications for the code generation phase. An overview of the process is given in Figure 10. The data organization design process was simplified by the adoption of hierarchical file organization and the network model for data base design and DML interface in code generation. Thus, for purposes of standardization of procedures and probability, a subset of the data structuring techniques of the CODASYL Data Base Task Group (DBTG) Report was selected to be used as the basis for data organizations generated by the system. Utilizing the set concept of data management of the DBTG makes it possible to define file design as a determination of how, within the constraints of the owner/ member concept, a given data organization may best be logically described. This eliminates the

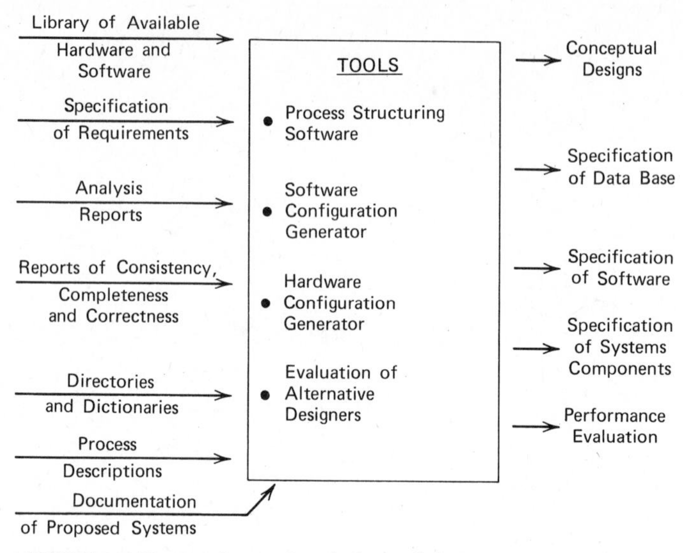

FIGURE 10 Phase 3: Logical and physical design.

task of selecting a particular data representation from among the total feasible class.

One of the primary responsibilities of Phase 3 is the determination of effective modularization of the system of processes into programs and routines. The purpose of the analysis of the logical data and process control flow is determination of grouping feasibility and evaluation of transport volume savings resulting from groupings. Transport volume is a measure that reflects the volume of data movement in the system. The processing time is a nondecreasing function of the transport volume; thus reduction of transport volume can result in considerable savings in processing time. Hardware constraints include the existing CPU speed, main and auxiliary memory and communications aspects of the target system, as well as the existence of unique features such as specialized dedicated processors, and so on.

One might wonder how many "processes" must be dealt with in a real world application. In a past project, 647 processes were defined and 62 program module specifications were generated through analysis of the interprocess relationships. The 62 program modules produce 79 formal reports and constitute the high level modules of the system.

Modularization, or a process grouping, is a factoring of the system into interacting modules such that the modules together perform as the system. In terms of a system of processes, modularization is the determination of subsets of the set of processes that satisfy some evaluation criteria such as reducing the intermodule interface. The subsets may be hierarchical and may overlap. For our present discussion, we will consider modularizations that form a cover for the set of processes and constitute a partitioning. This may not always be true.

The typical concerns of process grouping with respect to software design involve operating system features such as multiprogramming, multiprocessing, memory management, and so on. The importance of this aspect is often overlooked by manual designers until processes are procedurally bound, that is, when it is too late to take advantage or if so, only superficially. The nature of the analysis for design under such an environment necessitates a computer-aided approach.

Another, and initially significant, area of

concern is the operational aspect of the system use. A thorough analysis of operational constraints may significantly reduce the number of design alternatives, as the constraining features of operation are often absolute.

The necessity for operational (manual) intervention results in a primary decomposition of the system of processes. Such intervention often arises from factors such as having to await response from a management position or awaiting the arrival or production of necessary data. Further, the frequency of process invocation will frequently influence the design process.

One of the objectives in Phase 3 is the evaluation of the system design implicit in the PLEXSYS description. The intent is to provide the user with several measures of the "goodness" of alternative designs, enabling him or her to compare those measures with some prespecified criteria. The user may then choose to accept the system as it is or to redesign it. Note that redesigning may

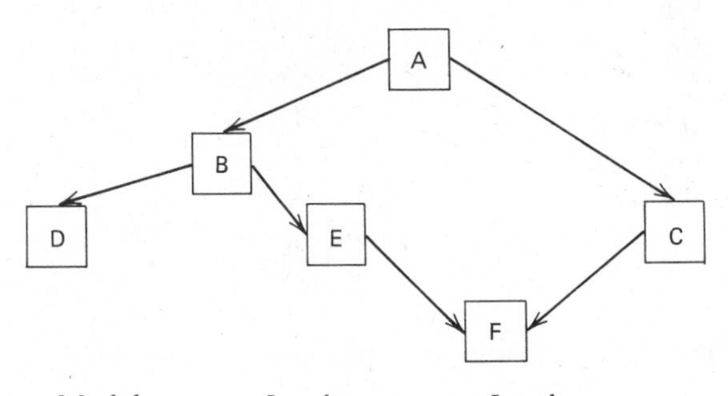

Module	Invokes	Level
HEAD	A	
A	B C	1
B	D E	2
C	F	2
D	γ	3
E	F	3
F	γ	4

FIGURE 11 Precedence table.

require a return to the earlier phase in which the initial PLEXSYS definition was formulated.

In order to facilitate logical analysis and design, several tables, matrices, and other intermediate forms are derived from the requirements data base. Two of these tables (Figures 11 & 12) are derived from the requirements data base for module invocation analysis. The subsystem indicates which module is to be performed first. Module characteristics consist of information as to other modules to be invoked, the conditions for invoking them, and the number of times each module is executed when invoked.

The incidence matrix for a subsystem is constructed module by module. For each module, *j*, the data sets, *i*, which are known to the module as inputs are denoted by a "-1" in the *i*th row, *j*th column. Those data sets that are outputs from the module are each denoted by a "1" in the appropriate row and column. Inputs are identified when data sets are read or used by a module without having been assigned a value previously. Outputs are identified when data sets are written or assigned a value. Similarly, an incidence matrix can be constructed for an entire design based upon data sets used by multiple process groupings.

The subsystem (process group) matrix is used for computation of the data utilization and information distribution measures. Transport volume is relevant when peripheral access is necessary. Because modules are grouped as subsystems, there is no transport between modules for external access. The locality of reference analysis addresses the interprogram communication. The incidence matrix for each subsystem is generated along with the "goodness" measures to be used in the redesign process if necessary.

Four design criteria are currently being used in program module determination (process grouping). These include:

Data Utilization—an analysis of data existence. Has significant impact on processing time which is a nondecreasing function of the

Invoking Module	Condition	Invoked Module	Exp. No. Executions	Probability
Head		A	1	1.0
A	$X_1 > X_2$	B	5	.6
A	$X_1 \leq X_2$	C	3	.4
	etc.			

FIGURE 12 Condition table.

transport volume, a consequence of the analysis of data utilization. Grouping and reorganization of processes can have a significant impact on data transport by eliminating multiple references and unnecessary creation of history items.

Information Distribution—deals with the knowledge of information by modules [Parnas]. By "hiding" information the complexity of development and change can be reduced considerably. A goal of this effort is to minimize the impact of change.

Control Transfer—analysis of the probability of in-line control transfer [Loew] between processes. Minimization of control transfers can have significant impact on both process and data organization.

Parallelism—through analysis of invocation procedures and data usage we can determine precedence relations and assess potential savings from parallel invocation of processes. This criteria takes on obvious importance when mixed multiple processors are available. A subanalysis includes determination of suitability for pipeline processability.

This list is only a partial set of the criteria that can be used; indeed, the authors have also used reference distribution (locality) and operational invocation as criteria in the past. The current list matches the software analysis capabilities and reflects our design priorities.

Data Utilization

Data utilization is a measure of the volume of data communicated between subsystems or groups of processes. The analyst retrieves from the database the data sets which are input to and/or output from each subsystem and must supply the volume for each data set, where it does not exist in system. An incidence matrix for the subsystem is derived, and the following formula is used to compute the transport volume.

Measure:

$$TV = \sum_{j=1}^{R} \left(V_j \times \sum_{i=1}^{K} |e_{ij}|\right)$$

where V_j is the volume for each data set j;
e_{ij} is a 0/1 matrix of the incidence of subsystem i using data set j,
R is the number of data sets, and
K is the number of subsystems.

Information Distribution

For determination of an information distribution (or information hiding) measure, the incidence matrix for the subsystem and a complexity vector are utilized. The analyst is presented with a list of data sets for which he or she supplies a weighting factor, if not in the data base, indicating the relative complexity of each data set.

Measure:

$$ID = \sum_{i=1}^{N} \sum_{j=1}^{K} (|e_{ij}| \times C_j)$$

where e_{ij} is the incidence matrix of subsystem i and data set j,
C_j is the complexity weight for data set j,
N is the number of subsystems, and
K is the number of data sets.

Control Transfer

Control transfer is a measure of the expected number of times a module will be executed when the program is run. It is based upon the probability that the invoking condition will be met and upon the expected number of executions of each module when invoked. The analyst is required to provide, where not available in the data base, invocation conditions, invoking and invoked modules, as well as expected-number-of-executions parameters. The analyst must provide missing probabilities concerning invoking conditions and expected number of executions of each module. We view the control transfer as a Markov model, using CT as a measure of effectiveness of our modularization with regard to control transfer.

Measure:

$$CT = 1 + \sum_{i=1}^{n} \sum_{j=1}^{n} a_i \cdot p_{ij} (1 - b_{ij})$$

where a_i is the expected number of times a process is referenced over a specified period, p_{ij} is the probability of a control transfer between processes i and j, and b_{ij} is a boolean matrix of consonance of module assignment (i.e., $b_{ij} = 1$ if both processes are in same module for that design).

Parallelism

Parallelism is measured in terms of the number of levels and the number of processes at each level. This analysis makes use of the precedence table. From this table the levels and processes can be determined using the following algorithm:

set all levels to zero
$level_{previous} = level_{head} + 1$
for each module invoked by $level_{previous}$
  if $level_{next}$ = zero
    then $level_{next}$ = (greater of $level_{next}$ or $level_{previous} + 1$)
    else $level_{next} = level_{previous} + 1$
end for
sum all modules for each level value.

Measure:

$$PA = \sum_{i=1}^{K} W_i \times N_i$$

where $W_i = i/N$, the weight at level i,
N_i is the number of modules at level i,
K is the number of levels, and
N is the number of modules.

The result of the interactive schema generation and process structuring is a formal specification of program module data and processing composition. In the event that appropriate code generation facilities are not complete, the system can generate readable specifications of program details for manual programming. The data requirements, procedural activities, basic control invocations, and relations to other programs are available in the data base.

PHASE 4: CONSTRUCTION, EVALUATION, AND OPERATION

The input to Phase 4 (as shown in Figure 13) is the program and data organization specifications needed to generate code on the target com-

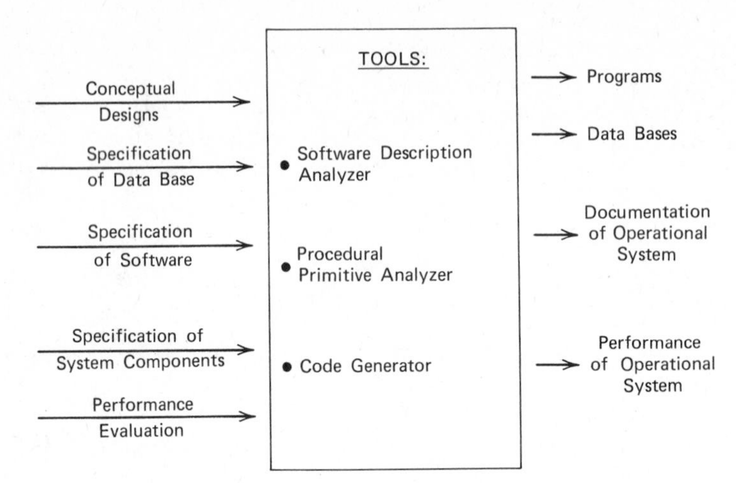

FIGURE 13 Phase 4: Construction, evaluation, and operation.

puter system. The specifications include the process descriptions complete with procedural descriptions as per the PLEXSYS procedural Definition Facility. The logical design of program and data structures is provided along with relevant specifications of the target hardware and systems software characteristics. If additional information is required, the PLEXSYS system prompts the analyst for whatever assumptions, defaults, or decisions are required for the code generation process.

The general approach to code generation was to determine certain primitive operations common to most languages, and to logically generate the target system code in terms of these primitives. The actual code is then realized utilizing a set of routines which translates this primitive "pseudocode" into actual instructions. A set of primitive routines is defined for each target language. Each set of primitives represents the class of data manipulations for the target system processing anticipated.

The purpose of the intermediate primitive representation is to facilitate portability. The portability benefit can be viewed from several positions. The first is that the portability of the design system is facilitated by making only a small portion of the automatic code generator system dependent on the specific low level characteristics of the target system. The second is that this procedure eases the portability of the system on a target system which has had modification that may call for instruction modification. A third reason for the intermediate form is that application of optimization analysis can be utilized to further optimize the code. The type of optimization that was not included in the earlier processing includes such operations as reorganizing multipass operations (loops) so as to take advantage of the process grouping in terms of processing efficiency. Finally, most of the current compiler level type optimization operations can be applied at this point to further improve the execution performance.

The automatic code generation phase also determines data structures required for processing and control sequences within the framework of the program module. The procedures presently incorporated include derivation of necessary declarations, initializations, input and history reference commands, output and updated history command generation, and location and other "bookkeeping" operations.

The structure and location of input, history input, output, and updated history output commands are a function of many variables, among them the schema and storage structures, report format requests, flow of control in the data manipulation statements, and distribution of data element references determined in Phase 3. Other "bookkeeping" operations include the opening and closing of files, the basic program flow outside the data manipulation control portion, the various data movements necessary for computations for input items, and movement of data set to set up for the output. In several information processing systems analyzed, this "bookkeeping" code accounts for approximately 40 to 60 percent of the code produced. It is important to note that these "bookkeeping" operations exist in the requirements statement only at the highest levels (i.e., GENERATE report, CONSISTS in the OUTPUT section, etc.), or are implied by the nature and organization of the processing. A part of our efforts includes an evaluation of the influence of bookkeeping code on the size and volume of software.

The goals of Phase 4 of the PLEXSYS effort include: (1) machine independent requirements definition; (2) automation of code generation; (3) target system portability; and (4) file analysis and process design.

Phase 4 is the last module in the life cycle support system of the PLEXSYS system. The output of Phase 4 is a set of programs, data base systems, performance reports of the target hardware and software system, and complete documentation of the operational system.

FUTURE DIRECTIONS

The first PLEXSYS prototype was implemented in 1979. The concept of "phased" languages and data bases was simulated by the use of several sets of languages appropriate for the different phases: PSL/PDL, SDL/MSL, and a Procedural Primitive Language (PPL). The analyzers and translators were discrete systems rather than the single model management control system of the current version of PLEXSYS that is operating on the Management Information System Department's VAX at The University of Arizona.

The subsequent PLEXSYS system has been implemented on the MIS Department DEC VAX. The configuration consists of 2.5 megabytes of main memory and 2 RMO5 Disk Drives with a total of 512 megabytes of memory. The VAX has been configured with 24 ports. The system includes color graphics and contains 5 DEC GIGI terminals with monitors from Barco Electronics. Additional graphics terminals are 5 DEC VT 100s with graphic retro-fit boards, 1 Tektronix 4006, and 1 Tektronix 4025.

Microprocessors in the MIS Laboratory include Apple IIs, TRS 80s, and S100 Z80 based systems. Graphics software has been supplied by DEC (GIGI Software), and Megatek (Template Software).

The current activities are centered on the incorporation of the color graphics facilities and integration of control structures into an effective "workstation" environment.

The information system software development process will continue to be a man-machine iterative process. With effective tools, the human and machine capabilities can best be exploited. Further, to the degree that we are able to facilitate user involvement in the development process, we will be better able to serve the organization by better addressing the ultimate goal—improved user productivity.

BIBLIOGRAPHY

Alford, M. W., "A Requirements Engineering Methodology for Real-Time Processing Requirements," IEEE Transactions on Software Engineering, Vol. SE-6, No. 1, January 1977, pp. 60–69.

Alter, Steven, *Decision Support Systems: Current Practice and Continuing Challenges*, Reading, Mass.: Addison-Wesley Publishing Company, 1980.

Ashenhurst, R. L., "Curriculum Recommendations for Graduate Professional Programs in Information Systems," *ACM Communications*, Vol. 15, No. 5, (May 1972).

Baker, F. T., "Chief Programmer Team Management of Production Programming," *IBM Systems Journal*, Vol. II, I, 1972, pp. 57–73.

Bell, T. E., Bixler, D. C., and Dyer, M. E., "An Extendable Approach to Computer-Aided Software Requirements Engineering," *Proc. 2nd International Conference on Software Engineering*, October 1976.

Bell, T. E., and Thayer, T. A., "Software Requirements: Are They Really a Problem?" *Proc. 2nd International Conference on Software Engineering*, October 1976.

Blosser, P., "An Automatic System for Application Software Generation and Portability," Ph.D. dissertation, Purdue University, 1975.

Boehm, B. W., "Software and Its Impact: A Quantitative Assessment," *Datamation*, May 1973, pp. 48–59.

Boehm, B. W., "Some Steps Toward Formal and Automated Aids to Software Requirements Analysis and Design," *IFIPS 74*, 1974, pp. 192–197.

Boehm, B. W., McClean, R., and Urfrig, D., "Some Experience with Automated Aids to the Design of Large Scale Reliable Software," International Conference on Reliable Software, Los Angeles, April 1975, pp. 105–113.

Boehm, B. W., "Seven Basic Principles of Software Engineering," *Infotech State of the Art Report on Software Engineering*, 1977.

Bracker, W., and Konsynski, B., "An Overview of a Network Design System," Proceedings NCC, 1980, pp. 41–47.

Carlson, Eric D., J. Bennett, G. Giddings, and P. Mantey, "The Design and Evaluation of an Interactive Geo-Date Analysis and Display System," *IFIP Congress 74*, 1045–1061. (Also described in Keen and Scott Morton, Addison-Wesley Pub. Co., 1978.) (1974).

Carlson, Eric. D., and J. A. Sutton, "A Case Study of Non-Programmer Interactive Problem Solving, San Jose, Calif.: *IBM Research Report RJ1382* (1974).

CODASYL Development Committee, "An Information Algebra Phase I Report," *Comm., ACM*, Volume 5, No. 4, April 1962, pp. 190–204.

Cotterman, William W., J. Daniel Couger, Norman L. Enger, and Frederick Harold, *Systems Analysis and Design: A Foundation for the 1980s*, New York: North Holland Publishing Company.

Couger, J. D., "Evolution of Business System Analysis Techniques," *Computing Surveys*, Vol. 5, No. 3, September 1973, pp. 167–198.

Davis, C. G., and Vick, C. R., "The Software Development System" *Proc. 2nd International Conference on Software Engineering*, October 1976.

Davis, C. G., and Vick, C. R., "The Software Development System," *IEEE Transactions on Software Engineering*, Vol. 3, No. 1, January 1977, pp. 69–84.

DeJong, P., "BDL," *IBM Systems Journal*, 1976.

Gain, C., and Sarsen, *Structured Systems Analysis: Tools and Techniques*, Englewood Cliffs, NJ: Prentice-Hall, 1979.

Grindley, C. B. B., "Systematics—A Nonprogramming Language for Designing and Specifying Commercial System for Computers," *Computer Journal*, August 1966, pp. 124–128.

Hamilton, M., and Zeldin, S., "Higher Order Software—A Methodology for Defining Software," *IEEE Transactions on Software Engineering*, Vol. SE-2, No. 1, March 1976, pp. 9–32.

Hax, A. C., and Martin W. A., "Automatic Generation of Customized Model Based Information Systems for Operations Management," *Proceedings of the Wharton Conference on Research on Computers in Organizations*, Philadelphia, Pa. October 1975.

Ho, T., and Nunamaker, J. R., Jr., "Requirements Statement Language Principles for Automatic Programming," *Proc. ACM National*

Conference, 1974, pp. 279–288.

IBM, "The Time Automated Grid System (TAG): Sales and Systems Guide," Publication No. GY20-0358-1, May 1971, pp. 1–12.

IBM, "HIPO: Design Aid and Documentation Tool," IBM, SR20-9413-0, 1973.

Konsynski, B., and Nunamaker, J. F., Jr., "Towards Computer-Aided Schema Generation," *10th Annual IEEE Asilomar Conference Proceedings*, November 1976.

Konsynski, B., "A Model of Computer Aided Definition and Analysis of Information System Requirements," Ph.D. dissertation, Purdue University, December 1976.

Konsynski, B., "Computer-Aided Process Structuring," MIS Working Paper, University of Arizona, Tucson, 1979.

Konsynski, B., and Nunamaker, J. F., Jr., "Automation of Systems Analysis, Design and Construction Process," *Proceedings of the National Canadian Computer Conference*, May 1980.

Konsynski, B., and Mannino, M., "Information Resource Specification and Design Language," *Entity-Relationship Approach to Systems Analysis and Design*, P. Chen, New York: North Holland Publishing Company, 1980.

Konsynski, B., and Bracker, W., "Software Systems to Solve Network Design Problems," *Data Communications*, July 1980, pp. 69–78.

Konsynski, B., and Bracker W., "Defining Requirements for a Computer-Aided Network Design Package," *Data Communications*, August 1980, pp. 75–84.

Konsynski B., and Nunamaker, J. F., Jr., PLEXSYS II, Working Paper, Department of Management Information Systems, University of Arizona, Tucson, July 1981.

Langefors, B., "Some Approaches to the Theory of Information Systems," *BII*, Vol. 3, 1963, pp. 229–254. Reprinted in *System Analysis Techniques*, J. D. Couger and R. Knapp, Eds., New York: John Wiley, 1974, pp. 292–309.

Lynch, H. J., "ADS, A Technique in Systems Documentation," *Database*, Vol. 1, No. 1, Spring 1969, pp. 6–18.

Mills, H., "Top Down Programming in Large Systems," *Computer Science Symposium*, New York University, Courant Institute, July 1970.

Mills, H., *Debugging Techniques in Large Systems*, Randall Rusten, Ed., Englewood Cliffs, N.J.: Prentice-Hall, 1971, pp. 41–55.

Morton, M. S. Scott, "Management Decision Systems: Computer Based Support for Decision Making," Cambridge, Mass., Division of Research, Harvard University (1971).

National Cash Register Co., *A Study Guide for Accurately Defined Systems*, Dayton, Ohio, 1968.

Ness, D. N., "Decision Support Systems: Theories of Design," presented at the Wharton Office of Naval Research Conference on Decision Support Systems (November 4–7, 1975).

Nunamaker, J. F., Jr., Swenson, D. E., and Whinston A. B., "Specifications for the Development of a Generalized Data Base Planning System," *Proceedings from AFIPS Conference*, Vol. 42 (1973).

Nunamaker, J. F., Jr., "A Methodology for the Design and Optimization of Information Processing Systems," *Proc. 1971 AFIPS SJCC*, Volume 38, AFIPS Press, Montvale, N.J., pp. 283–294; reprinted in *System Analysis Techniques*, J. D. Couger and R. Knapp, Eds., New York: John Wiley, 1974, pp. 359–376.

Nunamaker, J. F., Jr., Nylin, W., and Konsynski, B., "Processing Systems Optimization Through Automatic Design and Reorganization of Program Modules," *Information Systems*, Julius Tou, Ed., New York: Plenum Publishing, 1974, pp. 311–336.

Nunamaker, J. F., Jr., Ho, T., Konsynski, B., and Singer, C., SODA: Systems Optimization and Design Algorithm—An Aid in the Selection of Computer Systems and for the Structure of Computer Program Modules and Database," *Information Systems and Organizational Structure*, E. Grochla and N. Szyperski, eds., Berlin and New York: Walter de

Gruyter Publishing, 1975, pp. 127–150.

Nunamaker, J. F., Jr., Pomeranz, J., and A. Whinston, "Automatic Interfacing of Application Software in the GPLAN Framework," *Information Systems and Organizational Structure*, E. Grochla and N. Szyperski, eds., Berlin and New York: Walter de Gruyter Publishing, 1975, pp. 382–396.

Nunamaker, J. F., Jr., and Konsynski, B., "From Problem Statement to Automatic Code Generation," *Systemeering 75*, Studentlitteratur, Lund, Sweden, 1975, pp. 215–240.

Nunamaker, J. F., Jr., Ho, T., Konsynski, B., and Singer, C., "Computer-Aided Analysis and Design of Information Systems," *CACM*, December 1976.

Parnas, D. L., "On the Criteria to Be Used in Decomposing Systems into Modules," *CACM*, Vol. 15, No. 12, December 1972, pp. 1053–1058.

Prywes, N., "Automatic Generation of Software Systems—A Survey," Department of Computer and Information Sciences, University of Pennsylvania, 1974.

Prywes, N., "Model II—Automatic Program Generator User Manual," Department of Computer and Information Sciences, University of Pennsylvania, 1977.

Ramamoorthy, C. V., and So, H. H., "Software Requirements and Specifications: Status and Perspectives, *IEEE*, 1978.

Ramirez, J., "Automatic Generation of Data Conversion Programs Using a Data Description Language," Ph.D. dissertation, University of Pennsylvania, 1973.

Riddle, W., and Fairley, R., "Review of the Pingree Park Software Development Tools Workshop," *Proceedings of Computers in Aerospace Conference II*, Los Angeles, October 1979.

Rose, C. W., "LOGOS and the Software Engineer," *AFIPS Conference Proceedings*, Vol. 41, Part I, 1972, pp. 311–323.

Ross, D., "Structured Analysis (SA): A Language for Communicating Ideas," *IEEE Trans. on Software Engineering*, Vol. SE-3, No. 1, January 1977.

Ruth, G., "Automatic Design of Data Processing Systems," ACM Principles of Programming Languages, January 1976, pp. 50–57.

Ruth, G., "Protosystem I—An Automatic Programming System Prototype," AFIPS National Computer Conference, 1978, pp. 675–681.

Scott, James H., "The Management Science Opportunity: A Systems Development Management Viewpoint," *MIS Quarterly*, Vol. 2, No. 4 (December 1978), pp. 59–61.

Sprague, Ralph H., "A Framework for Research on Decision Support Systems," in *Decision Support Systems: Issues and Challenges*, Flick, G. and R. H. Sprague, eds., Oxford, Eng.: Pergamon Press, (1981).

Sutton, Jimmy A., "Evaluation of a Decision Support System: A Case Study with the Office Products Division of IBM, San Jose, Calif.: *IBM Research Report FJ2214*.

Teichroew, D., and Sayani, H., "Automation of System Building," *Datamation*, August 15, 1971, pp. 25–30.

Teichroew, D., "A Survey of Languages for Stating Requirements for Computer-Based Information Systems," *Afips Conference Proceedings*, Vol. 41, Part II, Fall 1972, pp. 1203–1224.

Teichroew, F., "Problem Statement Analysis: Requirements for the Problem Statement Analyser (PSA)," *Systems Analysis Techniques*, J. D. Couger and R. W. Knapp, eds., New York: John Wiley, 1974.

Thall, R., "A Manual for PSA/ADS: A Machine-Aided Approach to Analysis of ADS," ISDOS Working Paper No. 35, Department of Industrial and Operations Engineering, University of Michigan, Ann Arbor, October 1970.

Trembly, J. P., *PSL: Problem Statement Language Productions*, Ph.D. dissertation, Case Western Reserve University, 1969.

Yourdon, E. and Constantine, L., *Structured De-*

sign, New York: Yourdon Inc., 1976.

Young, J. W., and Kent, H. K., "Abstract Formulation of Data Processing Problems," Journal of Industrial Engineering, November/December 1958, p. 479.

Zelkowitz, M. V., Shaw, A. C., and Gannon, J. D., *Principles of Software Engineering and Design*, Englewood Cliffs, N.J.: Prentice-Hall, 1979.

Zilles, S. N., "Data Algebra: A Specification Technique for Data Structures," Ph.D. dissertation, Project MAC, MIT, 1975.

READING QUESTIONS

1. What is PLEXSYS?
2. What prior development techniques were incorporated into PLEXSYS?
3. Provide a one-sentence description of the four phases (steps) in the use of PLEXSYS.
4. Expand Figure 3 to show the detailed functions of PLEXSYS. Provide a separate sheet, annotating the activity performed in each block.
5. Combine Figure 4, 5, 10, and 13 to better show the interactions of tools in the four phases (steps).
6. Compare Figures 8 and 9 with Figure 7 in the Teichroew, Hershey, Yamamoto paper. How do the problem statement languages differ?
7. Revise Figure 8 to reflect the following changes:
   (a) Back order = Number ordered − (number on hand plus number in production)
   (b) Generate report on number in production
8. Revise Figure 9 to reflect the following changes:
   (a) Input: Number in production
   (b) Output: Number in production
9. Compare the problem statement language of PLEXSYS with its five predecessor languages described in Section III, "Third Generation Development Techniques for Computer-Based Systems." Develop a table to facilitate this comparison, comparing functions and commands.
10. How does PLEXSYS differ from PSA/PSL? Prepare a table to facilitate this comparison.
11. How does PLEXSYS differ from SREM? Prepare a table to facilitate this comparison.

SECTION 5

FIFTH GENERATION DEVELOPMENT TECHNIQUES FOR COMPUTER-BASED SYSTEMS

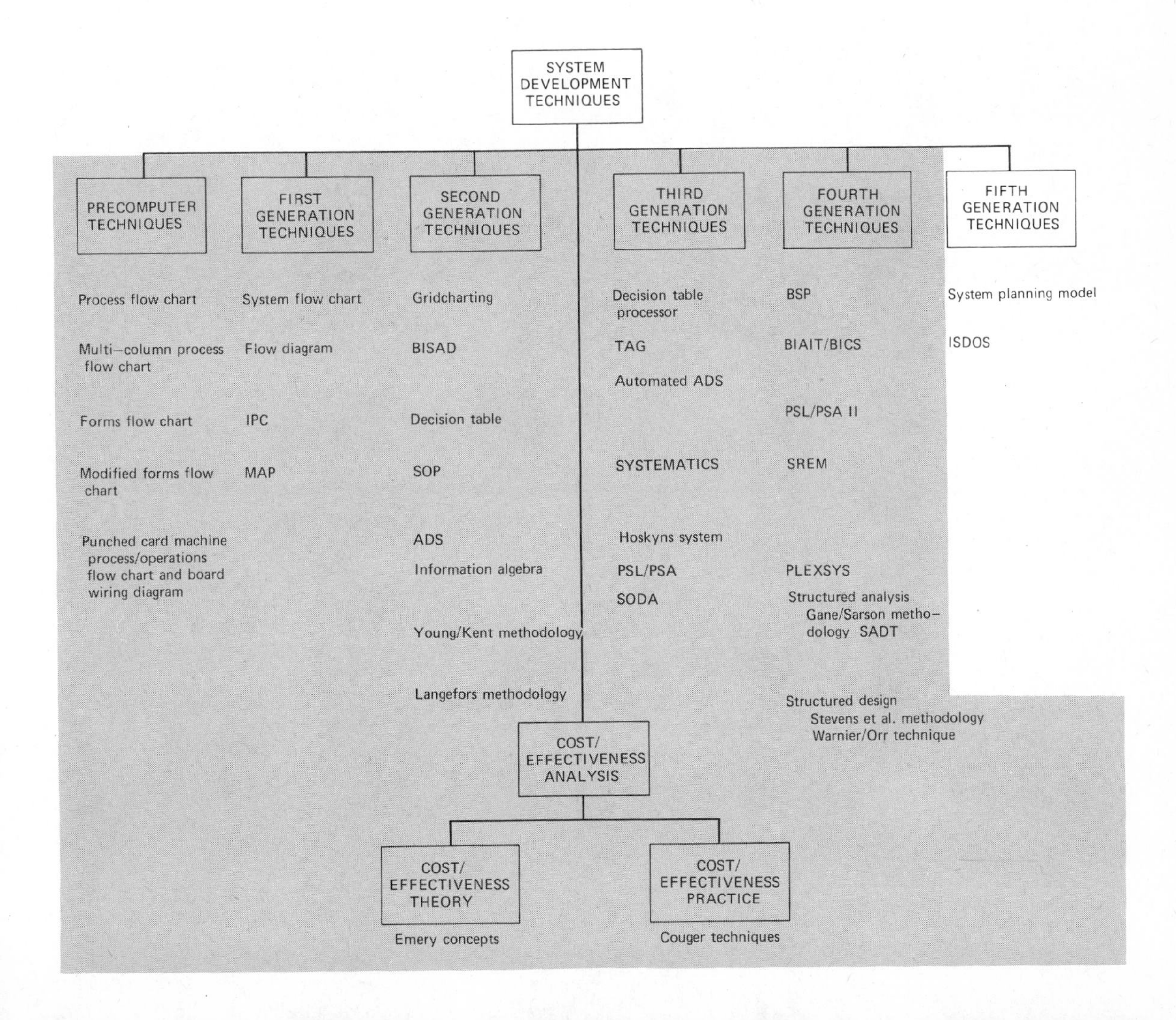

SECTION 5

FIFTH GENERATION DEVELOPMENT TECHNIQUES FOR COMPUTER-BASED SYSTEMS

J. Daniel Couger

OVERVIEW

In *System Analysis Techniques,* I projected for the fourth generation a linked set of techniques to be available to automate all phases of the system development cycle. That projection was overly optimistic.

The linked set of techniques is now projected for the fifth generation. Tools are already available for automating each phase of the development cycle. They have not yet been integrated into a uniform and comprehensive system. Just as companies have been slow to merge their computer applications into an integrated set of data bases, system developers have been slow in merging their techniques into a system that integrates analysis, design, programming, and testing.

The reasons for this delay are threefold:

1. Computer vendor personnel have been preoccupied with developing tools that are hardware/software specific.
2. Practitioner personnel have been preoccupied with developing tools that are application specific.
3. Academic personnel have been preoccupied with narrow scope research that is technique specific.

However, there is one multidiscipline research activity underway, the ISDOS project, coordinated by Professor Daniel Teichroew at the University of Michigan. ISDOS is an acronym for Information System Design and Optimization System.

The Teichroew-Sayani paper on ISDOS is reprinted in this section. However, a brief summary is needed to provide a frame of reference for comments on fifth generation techniques.

The ISDOS project is formalizing the design process along the lines of the mathematical approaches pioneered by Langefors, Cross, Turnburke, and Martin. Use of this multilevel approach, where the decision variables at one level become constraints at the next level, makes feasible evaluation of a large variety of design strategies.

ISDOS consists of four primary modules. The Data Reorganizer accepts: (1) specifications for the desired storage structures from the physical systems design process, (2) definition of data as summarized by the Problem Statement Analyzer, (3) the specifications of the hardware to be used, and (4) the data as it currently exists and its storage structure.

It then stores the data on the selected devices in the form specified. The third module, the Code Generator, accepts specifications from the physical design process and organizes the problem statements into programs recognizing the data interface as specified by the Data Reorganizer. The code produced may be either machine code, statements in a higher-level language (e.g., COBOL), or parameters to a software package. These two modules perform, automatically, the functioning of programming and file construction.

The final module of the ISDOS system is the Systems Director. It accepts the code generated, the timing specifications as determined by the physical design algorithm, and the specifications from the Data Reorganizer, and produces the target information system. This system then accepts inputs from the environment and produces the necessary outputs according to the requirements expressed in the problem statement.

The ISDOS concept meets fifth generation objectives. Whether the University of Michigan project will be the first to implement the concept is uncertain. There is little need for a crash effort, however. Based on past experience, it will be another decade before the majority of firms implement fourth generation system development techniques.

Figure 1 shows the evolution of techniques for all seven phases of the system life cycle. The number in the circle above each technique provides an index to references explaining that technique. For example, technique 1, processed flow chart, has four explanatory references. Appendix 1 lists the technique number and references. The references are cited in Appendix II, starting on page 433.

Reference to fifth generation techniques in Figure 1 reveals one that has not yet been discussed. Although a number of techniques such as SOP, BSP, BIAIT/BICS have been developed to facilitate Phase I of the system development, none are automated. Also, these techniques are static rather than dynamic. Their procedure for use does not include a mechanism to facilitate periodic updating as company objectives and resources change. To remain viable, they need to be linked to the corporate planning model, as shown in Figure 2. By redesigning Phase I techniques to produce a computerized systems/planning model, the linkage to the computerized corporate planning model would be facilitated.

Also, a better linkage to Phase II (logical design) is needed. Merging Phase I techniques into a system planning model would also facilitate this linkage. Figure 1 does not well connote one characteristic of the system development process—its iterative nature. The heading, system life cycle, properly describes the process. Systems are continually revised. The revision cycle should be initiated by the changes in corporate plans, which then are translated into revised system objectives. The next step is revised system requirements. Keeping system changes consistent and current with corporate objectives requires linkage of the corporate planning model, the system planning model, and the ISDOS-type model.

Figure 2 depicts the concept. Need for changes is generated by the corporate planning model, in a phase labeled Phase I minus one. Those requirements are translated to systems requirements for specific applications by the system planning model in Phase I. Phases II through IV are telescoped into one phase which automatically converts requirements to system specifications, revises the system design, and then generates revised program code.

Let us hope that in the near future as much emphasis is placed on front end activities as is now being placed on the subsequent activities of the development process.

CONCLUSION

Although ISDOS was not implemented in the fourth generation era as previously predicted, the concept is still viable. It provides for an integrated system development methodology. However, it starts with Phase II of the development cycle. A computerized system planning model is needed to link Phase I activities with the corporate planning model (at the front end) and with ISDOS (at the back end).

An impressive array of advanced system development techniques is now available for system designers. Merging those techniques into a *system* of techniques is the necessary next step in the evolution of development methodology. Such an approach will enable system designers to integrate efficiently and effectively complex company data bases and data communication systems.

APPENDIX 1

Index of System Development Techniques

1. Process flowchart: 39, 48, 53, 54
2. Geometry: 41
3. Linear algebra: 41, 54, 80
4. Probability theory: 39, 53, 54, 80
5. Truth tables: 41, 54
6. Gantt scheduling charts: 48, 53, 54
7. Accounting theory: 25, 54
8. Organization charts: 9, 48, 54
9. Simultaneous motion charts: 39, 48, 53
10. Time study (stop watch): 9, 39, 48, 53, 54
11. Punch card machine process chart: 22, 48
12. Punch card operations flowchart: 22, 48
13. Board wiring diagram: 22, 48
14. Information process chart: 28
15. MAP system charting technique: 51
16. System flowchart and flow diagram: 9, 12, 22, 31, 48
17. Point set theory: 41, 54
18. Linear programming (simplex method): 48, 53, 54, 80
19. Applied statistical analysis: 9, 39, 53, 54, 80
20. Precedence network technique: 31, 53, 80
21. Budgeting techniques: 39, 53, 54
22. Linear responsibility charts: 48
23. Work simplification: 48, 53
24. Synthetic time standards: 48, 53, 54
25. Graph Theory: 30

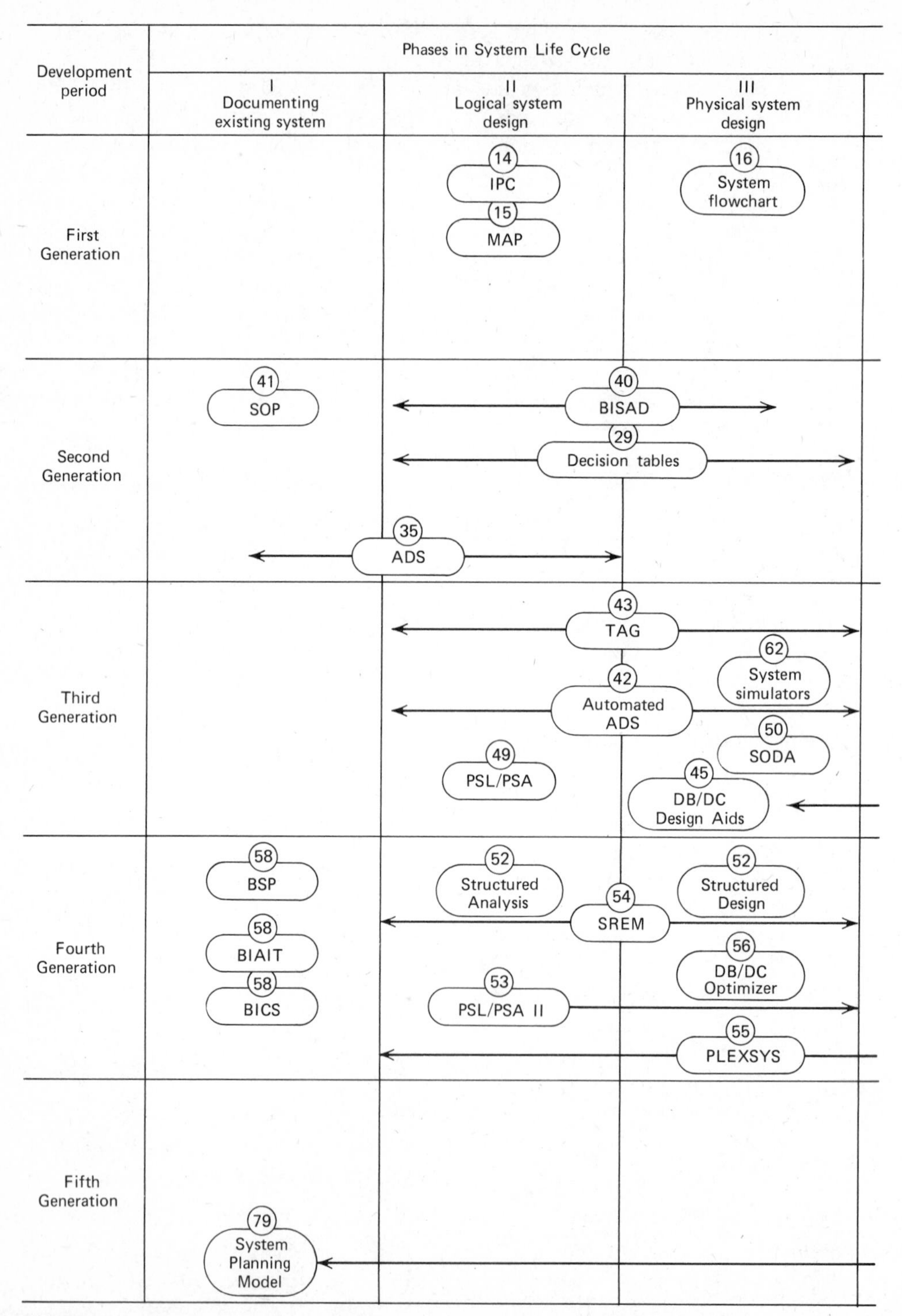

FIGURE 1 Evolution toward an integrated set of system development techniques.

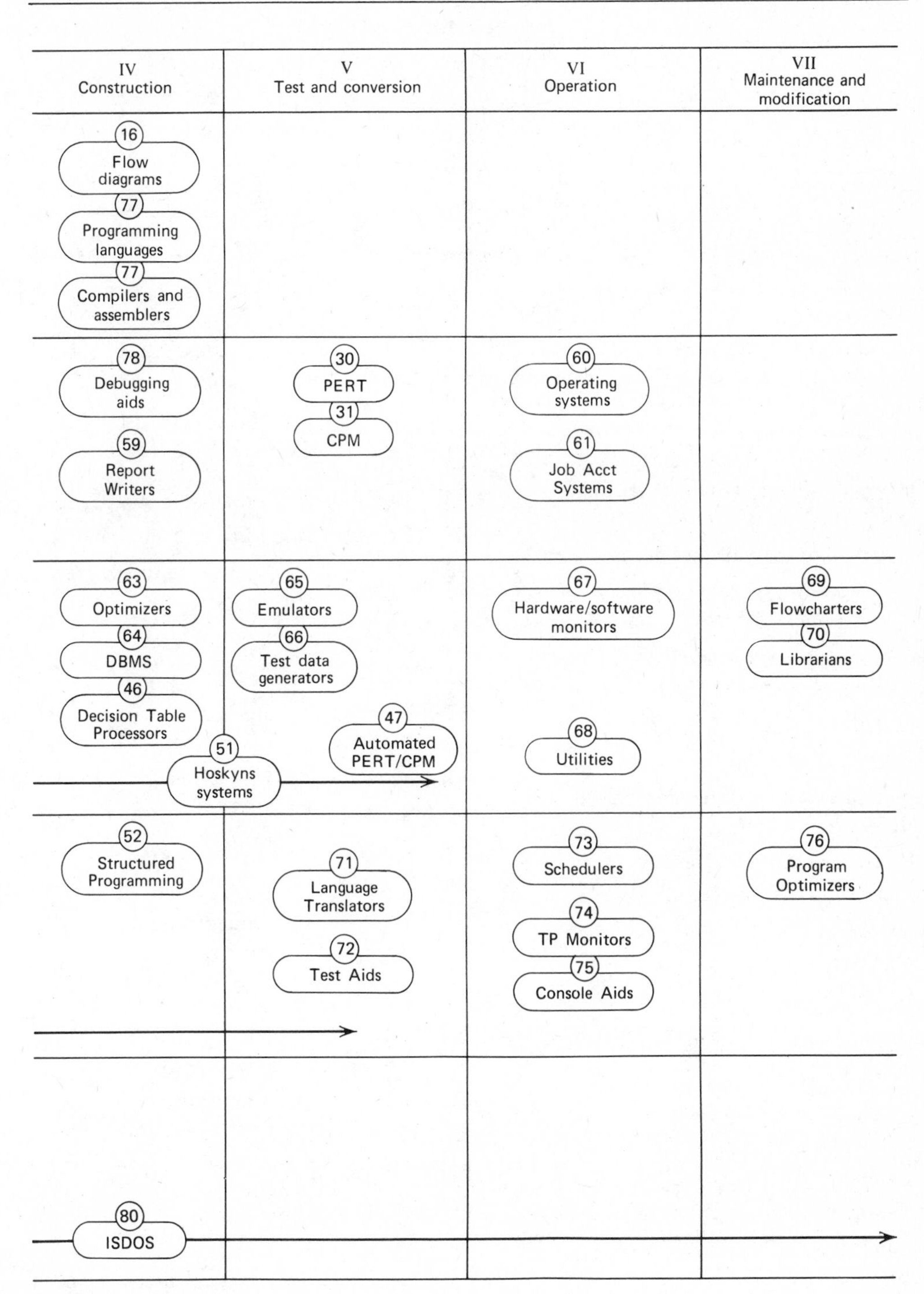

FIGURE 1 Continued.

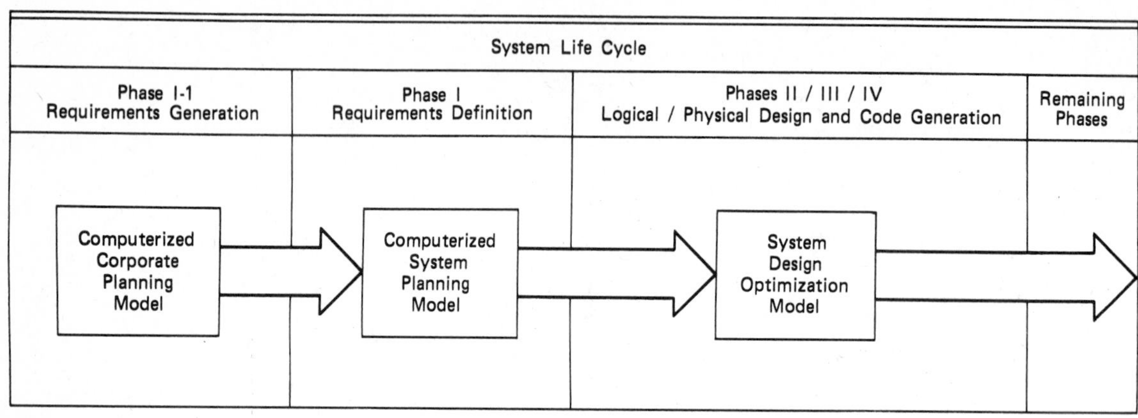

FIGURE 2 Revised system life cycle for fifth generation of development.

26. Mathematical programming:
    linear, nonlinear, dynamic methods:
    53, 80
27. Inventory model, waiting line models: 53, 54, 80
28. Computer simulation techniques: 53, 54, 80
29. Decision tables technique: 31, 56, 60
30. PERT network technique: 9, 31, 48, 54, 80
31. CPM network technique: 31, 48, 53, 54, 80
32. Clerical work sampling techniques: 48, 53
33. Clerical time standards techniques: 48, 53, 54
34. SOP activity analysis technique: 27, 70
35. ADS accurately defined systems: 69
36. Langefors methodology: 46
37. Young/Kent methodology: 83
38. Information albegra: 16
39. Large system optimization models: 47, 80
40. BISAD: 33a
41. SOP operation analysis and resource technique: 27, 70
42. Automated ADS: 29, 75
43. TAG time automated grid technique: 42, 77
44. DB specification languages: 6, 15, 13, 18, 17, 33b, 38
45. DB design tools: 3, 26, 34
46. Decision table processors: 10, 55, 60
47. Computerized PERT/CPM: 4, 57
48. Computerized planning models: 61
49. PSL/PSA: 71, 73
50. SODA: 58
51. Code generators: 62
52. Structured analysis and design: 7, 24, 59, 61, 64, 68
53. PSA/PSL II: 72
54. SREM: 1, 2, 5
55. PLEXSYS: 44a
56. Data base optimizer: 23, 82
57. Network modeling techniques: 8, 44b
58. BSP, BIAIT/BICS: 36, 11, 43
59. Report writers: 37
60. Operating systems: 21, 61, 63, 67
61. Job accounting systems: 37, 61
62. System simulators: 40, 67
63. Optimizers: 10, 37
64. DBMS: 17, 67
65. Emulators: 66
66. Test data generators: 10
67. Hardware/software monitors: 37, 67

68. Utilities: 37, 45, 61
69. Flowcharters: 10, 37, 67
70. Librarians: 37, 67
71. Language translators: 37, 81
72. Test aids: 32, 37
73. Schedulers: 37, 61
74. TP monitors: 37, 61
75. Console aids: 37
76. Program optimizers: 37
77. Programming languages, compilers/assemblers: 21, 63, 66, 67
78. Debugging aids: 10, 65, 66, 67
79. System planning model
80. ISDOS: 74

APPENDIX II

Bibliography

1. Alford, Mack W., "A Requirements Engineering Methodology for Real-Time Processing Requirements," TRW, Redondo Beach, CA, September 1976, pp. 1–28.
2. Alford, Mack W., "Software Requirements Engineering Methodology (SREM) at the Age of Two," TRW, Redondo Beach, CA, March 1978, pp. 1–48.
3. Aschim, F. and Braten, K., "THE SYSDOC Method and the SYSDOC Analyzer," Central Institute for Industrial Research, 1977.
4. Barnetson, P., *Critical Path Planning*, Princeton, NJ: Auerbach, 1970
5. Bell, Thomas E., Bixler, David D., and Dyer, Margaret E., "An Extendable Approach to Computer-Aided Software Requirements Engineering," TRW, Redondo Beach, CA, July 1976, pp. 1–48.
6. Berild, S., Bubenko, J. et al., "From Information Requirements to DBTG Data Structures," *Proceedings of the ACM SIGMOD/SIGPLAN Conference of Data: Abstraction, Definition, and Structure*, 1976, pp. 73–85.
7. Boehm, Barry, "Software Engineering," *IEEE Transactions on Computer*, December 1976, pp. 225–240.
8. Bracker, W. E. and Konsynski, B. R., "An Overview of a Network Design System," *Proceedings, AFIPS Conference*, 1980, pp. 41–48.
9. *Business Systems*, Cleveland, OH: Association for Systems Management, 1970.
10. Canning, R. D., "COBOL Aid Packages," *EDP Analyzer*, May 1972.
11. Carlson, Walter M., "Business Information Analysis and Integration Technique (BIAIT)—The New Horizon," *DATA BAS* , Vol. 10, No. 4, Spring, 1979, pp. 3–19.
12. Chapin, N., *Flowcharts*, Princeton, NJ: Auerbach, 1971.
13. Chen, P. (ed.), *The Entity-Relationship to Systems Analysis and Design*, North-Holland Publishing Company, 1980.
14. CODASYL, *Data Design Language*, Data Description Language Committee, National Bureau of Standards, No. 113, 1973.
15. CODASYL, "Data Description Language Report," *Journal of Development*, 1978.
16. CODASYL Development Committee, "An Information Algebra," *Communications of the ACM*, Vol. 5, No. 4, April 1962, pp. 190–204.
17. CODASYL, *Feature Analysis of Generalized Database Management Systems*, New York, NY: Association for Computing Machinery, 1971.
18. CODASYL, *Task Group Report*, Data Base Task Group, ACM, April 1971.
19. Codd, E. F., "Recent Investigations into Relational Data Base Systems," *Proceedings, IFIP Congress*, 1974.

20. Date, C. J., "Relational Data Base Systems—A Tutorial" in Ton, J., *Information Systems Science*, New York: Plenum Press, 1974, pp. 37–54.

21. Donovan, John J., *Systems Programming*. New York, NY: McGraw-Hill, 1972.

22. Feingold, C., *Fundamentals of Punched Card Data Processing*, Dubuque, IA: Wm. C. Brown Co., 1969.

23. Fry, J. P. and Teprey, T. J., *Data Base Design and Analysis*, Englewood Cliffs, NJ: Prentice-Hall, 1981.

24. Gane, Chris and Sarson, Trish, "Structured Methodology: What Have We Learned?" *Computer World/Extra*, Vol. XIV, No. 38, September 17, 1980, pp. 52–57.

25. Garner, Paul and Berg, K. B., *Readings in Accounting Theory*, Boston, MA: Houghton Mifflin, 1966.

26. Gerritson, R., "A Preliminary System for the Design of DBTG Data Structures," *Communications of the ACM*, October 1975, pp. 557–567.

27. Glans, T. et al., *Management Systems*, New York,: Holt, Rinehart and Winston, 1968.

28. Grad, B. and Canning, R., "Information Process Analysis," *Journal of Industrial Engineering*, November–December, 1969, pp. 470–474.

29. Gross, M. H., "Systems Generation Output Decomposition Method, "Standard Oil Company of New Jersey, July 1963.

30. Harary, F., *Graph Theory*, Reading, MA: Addison-Wesley, 1969.

31. Hartman, W. et al., *Management Information Systems Handbook*, New York,: McGraw-Hill, 1968.

32. Hetzel, William C. (ed.), *Program Test Methods*, Englewood Cliffs, NJ: Prentice-Hall, 1973.

33a. Honeywell, *Business Information Systems Analysis and Design*, 1968, Waltham, MA.

33b. Housel, Waddle and Yao, B., "The Functional Dependency Model for Logical Database Design," *Proceedings, 5th Annual Conference on Very Large Data Bases*, October 1979.

34. Hubbard, G. and Rover, N., "Automated Logical Data Base Design," *IBM Systems Journal*, 1977.

35. Huffer, J. A. and Severance, D. G., "The Use of Cluster Analysis in Physical Data Base Design," *Proceedings, 1st International Conference on Very Large Data Bases*, ACM, 1975, pp. 69–86.

36. IBM, *Business Systems Planning* (GE 20-0527-1), 1975, pp. 1–92.

37. *ICP Directory: Data Processing Management*, Indianapolis, IN, 1981.

38. Irani, K., Purkayastha, S. and Teorey, T., "A Designer for DBMS-Processable Logical Database Structures," *Proceedings, 5th Annual Conference on Very Large Data Bases*, October 1979, pp. 32–51.

39. Ireson, W. and Grant, E. (eds.), *Handbook of Industrial Engineering and Management*, Englewood Cliffs, NJ: Prentice-Hall, 1955.

40. Joslin, Edward, *Computer Selection*, Reading, MA: Addison-Wesley, 1968.

41. Kattsoff, L.O. and Simone, A. J., *Foundations of Contemporary Mathematics*, New York,: McGraw-Hill, 1967.

42. Kelley, Joseph F., *Computerized Management Information Systems*, New York,: Macmillan, 1970, pp. 364–400.

43. Kerner, David V., "Business Information Characterization Study," *DATA BASE*, Vol. 10, No. 4, Spring 1979, pp. 10–17.

44a. Konsynski, Benn R. and Nunamaker, Jay, "A System Development System," unpublished paper, University of Arizona, July 1981.

44b. Konsynski, Benn R. and Bracker, W. E., "Software Packages for Solving Network Puzzles," *Data Communications*, July 1980.

45. Lane, Ronn, *An Introduction to Utilities,* New York,: Petrocelli/Charter, 1975.

46. Langefors, B., *Theoretical Analysis of Information Systems,* 2 Vol., Lund, Sweden: Student Litteratur, 1966.

47. Lasdon, L. S., *Optimization Theory for Large Systems,* New York,: Macmillan, 1970.

48. Lazzaro, V. (ed.), *Systems and Procedures: A Handbook for Business and Industry,* 2nd Edition, Englewood Cliffs, NJ: Prentice-Hall, 1968.

49. Lum, V. et al., "1978 New Orleans Data Base Design Workshop Report," *Proceedings, 5th International Conference on Very Large Data Bases,* ACM, October 1979, pp. 328–339.

50. Lundeberg, Mats, "IA/1—An Interactive System for Computer-Aided Information Analysis," Working Report No. 14E, ISAC, University of Stockholm, 1972.

51. *MAP-System Charting Technique.* National Cash Register Company, Dayton, OH, 1961.

52. Martin, J., *Design of Real Time Computer Systems,* Englewood Cliffs, NJ: Prentice-Hall, 1967.

53. Maynard, H. B. (ed.), *Handbook of Business Administration,* New York: McGraw-Hill, 1967.

54. Maynard, H. B. (ed.), *Industrial Engineering Handbook,* 3rd Edition, New York,: McGraw-Hill, 1971.

55. McDaniel H., *Decision Table Software,* Princeton, NJ: Auerbach, 1970.

56. McDaniel H., *An Introduction to Decision Logic Tables,* New York,: John Wiley and Sons, 1968.

57. Miller, R. W., *Schedule, Cost and Profit Control with PERT,* New York: McGraw-Hill, 1963.

58. Nunamaker, J. F., Jr., "A Methodology for the Design and Optimization of Information Processing Systems," *Proceedings—Spring Joint Computer Conference,* Montvale, NJ: AFIPS Press, 1971, pp. 283–294.

59. Orr, Kenneth T., *Structured Systems Development,* New York,: Yourdon Press, 1977.

60. Pollack, S., Hicks, H. and Harrison, W. J., *Decision Tables Theory and Practice,* New York,: John Wiley and Sons, 1971.

61. Ralston, Anthony (ed.), *Encyclopedia of Computer Science,* New York: Petrocelli/Charter, 1976.

62. Rhodes, John, "A Step Beyond Programming," unpublished paper, copyright © J. Rhodes, September 1972.

63. Rosen, Saul, *Programming Systems and Languages.* New York,: McGraw-Hill, 1967.

64. Ross, Douglas T., "Structured Analysis (SA): A Language for Communicating Ideas," *IEEE Transctions on Software Engineering,* Vol. SE-3, No. 1, January 1977, pp. 16–34.

65. Rustin, Randall (ed.), *Debugging Techniques in Large Systems.* Englewood Cliffs, NJ: Prentice-Hall, 1971.

66. Sammet, Jean E., *Programming Languages: History and Fundamentals.* Englewood Cliffs, NJ: Prentice-Hall, Inc., 1969.

67. *Software Reports.* Philadelphia, PA: Auerbach Computer Technology Reports.

68. Stevens, W. P., Myers, G. J., and Constantine, L. L., "Structured Design," *IBM Systems Journal,* No. 2, 1974, pp. 115–139.

69. *A Study Guide for Accurately Defined Systems.* National Cash Register Company, Dayton, OH, 1968.

70. *Study Organization Plan.* International Business Machines Corporation (Form No. C20-8075), White Plains, NY, 1961.

71. Teichroew, Daniel, "Problem Statement Analysis: Requirement for the Problem Statement Analyzer (PSA), ISDOS Work-

ing Paper No. 43, Dept. of Industrial Engineering, University of Michigan, Ann Arbor, MI, 1971, pp. 20–53.

72. Teichroew, Daniel and Hershey, Ernest A., III, "PSL/PSA: A Computer-Aided Technique for Structured Documentation and Analysis of Information Processing Systems," *IEEE Transactions on Software Engineering,* Vol. SE-3, No. 1, January 1977, pp. 41–48.

73. Teichroew, Daniel, Hershey, Ernest A., III, and Yamamoto, Y., "The PSL/PSA Approach to Computer-Aided Analysis and Documentation," © Daniel Teichroew, 1979.

74. Teichroew, Daniel and Sayani, H., "Automation of System Building," *Datamation,* August 15, 1971, pp. 25–30.

75. Thall, R. M., "A Manual for PSA/ADS: A Machine-Aided Approach to Analysis of ADS," ISDOS Working Paper No. 35, Ann Arbor, MI, Department of Industrial Engineering, University of Michigan, 1971.

76. Thall, R. M., "A Manual for PSA/ADS: A Machine-Aided Approach to Analysis of ADS," *ISDOS Working Paper No. 35.* Ann Arbor, MI, Department of Industrial and Operations Engineering, University of Michigan, October 1970.

77. *Time Automated Grid System,* International Business Machines Corporation (Form No. GY 20-0358), 2nd Edition, White Plains, NY, 1971.

78. Turnburke, V. P., Jr., "Sequential Data Processing Design," *IBM Systems Journal,* March 1963.

79. Verrijn-Stuart, A. A., "Information Algebras and Their Uses," *Management Datamatics,* Vol. 4, No. 5, 1975, pp. 187–197.

80. Wagner, H., *Principles of Operations Research.* Englewood Cliffs, NJ: Prentice-Hall, 1969.

81. Weingarten, Frederick, *Translation of Computer Languages,* San Francisco, CA: Holden-Day, Inc., 1973.

82. Yao, S. B., Navathe, S. B. and Weldon, J. L., "An Integrated Approach to Logical Data Base Design," *NYU Symposium on Data Base Design,* May 1978, pp. 1–14.

83. Young, J. W. and Kent, H. K., "Abstract Formulation of Data Processing Problems," *Journal of Industrial Engineering,* November/December 1958, pp. 471–479.

Automation of System Building

Daniel Teichroew
Hasan Sayani

The building of computer-based information systems to serve the management and operation of organizations has become a large and visible activity. Furthermore, one need only note some of the recent news items regarding the system building process to be convinced that it does not always lead to satisfactory results.[1] In the past, the emphasis on improvement of techniques has been on methods to help the programmer. Programming is certainly an essential step in the process, but it is only one of the steps. The attention and emphasis it has received are completely out of proportion to its role in the building of systems, and this has resulted in insufficient attention paid to the improvement of other steps. It is becoming more generally recognized that the other steps in the system building process must also be improved.

What we call an "information system" consists of two subsystems: a management system and an information processing system. The management system consists of the organization, its objectives, the individuals or groups in it, and the rules and procedures under which they work. The information processing system is the subsystem which consists of hardware, programs, noncomputerized procedures, etc., that accomplish the storage, processing, and communication of information necessary for the functioning of the management system. An essential element in this view is that the information processing system (IPS) must, or at least should, be designed to serve the management system. This characterization of information systems is particularly relevant to management information systems.[2]

Source: D. Teichroew, and H. Sayani, "Automation of System Building," *Datamation*, August 15, 1971, pp. 25–30.

[1]See for example: "Burroughs Sued by Trans World for $70 Million," *Datamation*, Dec. 1, 1970, p. 47; "Bell's BIS: Bottomless Well," *Datamation*, July 15, 1970, p. 35; "Chrysler's Private Hard Times," Fortune, April 1970; "Computer Classic," The Economist, Oct. 24, 1970, p. 94.

[2]Emphasis in this paper is on the use of the information processing system which serves management because this is the most important and the most difficult type of system to construct today. The techniques discussed, however, are applicable with only minor changes to other types: routine business data processing, command and control, information storage and retrieval, message switching, and process control.

SYSTEM BUILDING

Organizations normally go through a number of phases in building information systems. Initially, the potential use of the computer is treated as a one-time task for a few programmers. Soon it becomes obvious that the task is much bigger than first suspected, and during the second phase more manpower is assigned to the project. The third phase begins when it is recognized that a series of systems will have to be built and that a procedure will have to be developed. This results in the establishment of a systems department. A fourth stage is reached when it becomes apparent that the systems being built have many features in common and considerable reduction in effort might result from using standard building blocks.

Most medium and large organizations have reached at least the third phase; they recognize that information processing system building will be a continuous activity and have established systems departments. System building therefore usually involves three groups; top management, users, and system builders. The users, in general, are the functional divisions such as manufacturing, finance, personnel, etc.; however, a system may frequently be designed to serve more than one function. The system builders are centralized in a systems department. Top management is involved because it must define the responsibility of the user and the builders and adjudicate differences. It must also allocate resources and assign priorities where resources are scarce.

Frequently, one of the activities that the system departments undertake early in their existence is the development of a set of procedures and standards to be followed in the building process. The number of papers and books describing such procedures and standards has increased very rapidly in the last few years. Most organizations, however, prefer to develop their own. Review of the published methods and the manuals developed by organizations for their own use indicates that the procedures are basically similar though they may differ in details. Here a brief outline of the major steps will be given in order to illustrate the need for, and potential scope of, automation.

The first step in the process is a request which indicates the need for a new system or the modification of an existing one. This request, ideally, is originated by the user, though it may come from the systems department if many users will be served. The request should contain sufficient information to initiate the next step.

Feasibility, or impact analysis, is the second major step. It consists of estimating the potential benefits of a system to satisfy the expressed request. A proposed system must be developed in sufficient detail to estimate the costs. The impact on the organization and the existing systems is evaluated since this may affect both benefits and costs.

While the analysis in the second step may frequently be extensive and time-consuming, neither the statement of user requirements nor the description of the proposed system is in sufficient detail to proceed with the construction of the proposed system. The third step is to determine the user needs in full detail and to describe them in a form which the users can agree to, and which is also suitable for the design and construction phases that follow. In some cases extensive analysis is required to verify that the detail requirements, as stated, do in fact satisfy the more general needs stated in the first phase. It may in fact be necessary to simulate the management system. This phase is sometimes referred to as the "functional specifications" or "logical system design" phase.

The fourth phase—the physical systems design—is concerned with developing the specifications for the proposed computer-based information processing system that will accomplish the logical requirements detailed in the previous phase. Ideally, this is an elaboration of the proposed system used in the feasibility phase to estimate potential costs; if not, the feasibility results may have to be modified. The design of the proposed system consists of selecting, within whatever constraints that may exist, the processing organization (real-time, batch, etc.) and hardware, and designing the programs and data base. The result of this phase is a set of specifications.

These specifications are used in the system construction phase to build the actual target system. The new hardware requirements, if any, go to the hardware acquisition group for procurement. The program specifications go to the programming department which writes and tests the various programs. The data base specifications go to the group which has the responsibility for constructing the data base that will be needed. Normally, all these activities are the responsibility of the systems department. Other specifications may go to the personnel department for training and educational requirements.

In the sixth phase, all the components, already tested individually, are brought together and tested as a system. Errors discovered in this phase must be corrected; this may require going back several phases.

During the seventh phase, the system is in operation. Since requirements may change or errors may be discovered, a change control procedure must be established to ensure that changes are appropriately recorded. The performance of the system must be compared with the estimates made in the feasibility phase.

The amount of attention paid to the steps in the process depend, of course, on the size of the system. If the system can be built by one person in a short period of time, he can usually build it satisfactorily without explicitly following the procedure. As the number of individuals involved increases, the formal procedure and complete documentation become essential. Unfortunately, documentation is usually neglected because the system building process is essentially manual. Formal techniques are not widely used; the steps are carried out by individuals using pencil and paper, and the documentation consists of descriptions in English supplemented by flowcharts, tables, etc. Formal, computer-aided techniques are used only in that part of the construction process in which higher level statements are compiled into object code.

Another consequence of the present system building methods is that the process from the user request to successful operation takes a long time. The elapsed time is a function of the size of system, but several years is not unusual. The elapsed time can be reduced by carrying out some subphases in parallel; but this must be planned very carefully or it may result in inconsistencies and require more time rather than less.

A major problem in system building occurs right at the beginning in determining what the user wants. In fact, it is not even clear that this is the right way to state the objective. The user may not be able to articulate what he wants and usually is not the appropriate person to decide what he should have. The situation is further aggravated by the fact that the user usually is not accustomed to describing what he wants in sufficient detail to translate it into computer programs, a point that has been very well stated by Vaughn.[3]

There is no doubt that, ideally, it is very important to start with the "correct" requirements. Building a system to accommodate wrong requirements is a waste of time and effort. However, it is our view that in the present circumstances it is more important to develop methods to reduce the time to build systems *once the requirements are given*. The major reason for this is that it does not do much good to produce the "absolutely perfect" set of requirements if it will then take a long time (six months to several years) to produce the system that will accomplish the requirements. The absolutely perfect requirements are not constant; they change as the environment in which the organization exists changes. The organization itself changes, and the individual users change or learn to use the outputs from the computer-based system. We have therefore adopted as our basic objective the need to reduce the length of time from the point where requirements are first stated until the target system to accomplish the requirements is in opera-

[3]Vaughn, P. H., "Can COBOL Cope?" *Datamation*, Sept. 1, 1970, pp. 42–46.

tion. Obviously the major tool to accomplish this reduction must be the computer itself.

Once it is decided to use the computer in the system development process, the next step is to decide where to begin recording data in machine processible form. Here we try to apply the first principle of automation: record the input data in machine-readable form as close to the source as possible and thereafter process it with as little human intervention as possible. For the reasons mentioned above, we have decided to start at the point at which the requirements of the management system have been determined and the specification of individual inputs and outputs can begin. In the future we hope to extend our techniques to aid the process of determining what the requirements of the management system *should* be. There is no reason why any piece of data about requirements should not be recorded in machine-readable form the first time it appears in the system building process. (The proposed format for capturing the specification of requirements at this point and the software to process it are outlined later.)

Once it is decided to base the system building process on the use of the computer, there are other potential benefits than just the reduction in elapsed time. It should be possible to accommodate changes in requirements more easily both during the design process and during system operation. The computer can also be used as the basis for coordinating the activities of many analysts and to relieve them of many tedious and laborious clerical tasks which they now must do manually.

METHODS OF IMPROVEMENT OF SYSTEM BUILDING

The conclusion reached in the previous section is that the system building process itself should be automated, or at least computer-aided. Before describing our approach on how this might be done it is worthwhile examining some other ways to improve the process. The alternative methods may be grouped into four major categories: improve education and training of system builders, provide aids (computer based and others) for the system builders, use application packages, and use generalized software.

System building, as a profession, is still in its infancy and most practitioners were trained in other fields. In the early days, practice was relatively simple and required little more than programming. Now, however, the practice is becoming more professional and educational programs for the "information engineer" are being developed. However, it is extremely unlikely that it will be possible to build the number of systems of the size and complexity desired by manual methods; there will not be enough people. It will be necessary to use the trained professionals more effectively by moving from "handcrafted" systems to "mass-produced" systems.

Many aids designed to facilitate individual tasks in system building have been proposed. Probably the most generally used are the general-purpose programming languages. Less widely used are programs for other aspects of the process such as flowcharters and system simulators (SCERT, CASE, etc.). Space does not permit a detailed analysis of these aids here; however, our conclusion from such analysis is that these aids tend to be useful in only one particular (and usually narrow) aspect of the whole system development cycle. Manual intervention and manual preparation of input is required at each stage. What is needed instead is a coherent system that covers all phases of the life cycle in which the output of one phase is automatically an input to the next.

Application packages have been available since the early days of computers. Their use has been limited primarily because the user needs are continuously changing and attempts to provide flexibility usually result in high processing cost. There is a spectrum of methods to build application packages so that they can be tailored for a specific set of requirements ranging from applications in which the user has no alternatives, to ones in which he has complete freedom. In the most completely specified packages the user can only enter data values. This approach tends to be satisfactory only where the problem is relatively small and very well defined. Some packages allow more freedom through the use of parameter values as well as data values. Another level of generality is reached by providing for a number of options which the user specifies by filling out a questionnaire or by completing a form. This method has been used to generate simulation programs[4] and is the basis of generating software for the IBM System/3. An even more general approach to application packages is represented by user-oriented languages. These give the user a relatively flexible method of specifying his problem but require less effort than would be required to write a program in a general-purpose language. To cover all user needs would require many different languages and maintenance of the associated software. Some standardization clearly is desirable.

Generalized software started from input/output subroutines, and packages such as sort, merge, report generators, etc., are now in common use. Generalized file maintenance packages, however, have only fairly recently evolved into "data base management systems." These systems differ from application packages in that they "generalized" in terms of operations inside the computerized system rather than in terms of view of the user from the outside. Sorting, for example, is a processing operation that is not dependent on the particular application. Data base management systems will undoubtedly achieve a major role in the next few years. They are attractive, despite their high processing cost, because they relieve the programmer of the need to program frequently used operations such as access methods for complicated data structures and variable-length items, records, and files.

All of the methods of improving system building described above have been used and will continue to be used in the future. What we are concerned with is the next major plateau. There has been a progression in which general-purpose programming languages have replaced assembly languages and general-purpose languages themselves have had to be augmented by data base management systems to provide the framework for the programmer to communicate with the machine. In turn, the limitations of data base management systems will be overcome through automation of the system-building process. The effectiveness of trained professionals can be amplified and the computer-based aids to system building integrated into a software factory that can produce user programs tailored to user requirements.

AUTOMATION OF THE SYSTEM-BUILDING PROCESS

The need to automate the whole system-building process, as contrasted with the development of aids for parts of the process, has been recognized. For example, this is the expressed goal of the CODASYL Systems Committee.[5] So far, however, the committee has been primarily concerned with data base management systems. A computer-aided approach, the TAG (Time Automated Grid) System, has been developed by IBM. A number of other systems have been proposed.[6] Many concepts from these systems have been incorporated into ISDOS.

ISDOS (Information System Design and Optimization System) is the name of a software

[4] Ginsberg, A. S., H. M. Markowitz, and P. M. Oldfather, "Programming by Questionnaire," AFIPS Conference Proc. Vol. 30, 1967, SJCC, pp. 441–446 [CR 12764, 12149].

[5] The CODASYL Systems Committee states its objectives as: ". . . to strive to build up an expertise in, and to develop, advanced languages and techniques for data processing, with the aim of automating as much as possible of the process currently thought of as system analysis, design, and implementation."

[6] For a discussion and comparison, see Teichroew, D., "A Survey of Languages for Stating Requirements for Computer Based Information Systems," ISDOS Working Paper No. 42.

package being developed by faculty, students, and research associates in the Department of Industrial Engineering at the University of Michigan. It consists of a number of major components which are shown in Figure 1; this section gives a description, and purpose, of each component.

As mentioned earlier, ISDOS begins with the user requirements recorded in a machine-readable form. The problem definer (i.e., the analyst or the user) expresses the requirements according to a structure format called the Problem Statement Language. This language can be considered a generalization of those of Young, and Kent;[7] Information Algebra,[8] SYSTEMATICS,[9,10] TAG Input/Output Analysis Form; and ADS.[11] All of these languages are designed to allow the problem definer to document his needs at a level above that appropriate to the programmer; i.e., the problem definer can concentrate on *what* he wants without saying *how* these needs should be met.

It is very important to note that a problem statement language is not a general-purpose programming language or, for that matter, any programming language. A programming language is used by a programmer to communicate with a machine in the fifth phase of the system building process. A problem statement language, on the other hand, is used to communicate the needs of the user to the analyst and therefore is needed in the third phase. The problem statement language consequently must be designed to express what is of interest to the user; what outputs he wishes from the system, what data elements they contain, and what formulas are to be used to compute their values. Analogous information must be given for inputs. In addition, the user must be able to specify the parameters which determine the volume of inputs and outputs and the conditions (particularly those related to time) which govern the production of outputs and acceptance of inputs. The Problem Statement Language is designed to prevent the user from specifying processing procedures that should be selected in the fourth or fifth phase; for example, the user cannot use statements such as SORT and he cannot refer to physical files.

The Problem Statement Language has sufficient structure to permit a Problem Statement to be analyzed by a computer program called a Problem Statement Analyzer. This program is intended to serve as a central resource for all the various groups and individuals involved in the system building process as shown in Figure 2.

Since the problem definer may be one of many, there must be provision for someone who oversees the problem definition process to be able to identify individual problem definitions and coordinate them; this is done by Problem Definition Management. One desirable feature of a system building process is to identify system-wide requirements so as to eliminate duplication of effort; this task is the responsibility of the System Definer. Also, since the problem definers should use common data, there has to be some standardization on their names and characteristics and definition by computations (these are referred to here as "functions"). One duty of the data administrator is to control this standardization. If statements made by the problem definer are not in agreement as seen by the system definer or data administrator, he must receive feedback on his "errors" and be asked to correct these.

---

[7]Young, J. W. and H. Kent, "Abstract Formulation of Data Processing Problems." J. of Ind. Engr., Nov.-Dec. 1958, pp. 471–479. Reprinted in Ideas for Management, Internat. Systems-Procedures Assoc., 1959.

[8]CODASYL Development Committee. "An Information Algebra-Phase I Report," Communications of the ACM, 5, 4, April 1962, pp. 190–204.

[9]Grindley, C. B. B., "SYSTEMATICS—A Non-Programming Language for Designing and Specifying Commercial Systems for Computers," Computer Journal, Vol. 9, August 1966, pp. 124–128.

[10]Grindley, C. B. B. and W. G. R. Stevens, "Principles of the Identification of Information." File Organization, IAG Occasional Publication, No. 3, Scolts and Zeitlinger N.V., Amsterdam, 1969, pp. 60–69.

[11]National Cash Register Company, Accurately Defined Systems, 1967.

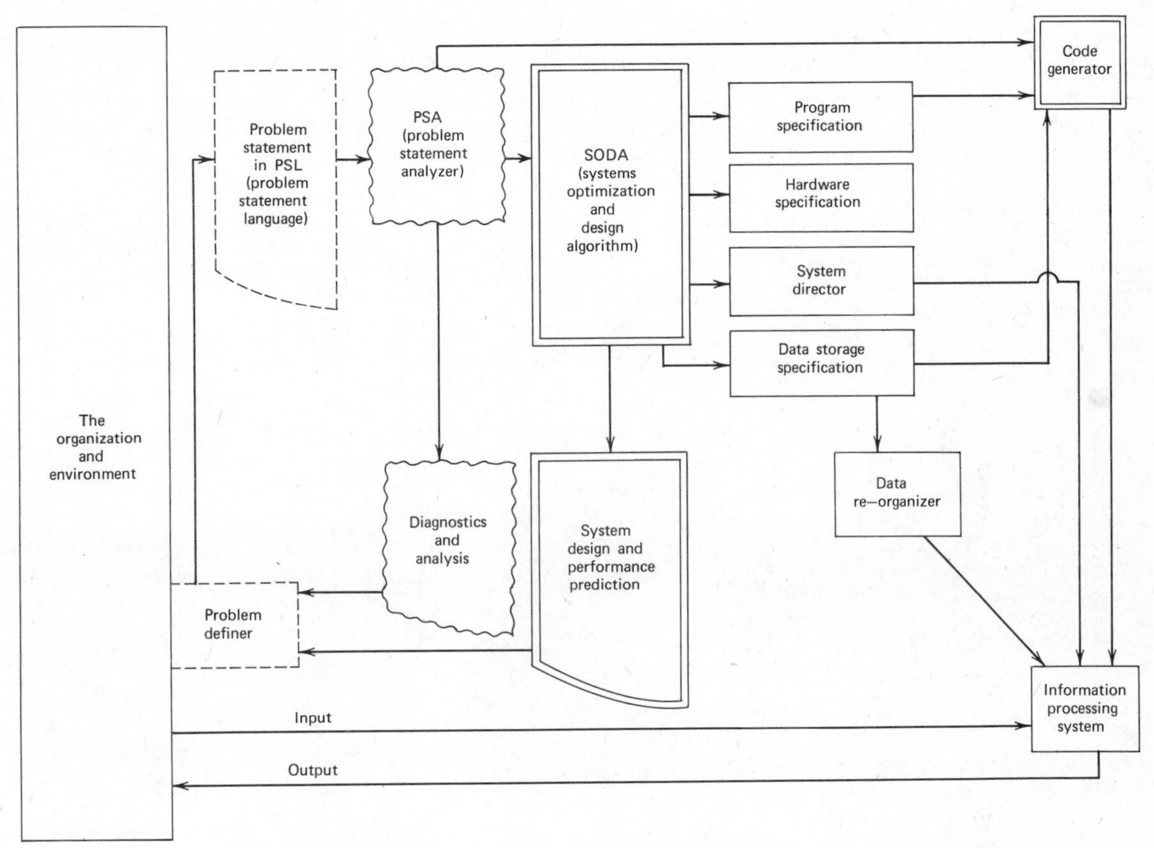

FIGURE 1 Information system design and optimization system (ISDOS).

All of these capabilities are being incorporated in the Problem Statement Analyzer, which accepts inputs in the Problem Statement Language, analyzes them for correct syntax, and produces, among other reports, a comprehensive data dictionary and a function dictionary that are helpful to the problem definer and the data administrator. It also performs static network analysis to ensure the completeness of the derived relationships, dynamic analysis to indicate the time-dependent relationships of the data, and an analysis of volume specifications. It also provides the System Definer with a structure of the problem statement as a whole. All these analyses are performed without regard to any computer implementation of the target information processing system. When these analyses indicate a complete and error-free statement of the problem, it is now available in two forms for use in the succeeding phases. One, the problem statement itself, becomes a permanent, machine-readable documentation of the requirements of the target system *as seen by the problem definer* (not as seen by the programmer). The second form is a coded statement for use by the physical systems design process and other modules of ISDOS.

In the conventional approach, the physical systems design phase (phase four) is concerned with accepting a consolidated statement of the requirements from the system analysts and outlining specifications for the actual construction of programs, files, the relevant schedules, etc. The number of alternatives available is usually so large that the manual approach does not per-

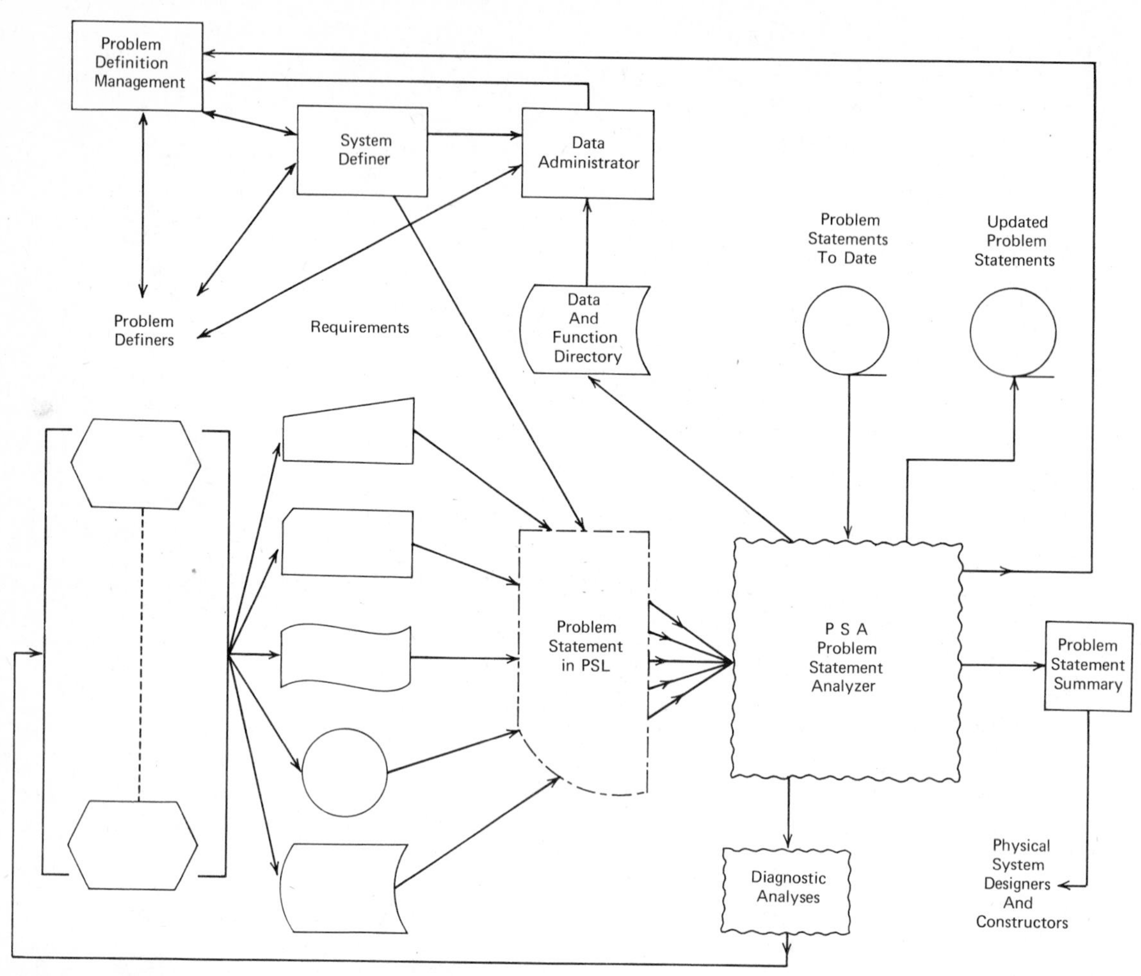

FIGURE 2 Information flows in problem statement analysis.

mit the examination of more than a handful of these. An objective of ISDOS is to formalize the physical design process along the lines pioneered by Langefors,[12] Grosz,[13] Turnburke,[14] Martin,[15] etc. The design problem is formulated mathematically. Operations research methodology is used to develop methods to search over the range of alternatives. A multilevel approach, where the decision variables at one level become the constraints at the next level, is required. This makes it possible to evaluate various design strategies and aids the hardware acquisition group in the selection and justification of appropriate hardware. It also gives the performance officer (who

[12]Langefors, B., "Theoretical Analysis of Information Systems," 2 Vol. Studentlitteratur, Lund, 1966. (Also available from National Computing Centre Ltd., Quay House, Quay Street, Manchester, England.)

[13]Grosz, M. H., "Systems Generation Output Decomposition Method," Standard Oil Company of New Jersey, July 1963.

[14]Turnburke, V. P., Jr., "Sequential Data Processing Design," IBM Systems Journal, March 1963.

[15]Martin, J., Design of Real-Time Computer Systems, Prentice-Hall, Englewood Cliffs, N.J., 1967.

is responsible for the efficient use of resources in computer operations) and the physical system designers a good indication of the expected performance of the system. In addition to the requirements as prepared by the Problem Statement Analyzer, a description of hardware characteristics is required. The outputs are specifications for program modules, storage structures, and scheduling procedures which are in a form suitable for processing by the next two ISDOS modules.

The Data Re-organizer accepts specifications for the desired storage structures from the physical systems design process, definition of data as summarized by the Problem Statement Analyzer, the specifications of the hardware to be used, and the data as it currently exists, and its storage structure. It then stores the data on the selected devices in the form specified. The Re-organizer also produces information for the data administrator and the performance officer. The other module, the Code Generator, accepts specifications from the physical design process and organizes the problem statements into programs recognizing the data interface as specified by the Data Re-organizer. The code produced may be either machine code, or statements in a higher level language (e.g., COBOL), or parameters to a software package. These two modules perform, automatically, the function of programming and file construction in the fifth phase of the system building process.

The final module of the ISDOS system is the Systems Director. It accepts the code generated, the timing specifications as determined by the physical design algorithm, and the specifications from the Data Re-organizer and produces the target IPS. This IPS is now ready to accept inputs from the environment and produce the necessary outputs according to the requirements expressed in the problem statement.

The central concept which makes possible the automation of design and construction is the separation of user requirements from decisions on how these requirements should be implemented. This philosophy is incorporated in the design of the Problem Statement Language. From then on the problem statement can be manipulated by the Problem Statement Analyzer. The decisions which are made in the physical systems design are basically "grouping" decisions, which theoretically can be represented as combinatorial problems. In practice, of course, the number of combinations is very large; and therefore a major research task is to develop efficient algorithms.

ISDOS Development Plan

If a system such as the one outlined in the previous section were available it would go a long way towards improving the effectiveness of computer-based information systems. Since new requirements or modifications to existing requirements could be implemented at computer speeds, management would be able to get the information it asked for in a much shorter period of time. It would therefore be much less important to get the requirements right the first time since a change could be incorporated more easily than at present. The user would be closer to the requirement specifications since the language is closer to the one he is familiar with. Hardware could be used more effectively since design would be based on a formalized procedure using latest available parameters which specify volume of system inputs and outputs.

The system described is itself an information system; and the development of functional specifications, design, and construction is a substantial task which involves three major subtasks:

1. The specification of man-machine communication problems encountered by the analyst in acquiring and recording the requirements for the target system; in other words, in the design of the Problem Statement Analyzer we must ask, "What type of information, in what form, would most aid the analyst?"

2. The specification of the system development cycle with sufficient detail of subtasks to indicate what functions must be performed, and their interrelationships.
3. The development of algorithms using decision-making (operations research) methodology where possible, synthesizing wherever appropriate, the various "micro" decision models already available.

These tasks are being undertaken in the ISDOS Project. Basic engineering philosophy is followed: development of subsystems, evaluation, and validation in real life situations, and eventually, demonstration of the feasibility of the whole concept.

READING QUESTIONS

1. Describe the two subsystems within an "information system."
2. What are the seven major steps (phases) in system building?
3. The Teichroew/Sayani technique concentrates on which steps (phases)? Why?
4. Explain the four major alternatives to methods of improving the system building process.
5. What is ISDOS and what theory serves as its foundation?
6. What is unique about ISDOS?
7. Summarize the ISDOS functions (reduce the section on Automation of the System-Building Process to one page of explanation).
8. The Hoskyns System is being incorporated into ISDOS. What modules in Figure 1 will be accomplished by the Hoskyns System?
9. Based on prior papers, which modules of ISDOS are operational?
10. In your opinion, what is the most difficult ISDOS module to implement? Why?
11. What constraints do you visualize (if any) in widespread use of the ISDOS technique?

SECTION 6

COST/EFFECTIVENESS ANALYSIS TECHNIQUES

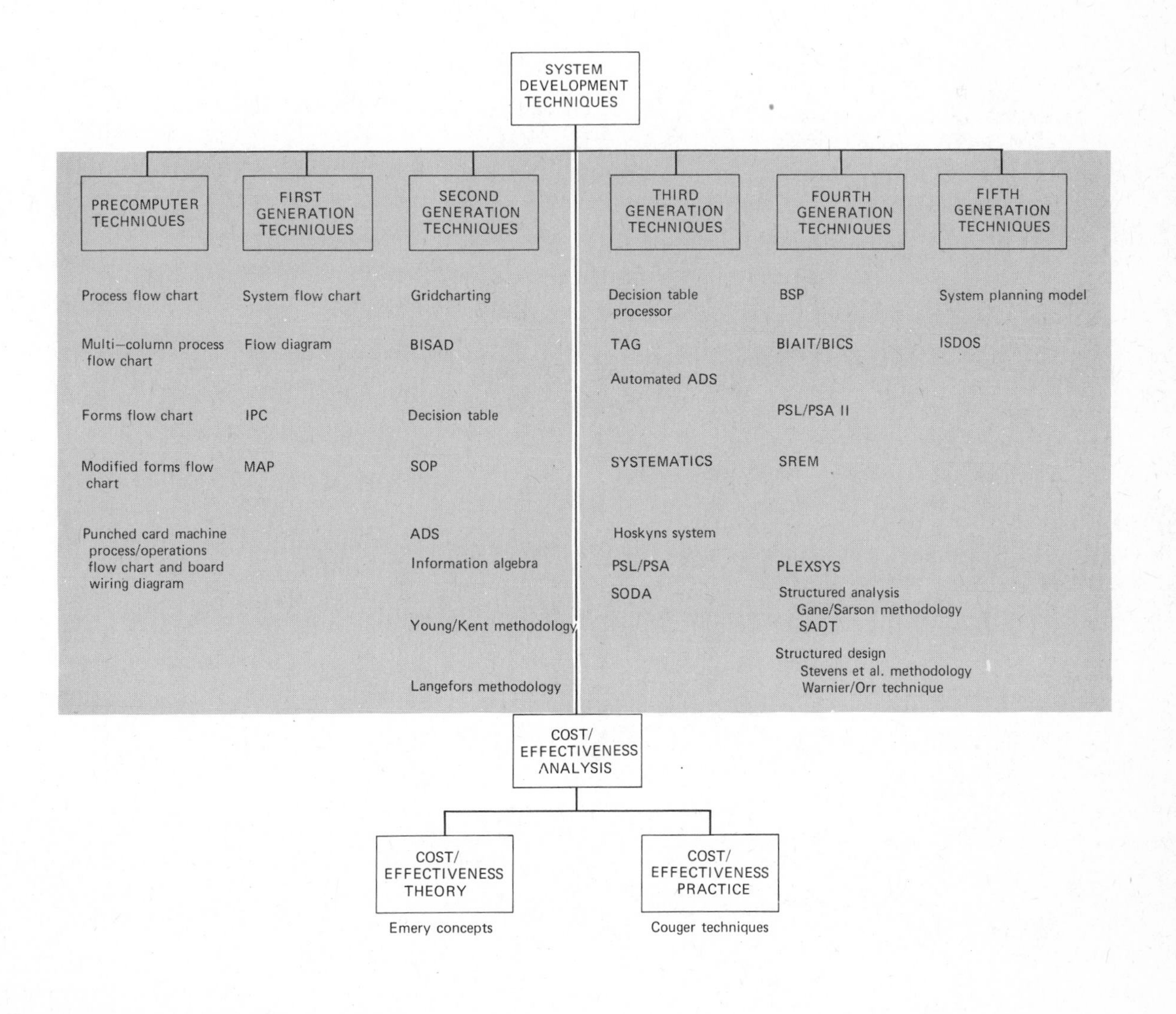

SECTION 6

COST/EFFECTIVENESS ANALYSIS TECHNIQUES

J. Daniel Couger

OVERVIEW

The approach to cost/benefit analysis has changed significantly in recent years. Foremost among the causes is the increased emphasis on the integration of systems. The synergistic effect (the whole being more than the sum of the parts) of integration forces a new perspective on cost/benefit studies. Two applications—integrated—produce more benefits than they produce operating independently. However, the measurement of these synergistic benefits is often difficult; for example, what is the prime benefit of improved information for management? Another example, the perpetual conflict between sales and finance has a better basis for resolution through integration of many of the subsystems of these major functions. The sales manager desires high finished-goods inventory levels to enable quick response to varying customer demands. The controller is charged with the responsibility of keeping inventory levels as low as possible.

In the era of independent systems, each manager could optimize the set of subsystems in his/her area of responsibility—yet produce a suboptimal effect for the company. Integration of systems provides information for management to facilitate optimization for the organization as a whole. Such information is unquestionably better than that produced before integration, but determining how to quantify the improvement in terms of cost/benefit analysis is not simple.

Nor are costs easy to derive. Before the multiprogramming capability existed, projecting processing cost was straightforward. With today's computer configuration, job mix and multiple device usage influence cost significantly, and a more sophisticated cost estimating approach is required.

Total cost of data processing (personnel, equipment, supplies) is *increasing* despite the improvement in hardware/software and the availability of well-trained personnel. Management is justifiably concerned with the increased expenditure levels for data processing. However, total cost is higher principally because more applications are being processed. Previously, only operating level applications were computerized. Now, higher level systems—the management systems—are being computerized. Nevertheless, a negative synergistic effect is possible. As more activities in the firm are computerized, the firm's level of dependence on the computer grows out of proportion to the benefits received. A rising level of computer use should produce a corresponding increase in the number of benefits. Yet, the new types of benefits are not as easy to quantify.

This combination of circumstances has led to an environment where management is much more involved in cost/benefit studies. New techniques have been developed for conducting cost/benefit analysis and for reviewing the results with management.

MARGINAL COST, MARGINAL VALUE

In his landmark publication, *The Economics of Computers*, William Sharpe illustrated an initial attempt to apply microeconomic theory to computer science. Figure 1 shows his representation of the marginal cost/marginal value concept for computer use charging schemes.[1]

He demonstrated that demand (marginal curves) could be applied to the problem of determining optimal computer utilization. Figure 1 depicts the situation for a customer of a service bureau which charges a fixed fee per month and an hourly charge (T^*). The imposition of the fee does not affect the *marginal cost* of computer time. If any time is used, the fee must be paid.

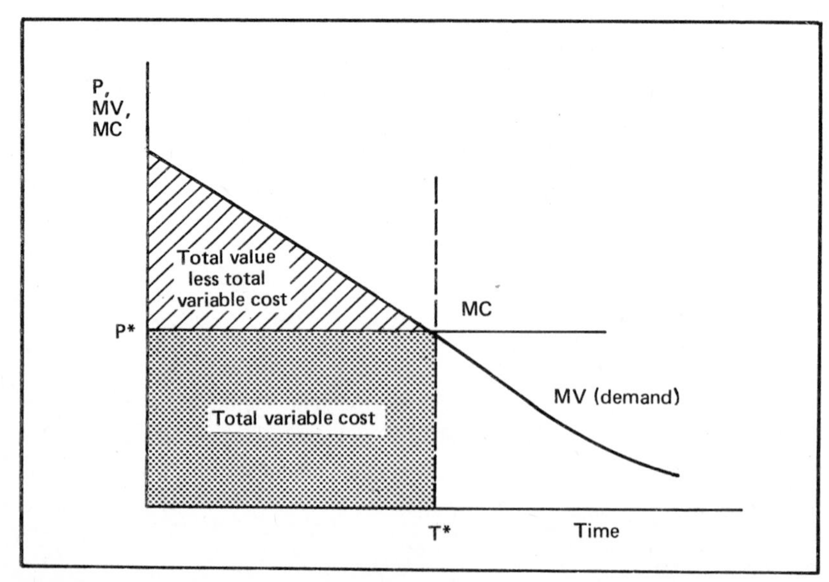

FIGURE 1 Marginal cost/marginal value considerations in service bureau utilization.

Given the decision to use computer time, it is desirable to maximize total value less total *variable* cost, where the latter refers to all costs that vary with utilization once the fixed fee has been paid. This figure can be used to find the optimal utilization regardless of the fee charged. But the net value (upper shaded area) is now total value less total variable cost. The true net value exceeds this amount less the fixed fee. If the fee exceeds the upper shaded area, no time should be used; if it does not, T^* hours should be used.

The practical implication of this case is quite important. When all-or-nothing decisions are required (e.g., whether or not to pay a fixed fee, whether or not to buy a computer, whether or not to add a second-shift operator), it is often useful to analyze the problem by asking the following questions:

1. If the step is taken, what is the optimal way of utilizing the added capability?
2. Is the value that will be obtained if the step is taken and the added capability is used optimally sufficiently larger than the variable costs to justify the expense of taking the step at all?

COST PER UNIT OF EFFECTIVENESS

Herbert Grosch's contribution to cost/benefit analysis was his assertion that, "for computer equipment, average cost decreases substantially as size increases." Grosch's Law, well known in the computer industry but not published, was formalized by Sharpe.[2]

$$C = K\sqrt{E} \text{ or } E = \left(\frac{1}{K^2}\right)C^2$$

where C = the cost of a computer system.
E = the effectiveness (performance, speed, throughput) of the system, and
K = some constant.

Concerning average cost (C/E), the law asserts:

$$\frac{C}{E} = \frac{K^2}{C} \text{ or } \frac{C}{E} = \frac{K}{\sqrt{E}}$$

where K = some constant.

These relationships are shown in Figures 2a and 2b for systems with cost and effectiveness normalized to $C = 1$ when $E = 1$. Economic theory usually relates average cost to output (as does Figure 2b); however, it is often convenient to use the cost as an independent variable (as in Figure 2a) when comparing results based on different measures of effectiveness. Kenneth Knight[3] proved Grosch's Law (Figure 3) in developing similar curves to measure the effectiveness and cost of five IBM System 360 computers (Models 30, 40, 50, 65, and 75).

INDIFFERENCE CURVES

Jacob Marschak applied to economic concept of indifference curves to analyze the value of information to a decision-maker. Marschak's "Economics of Information Systems"[4] was a major theoretical contribution to cost/benefit analysis. Figure 4 shows the indifference set reduced to two points (1, 1) and (0, 0): perfect information. By eliminating the half-square below the diagonal, the likelihoods of p_1 and p_2 are represented by strictly concave curves. When the value (V) of information rises to the point $p > \frac{1}{2}$, it is "good" information.

Marschak's principal contribution was a demonstration of an approach to quantify the value of information, opening the way for the pragmatic analytical techniques in use today.

EFFICIENCY FRONTIER

Each of the above-described analytical approaches was important in laying the foundation for a unified theory of cost/benefit analysis.

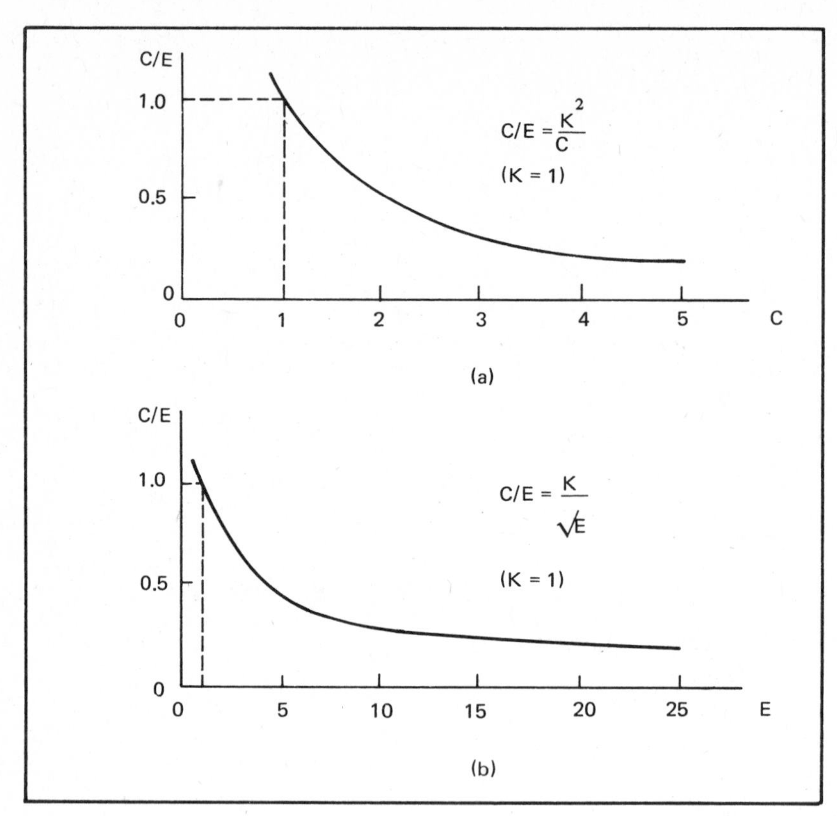

FIGURE 2 Cost per unit of effectiveness versus *(a)* cost and *(b)* effectiveness of system.

However, it was the work of James Emery that served as the cornerstone to the foundation. In *Cost/Benefit Analysis of Information Systems,*[5] Emery introduced a process for analyzing the value of information.

Figure 5 shows Emery's concept of efficiency frontier. Information is attained from a specific system; alternative systems vary in their efficiency of the design used. The efficiency frontier represents the set of systems that provides each level of quality at the lowest cost. For a given level of efficiency, cost rises with increased quality.

The effect of an advance in information processing technology on the optimum information quality is shown in Figure 6. The optimum quality of information changes when an advance in technology lowers the cost of information processing. The new optimum is almost always at a higher level of quality. In this illustration, the new optimum dominates the old (i.e., provides higher quality information at lower cost). In other cases, the new optimum could result in a higher cost than before but provide benefits that more than offset the cost.

Emery uses these analytical devices to facilitate the study of system benefit areas where dollar value is difficult to determine. Examples are response time and accuracy.

Whether at the operational or strategic level, decision making requires a prediction for each of the input variables used in the decision process (e.g., sales, forecasts, inventory levels, aircraft positions, etc.). The prediction span for

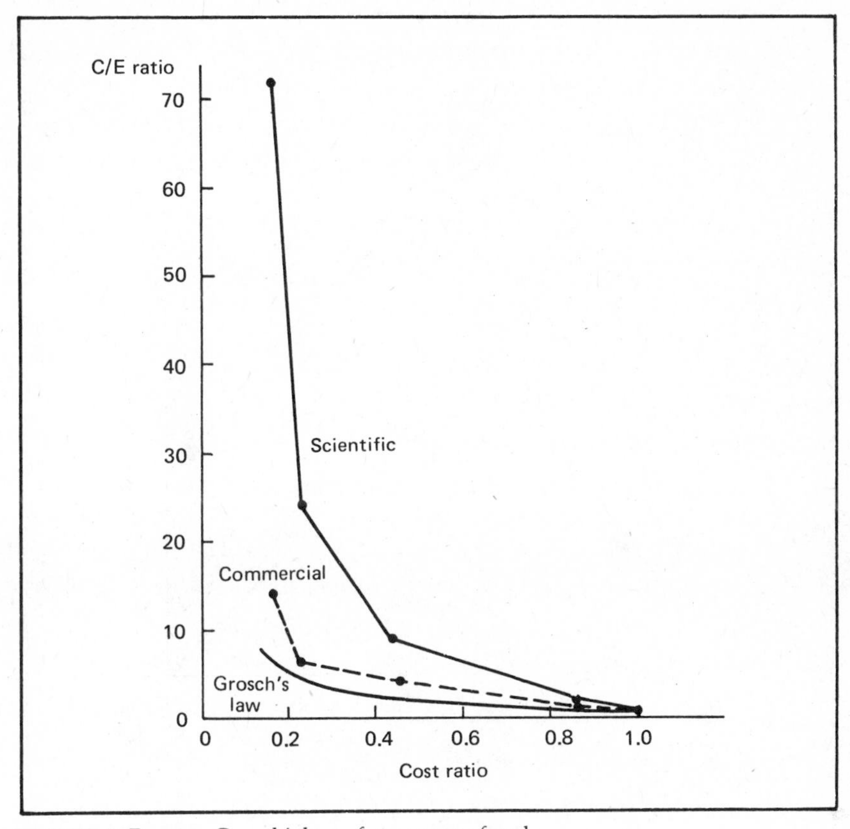

FIGURE 3 Proving Grosch's law of economy of scale.

a given variable extends to the decision horizon from the point in time at which the data base was last updated (Figure 7). The quality of a decision depends in part on the accuracy of the predictions. Accuracy, in turn, depends on the length of the prediction span and the inherent variability of events within the span. A rapidly changing and unpredictable environment requires a short prediction span in order to maintain control; accordingly, information must be very current in order to provide suitable accuracy. Strategic decisions, on the other hand, tend to have long-term effects, and so the decision horizon is well into the future; in this case reducing the age of information adds little to predictability over the long span.

PERSPECTIVE FROM CLASSICAL ECONOMICS

The perspective gained from the use of classical economics is important for cost/benefit analysis. First, marginal cost/marginal value analysis facilitated *delineation* of the attributes of a system. Then, the use of indifference curves led to a process for *quantifying* system attributes.

These approaches proved that system benefits heretofore categorized solely as qualitative could now be quantified, though in nonfinancial terms. The next step was to develop an approach for converting all quantitative measurements to a financial basis.

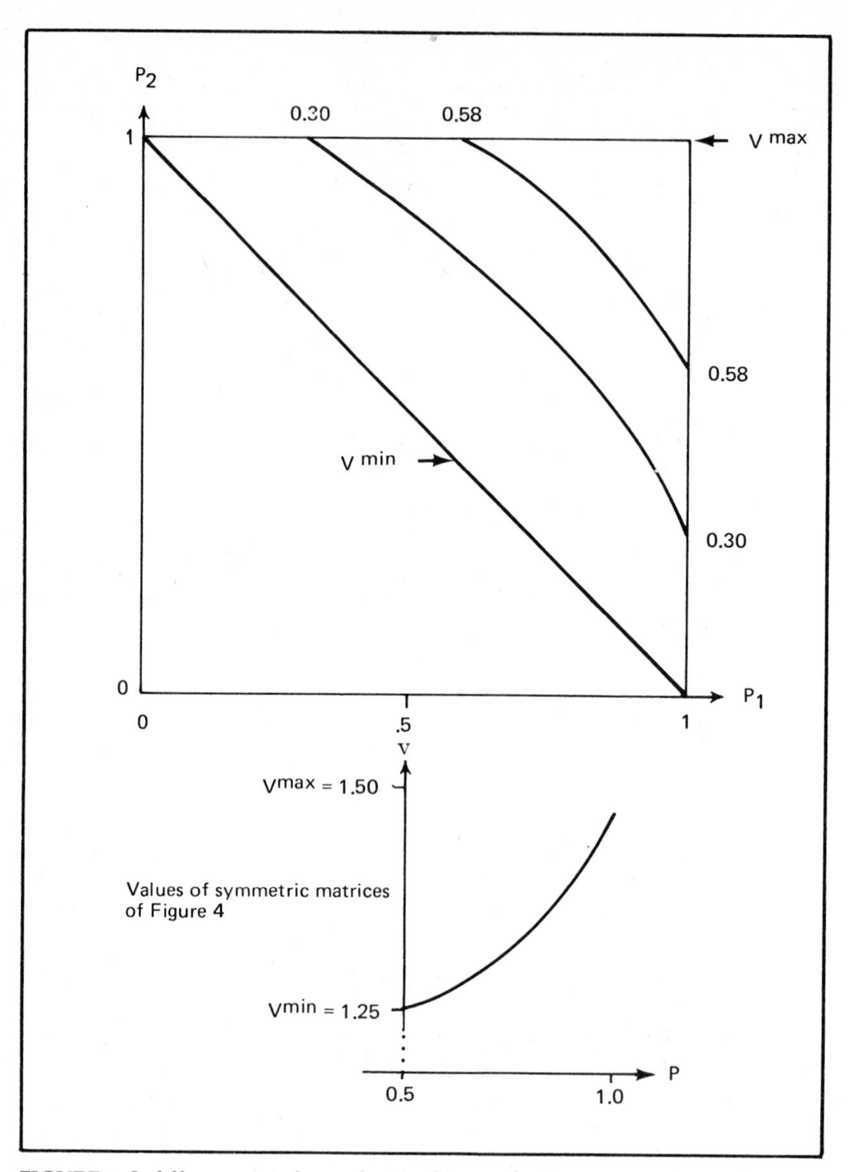

FIGURE 4 Indifference sets for evaluating binary decisions.

BENEFIT IDENTIFICATION

One of the shortcomings of previous approaches to benefit determination was the use of the "intangible" category. It displayed sheer laziness on the part of analysts who classified some benefits in this manner.

The proper approach is to recognize that benefit attainment is a stochastic process. There is a probability associated with attainment of

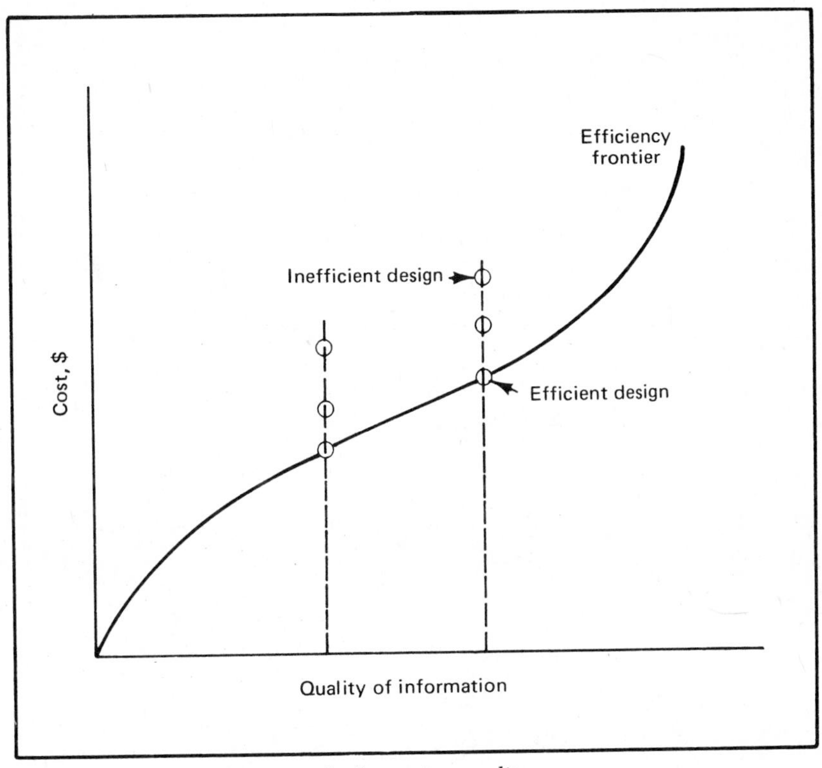

FIGURE 5 Cost as a function of information quality.

each benefit. Present-day benefit analysis concentrates on identifying each benefit and determining the probability of its occurrence.

Figure 8 provides a list of possible benefits, classified in four categories. More than half of the benefit areas listed are straightforward, as far as benefit determination is concerned. They primarily concern labor displacement. In early-day computerization, most computer applications were designed for that purpose; today's systems are complex in development. Earlier, only subsystems such as the payroll system were computerized. Today, in the era of integrated systems, the scope of the system is enlarged many times. Payroll is a module in the accounting subsystem, which is only one of several subsystems in the finance system.

Previously, *in*dependent subsystems were designed for *inter*dependent activities. The payroll application was designed as an entity when, in reality, it was a part of both the finance and personnel systems of the firm. The payroll module of this era is redesigned to feed both of these major systems.

Those early-day systems were largely operational-level systems. They provided the information needed by first-level supervisors and their subordinates. Today's systems include the tactical (control) and strategic (planning) levels, as well. The thrust of system analysis/design efforts in the 1980s has been to expand systems horizontally and vertically.

The expansion in scope and sophistication of systems increases the complexity of system analysis and design. There are more "front-end" costs in designing integrated subsystems.

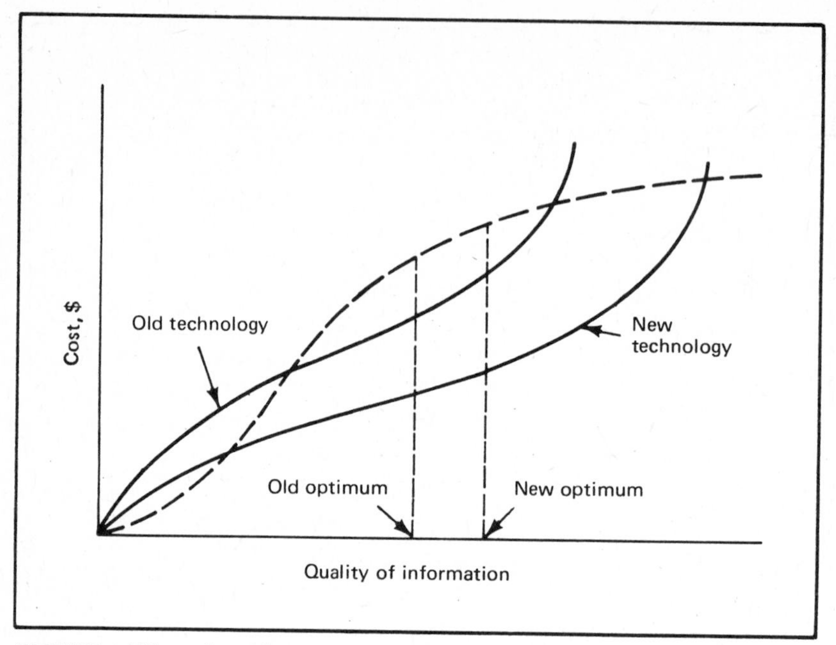

FIGURE 6 Effect of an advance in technology on optimum information quality.

PRIORITIZING SYSTEMS TO ENHANCE RETURN ON INVESTMENT

Another characteristic of the present-day system approach is the emphasis on systems that produce high return on investment (ROI).

Many first-time users of computers concentrate on accounting applications. The same level of effort in system development produces significantly higher ROI when applied to those activities directly related to the product or service. Figure 9 illustrates this point. Before-tax profit is 5 percent higher when the same level of effort is applied to systems related to the product rather than to G&A-type systems.

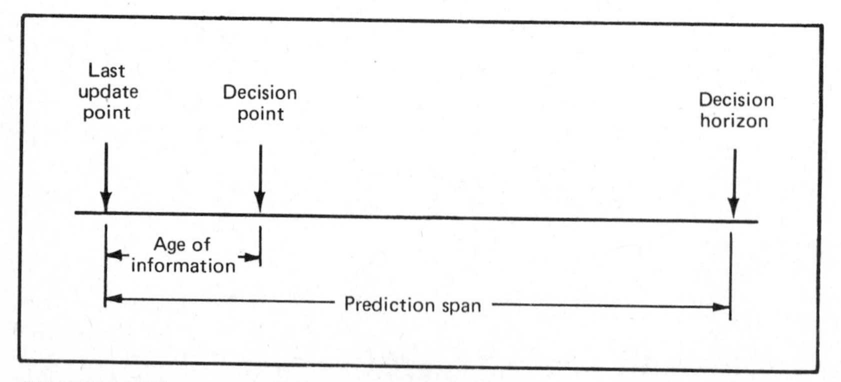

FIGURE 7 Effect of age of information on decisions.

1. **Lower Costs**
   a. Reduction in clerical operations.
   b. Savings in space required for personnel, desks, and files.
   c. Reduction in redundant files.
   d. Reduction in duplication of operations.
   e. Detection of problems before they become costly.
   f. Reduction in the routine, clerical elements in high-caliber jobs.
   g. Reduction in amount of paperwork by utilizing exception principle.
   h. Reduction in inventory.
   i. Combination of like functions in several departments.
2. **Faster Reaction**
   a. Improved ability to react to changing external conditions.
   b. Larger reservoir of information for producing realistic operating plans and forecasting market conditions.
   c. Closer monitoring of operations and utilization of feedback principle to produce corrective actions.
   d. Assessing impact of problems of one area on the other activities of the firm.
   e. Faster turnaround time for processing jobs due to less clerical activity.
   f. Ability to compare alternative courses of action more comprehensively and rapidly.
3. **Improved Accuracy**
   a. Mechanization of operations, permitting more checks and less error possibilities.
   b. Sharing of information between files, reducing the errors resulting from manual intervention.
   c. Ability to raise confidence limits on activities due to more information for measuring performance and to permit more accurate forecasts.
   d. Integrity of information maintained through improved validation techniques.
4. **Improved Information for Management**
   a. Higher quality information through feasibility to employ management science techniques.
   b. Capability to utilize management-by-exception principle to a greater extent.
   c. Capability of developing simulation models for inclusion of all factors in forecasting and developing alternative management plans.
   d. Improved performance indicators through more quantitative data and faster response on performance of all functions.

FIGURE 8 Four categories of benefits of computerization. (Source. *First Course in Data Processing with BASIC,* Copyright 1981, J. Daniel Couger and Fred R. McFadden, John Wiley & Sons, Inc., p. 382.)

SUMMARY

In this section of the book, two papers are reprinted. James Emery's classic paper is taken from his book *Cost/Benefit Analysis of Information Systems*. Few would question that it is the leading paper on the theory of cost/benefit analysis.

The second reprint is my paper on "The Techniques for Estimating System Benefits." It provides a practical approach to quantifying benefits of computer-based systems. I've used the technique in feasibility analysis studies for more than 40 organizations, both government and industry.

Although the cost side of feasibility analysis has received a great deal of attention over the years, the benefit side has been neglected. The high cost and impact of today's system demands rigor on all aspects of feasibility analysis.

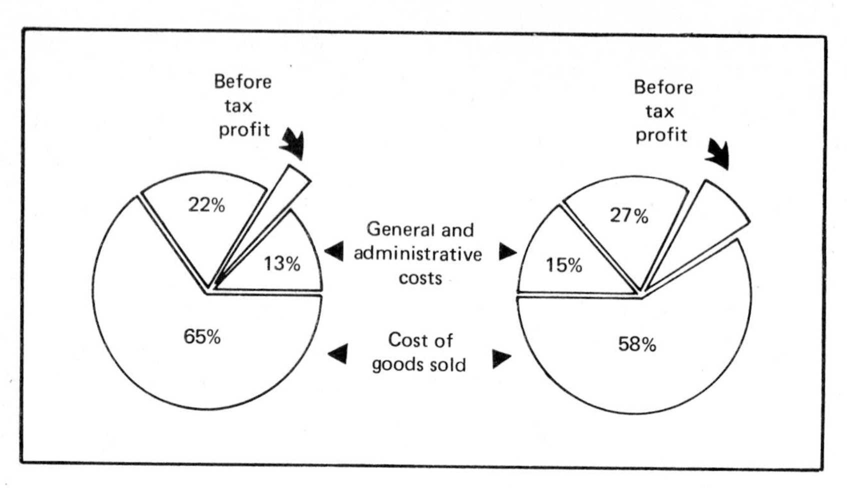

FIGURE 9 Difference in benefits depending upon application area.

REFERENCES

Most of this material was published originally in:
Couger, J. Daniel, "The Benefit Side of Cost Benefit Analysis—Part I," Data Processing Management Portfolio, 1-01-07, Pennsauken, N.J.: Auerbach Publishers, Inc., 1975, pp. 1–11.

1. Sharpe, W. F., *The Economics of Computers*, New York: Columbia University Press, 1969, p. 54.
2. Ibid., pp. 315, 316.
3. Knight, Kenneth E., "A Study of Technological Innovation—The Evolution of Digital Computers," Doctoral dissertation, Carnegie Institute of Technology, Nov. 1968, pp. IV-1–IV-6.
4. Marschak, Jacob, "Economics of Information Systems," *Journal of the American Statistical Assoc.*, March 1971, pp. 192–219.
5. Emery, James C., *Cost/Benefit Analysis of Information Systems*, Chicago: The Society for Management Information Systems, 1971.

SECTION QUESTIONS

1. What has caused the change in approach to cost/benefit analysis in recent years?
2. Define synergism. Give an example.
3. Why is system cost more difficult to estimate today than in earlier days of the computer era?
4. What causes the negative synergistic effect?
5. What are the four categories of benefits to be derived from computerization? Are they in order of difficulty of attainment? Explain.
6. What is the impact of the emphasis on high return systems?
7. How is cost/benefit analysis facilitated by the study of:
   (a) Sharpe's marginal cost/marginal value analysis
   (b) Grosch's law
   (c) Marschak's indifference curves
   (d) Emery's efficiency frontier analysis
8. Grosch's law is being affected by the reduced cost and increased capability of micros and minis. How would the curves in Figure 2 be affected? Explain.
9. See Question 8. Would Figure 6 be affected by this same trend? Explain.
10. See Question 8. Would the other curves (Figures 1, 4) be affected? Explain.

Cost/Benefit Analysis of Information Systems

James Emery

VALUE AND COST AS A FUNCTION OF INFORMATION QUALITY

A system can be perfectly efficient (in the sense of meeting specifications at the lowest possible cost) and still be a very bad one. Efficiency is not enough; the desirability of a system depends on both the cost of meeting the specifications and the value derived from the information.

Thus, any attempt to define an optimum system must consider alternative specifications. Each specification defines information in terms of such characteristics as content age, accuracy, and so on. For the time being, however, we can enormously simplify the discussion if we artificially think of information as having only a single characteristic, which we will call its *quality*.

In effect this assumes that it is possible to trade off all of the detailed characteristics of information into a single, overall index. The justification for this totally unrealistic assumption is that it permits us to discuss some important conceptual issues without the burden of unnecessary details. Our immediate goal is to determine the relationship between the quality of information and its value and cost. This will in turn allow us to consider the optimum balance between value and cost.

Let us first consider value as a function of quality. As we have already seen, it is usually not possible to determine this relationship. Nevertheless, we can still discuss its general characteristics. The most important one is the declining incremental value of information as its quality increases.

The gross value of information continues to increase as quality goes up. Beyond a certain point, however, increased quality may add very little to value. For example, increasing the accuracy of invoices issued to customers from 98 to 99 percent reduces the number of errors by a half. This may be viewed as a highly worthwhile improvement. A further reduction by a half, to 99.5 percent, is of less value. Eventually the

Source: J. Emery, *Cost/Benefit Analysis of Information Systems*, The society for Management Information Systems, © 1971, pp. 16–46.

point will be reached at which continued improvement provides very little benefit. The same thing can be said of increases in level of detail, timeliness, or any other desirable characteristic of information. Figure 1 shows this general phenomenon.

We need a similar relationship between cost and quality. Each level of quality represents a different set of detailed specifications. For each set we are interested in finding the efficient system. The curve connecting the efficient points shows the tradeoff between cost and quality provided by current technology. In the terminology of the economist, this curve is the *efficiency frontier*. It is shown in Figure 2.

Of course, we do not actually prepare curves of this sort. At most we may look at a few alternative levels of quality—a "real-time" versus a batch processing system, for example. For each level of quality we may then evaluate a few alternative designs. The smooth curve shown in Figure 2 thus represents a considerable abstraction from reality.

It is difficult to defend any particular cost curve. Nevertheless, certain characteristics are probably fairly general. For example, over a considerable portion of the curve, economies of scale are common—that is, as quality goes up, costs go up less than proportionally. Eventually, however, quality increases to the point that costs start to rise very rapidly as the limits of current technology are approached.

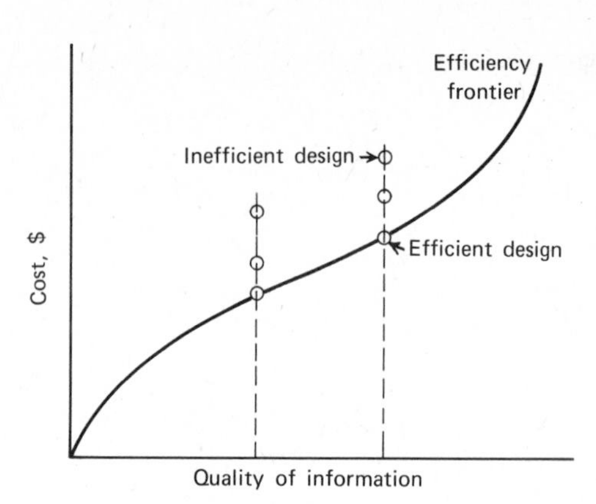

FIGURE 2 Cost as a function of information quality. Information is obtained from a specific system. Alternative systems vary in their efficiency, and so the cost of a given quality of information depends on the efficiency of the design used. The *efficiency frontier* represents the set of systems that provide each level of quality at the lowest cost. For a given level of efficiency, cost rises with increased quality.

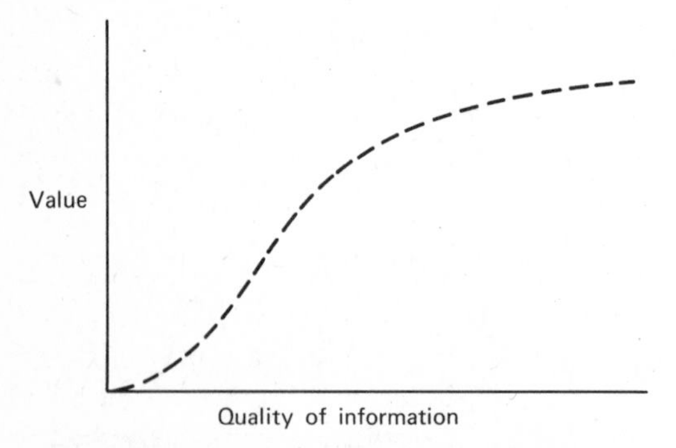

FIGURE 1 Value as a function of information quality. The value of information goes up as its quality increases. At high levels of quality further improvements yield relatively small incremental benefits.

BALANCE BETWEEN VALUE AND COST OF INFORMATION

Finding the Optimum System

Having discussed (abstract) curves relating value and cost to the quality of information, we are now in a position to consider the design of the optimal system. We should aim at finding the design that maximizes the *net* benefits—that is, the difference between gross benefits and cost. Equivalently, the optimum occurs at the point where incremental value just matches incremental cost. This is shown in Figure 3.

Since in practice such curves rarely exist, it is not possible to actually find the optimum design. Even if the curves were known, the optimum would tend to shift during the time span required to implement the system.

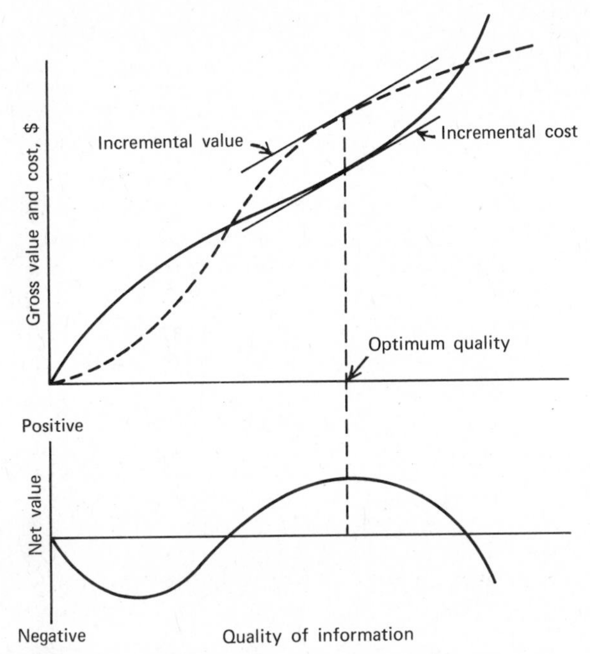

FIGURE 3 Determining the optimum system. The optimum level of quality occurs at the point at which net value (i.e., gross value minus cost) is maximum. This can also be viewed as the point at which incremental value equals incremental cost. Obviously, the optimum system does not provide all useful information; there will always remain unfulfilled information "requirements" that cost more to satisfy than they contribute in benefits.

Nevertheless, an important and valid conclusion emerges from this simplified view of reality. The optimum system does not supply all useful information, since some information costs more than it is worth. Therefore, the specifications of systems requirements must simultaneously consider both cost and value of information.

Effects of an Advance in Information Technology

The (gross) value of information does not depend on the technical means of obtaining it, but costs do. Suppose that an advance takes place in the technology of processing information, such as occurred, for example, between the early 1950's (when punched card technology prevailed) and the current computer era. The advance may come from either hardware or software improvements. The effect is to drop the current cost curve below the earlier one, as shown in Figure 4.

The organization can respond in different ways to a technological advance. It can, as one alternative, choose to exploit the new technology primarily by lowering the cost of producing information. This presumably is the motivation behind projects that merely convert an old system to the new technology, without making any basic changes in information outputs. If outputs remain essentially constant, so must value; the justification must therefore come solely from the lower cost of information.

Alternatively, the system can be redesigned in a more fundamental way that enhances information value. Benefits in this case might come from lower operating costs, improved service, or better decision-making information. Figure 5 shows the alternatives available.

It is difficult to lay down hard and fast rules about the best strategy to follow. Clearly, how-

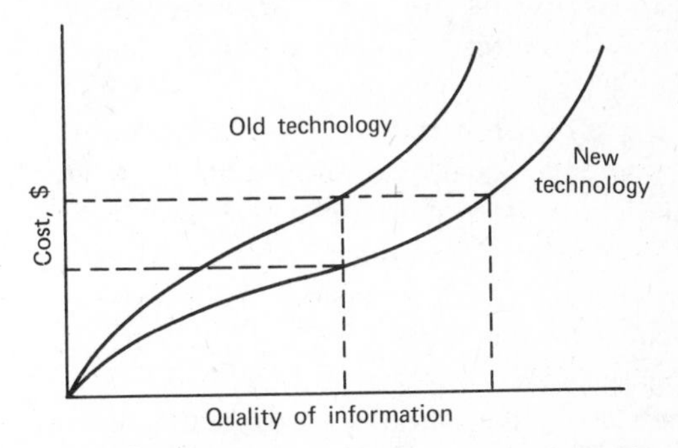

FIGURE 4 Lowering the cost curve with an advance in technology. As information processing technology advances, it becomes possible to obtain a given quality of information at lower cost than before. Alternatively, higher quality can be obtained at the same cost. The heavy line represents *dominant* systems that provide some combination of both higher quality and lower cost.

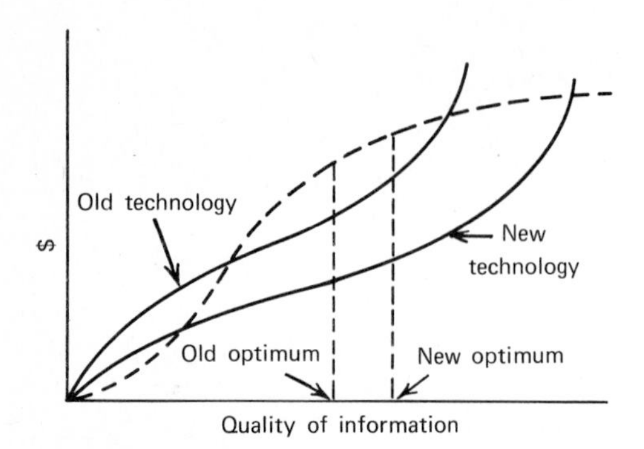

FIGURE 5 Effect of an advance in information processing technology on the optimum information quality. The optimum quality of information changes when an advance in technology lowers the cost of information processing. The new optimum is almost always at a higher level of quality. In this illustration the new optimum dominates the old (i.e., gives higher quality information at lower cost). In other cases, the new optimum could result in a higher cost than before but give benefits that more than offset the higher cost.

ever, an advance in information technology tends to shift the optimum design toward higher quality information. This is simply a manifestation of the general economic principle that a reduction in the price of a resource (relative to other resources) should normally lead to its greater use.

But even if the best long-run strategy is to upgrade the quality of information through a redesign of the system, attractive short-run benefits may also be possible through a relatively straightforward conversion of the existing system. The two approaches are not necessarily in conflict; they may proceed more or less concurrently. By the time the fundamental redesign is ready for implementation, an earlier short-term conversion may have already paid for itself handsomely through cost reductions. Unfortunately, too many organizations appear to pursue short-term savings at the exclusion of any long-term benefits.

Important Characteristics of an Information System That Govern its Value and Cost

In discussing the balance between value and cost, we found it convenient to use a composite characteristic of information called quality. In practice we cannot deal with information in this way; instead we must consider each of its individual characteristics. Although there are tradeoffs among the characteristics—between detail and timeliness, for example—it is useful to consider their separate effects on the overall value and cost of a system.

Allocation of Tasks Between Man and Machine

An information system includes both human and automatic components. A critical characteristic of a system is the way in which these tasks are divided between human and computer. Certain tasks clearly belong to one or the other, but this is by no means always the case.

Complex decision-making that deals with ill-structured goals and relationships is typically best handled by man. So are tasks that occur rarely and do not involve major risks. It is exceedingly difficult (or impossible) for a computer to duplicate man's flexibility and ability to generalize, recognize complex patterns, and deal with unexpected or unusual situations. On the other hand, the computer enjoys an obvious edge over man in a number of respects—in speed, accuracy, volume of data, and the ability to draw inferences from complex models.

A system designer faces the job of allocating tasks between man and machine in the way that leads to the best overall performance. Problems of allocation arise at all levels in the system. At the operating level the designer must determine the extent to which clerical tasks should be replaced by the computer. Typical examples of such questions are:

- Should freight or passenger rates be calcu-

lated automatically within an airline information system?
- Should premiums be calculated automatically within an insurance system?
- Should detected errors be corrected automatically?
- Should a rare combination of circumstances be handled automatically (or as an "exception" dealt with by a clerk)?

Similar types of issues arise in connection with the design of decision-making systems:

- Should inventory order points and order quantities be calculated automatically?
- Should buy and sell orders be generated automatically in a trust management system?
- What thresholds should be set to require automatic decisions to be reviewed by a human?
- What tasks can the computer perform to aid human decision making?

It is difficult to provide many hard and fast generalizations about such questions. One generalization, however, is inescapable: the optimal system falls far short of complete automation. Insofar as possible, each task should be dealt with on its own merits, considering both the value and cost of performing it automatically.[1]

Content of the Data Base

The data base provides an organization with an image or analogue of itself and its environment. The more detailed the image, the greater its realism. This may improve the decisions that rely on the data base as a source of information, but it also increases costs.

The content of the data base depends on the data that enter the system and how long they are retained. High-volume data are usually captured in the form of transactions that feed some operational system. Sales data, for instance, are collected as part of order processing. Once immediate needs have been met—after an order has been shipped, for example—transaction data can be retained in detailed form in an accessible storage medium, or they can be retained only in aggregate form.[2] The level of aggregation and the length of retention are important system characteristics.

Let us examine this issue in more detail. Suppose we are designing an order processing system for a supermarket chain. Replenishment orders for each store are processed daily at a central location. The system generates shipping schedules and thus has access to data about the current day's shipments of each stocked item to each store (three cases of Campbell's tomato soup shipped to Store 53, say).

Now, it is highly unlikely that each individual replenishment order will be worth saving after it has been processed. The real issue is the level of aggregation to be retained. One alternative is to retain individual item data aggregated across all stores within a given week. Shipments to each individual store could be aggregated across all items.

Aggregation of this degree washes out information about the movement of a given item at a given store. Suppose the buying department wishes to analyze the sales of each product to determine if it generates enough gross margin to justify its use of shelf space. If only total figures are known, the decision must be based on an

[1]The interdependencies among tasks make this approach difficult to apply in practice. For example, an on-line system for printing railroad passenger tickets would probably not be feasible if rates were calculated manually. We can broaden the definition of a "task" to include both printing and rate calculation, and then decide whether the combined task should be automated. When a task is defined too broadly, however, the designer may overlook some subtasks that could be better handled independently as a manual operation.

[2]Detailed transactions may also be retained in an inexpensive but relatively inaccessible from for archival purposes (such as microfilm). We are concerned here with data stored in a way that permits easy access for further processing.

item's *average* sales per store. But an average can be very misleading because it hides all variation among the stores.

Sales of some items may vary greatly among stores. One supermarket firm found, for example, that virtually all of the movement of 25-pound sacks of flour occurred in a small number of rural stores serving women who were accustomed to baking their own bread. For these few stores it was essential to carry the item. However, if the decision to stock it were based solely on average sales per store, large sacks of flour would not be retained. Only a data base that retains item-store data would lead to the proper conclusion. See Figure 6.

Even if detailed data find little direct use, their retention may still be very desirable in order to allow aggregation in unanticipated ways. Suppose, for example, that an analyst performing a distribution study wishes to find the total weight of products shipped to each store. Unless item-store shipment data are maintained (as well as the unit weight of each product), it would be very difficult to retain reliable estimates of shipments. To be sure, this information could be accumulated as part of the periodic reporting scheme, but it is impossible to anticipate all possible aggregations of detailed data. The retention of relatively disaggregated data is the best way to overcome this difficulty. See Figure 7.

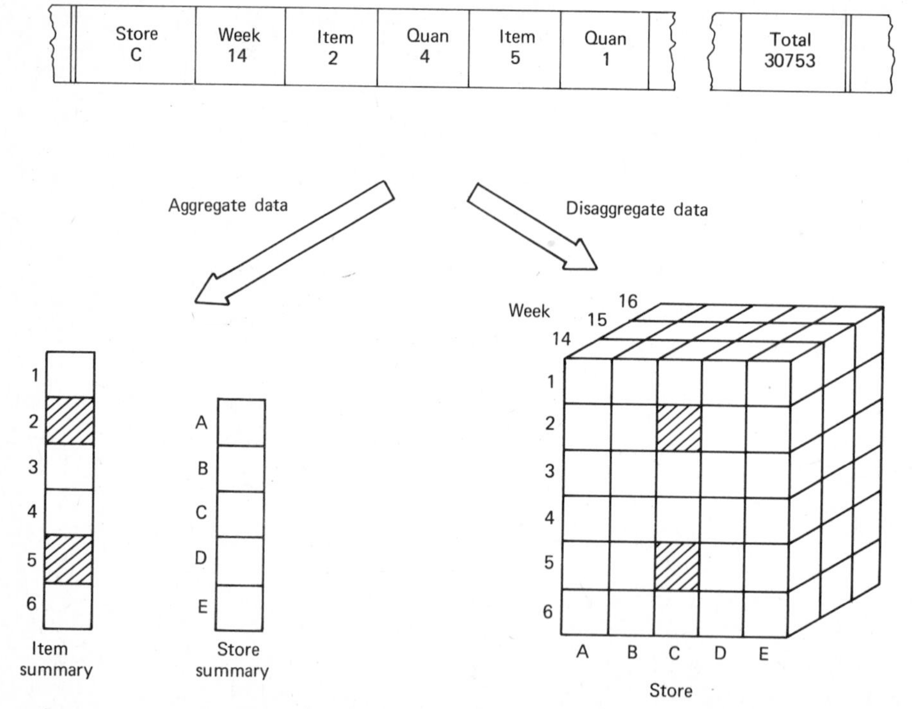

FIGURE 6 Retention of aggregate versus disaggregate data. Transaction data, such as the shipment transactions shown above, are rarely kept very long in complete detail; the question is, how much detail should be retained (and for how long). Maintaining only aggregate data—for example, total units shipped of each item and total dollar shipments to each store—greatly reduced storage requirements. Retaining item-store shipment by week, say, drastically increases the size of the file, but allows a much greater variety of information to be obtained.

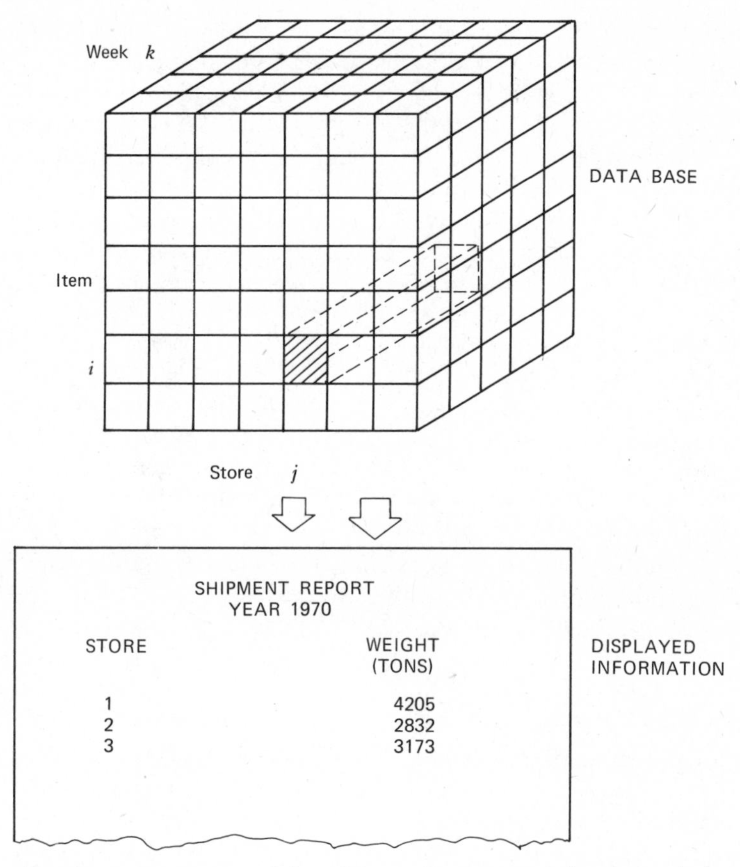

FIGURE 7 Aggregation of data from a disaggregated data base. A detailed data base allows great flexibility in calculating specific aggregations. For example, the total weight of product shipped to a store can be calculated if item-store shipment data are available. The item-store shipment data may, in turn, be aggregated from weekly shipment data. Thus,

$$\text{total weight shipped to store } j = \sum_i w_i S_{ij} = \sum_j w_i \sum_k s_{ijk}$$

where

w_i = unit weight of ith item
S_{ij} = total unit shipments of ith item to ith store
S_{ijk} = total unit shipments of ith item to jth store during kth week.

The incremental value of information diminishes as detail grows. This is so because it becomes increasingly likely that information needs can be satisfied without resorting to still finer detail. In the supermarket example, storage of item-store data by *week*, say, would probably meet most information needs; it would not be necessary to retain *daily* item-store shipments.

The incremental value of data declines with age as well as with detail. Retention of complete transaction data may be justified for a short time following a shipment to a store. After a few months, however, the details should be summarized into, say, item-store data by week; the transaction data can then be discarded (or stored in a low-cost medium for archival purposes). When older than a year, the data might be further summarized into sales by product groups (canned soups, for instance) and month. This process of increasing the degree of aggregation might continue for several years. Usually only highly aggregated data are retained permanently.

The size of the data base obviously grows as the level of detail and length of retention increase. In fact, it can grow explosively. For example, if sales data are maintained on a daily instead of a weekly basis, the size of the file may increase by a factor of six (assuming a six-day week). A similar factorial growth occurs as other dimensions of classification are added. Differentiating between credit and noncredit sales, for example, potentially doubles the size of the file. The length of retention also has an important effect on size. The data base can grow very rapidly if steps are not taken to cull out old data.

The cost of storing data naturally increases as the size of the data base grows. The cost of the storage medium increases, but usually less than proportionately (i.e., some economies of scale are exhibited, at least over a considerable range of volume).

Far more significant is the cost of retrieving desired information. This cost depends on the way in which data are organized and accessed. Retrieval may require a sequential scan of a portion of the data base or, alternatively, relatively direct access by means of indices or linked records. In either case cost of retrieval may increase more than proportionately as the size of the data base grows.

The decision regarding the proper level of detail and length of retention should be based on the expected value of stored data. The expected value of a given data element equals its value, if it is required, multiplied by the probability that it will be required. Thus, the storing of information may be justified on the grounds that (1) it will be extremely useful if asked for, even if this is fairly unlikely (such as a cancelled check); or (2) it is very likely to be needed and its value upon retrieval exceeds the cost of retaining it. In practice one must usually rely on subjective (and somewhat vague) estimates of value and probabilities, but the person making these judgments should at least have his objective clearly in mind.

We may summarize this discussion in the following way:

- Increasing the degree of detail and the length of retention of data may drastically increase the size of the data base.
- The value of the data base tends to increase with its size, but at a diminishing rate.
- The cost of maintaining and retrieving from the data base grows rapidly as its size increases.
- The optimum size occurs at the point where incremental value equals incremental cost. Typically this point falls considerably short of retaining (for very long, at least) complete transaction data.

Selectivity of Displayed Data

It is not enough merely to keep useful information in the data base; the information must be displayed if it is to be used for human decision-making. As the size of the data base grows, it becomes all the more critical to display only highly selective information. Ideally, the information displayed should always have some surprise content and lead to a better decision than would otherwise be made. In practice we can only approach this ideal.

Let us consider the total variety of ways information can potentially be obtained from the data base. This includes all of the individual data elements, as well as any arbitrary transformation performed on these elements. The transformations include simple aggregations, standard statistical analyses, preparation of graphical output, and even calculations using complex decision models.

Now, let us look at the information needs of a given user. The very large (in fact, infinite) set of potential information available from the data base can be partitioned in two independent ways: displayed versus nondisplayed information, and relevant versus irrelevant information.

Displayed information is that portion presented to the user. It may be displayed either in hard copy form or as a transient display on a CRT or some similar device.

The relevance of information is not as easy to define; it is a matter of degree. We will define information to be relevant if its value exceeds the incremental cost of using it—in short, if the user would benefit by having it displayed. This means that relevant information offers some surprise, leads to a decision that would otherwise not be taken, and improves payoff.

The total set of potential information can be broken down into four subsets:[3]

1. Relevant and displayed.
2. Irrelevant and not displayed.
3. Relevant and not displayed.
4. Irrelevant and displayed.

The designer of a system would naturally like to limit the information included in the third and fourth categories, since they represent "errors" of omission or commission. When relevant information is not displayed, decisions will not be as good as they otherwise would be. When irrelevant information is displayed, the user himself must select the useful information from the useless. The greater the proportion of irrelevant information, the greater the effort on the user's part to cull it out and the greater the risk that he will overlook valuable information. Like the gold in the ocean, relevant information too diluted with useless information ceases to have any value.

Errors in information selection are unavoidable. In order to display *all* relevant information, and *only* relevant information, the system would have to determine what each user already knows and what his decision process is. This is clearly impossible, and so the choice of information to display must be a compromise between displaying too much or too little. The probability of displaying relevant information goes down as the degree of filtering increases, but this is accomplished by an increase in the probability of overlooking relevant information.

The optimum degree of selection depends on the relative penalties of the two types of errors. If failure to display critical information carries a very high penalty relative to the cost of displaying irrelevant data, then it is advantageous to err on the side of displaying too much. This is the implied motivation behind many existing information systems.

Both types of errors can be reduced simultaneously if sufficient resources are spent in implementing more effective selection.[4] It is primarily the cost of design effort and information processing that imposes an upper limit on the extent to which selection errors should be avoided.

Increased selectivity can be achieved through a variety of means. Some of the techniques have been used for many years, while others have become feasible only with relatively recent advances in information technology. Let us examine the more important techniques.

Appropriate Aggregation of Details. Virtually no one needs transaction data in complete detail (except for handling the transaction itself, of

[3]The four-way breakdown is analogous to that faced in quality control, in which one is concerned with whether a product should be accepted as good or rejected as bad. The possibilities are: (1) satisfactory quality and accepted, (2) unsatisfactory and rejected, (3) unsatisfactory and accepted (the so-called "buyer's risk"). and (4) satisfactory and rejected (the "producer's risk"). This analogy is worth pointing out because the issues are much the same in the two contexts.

[4]This is analogous to taking a larger sample in quality control applications in order to simultaneously reduce both the producer's and buyer's risks.

course); it is almost always necessary to aggregate the details before they can be used. The aggregation may be by product group, organizational unit, cost category, time period, or some similar dimension. The intent is to aggregate in a way that washes out irrelevant dimensions and preserves the relevant ones. For example, sales data used by the marketing vice president may be aggregated by major product groups and sales regions. Finer detail (sales of a specific product, say) or some alternative aggregation (sales classified by industry, perhaps) would probably not be particularly useful in assessing the performance of regional sales managers (although obviously such information might be entirely relevant for other purposes).

Simple aggregation, in which each data element is added to its appropriate category, implies that each element carries the same relative value. A summary sales report, for example, implicitly assumes that a dollar of sales for Product X is equivalent to a dollar of sales for Product Y if they both fall within the same product group. This is often valid, but it need not be. A production report that shows delivery performance may give quite misleading information if it merely provides a count of late jobs. A much more meaningful figure would be one that weights each late job by a measure of its lateness and importance (man-hours of labor applied, for instance).

Although aggregation is essential to reduce the volume of displayed information, it always carries some risk of washing out relevant information. A sales report may hide significant trends within a product group or sales territory, for example. Similarly, an inventory report that aggregates across all items may cancel out a serious imbalance in which some items are in critically short supply while an offsetting group has a large surplus. The remedy for this is to aggregate within finer (and hopefully more relevant) categories. For example, the inventory report can aggregate according to the current status of each item, using the three categories "in control," "short," or "surplus." See Figure 8. Unfortunately, increasing the fineness of aggregation also increases the likelihood of displaying irrelevant information.

The risk of overlooking relevant information can be reduced by providing the user with backup details that explain each aggregation. For example, a report giving sales by product group and region might be supported with more detailed reports by sales office and product subgroups. Each detailed report may, in turn, be supported with still more detailed reports that show sales by individual salesmen and product.

INVENTORY ANALYSIS

STATUS	NUMBER OF ITEMS	BALANCE ($000)	STANDARD ($000)	PERCENT OF STANDARD
IN CONTROL	9242	2530	2400	105
SHORT	1025	135	420	32
SURPLUS	779	832	205	406
TOTAL	11046	3497	3025	116

FIGURE 8 Aggregation of displayed information to reduce detail while preserving essential information. Aggregations are necessary to increase selectivity, but run the risk of washing out significant details. A complete aggregation of the above inventory information would give the misleading impression that inventory is in control (only 16 percent above standard). Aggregating the inventory items into three categories according to their current status gives a much more realistic picture at little increase in the amount of information displayed.

A detailed report should clearly show the relation between its data and the next higher aggregation. See Figure 9. Similar backup can be provided to support aggregations across cost categories, multiple time periods, and the like.

This hierarchical linking of reports provides an effective means of reducing unnecessary detail while still allowing the user to penetrate into the details when this appears to be warranted. An obvious requirement of a complete hierarchical linking of reports is the proper nesting of data elements in terms of data definitions, reporting periods, and classifications. In other words, a lower level data element must be identified uniquely with a higher level aggregation for a given classification scheme.[5]

MONTHLY SUMMARY SALES REPORT ($ MILLIONS)

		REGION					
		NORTH EAST	SOUTH EAST	CENTRAL	SOUTH WEST	WEST	TOTAL
PRODUCT GROUP	A	25	12	32	8	10	87
	B	17	9	35	4	12	77
	C	12	5	22	2	3	44
	D	(20)	7	15	15	18	75
	TOTAL	74	33	(104)	29	43	(283)

MONTHLY REGIONAL SALES REPORT ($ MILLIONS)
PRODUCT GROUP D NORTHEAST REGION

		OFFICE			
		NY	BOS	PHIL	TOTAL
PRODUCT SUBGROUP	D1	4.2	1.5	2.1	7.8
	D2	1.8	.8	2.6	5.2
	D3	1.1	1.7	.3	3.1
	D4	3.0	.3	.8	4.1
	TOTAL	10.1	4.3	5.8	(20.2)

FIGURE 9 Hierarchic relationship among reports. It is desirable to provide detailed backup reports for all aggregate data. A figure on a high-level report should appear as a total on the next lower level report.

Good Human Factors in the Design of Display Formats. Information must be perceived before it can have value for human decision-making. The effectiveness with which a user perceives information is largely governed by the way in which it is displayed. The interface between the system and user is one of the more critical design factors.

Some of the general principles of good display are as follows:

- Use standard report formats, headings, and definitions whenever possible. This permits a user to scan a display without having to interpret each item.
- Each item displayed should be labeled or have an obvious interpretation.
- Avoid unnecessary precision. Since an aggregation inevitably represents an approximation of reality, excess precision adds little value while it clutters up the display.
- Use graphical display when feasible. A graphical display reduces unneeded precision while often revealing relationships among variables much more perceptibly than a tabular display.
- Provide a basis for interesting information. A given piece of information seldom has value by itself; it must be assessed relative to some standard or anticipated result. It is therefore important that a user be provided sufficient information to comprehend the significance or surprise content of new information. This can be done by displaying the new information (e.g., actual current results) in juxtaposition with the existing plan, standard, or past results.
- Provide links among separate displays. Each display should contain relatively little information in order not to swamp the user. It is therefore necessary to use multiple displays if much information is to be conveyed. The user should be able to relate one display to another. Hierarchical relationships among displays, as discussed in the previous section, provide one of the basic means of doing this.

[5]Geographical boundaries provide a useful analogy. An example of proper nesting is the aggregation of United States countries to form states and the aggregation of states to form the United States. On the other hand, metropolitan areas do not nest within state boundaries. Proper nesting for a given classification does not preclude alternate nesting for other classifications. For example, the sales of a given item can be aggregated by geographical boundaries, product groupings, or industries.

Use of the Exception Principle. The exception principle is by no means a new idea. Its intent is to identify "exceptions" that require human attention. Information about exceptional conditions is displayed while all other information is filtered out. An exception is deemed to have occurred when actual results deviate from a standard by more than an established threshold value. See Figure 10.

Threshold values are ideally set at a level that correctly distinguishes between relevant and irrelevant information. This means that information about conditions outside control limits should lead to a new decision, such as altering an existing plan. The deviation from the plan may be either favorable or unfavorable. A favorable deviation opens new opportunities that would be lost if the current plan were not revised—an increase in sales, for example, may call for increased output. An unfavorable deviation may require a change in plan in order to minimize the cost of the deviation—a renegotiation of delivery schedule when difficulties are detected in an engineering development program, for instance.

The effectiveness of an exception reporting system obviously hinges on its ability to distinguish between relevant and irrelevant information. The ideal system cannot generally be achieved in practice, since this requires complete formalization of the decision process. Suppose, for example, that we wish to report inventory status according to the exception principle. In order to identify items of inventory requiring attention (to change the existing production schedule or the order point and order quantity, say), the system must be capable of comparing the existing plan with the new optimum based on the latest known conditions. Only when there exists a significant disparity between the two is it necessary to signal an exception.

Thus, the identification of a true exception requires (1l a prediction of the likely outcome if no change in plan is made, and (2) the penalty that this outcome would entail in comparison with the current optimum plan. Only if the penalty exceeds the cost of changing the plan should the situation be labeled an exception. Such formalization rarely exists, and so it is necessary to strike a balance between the risk of displaying too much or too little information.

The control limits used to identify exceptions can be based on any (or all) of the variables

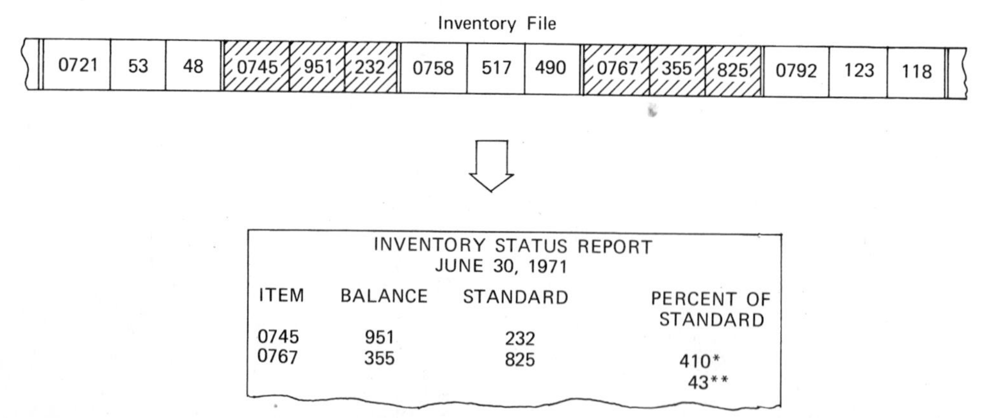

FIGURE 10 Exception reporting. An exception report displays only the conditions that fall outside of defined control limits. An inventory status report, for example, may show only the items that are over 200 or less than 50 percent of their standard. The system should readily handle changes in control limits in order to allow users to redefine their information requirements. The inventory standard used in the report may be set externally by management, or it could be computed periodically from an inventory model.

used in a plan. In the case of an operating budget for example, manufacturing overhead might be monitored in order to signal an alarm when costs become excessive. The deviation allowed each variable should be set according to its relative importance and its normal random fluctuations. Thus, a variation of 50 percent from standard may be allowed for a minor variable subject to considerable routine fluctuation, while a 5-percent variation in a well-behaved aggregate variable may be deemed worthy of management attention.

An exception can be defined in terms of trends as well as deviations. For example, an unfavorable deviation in raw material costs of, say, 8 percent may normally be considered unexceptional; but if it follows a month with a favorable variance of 6 percent, the unfavorable trend could be tagged as an exception.

Because of the imprecise nature of control limits, they should be viewed as parameters subject to change as users see fit. If too many exceptions are currently being displayed, a user may wish to broaden the limits in order to display only the most severe exceptions. If only a few exceptions occur, a user should be able to tighten limits in order to reveal the situations that may most benefit from his attention. The extreme case of zero limits should be allowed when an exhaustive display is desired, as well as "infinite" limits when no information is wanted.

A well-designed exception reporting scheme can add greatly to the value of an information system. To be sure, it carries some risk of overlooking important information. It is a profound mistake, however, to assume that this is a risk unique to formal exception reporting; a user may stand a much greater risk of overlooking significant information if it is immersed in a huge report containing mostly irrelevant data. Selection of relevant from irrelevant information must be performed either by the user himself or within the system. The capability of the computer to apply sophisticated selection criteria often gives it a tremendous advantage over the user in identifying likely candidates for closer inspection.

There are obvious costs of providing exception reporting services. The design and maintenance of a sophisticated system is by no means trivial. Users must be trained to use the system well. Some additional processing costs are usually incurred to filter out irrelevant data (although savings in display costs may in some cases more than offset this). However, when these costs are compared to the (largely hidden) costs of having users perform their own selection, the economies almost always favor formal exception reporting.

Use of ad hoc Inquiries. The typical periodic report is based on the anticipated recurring needs of a group of users. It therefore necessarily contains a great deal of information that is not relevant to a given user at a given point in time.[6]

An alternative to periodic reports is an ad hoc inquiry system that provides a response (within a reasonable time lag) to each specific request for information. See Figure 11. If information is supplied only on demand and is tailored to a specific user's needs, it stands a much higher probability of being useful.

The capabilities offered by inquiry systems vary widely. Some systems only allow data to be extracted from the data base, without any further manipulation. Of much greater use is a system that can perform appropriate aggregation of detailed data. Some systems also have built-in standard transformations of data, such as the calculation of an average, range, or standard deviation. A still more advanced capability is the ability to handle transformations defined by the user in the retrieval language of the system. Fi-

[6]A specialized example of a periodic report is the standard telephone directory. It contains a very low density of useful information for any given subscriber. Nevertheless, economies favor such an approach because the cost of providing each subscriber with selected information (through a directory assistance operator or a tailor-made directory) would be prohibitive with existing technology. Besides, the cost of selection from the telephone book is relatively small (if one knows the name of the person whose number he is seeking) and is borne by the subscriber.

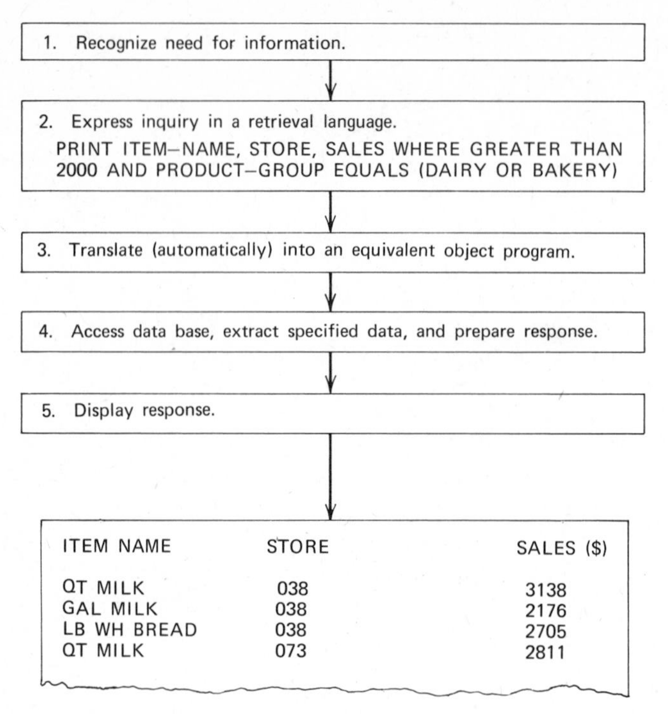

1. Recognize need for information.
2. Express inquiry in a retrieval language.
PRINT ITEM–NAME, STORE, SALES WHERE GREATER THAN 2000 AND PRODUCT–GROUP EQUALS (DAIRY OR BAKERY)
3. Translate (automatically) into an equivalent object program.
4. Access data base, extract specified data, and prepare response.
5. Display response.
ITEM NAME
STORE
SALES ($)
QT MILK 038 3138
GAL MILK 038 2176
LB WH BREAD 038 2705
QT MILK 073 2811

FIGURE 11 Processing of an ad hoc inquiry. A specific display is prepared in response to an ad hoc inquiry. The resulting information can therefore be highly selective.

nally, in the most sophisticated systems the user can define transformations with any available language (such as FORTRAN, COBOL, or PL/1); these are properly termed data base management systems, since they serve as the basic interface with the data base. Olle (1970) discusses some of these issues.

The costs of handling ad hoc inquiries depend greatly on the particular system used, the size of the data base, and response time requirements. Costs include the effort required by the user to specify requested information, as well as the processing of the inquiries. A sophisticated system providing a powerful retrieval language and fast response may be quite expensive indeed. On the other hand, handling batch-processed inquiries of a standard nature may be quite inexpensive, particularly if the processing is combined with routine file updating.

It is sometimes feasible to maintain certain commonly used information, such as standard financial data, as a separate subset of the total data base. Relatively standard inquiries can then be handled by retrieving data from this subset. Its size may be very much smaller then the entire data base, permitting substantial economies.

Use of a Decision Model. A decision model can be viewed a a particularly effective information filter. The user of a model is presented with the output of the model, rather than its detailed inputs. Output normally is very much less voluminous than input, which drastically reduces the amount of information displayed.

The degree of filtering depends greatly on the type of model used. A decision model fully embedded within the information system, in which all input data are obtained automatically from the data base and all decisions are fed automatically into operations, provides the most extreme example. In this case only aggregate results need to be displayed for human monitoring.

This is obviously not the common situation. In most cases even "optimizing" models require human intervention in providing inputs and in reviewing and modifying outputs. Simulation models usually require active human participation in proposing alternatives and selecting the best one among them. Nevertheless, the amount of information presented to the user is usually much less than would be required if he made the decision unaided by a model.

A decision model can be combined effectively with exception reporting. The model can be run periodically to test if the latest conditions—e.g., sales, costs, etc.—should cause a revision in existing plans. If so, this fact can be displayed for human review; otherwise, nothing is reported (except perhaps aggregate performance).

Response Time

The response time of an information system is the time it takes to respond to a significant stimulus. The stimulus may be the occurrence of an event that is to be reflected in the data base, such as the arrival of a sales order from a customer. Alternatively, the stimulus may be a request for information already stored in the data base. Thus, in discussing response time we are concerned with two aspects of time lag: (1) the time it takes to update the data base, and (2) the time it takes to retrieve desired information from the data base.

It is useful to break down response time in this way because value and cost may depend heavily on which aspect is being considered. Some applications may benefit significantly from quick updating and retrieval, others may call for only quick retrieval, while still others may require neither quick update nor retrieval.

Value of Short Update Time. Let us first consider the value of information as a function of its recency. The timeliness of the data base depends on the update response time. In an environment that changes both rapidly and unpredictably, the data base must be updated quickly in order to keep a faithful representation of current reality. If decisions are highly sensitive to changes in the environment, substantial penalties may be suffered if the update time is not kept short relative to the rate of these changes.

Some applications clearly benefit from online updating. These are nearly always found at the operating level of the organization. It is only here that events are likely to take place frequently enough to justify very short update lags. Such "real-time" applications as air traffic control, industrial process control, and stock quotation services would scarcely be feasible without a data base kept current within a matter of seconds. Airline reservation systems begin to incur significant penalties (in the form of underbooking or overbooking, for example) as updating lags exceed an hour or so. High-volume production and inventory control systems similarly benefit from rapid file updating. All of these examples exhibit the same essential characteristic: the average time interval between (unpredictable) events is short, and so effective control demands a correspondingly short update lag. Failure to revise current plans in the face of unanticipated events may lead to significant loss. See Figure 12.

This is not a common characteristic of higher level decision processes. Individual events have little relevance at this level; all that usually matters is aggregate behavior. Aggregate variables, such as total sales and total capacity, exhibit significant change only slowly and often in a fairly predictable way. It is not necessary to provide rapid updating of the data base in order to track sluggish and well-behaved changes. Furthermore, the accuracy of predictions over the relatively long planning horizon typically associated with higher level decisions would not be sig-

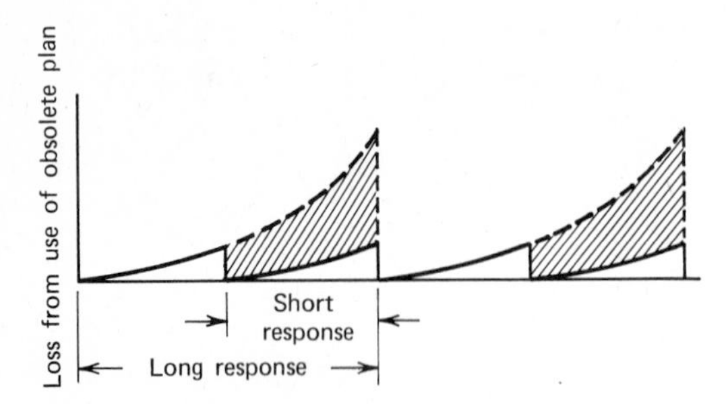

FIGURE 12 Loss due to use of old information. A plan, once made, begins to decay with time when unexpected events occur. Replanning brings the plan back to the new optimum based on current information. The more frequent the planning and current the information, the less the loss due to use of obsolete plans. The shaded area shows the incremental loss when response time is doubled; it therefore represents the incremental value of the shorter response time.

nificantly reduced by update lags of days or even a few weeks. We can therefore conclude that quick updating of the data base usually adds little value to higher level decision making. See Figure 13.

Value of Quick Retrieval Time. The value of quick retrieval is often quite independent of the age of information obtained. Thus, under some circumstances we may want fast access to relatively old information. A short retrieval time enhances value under two circumstances: (1) a decision (and the resulting action) must be made quickly, or (2) the decision process benefits from a series of accesses performed in "browsing" fashion.

The typical real-time system requires fast action (as well as fast file updating) in order to control a dynamically changing environment. Air traffic control, to use this example again, must deal with the problem of collision avoidance. A speedy decision is warranted when two aircraft on a collision path approach one another at Mach 2. The information on which the decision is based must therefore be both current and rapidly accessible. See Figure 13.

In less dynamic cases it may be important to make a fairly quick decision even if the information supplied is not particularly current. In an order entry application, for example, it may be quite valuable to respond immediately to a cus-

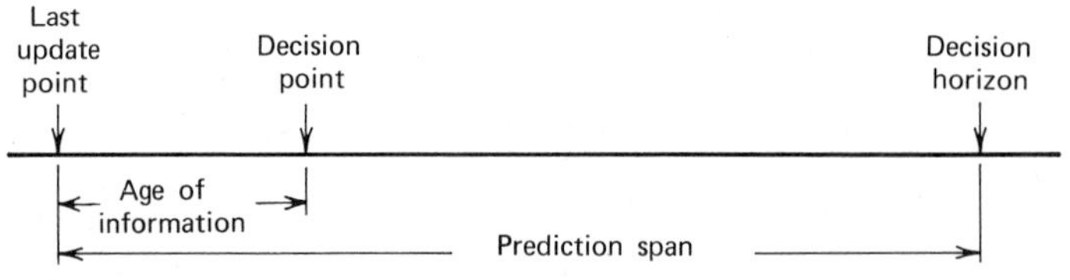

FIGURE 13 Effect of age of information on decisions. Decision-making—whether at the operational or strategic level—requires a prediction for each of the input variables used in the decision process (e.g., sales forcasts, inventory levels, aircraft positions, etc.). The prediction span for a given variable extends to the decision horizon from the point in time at which the data base was last updated. The quality of a decision depends in part on the accuracy of the predictions. Accuracy, in turn, depends on the length of the prediction span and the inherent variability of events within the span. A rapidly changing and unpredictable environment requires a short prediction span in order to maintain control; accordingly, information must be very current in order to provide suitable accuracy. Strategic decisions, on the other hand, tend to have long-term effects; and so the decision horizon is well into the future. In this case reducing the age of information adds little to predictability over the long span.

tomer's inquiry concerning stock availability (while the customer waits on the telephone, say). In most cases there would be no great value in having an up-to-the-minute picture of inventory status; normally it would be sufficient to base the decision on status as of some earlier cut-off point[7] (e.g., at the end of the previous day). This allows file updating to be handled on an economical batch basis. Inquiries between file updates can be processed through an on-line system or a manual operation that uses a daily status report (the choice depending on speed and cost of processing an inquiry). Most on-line reservation systems are motivated much more by the requirement for fast confirmation than by the need to keep files current within a few seconds.

Data entry applications can also benefit substantially from quick access to the data base. An interactive system can perform validity checks while a clerk remains at a remote terminal. A detected error usually can be corrected immediately, thus avoiding the serious complications involved in off-line error correction (as well as speeding up the updating process). Data accepted as valid may then be used for on-line file updating (if a short update lag is required), or may simply be stored temporarily for later batch updating.

Although the above examples are representative of important quick response applications, it certainly is true that the bulk of decision-making within an organization cannot justify quick retrieval on the grounds that hasty action is called for. On the contrary, a delay in reaching a decision of days or weeks may not carry a serious penalty, especially at the strategic level. Nevertheless, even in these cases there may still be considerable value in providing information within a short response time.

[7]If inventory levels change rapidly, but fairly predictably, the system can adjust the inventory balance by the predicted withdrawals from the cut-off point up to the time of the decision. In any case, one can always increase the probability of having an item in stock by simply maintaining a larger safety stock.

Quick response tends to be especially valuable when one deals with an ill-structured problem. Typically one cannot specify in advance all required information. Instead the problem is best approached through a sequential examination of responses. Each response may suggest new information that would shed additional light on the task at hand. This "browsing" process can continue until the problem solver feels that further probes are not justified.

A quick response allows more alternatives to be examined, which normally results in an improved decision. Even though there may be no great urgency in reaching a decision, there is always some upper limit on the time available. The number of sequential probes is therefore limited by response time.

If the response is fast enough, in the order of seconds or perhaps at most a minute,[8] the decision can be reached through the uninterrupted participation of the human problem solver. This allows him to retain a grasp of the problem that would otherwise be lost if he were forced to switch to some other activity while he waits for a response. Although the evidence is by no means overwhelming, it appears reasonably certain that interactive man-machine problem solving of this sort can be very effective in dealing with many types of complex tasks.

A capability of this sort may be particularly useful when a decision is reached through group cooperation. For example, setting a quarterly production schedule may require the participation of the managers of marketing, manufacturing, purchasing, and personnel. The value of a man-machine model may be fairly limited unless it allows the group as a whole to explore alternative schedules. This probably is feasible only if the model provides a response quick enough to make a decision in a relatively short meeting among all participants.

[8]Humans become extraordinarily impatient in an interactive environment if they have to wait very long for a response.

Cost of Short Response Times. Having discussed the value of information as a function of response time, we can now turn our attention to the matter of costs. Cost is governed largely by the frequency with which the data base is accessed for updating or retrieval.

Consider first the case of batch processing. Response time is clearly a function of the interval between successive batches. Suppose that this interval is I, and that there is a processing lag L between the cut-off point at the end of the batch cycle and the availability of output.

Update time depends on the timing of an event. If the event occurs just before the end of the cycle, update time is the processing lag L; if it occurs just after the cut-off point, update time is I + L. The average time, assuming that events are spaced uniformly over the processing cycle, is I/2 + L. This is shown in Figure 14. The same minimum, average, and maximum response times apply to retrieval, whether information is obtained from a periodic report or through an ad hoc inquiry.[9]

Thus, response time is a function of the processing cycle and processing lag. The processing lag can be reduced primarily through the use of more rapid means of collecting and transmitting data, such as an on-line data collection system tied directly to a central processor. The lag can also be reduced somewhat by reducing the average time a batch must wait in queue prior to processing. This is achieved by increasing surplus capacity to absorb fluctuations in demand. The actual run time taken to process a batch is rarely a significant portion of overall response time.

The most significant component of response time is typically the processing interval I. The number of runs is, of course, inversely proportional to this interval—e.g., twice as many runs are required if the interval is cut in half. Sequential processing runs tend to be input/output bound; this tendency increases all the more as the interval shrinks (since the number of transactions goes down when they are accumulated over a shorter interval). Therefore, processing time—and hence cost—is inversely related to response time.

As the processing interval shrinks, costs of conventional sequential processing begin to grow very rapidly. Eventually the point is reached where it becomes less expensive to use indexed sequential file organization, since this allows inactive portions of the file to be skipped. As the batch interval goes down still further, random processing—i.e., handling each transaction separately—eventually becomes the least expensive method. Figure 15 illustrates how cost varies as a function of response time.

Because of the dominant nature of input/output time for batch processing, it is usually advantageous to perform both file updating and retrieval during the same run. Retrieval may consist of handling ad hoc inquiries as well as preparing periodic reports. Since all records have to be accessed with each run, the only incremental cost of combining file updating and retrieval is the added size and complexity of the combined program.

Only active portions of the file need to be accessed with either indexed sequential or random file organization. Therefore, the number of active records becomes a principal determinant of cost (and not just the number of processing runs). It may be possible to substantially reduce the number of random accesses if updating is separated from retrieval. For example, all updating can be handled in batch fashion, while ad hoc in-

[9] These times apply only to *status* information, which gives the value of a particular variable, such as inventory level, at the cut-off point. *Operating* information deals with the series of events that occur over a reporting interval, such as the orders shipped during the past month. Immediately before an operating report is produced, the latest information available covers events that occurred as long as R + I + L time units earlier, where R is the reporting interval (usually, but not always, the same as I). The average age of operating information is R/2 + I/2 + L. Gregory and Van Horn (1963) discuss these matters in some detail (pp. 576–580).

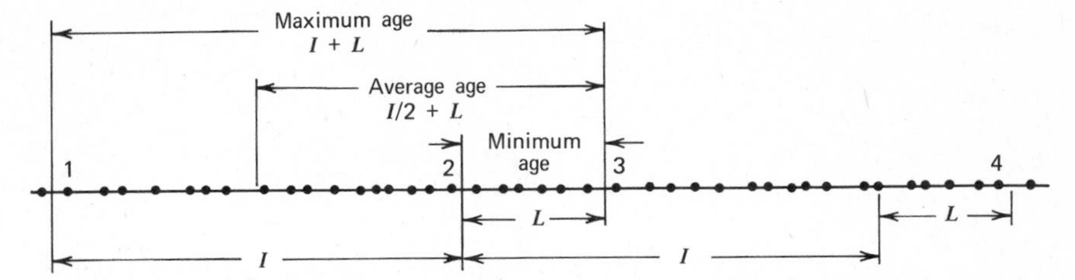

FIGURE 14 Response time as a function of the batch processing interval and processing lag. The time required to update a sequential file depends on the processing interval I and the processing lag L. All events that occur during a given interval are processed together in a batch. The first event within the interval (at time 1 in the figure) is not reflected in the file until $I+L$ time units later (at time 3). Information about the last event (at time 2) enters the file in only L units. The average, assuming a uniform distribution of events, is $I/2+L$. The minimum, maximum, and average *retrieval* times are the same as the *update* times. The information retrieved from a periodic report or through batch processing of ad hoc inquiries is L time units old immediately upon completion of the processing cycle (at time 3); it is $I+L$ units old at the end of the cycle (at time 4). In some applications the minimum response time may be the most important design consideration (e.g., in the case of a financial statement reviewed by management only when it first becomes available); in others, the maximum (e.g., the handling of customer orders); and in still others, the average (e.g., inventory updating).

quiries are processed randomly. Batch updating cannot only lower processing costs, but it can also significantly reduce the problems of file security and reliability that always attend on-line file updating.

Accuracy

Information is accurate if the image it provides conforms closely with reality. Thus, an inventory report is accurate if the stated levels agree with actual inventory status. A sales forecast is accurate if it correctly estimates future sales.

The value of accuracy is derived from the improvements it brings in decisions or operating actions. Any decision or action is based on the information available. To the extent that this information does not accurately portray reality, outcomes will be less desirable than would otherwise have occurred.

The most stringent requirements for accuracy generally come from operational activities. A considerable loss may ensue from an error in a paycheck, purchase order, invoice, or bank balance. A good deal of effort is justified to avoid the unrecoverable losses, the confusion, and the loss of goodwill usually associated with errors of this type.

Higher level decision making generally imposes relatively mild requirements for accuracy. Most decisions and payoffs are fairly insensitive to moderate errors, and so great accuracy adds little value.

Decision-making information that comes from the aggregation of transaction data usually is accurate by virtue of operational needs. Any errors that exist in the details tend to cancel out in the aggregation process.

Some data are collected specifically for decision-making purposes—for example, demographic data used in store location studies, consumer survey data used in marketing, business intelligence data used in competitive evaluations, and economic indicators used in forecasting. It usually is not necessary to subject these data to

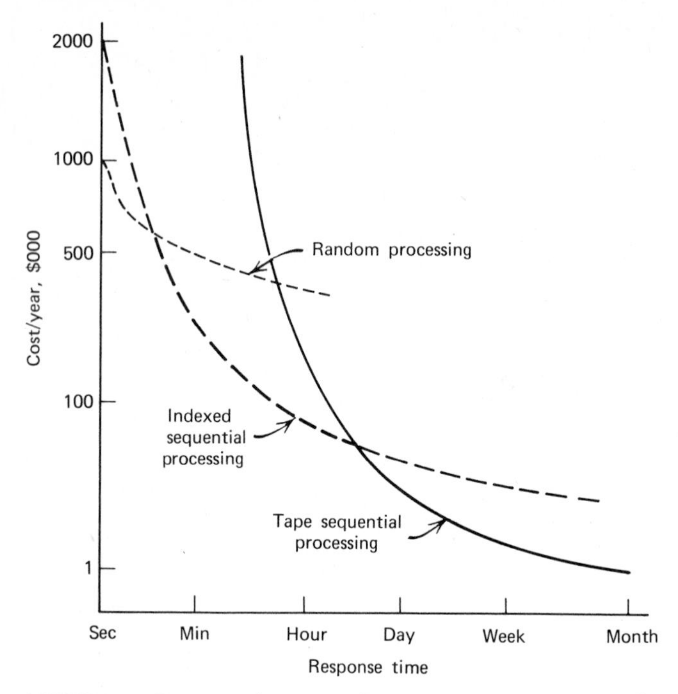

FIGURE 15 Cost as a function of response time. Economical batch processing with magnetic tape can be used when response time is not critical. As response time is reduced—and hence run frequency is increased—total input/output time grows rapidly. Eventually batch processing with an indexed sequential file becomes less expensive, since it allows skipping over inactive portions of the file and thus reduces input/output time. If response times are reduced still further, eventually random processing of a randomly organized file becomes the least expensive approach.

the same close control applied in collecting operational data.

Accuracy, like all desirable characteristics of information, has a cost. It is achieved by collecting error-free data that are then processed by suitable routines. Accuracy in data collection comes from error control procedures that both discourage creation of errors and provide the means for detecting and correcting any errors that enter the system. A great deal of sophistication and ingenuity can go into these procedures. Almost any desired degree of accuracy in data collection can be attained, but at an increasingly high cost as perfection is approached. Cost is incurred in the form of collecting and maintaining redundant data for error detection and correction purposes, as well as in the implementation and operation of all routines for performing error control functions.

Accuracy depends not only on the quality of data inputs, but also on the routines use in processing information. An obvious requirement for assured accuracy is that the routines be bug free. It virtually is impossible to eliminate all bugs from a large program, but careful (and expensive) testing and maintenance procedures can reduce the probability of an error to a manageable level.

A less obvious way to increase accuracy is to apply improved estimating procedures. Decision-making is concerned with predicted future conditions rather than historical accuracy. Sales data used in forecasting may be perfectly accurate: but if the estimating procedures are not suitable, the resulting errors may cause substantial penalties of stockouts or excessive inventory.

Any prediction has inherent errors, but the average size of error can be reduced through the use of appropriate estimating procedures. Accuracy is governed to some extent by the type of input data available for calculating predictions. Inputs include data about past transactions, past forecasts, economic indices, and the like. These inputs may be subjected to relatively simple procedures, such as exponential smoothing or other averaging techniques. On the other hand, accuracy may be improved significantly by using some of the more elaborate forecasting methods. Cost of such accuracy includes the maintenance of data inputs and the design and operation of the forecasting routines.

TECHNIQUES FOR MAKING COST/BENEFIT ANALYSES

We have seen that cost/benefit analysis of information systems faces some extremely complex issues. We are a long way from developing fully satisfactory approaches. Nevertheless, an organization that sinks vast sums of money into the development and operation of an information system cannot ignore the complexities; it must deal with them in the best way it can.

Analysis of Tangible Cost Reductions

Tangible costs and benefits are those that can be expressed in *monetary* terms. This clearly is possible in the case of projects aimed at clear-cut cost reductions in information processing. Elimination of clerical operations or lowering the cost of equipment rental (through greater efficiency or use of a later generation computer, say) are common examples of such projects. They are often accompanied by little or no change in information quality, essentially the same information is provided (hopefully) at lower cost. Under these circumstances the analysis need not concern itself with determining benefits, since they remain the same as before.

A cost reduction project thus can be viewed as a straightforward investment. As such, traditional methods of analysis can be applied. Certain expenditures must be made to implement the modified system. These are then followed (if all goes well) by reduced future operating expenditures. See Figure 16. One can analyze the investment in terms of net present value, internal rate of return, or payback period. Project selection can be dealt with in the same way used to set priorities on other forms of investment.

There lurks a hazard in this: it is too easy. Because cost reduction projects can be analyzed in traditional ways, the organization is often tempted to concentrate on them. This avoids the problem of having to assess benefits of improved information but runs the risk of misdirecting efforts away from projects that can make more fundamental improvements in organizational performance. This is not to say that worthwhile

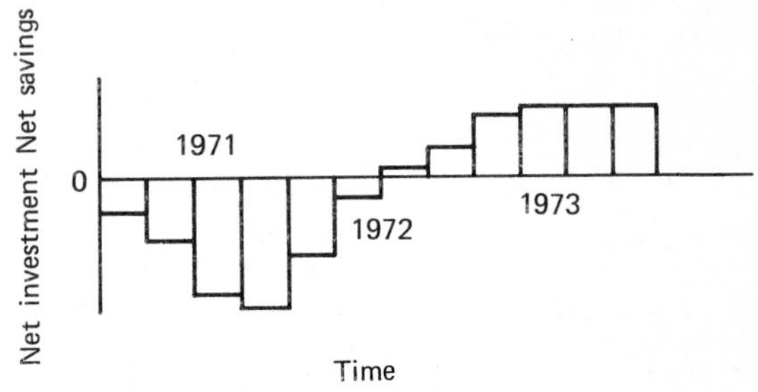

FIGURE 16 Analysis of tangible savings from an investment in an information system project. Tangible costs and benefits can be handled using traditional methods of investment analysis. The net cash flows that occur in different time periods can be translated into a single index by calculating the present value of the investment, its rate of return, or its payback period.

cost reductions should not be pursued, but the bias is often too heavily weighted in their favor.

Analysis of Tangible Benefits

Really significant contributions normally come through enhanced information quality. Many of the benefits from improved quality may be perfectly tangible. The benefits are certainly tangible, for example, when faster customer billing reduces cash requirements. The translation into a monetary savings requires an estimate of both an annual rate of return and the total expected cash reduction. If the rate of return is estimated to be, say, 20 percent (based perhaps on the opportunity cost of internal investments), a $100,000 reduction in cash is worth $20,000 per year.

The principal difficulty in assessing benefits of this sort is the estimation of the effects of improved information. In the above example the analyst must estimate the relation between cash requirements and billing time. If it can be assumed that faster billing will not change the distribution of time lags between the receipt of an invoice by a customer and the receipt of his payment, then the relation can be estimated quite easily. For instance, a two-day reduction in billing, with average daily sales of $50,000, will result in a $100,000 reduction in cash requirements.

The use of a formal decision model often greatly facilitates the estimation of benefits. Suppose, for example, that we would like to estimate the inventory reduction stemming from more frequent order processing. The reduction comes from lower inventory safety stocks made possible by the shorter reorder lead time (since more frequent order processing reduces the time from the breaking of an order point to the preparation of a replenishment order). The relation between processing cycle and inventory level is a very complex one, but it can be estimated easily if a suitable inventory model exists. Sensitivity studies of this sort can be used to consider the effects of altering any of the model's input data. Such studies should be an important part of implementing any decision model.

In the absence of a formal model, it becomes considerably more difficult to estimate the effects of changes in information quality. Formal models often do not exist, of course. In particular, they do not exist when the information project under consideration is the implementation of such a model.

One way to deal with this situation is to develop a "quick and dirty" model that gives a gross estimate of possible savings. For example, in estimating the benefits of more frequent order processing, we might analyze a typical inventory item using a standard inventory model. This will not give the same accuracy as an analysis of all items using a tailor-made model, but when used with discretion this approach gives a satisfactory first-order approximation. Often even a cursory study allows management to reject or accept clear-cut cases. Attention can then be focused on projects that appear to be borderline cases.

Sometimes a simple model can be used to provide boundary estimates of benefits—i.e., an upper or lower limit on possible benefits. An upper limit can be used to reject unworthy projects, while a lower limit can identify worthwhile projects. For example, in considering the implementation of an improved forecasting system one can establish an upper limit by assuming that the system will give *perfect* forecasts. In a similar fashion an upper limit on the improvement in production scheduling can be obtained by assuming 100 percent capacity utilization and no interference among jobs. A proposed system could be rejected out of hand if the estimated cost of implementation exceeds the upper limit on benefits. If this is not the case, more refined analysis can often lower the upper limit in order to bring it closer to the actual value.

Lower limits can be estimated in a variety of ways. One means is to put a value on only the most easily determined benefits. In analyzing a customer billing project, for example, estimated savings from a reduction in cash requirements gives a lower limit on total benefits. Actual benefits might exceed this limit by the (unknown) value of fewer errors in billing, better by-product information for market analysis, and the like. Another technique is to calculate benefits by using *worst case* estimates for probabilistic variables. If a project is acceptable using the most pessimistic estimates, actual benefits are highly likely to prove even more attractive. See Figure 17.

A similar approach is to estimate the break-even improvement necessary to balance the cost of implementing a project. Suppose, for example, that a firm spends $1 million per year on advertising in magazines and newspapers. The cost of obtaining and analyzing readership data is estimated to be, say, $10,000. Thus, a one-percent reduction in advertising expenditures (while holding exposure constant) would justify the cost of the analysis. Similarly, a supermarket firm might determine that the cost of maintaining detailed sales statistics could be balanced by an increase in sales of 0.3 percent. Given such an analysis, an experienced manager can often judge whether or not the likely improvement exceeds the break-even point.

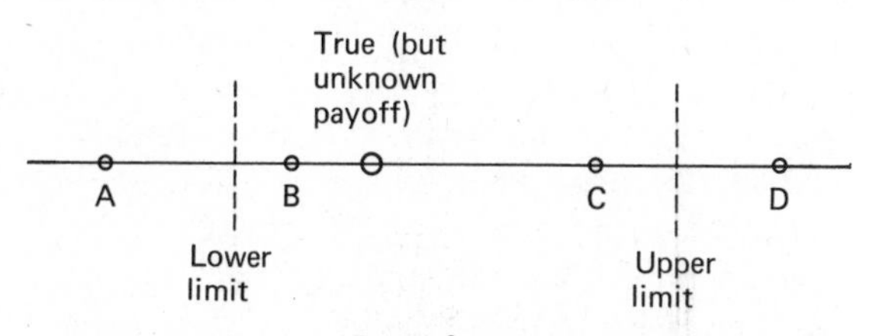

FIGURE 17 Use of upper and lower limits in analyzing benefits. Suppose that the upper and lower limits shown in the diagram have been estimated for a given project. The project must provide sufficient benefits to justify its estimated cost. If point A represents the minimum acceptable payoff, the project is worthwhile by virtue of the fact that the lower limit on benefits exceeds the acceptable payoff. Similarly, the project can be rejected if point D is the minimum acceptable payoff, since this value exceeds the upper limit on benefits. The acceptance of point B and the rejection of point C require further analysis to refine the estimated upper and lower limits to bring them closer to the true payoff.

Analysis of Intangible Benefits

The benefits discussed so far are tangible enough, even though they may be difficult to estimate. Some benefits, however, are especially difficult to translate into a monetary value. For example, it would be very difficult indeed for General Motors to put a dollar value on the customer goodwill brought about by a reduction from 10 percent to 5 percent in the probability that a dealer will have a stockout of a needed repair part. It would be equally difficult for the Southern Railway System to attach a dollar benefit to the improved service achieved by a one-day reduction in average delivery time of freight shipments.

Almost any benefit can be assigned a tangible value if sufficient effort is devoted to the task; the difference between a "tangible" and an "intangible" benefit thus lies in the difficulty of estimating monetary value. Even a benefit such as customer goodwill could be translated into a reasonable estimate of monetary value if the effort were justified. If this is not the case, however, we must deal with the problem in other ways.

At the outset it should be pointed out that difficulty in expressing a benefit in *monetary terms* does not imply that the benefit cannot be *quantified.* Failure to appreciate this fact has often resulted in an unnecessary lack of specificity in describing intangible benefits. Thus, in the above examples it was possible to quantify benefits (e.g., stockout probability reduced from .10 to .05), even if no dollar value was attributed to them. In some cases very little quantification may be possible (and so a narrative description must suffice), but this is the exception.

Under some circumstances the analysis of intangible benefits becomes fairly straightforward. This is the case when a proposed system can be justified on tangible grounds alone (i.e., through some combination of cost reductions and tangible benefits), while also contributing significant intangible benefits. The proposed system is said to *dominate* the existing one, since it is superior in terms of both its tangible and intangible characteristics.

Opportunities of this sort are not as uncommon as one might suppose. Existing systems are rarely as efficient as they could be, and therefore offer considerable potential for cost reductions. Such savings are made all the greater when technical advances permit new economies. Rather than concentrating solely on tangible effects, however, an organization usually finds it worthwhile to take some of its gains in intangible form.

Advanced "real-time" systems often provide a variety of benefits of this sort. For example, a comprehensive airlines reservation system may be justified on such tangible grounds as reduction in the salaries of reservation and ticketing clerks, higher seat bookings, and more efficient routing and scheduling of flights. Improved customer service may be as valuable as any of these, but it is probably not possible to assign a dollar estimate to it.

If a formal model exists, it may be possible to establish an explicit tradeoff between an intangible benefit and a tangible one. Suppose, for example, that an improved inventory model is implemented. Benefits can be taken in the form of lower inventory cost, lower stockout probability, or some combination of both. Even if a monetary value cannot be placed on fewer stockouts, its cost can be expressed in terms of foregone opportunity to reduce inventory costs. See Figure 18. An experienced manager presumably can resolve questions of this sort when he is presented with explicit tradeoff information.

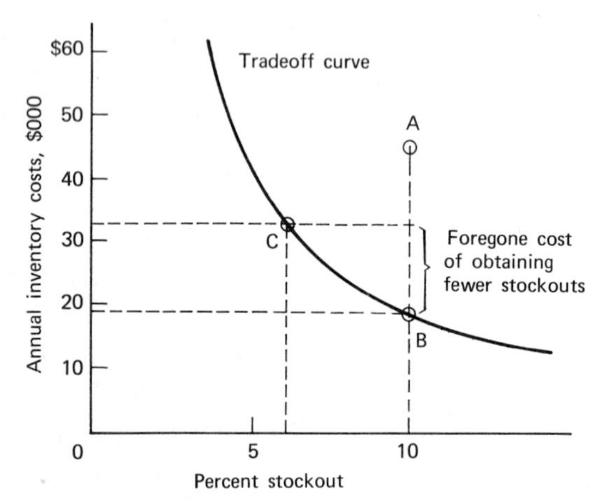

FIGURE 18 Tradeoff possibilities offered by increased efficiency. Changes in technology (or simply discovery of more efficient ways to exploit existing technology) allow new alternatives that provide tangible or intangible benefits. Such changes might permit, for example, the replacement of inventory system A by system B or C. All the benefits of system B are in the form of a tangible cost reduction. The selection of system C in preference to B implies that the intangible benefit of fewer stockouts is worth *at least* as much as the resulting reduction in cost savings.

USERS' ROLE IN COST/BENEFIT ANALYSIS

As information systems become more comprehensive and integrated, formal methods of analysis become increasingly difficult to apply. Joint costs and joint benefits often make it impossible to determine the payoff from any one subsystem. As the system begins to pervade day-to-day operations and higher level decision making, benefits become increasingly difficult to evaluate in monetary terms. Inevitably, then, we must rely on experienced judgment as well as technical analysis, to make cost/benefit decisions.

Communication between User and Technical Staff

One of the complications of making a cost/benefit decision is that it requires close coordination between two disparate groups, the users and the technicians. A rational decision requires knowledge of both costs and benefits. Information acquires value only through use, and so it is the users who are in the best position to judge the

Response Time	Incremental Benefit	Incremental Cost	Net Incremental Value
10 seconds, 90% confidence	$100,000	$50,000	$50,000
10 seconds, 75% confidence	98,000	45,000	53,000
60 seconds, 75% confidence	75,000	30,000	45,000
2 hours	72,000	10,000	62,000
Overnight	70,000	5,000	65,000

value of information. On the other hand, users very often have only a hazy idea of the cost of information. As a result they may specify an information "requirement" that is very expensive to satisfy, or they may fail to ask for useful information that could be provided at very little extra cost. To get costs into the picture, the technical manager and his staff must play the vital role of (1) analyzing costs and benefits to the extent possible, and (2) providing users with tradeoff information so that they will know the cost implications of their specifications.

The knowledge necessary to strike a reasonable balance between the cost and value of information is thus split between users and technicians. Clearly, then, this knowledge should be shared in some way. The dialogue between users and technicians may take several forms. One important way is by means of an iterative revision of requirements.

This process can begin by having the users specify desired characteristics of information outputs. A user might, for example, request that the system provide a response time of ten seconds in handling a given type of inquiry.[10]

With this and other specifications as a starting point, the technicians can then design, in gross terms, a system that meets (most of) the stated requirements. The designers may find that some few of the requirements are wholly infeasible; in these cases an alternative should be proposed. Even when a specification is perfectly feasible, the designers should provide tradeoff information so that users have some indication of the cost consequences of their specifications. For example, users may be given the following tradeoffs:

Response Time	Incremental Annual Cost
10 seconds, 90% confidence	$50,000
10 seconds, 75% confidence	45,000
60 seconds, 75% confidence	30,000
2 hours	10,000
Overnight	5,000

The great value of such tradeoff information is that it gives users a basis for balancing benefits against costs. A user may very well choose to alter his initial specification in light of its cost implications; at least he should be presented with this option.

Let us continue with the example of the inquiry system. Suppose that benefits are estimated as shown at the top of the page. The *net* value of information can then be determined. Overnight service turns out to be the optimal re-

[10]Since the load on the system varies minute by minute, response time must be specified in probabilistic terms. The user may, for instance, specify that at least 90 percent of the transactions receive a response within ten seconds.

sponse time in this example, and so the original specification should be changed.

Tradeoffs should be expressed in monetary terms whenever feasible. Ultimately, however, the final judgment usually involves considerable subjective evaluation of intangible benefits. This should not be viewed as a serious limitation—after all, managers are paid to exercise judgment. When presented with sufficient tradeoff information, a user is probably able to strike a satisfactory balance between cost and value.

An important question remains, however: Is the user motivated to choose what he considers to be the proper balance between cost and value? He may well not be if he receives benefits while not bearing the incremental cost of obtaining them. Question: In the above example if the user does not pay incremental costs, what response time is he likely to specify? A policy of charging users for incremental costs raises all sorts of costing problems, but it is very difficult otherwise to motivate users to make wise judgments.

Determination of Tradeoffs

The preparation of tradeoff information is obviously expensive. Each tradeoff point represents an alternative design in terms, say, of different capacities of main or auxiliary storage, number of I/O channels, or communication network configurations. The designers should therefore limit their consideration to a relatively few alternatives that stand some chance of being acceptable to users. For example, it serves no useful purpose for the designer to determine the cost of an order processing system that provides invoices of 95 percent accuracy if the marketing vice president insists on a minimum of 99 percent accuracy.

Some alternatives can be dismissed out of hand by having technical personnel work with users in preparing initial specifications. This will better insure that the alternatives considered will be limited to those that are reasonable from a technical point of view. An experienced designer can establish, without much analysis, fairly good boundaries on the range of feasible alternatives.

It is quite unlikely that users will be able to supply explicit estimates of value for alternative levels of information quality. A user may be willing to choose an alternative when presented with tradeoffs, and thereby establish an implicit estimate of relative values, but this is a considerably easier matter than giving explicit values. Nevertheless, the user should at least be able to provide *qualitative* tradeoff information in order that the designers can confine their attention to acceptable alternatives. In the case of the inquiry system, for example, users might be able to state in advance that overnight service is fully acceptable, interactive response is marginally useful, and any response time in between contributes very little extra benefit compared with overnight response. Designers then need only consider overnight batch processing and an interactive system.

Hierachical Nature of Cost/Benefit Analysis

Any cost/benefit analysis takes money to perform. The more money spent, the better the estimates that can be provided. In other words, one can reduce uncertainty about the economic payoff of a project by devoting resources to the task of analysis.

The organization thus faces a typical resource allocation problem in trying to decide how much to spend on cost/benefit studies. Like other information expenditures, resources used for analysis should be spent where the results have the greatest surprise content, will lead to the most significant modifications in decisions, and offer the greatest potential payoff from improved decision making. These conditions are not met if the payoff from a project is fairly obvious, if the level of benefits expected from a project cannot justify an elaborate study, or if the benefits are intangible enough so that trying to place a monetary value on them is not worth the effort.

Very often a project can be dismissed without a great deal of analysis. Although a superficial analysis may be subject to considerable uncertainty, the estimated cost/benefit performance for a given project may be unattractive enough that further refinement is not necessary. An experienced information specialist can be very effective in identifying these marginal projects.

The remaining projects require further study. Uncertainty is reduced as implementation proceeds through gross design, detailed design, and programming. Periodically during this process management should review the project to determine if it should be abandoned, accepted, or subjected to still further analysis. See Figure 19.

Thus, cost/benefit analysis should continue throughout the life of a project. During the early, uncertain stages of a project, the amount of money spent is relatively modest. The ante goes up significantly when it reaches the stages of de-

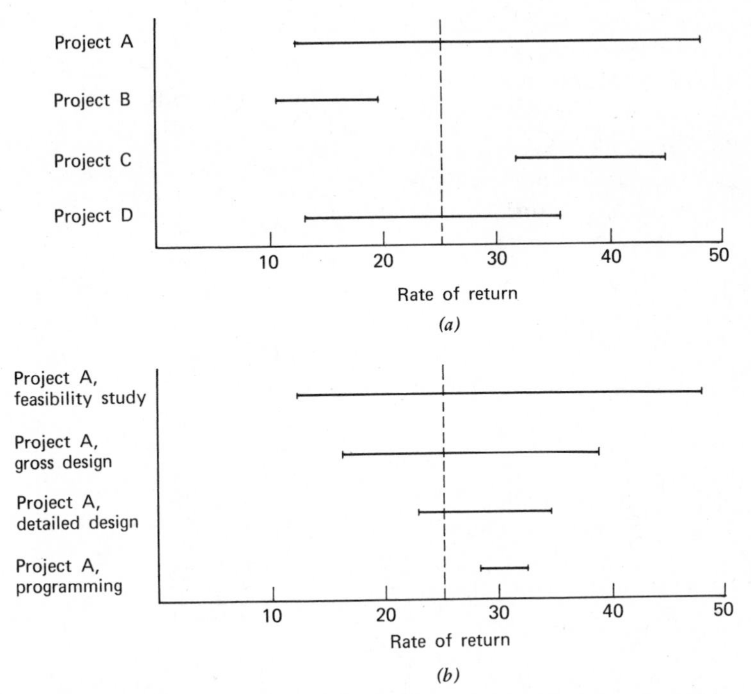

FIGURE 19 Uncertainty in estimating return from different projects. (a) shows the estimated return from different projects, based on preliminary feasibility studies. Such estimates are subject to a considerable range because they are made without detailed information about the proposed systems. Nevertheless, even a superficial analysis can identify projects that appear worth pursuing in greater detail (e.g., Project A). It can also screen out projects that clearly fail to meet the minimum acceptable return of, say, 25 percent (e.g., Project B). Uncertainty is reduced as the analysis proceeds through the various stages of implementation, as is shown for Project A in (b). More detailed analysis and design narrows the range of estimated payoff. The estimate may stay within the acceptable region, or it may fall below the limit and call for the project to be modified substantially or even abandoned.

tailed design and programming, but by this point most of the uncertainty concerning payoff should have been eliminated.

Subjective Estimation of Benefits

Since it is usually necessary to rely at least partially on subjective evaluation of information benefits, it is important that users have a good conceptual grasp of the factors that contribute to value. The evidence is that these concepts are not at all universally understood. More than a few systems have been implemented in which there is an obvious imbalance between cost and value; either the system falls far short of providing the information quality that it should, or it supplies information at a cost drastically exceeding any possible value. Even a primitive understanding of the issues involved would have avoided many of the problems of this sort.

Surprise Content

A user, along with the technicians who aid him, should ask himself questions of the following sort:

- Does the contemplated information tell me something that I did not already know or strongly suspect?
- How frequently will I obtain a significant surprise?
- Is the information selective enough that I am likely to perceive significant facts?
- Would less detailed, accurate, timely, or reliable sources provide essentially the same surprise content?

Effect on Decisions

- Will any surprise information cause me to take an action that I would otherwise not take?
- Would my decisions be significantly altered by less accurate or timely information?

Effect on Performance

- What are the benefits from any change in action that results from receiving surprise information?
- Are the benefits sensitive to moderate deviations from "optimal" actions?
- Do benefits decrease significantly with delays in taking action?

REFERENCES

1. James C. Emery, 1969. *Organizational Planning and Control Systems.* Macmillan, New York. Chapter 4 discusses the economics of information, relying heavily on a theoretical model.
2. Robert H. Gregory and Richard L. Van Horn, 1963. *Automatic Data-Processing Systems,* second edition. Wadsworth Publishing Company, Belmont, California. A classical treatment of the technical and managerial issues of information systems. Chapters 14 and 15 deal with economic questions.
3. E. Gerlad Hurst, Jr. 1969. "Analysis for Management Decisions." *Wharton Quarterly,* Winter, 1969. This paper presents some simple examples of Bayesian decision making and shows how the value of information can be calculated.
4. Börje Langefors. 1970. *Theoretical Analysis of Information Systems,* Volumes 1 and 2. Studentlitteratur Lund. Available from Barnes and Nobel, New York. A valuable reference for anyone willing to wade through fairly heavy theory. Chapter 3 deals with the economics of information. Volume 2 is devoted mostly to questions of file organization.
5. Jacob Marschak, 1959. "Remarks on the Economics of Information." In *Contributions to Scientific Research in Management.* Graduate School of Business Administration, U.C.L.A., pp. 79–98. One of the fundamental theoretical contributions to an understanding of information economics.

6. Norman R. Nielsen, 1970. "The Allocation of Computer Resources—Is Pricing the Answer?" *Communications of ACM*, 13,8 (August 1970), pp. 467–474. An excellent general discussion of pricing for computer services. It contains useful references to other papers on the subject.
7. T. William Olle. 1970. "MIS: Data Bases." *Datamation*, November 1970. A survey article on data base management systems.
8. Howard Raiffa. 1968. *Decision Analysis—Introductory Lectures on Choices Under Uncertainty*. Addison-Wesley, Reading, Massachusetts. A very readable discussion of Bayesian decision making by one of the leading authorities in the field.
9. William F. Sharpe, 1969. *The Economics of Computers*. Columbia University Press, New York. The most complete discussion of economic issues connected with the use of computers. Chapters 2, 5, and 9 are particularly relevant to a discussion of cost/benefit analysis of information systems.

READING QUESTIONS

1. What overall index does Emery suggest be used for a trade-off of all the detailed characteristics of information? Why?
2. What does he call the curve connecting the efficient points for the trade-off between cost and quality of information? Explain it.
3. Explain Figure 5.
4. In what tasks does man excel over the computer and vice versa?
5. The designer's task is to produce a synergistic effect. Although that term is not used in Emery's paper, you should be able to give an example, based on his discussion in the section "Allocation of tasks between man and machine."
6. What factors determine the degree of aggregation in a data base?
7. Explain Emery's concept of surprise content and its effect on the design of systems.
8. Discuss the two types of errors that result from varying approaches to information selection in a M.I.S.
9. How is the exception principle implemented in M.I.S.?
10. Why is the index sequential access method often specified system design?
11. What factors must be considered in design of data bases to meet the need for ad hoc inquiries?
12. Discuss the criteria considered in the design of response time capability of a system.
13. What are the considerations in determining accuracy objectives of a system?
14. Describe the process of determining benefits of a proposed system:
    (a) Tangible benefits
    (b) Intangible benefits
15. What are the responsibilities of the two "disparate" groups determining cost and benefits of potential systems:
    (a) the role of the system analyst (technician)
    (b) the role of the user
16. Figure 18 requires more explanation than is provided in the material below the figure. Clarify it by drawing a cost versus savings curve which shows the effect of developing a more elaborate system to produce both types of benefits shown in Figure 18.
17. Discuss the effects of various aggregation approaches in a college student record system where data will be accessed for three levels of management (by counselors, by the Director of Admissions, and as input to the 5-year planning model used by college executives).
18. What tangible and intangible benefits might be produced by the system described in Question 17?

19. Use the approach suggested by Emery in developing a cost/benefit analysis for the information derived in Questions 17 and 18.

Techniques For Estimating System Benefits

J. Daniel Couger

INTRODUCTION

In the past, a number of potentially high-benefit systems were given low priority for computerization because of the difficulty in quantifying benefits. The benefits of these systems were classified as intangible. The best illustration is the general category of computer-based management information systems. Due to the difficulty in quantifying the benefits of improved quality or timeliness of information, those systems were rarely computerized. Instead, cost reduction systems were emphasized—where benefits were easily measurable.

This situation is unfortunate because the computer can significantly enhance managerial decision making. Computers are not utilized at their full potential when they are used merely to do things faster than a human can do them; computers are used most beneficially when they perform tasks that are too complicated for humans to accomplish.

The complex process of decision making has the most to gain from computer use because it exploits the true potential of the computer. Therefore, present-day emphasis in feasibility analysis is on quantification of benefits in this category. The potential justifies the expenditure of additional effort to determine a satisfactory approach for quantifying benefits.

Another area of difficulty in benefit determination is that of customer-oriented systems. In contrast, a production scheduling system involves resources (personnel, materials, and equipment) that are entirely under the control of the company. A system allocating these resources is deterministic—an optimal solution can be derived and enforced because the factors for success are controlled by the production manager. Therefore, benefits are easily quantified and measured. The impact of a new marketing system is considerably more difficult to predict, since many of the factors are outside the control of the decision maker, the marketing manager.

Source: J. Daniel Couger, "The Benefit Side of Cost Benefit Analysis, Part II," Data Processing Management Portfolio 1-01-08, Auerbach Publishers, Inc., Copyright© 1975, pp. 1–13.

Both economic conditions and customer response are difficult to predict. Because of this difficulty, some companies place low priority on computerization of these potentially high-yield systems.

Despite the difficulty in quantifying benefits, computerization of marketing and management systems has been authorized in some organizations by managers who intuitively recognized its potential. These organizations proved that systems of this type can produce high return on investment. These results induced research on new approaches to benefit determination for application to benefit areas previously classified as intangible.

TECHNIQUES FOR QUANTIFYING BENEFITS

From the above described research two techniques, Bayesian analysis and statistical sampling, were identified as particularly advantageous. These techniques for benefit determination have proven effective for a large variety of organizations and applications. They are explained and illustrated in the following sections.

Bayesian Analysis

The eighteenth century theories of Thomas Bayes were the forerunners of the present-day approach to establishing the probability of occurrence of various possible benefits. Harvard's Robert Schlaifer is generally credited with expanding Bayes' work to apply to managerial decision-making applications.

A basic assumption made in using subjective probability estimates is that the manager, by virtue of past experience and knowledge of the situation, can quantify the relative likelihood of events.

Applied in the system benefit area, Bayesian analysis takes the approach that when there are no historical data or other bases for determining benefits, it is best to rely on the judgment of the individual who is most knowledgeable about the potential benefits—the person for whom the system is being designed.

The Bayesian approach also provides a framework for deriving probabilities. Probabilities are established for a range of benefit values. For example, assume that a new competitive analysis system has been proposed. The specifications are detailed enough that the marketing manager can determine precisely what kinds of information will be produced. With the new system, marketing efforts can be better directed toward competitors' weaknesses. The system will undoubtedly be beneficial; however, determination of benefits is not as precise as it is in an area under complete control of a manager, such as inventory. Nevertheless, the experience and expertise of the marketing manager can be utilized in deriving the probability of benefits occurring.

The manager is guided through the benefit determination process by the system analyst. Probabilities are established for various increases in the sales level resulting from the improved marketing information. The result might appear as:

Possible Improvement	Manager's Estimate of Probability
1% increase in gross sales	80%
3% increase in gross sales	15%
5% increase in gross sales	5%

The next step is to calculate the expected return from the system, based on the manager's estimates. Figure 1 provides the results of these calculations, illustrated for a firm with $75 million in annual sales. Assuming a 15 percent before-tax profit on sales, the new competitive analysis system should achieve an annual profit improvement of $168,750 (15 percent of $1,125,000).

In addition to illustrating the process of Bayesian analysis, this example demonstrates the

Possible Increase in Gross Sales	Dollar Amount Based on $75 Million Sales Level	Probability of Occurrence	Expected Return, $ (Col 2 x Col 3)
1%	750,000	0.80	600,000
3%	2,250,000	0.15	337,500
5%	3,750,000	0.05	187,500
		1.00	1,125,000

FIGURE 1 Determination of expected benefits.

high-yield potential of areas previously given low computerization priority because they were classified as intangible benefits. The example shows that a slight improvement in sales can produce significant benefits. Progressive organizations are utilizing Bayesian analysis to eliminate the intangible benefit category.

Statistical Sampling

Statistical sampling provides a means to quantify intangible system benefits to customers. For example, the material requirements planning system integrates several manufacturing systems previously processed separately. One of the resultant benefits, finished-goods inventory reduction, is fairly easily calculated. Another benefit, reduced in-process inventory due to improved scheduling, is relatively more difficult to ascertain. However, a Bayesian analysis can be conducted by the affected managers to quantify these benefits. As further integration occurs, benefit potential increases—but at a progressively more difficult level of prediction. By linking the inventory system to the sales order processing system, customer service will be improved. For example, the integrated system is capable of producing an order confirmation to be sent to the customer with a schedule for shipment of the product.

Assume the system analyst wants to determine if it is cost-effective to add the confirmation function to the proposed integrated system. One approach is to conduct a statistical survey of customers to determine the extent to which this feature might affect their decision to purchase. The sample might also be stratified by order volume, customer type, etc. Specifically, customers would be asked, "All other things being equal, would you give the order to the firm that provides the confirmation procedure?". With this information, it is possible to project the impact on sales.

The same sampling procedure and the same question would be used if (1) the company were deciding if it should be the first in the industry to include the confirmation subsystem *or* (2) if it were trying to determine the feasibility of adding the function to make its product more competitive.

Use of the statistical sampling procedure ensures representativeness and is a means of obtaining the information at a reasonable cost.

BENEFIT COMPUTATION

Figure 2 illustrates a useful format for presenting benefits to management to gain approval of system priority. The figure emphasizes two elements: (1) priority according to impact on the firm (lifestream systems) and (2) anticipated benefits, classified in three groups—cost reduction, cost avoidance, and profit improvement.

Calculation of benefits will be illustrated and then portrayed in Figure 2 format. Assume that the marketing function has been determined to be a lifestream system. One might think that the marketing system is a lifestream system for

LIFESTREAM SYSTEMS	BENEFITS		
	Cost Reduction	Cost Avoidance	Profit Improvement
SYSTEM 1 Probability Sub-Total			
SYSTEM 2 Probability Sub-Total			
SYSTEM 3 Probability Sub-Total			
TOTAL			

FIGURE 2 Format for presentation of benefits.

every firm. It frequently is; however, the system may not justify computerization. For example, a firm designing products for airlines may not need the computer for market analysis—there are so few customers that the data can easily be processed manually. So for our demonstrations the mail order industry will be used; marketing here is unquestionably a lifestream system.

Sales order processing is a subsystem within the marketing system. Benefits accrue in each of the three categories through computerization of sales order processing. Not only is the large volume of orders processed more quickly and efficiently, but data is captured simultaneously for the marketing data base. Examples are provided for each of the three benefit categories.

Cost Reduction

Sales order processing consists of capturing data from orders, sorting items into efficient filling sequence, and printing picking tickets. The cost of performing these functions manually is significant for a mail order firm, and computerization is easily justified.

Assume that processing cost can be reduced by $0.04 per order through conversion to the computer. The cost reduction calculation is given in Figure 3.

The probability estimate came from the manager of order processing, since he is the individual most knowledgeable concerning possible benefits. In this situation, it was unnecessary to develop an array of possible benefits. It is possible to rather precisely determine what activities will be performed on the new system. All of the factors are under the control of the decision maker. A very high probability exists that the projected level of savings will be attained.

Cost Avoidance

In a fast-growing organization, computerization permits resource usage in a linear rather than stairstep fashion. The volume of orders processed can be increased significantly without requiring additional equipment or second-shift operation (both of which would constitute stairstep cost increase). In contrast, manual processing of increased volume would result in many lower-magnitude stairstep cost increases, one for each additional person hired.

Assume that order volume is growing at a rate of 35 percent per year. The benefit calcula-

$0.04/order x 500,000 orders/year = $20,000
@ 95% probability of attainment, annual cost reduction = $19,000

FIGURE 3 Illustration of cost reduction.

tion shown in Figure 4 is representative of such a situation. The probability of achieving the cost avoidance is very high for the same reasons that the probability of the cost reduction is high, as given above.

Profit Improvement

Several other benefits result from the sales order processing system. Some of these would be listed as intangible if previous feasibility approaches were followed. Yet the potential is significant and justifies the additional effort for quantification. For example, consider the impact of faster order processing on reorders. Common sense indicates that customer satisfaction will increase due to faster turnaround of orders. The benefits of customer satisfaction are not intangible; they can be estimated.

Consider the actual situation depicted in Figure 5 where the result of computerized order processing is plotted. The mail order company separated its sales order processing system into three phases; the results of the first two phases are shown on the graph. Sales volume was increasing at a rate of approximately 35 percent per year. In the three years before the system was computerized, in-plant turnaround of orders increased from less than 12 to almost 22 days. After implementation of the second phase of the sales order processing system, turnaround time was reduced to 9 days.

Benefits from cost reduction and cost avoidance (primarily labor savings) resulted as calculated above and shown in Figure 4. The firm also anticipated increased sales due to the improved service level. System analysts decided to use the Bayesian approach to benefit determination.

The mail order industry experiences a peak during the Christmas season. the company sells stationery and other paper products suitable as gifts. Customers order from a catalog, typically in the early fall. When the order is received the customer may find a product especially appealing and wish to order more as additional Christmas gifts. However, with postal delays and the order filling delay, initial orders were usually received by the customer too late to reorder for Christmas gifts. With the new sales order processing system, the company expected to return initial orders soon enough to allow time for delivery of reorders. Using the Bayesian approach, benefits (in terms of additional profit) could be calculated as shown previously in Figure 1. So, in addition to cost reduction and cost avoidance, profit improvement occurred.

Other profit improvement benefits also resulted. By capturing customer information from the order as a by-product of sales order processing, the marketing data base could be established. Since zip code information can be corre-

35% increase in orders = 20 additional personnel if the manual system is continued.
20 persons @ $7,000 per year salary and benefits = $140,000/yr
Less cost of computer-processing additional 35% volume = $ 27,000
Cost avoidance = $113,000
@ 95% probability of attainment, annual cost avoidance = $107,350

FIGURE 4 Illustration of cost avoidance.

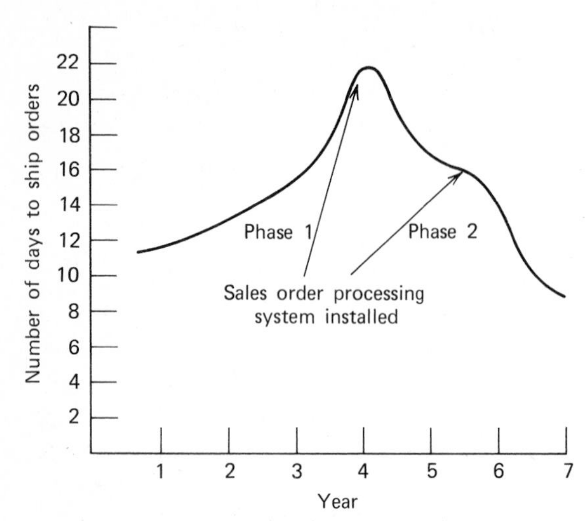

FIGURE 5 Reduction in order cycle time.

lated with sales volume and product preferences, demographic studies are possible.

Based on the resultant market analysis, the mail order company established an objective of doubling the dollar volume per order during the ensuing three years. Marketing management believed that such a goal could be attained by extending the product line to respond to the customer preferences revealed in the market analysis. Profit improvement projections, resulting from the establishment of the marketing data base, are given in Figure 6.

However, this approach involved projections of customer response, and customers are not perfectly predictable. A probability was associated with realization of the new sales objective. Marketing management selected the 75 percent probability level, the results of which are shown in Figure 7, along with previously calculated benefits. It is important to note that actual benefits exceeded the estimates. These results will be analyzed in more depth after calculating other system benefits.

Also shown in Figure 7 are benefits resulting from computerization of the inventory system. Those calculations were straightforward, based on estimates of reduced inventory levels. Again, actual benefits were higher than those projected. The most important benefit was improved control of inventory, which is shown for the mail order firm in Figure 8. In earlier years, the manual system kept inventory growth below the sales growth; however, the figure indicates that the manual system lost control in year 3. The inventory level climbed rapidly until the computerized system was installed. During the first year the computerized system was in operation, inventory was reduced by $500,000 despite a 35 percent increase in sales. The cost of carrying inventory for this firm (labor, material handling, stockroom, obsolescence, etc.) was 20 percent of the value of inventory; the cost reduction was therefore $100,000 (20 percent of $500,000). However, management estimated a saving of $80,000 as shown in Figure 7.

Cost avoidance also occurred. If the system had been installed, the inventory level would have risen to $1,900,000 (based on continuation of the trend shown in Figure 8). The actual inventory cost avoided was $600,000 (projected level of $1,900,000 minus actual level of $800,000 equals a $1,100,000 reduction minus $500,000 already listed as cost reduction). With a carrying cost of 20 percent, cost avoidance amounted to $120,000. Again, cost avoidance estimates were below benefits actually achieved ($100,000 was the estimate, as shown in Figure 7). The author's experience is that benefit projections are frequently conservative.

Projected sales/order = $13.00
Current Sales/Order = $ 6.50
Increase/Order = $ 6.50 ÷ 3 (years) = $2.17/year
$2.17/order x 500,000 orders = $1,085,000 sales increase per year
$1,085,000 @ 15% before-tax profit = $162,720 improvement profit

FIGURE 6 Illustration of profit improvement.

LIFESTREAM SYSTEMS	BENEFITS		
	Cost Reduction	Cost Avoidance	Profit Improvement
SYSTEM 1 Sales Order Processing	$20,000	$113,000	
Probability	95%	95%	(Varied, see Figure 1)
Sub-Total	$19,000	$107,350	$168,750
SYSTEM 2 Market Analysis			$162,720
Probability			75%
Sub-Total			$122,040
SYSTEM 3 Inventory	$80,000	$100,000	
Probability	95%	95%	
Sub-Total	$76,000	$ 95,000	

FIGURE 7 Benefits from three lifestream systems.

Conservatism in Estimates

Contrary to what might be expected, benefit projections tend toward conservatism. Some companies have precipitated this result by holding managers responsible for their estimates with the policy that budgets will be reduced by the amount of cost reduction forecast.

Policies of this type aside, experience shows that managers fail to anticipate some of the benefits that accrue. The unanticipated benefit of the mail order company market data base application is a good illustration. The market analysis revealed demographic patterns in buying. It also showed high mobility. A bright marketing manager capitalized on this information by designing a mailing brochure to be sent to "The New Folks" at address such and such. Sure enough, the family that moved into the residence vacated by a customer proved much more responsive to the company's catalog than persons whose names were purchased from some other organization's mailing list.

Another example occurred with an airline's customer data base. The original feasibility study did not include the possibility of using the computerized billing system as a marketing device. Analysis of the customer data base showed positive correlation between consumer buying habits and travel patterns. Customers who periodically traveled overseas with their families were susceptible to the purchase of quality luggage. Individuals who traveled as frequently as twice per month were susceptible to the purchase of pliable under-the-seat luggage (to avoid wait-

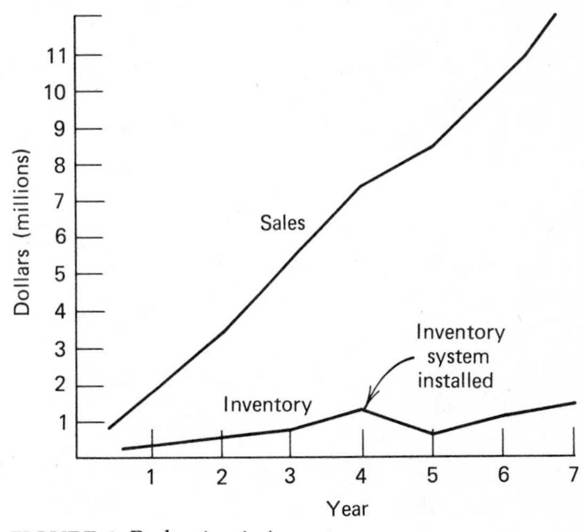

FIGURE 8 Reduction in inventory.

ing in baggage claim areas). Customers in the same category whose trips averaged more than three days in length were susceptible to the purchase of garment bags. As a result of analyzing the customer data base, the airline company entered the mail order luggage business and began enclosing brochures in the monthly billing envelopes. This approach produced higher readership than separate mailings of a typical mail order brochure.

These brief examples illustrate reasons for giving attention to benefit determination in areas other than cost reductions and cost avoidance. They also demonstrate why benefit projections in the profit improvement area have tended to be conservative.

Benefits of Integration

It was stated earlier that present-day emphasis is on the benefits of integrated systems and data bases, as well as individual applications. The previous examples of benefit projections for mail order companies illustrated determination of data base benefits. Implicitly illustrated were the benefits of integrated systems. Those benefits will now be made explicit.

As stated earlier, inventory system benefits are largely a result of improved control. Such improvement can result from computerization of this application as an independent system. The system's discipline, imposed on all individuals involved in inputting data, produces inventory integrity. The system outputs enable better purchasing and storage patterns. However, control is not assured until the inventory system is integrated with those systems which most affect it. The sales order processing system is one of these systems as are the purchasing and production control systems.

If the inventory and sales order processing system are integrated, the same transaction is used to deplete (update) finished-goods inventory each time an order is processed. If the purchasing system is also integrated, each transaction automatically updates raw-material inventory. If the production control system is integrated, each completion of a step in manufacturing is a transaction for both that system and for in-process inventory.

Not only does integration of this type reduce cost of input, but more importantly, it assures consistency. The accounting and production systems are in agreement, since both utilize the same input. Thus, more resources can be applied to assuring validity and accuracy of input, instead of being spent in costly reconciliation of the physical and book inventories that are perpetually out of balance when the two systems are not integrated.

The procedure for determining benefits of integration is the same as that illustrated previously. The important difference is that the manager responsible can usually expect a higher probability of attainment of benefits. The manager who knows that, once captured, the input transaction is shared among all affected systems has much more confidence in system integrity.

There are other, more easily discernible benefits. Manual recordkeeping is eliminated. Data redundancy is reduced. But the important benefit is the fact that everyone in the organization is now using the same information, which eliminates the disparity associated with independent systems.

RESULTS FROM IMPROVED LEVELS OF SOPHISTICATION

Few companies can afford to design the "ultimate" system in the first implementation effort. Most prefer to design several "first-level sophistication" systems simultaneously and improve each of the systems at a later date. More benefits are typically accrued by staging systems, as shown in Figure 9, than by using all resources to implement a highly sophisticated system for one function of the firm. There are exceptions, of

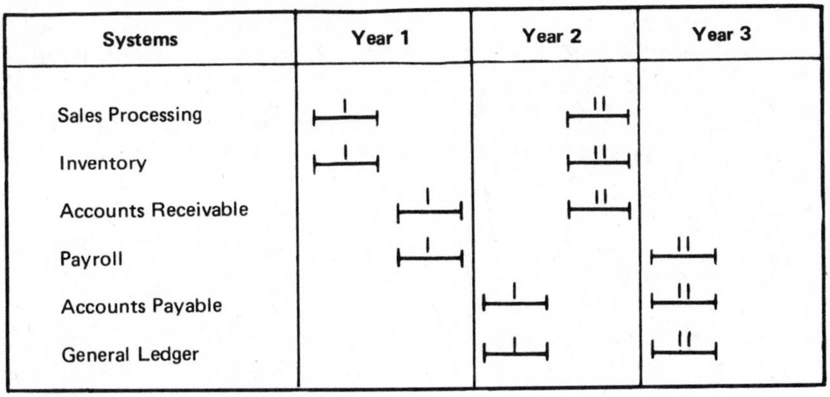

Systems	Year 1	Year 2	Year 3
Sales Processing	I	II	
Inventory	I	II	
Accounts Receivable	I	II	
Payroll	I		II
Accounts Payable		I	II
General Ledger		I	II

FIGURE 9 Staging systems to facilitate integration.

course. The policy accounting system of an insurance company is so important (e.g., a lifestream system) that all but nominal budget allocations for other systems are precluded. A high level of sophistication is typically designed into the policy accounting system during the first stage.

Benefit analysis is typically separated by stages of system implementation. Staging often occurs as shown in Figure 10. In this illustration, one level of sophistication is achieved in each stage. In actuality, several levels of sophistication may be designed into the system at one stage. However, such an approach would lengthen the design process considerably or would require a large amount of resources. The firm may be in a competitive situation which requires the telescoping of levels of system design sophistication. For example, an airline may find it imperative to combine all four levels of sophistication to meet the competition of other airlines that began computer use much earlier.

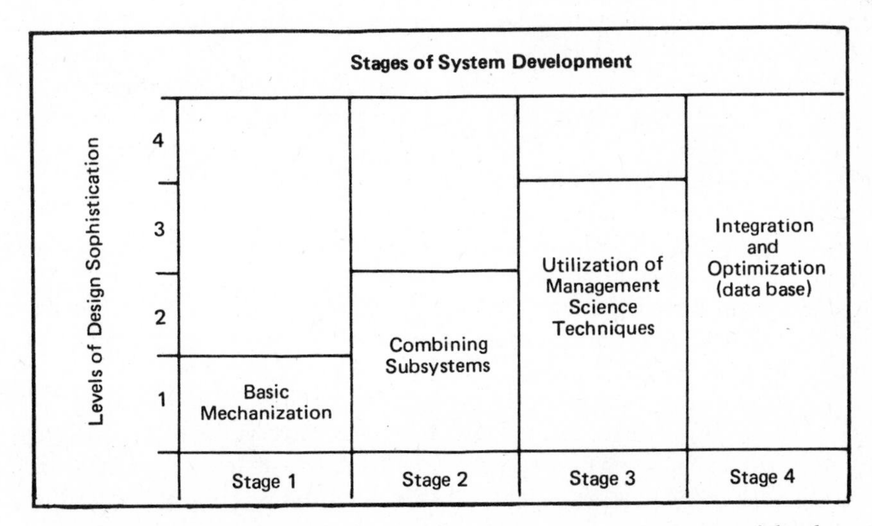

FIGURE 10 Designing four levels of sophistication in successive stages of development.

Circumstances of organizations vary and some firms may reach the fourth level of sophistication in only three stages of systems refinement. For example, many firms are combining levels one and two to take advantage of the benefits of integration. Utilizing the multi-stage approach may require four to five years before the fourth level of sophistication is achieved; however, benefits of computerization are occurring within one year. Benefits from the first stage are underwriting development of successive stages.

We will now illustrate benefit analysis for the four levels of sophistication. The sales order processing system previously illustrated included benefits attained at stage 1. Essentially, at stage 1 the system captured order data, produced picking tickets (sorted into the optimum sequence for manually filling the order), and provided sales statistics. Stage 1 reduced order processing cost and increased order turnaround (Figures 3 and 5). Stage 2 was designed, programmed, and implemented the following year. Normal improvements in processing efficiency resulted in continued improvements in turnaround (Figure 5). In this stage, the sales order processing system was integrated with the inventory system, enabling the finished-goods inventory to be updated automatically. The accounts receivable subsystem was also updated automatically, producing further clerical cost reductions.

During stage 3, which was implemented the following year, decision tables were utilized to optimize shipping container size and to calculate postage. Orders were sorted into groups by carton size, enabling reduction in the number of carton types and permitting efficiency in boxing by specializing shipping lines according to carton sizes. The ability to calculate precise carton sizes resulted in annual savings of $165,000 by eliminating the need for fillers (sleeves). Previously, the sleeve was required to fill the carton and prevent damage to the goods during shipment. Also, a forecasting system with scheduling algorithms was integrated with the sales order processing/inventory/accounts receivable system, as was a purchasing system.

Stage 4 provides for integration of the firm's major systems into a common data base. Although cost reduction occurs through elimination of separate inputs and data redundancy, the significant benefit is improvement in management information. Managers can now ask questions which relate to several systems. For example, a manager might want to know which production departments were behind schedule, how these departments were performing against their budgets, and the qualification level of personnel. Three systems are affected: production control, finance, and personnel. A data base approach is required for quick response to such questions. With a stage 4 system, management can also ask "What if" kinds of questions. The marketing manager can use a terminal for an inquiry concerning the impact of a change in product mix on profit. Benefit determination for improved management information of this type utilizes the Bayesian analysis and statistical sampling procedures illustrated previously.

RECOMMENDED COURSE OF ACTION

The author has specialized in cost/benefit analysis during the past seven years. One very important pattern has been noted in the many organizations analyzed during this period. There is a universal tendency toward conservatism in projecting benefits—particularly in the profit improvement category. This situation is particularly significant when one recognizes that these benefits previously were left unquantified, i.e., considered intangible. Even more important, as shown by the figures cited in the illustrations, is the fact that the profit improvement category often yields the greatest benefits.

Progressive firms today are utilizing techniques of Bayesian analysis and statistical sampling, analytical approaches evolved from classical economics, to quantify benefit areas previously considered intangible. In these organizations, the computer is being utilized at its true potential—not just to perform tasks faster than humans can, but to produce comparisons and analyses that are difficult for humans to perform.

With this new emphasis on the benefit side of the cost/benefit equation, firms are experiencing a greater return on investment from computer applications. Systems key to the continuance of the firm, lifestream systems, are now being given priority for computerization.

READING QUESTIONS

1. Who makes the benefit estimate? Why?
2. What factors influence the attainment of the fourth level of system sophistication?
3. What value accrues from the application to benefit determination of the following techniques:
   (a) Bayesian analysis
   (b) Statistical sampling
4. How would you go about estimating benefits for a system that permitted "what if" kinds of questions to be processed?
5. Why are benefits typically underestimated?
6. What special benefits accrue from integration of systems?
7. Why stage systems as shown in Figure 9?
8. Recalculate the system benefit example for conservative management that places the highest probability of achievement at 70 percent on the four categories that now show a 95 percent probability. If the development cost of these systems is $600,000 and the annual operating cost is $100,000, what is the payback period? (Assume that cost reduction occurs in year 1 and that cost avoidance and profit, management occur each year thereafter.
9. Recalculate the system benefit example, for a firm with inventory value of $750,000. Assume an inventory reduction of 25% and a carrying cost of 20% and probability estimate at the two sigma level.
10. Recalculate the system benefit example, where the company installs a decision support system (DSS) to provide "what if" responses. As a result management expects a 3% increase in gross sales but conservatively estimates probability of achievement at the original level.
11. Appendix E of the paper entitled "Business System Planning" (BSP) shows a column for "intangibles." Based on your reading of the Couger paper would you agree that these benefit categories are not capable of being quantified? Record your evaluation of each of the factors under:
    (a) Advanced Administrative System
    (b) Common Manufacturing System
12. Appendix E of the paper entitled "Business System Planning" (BSP) shows a column for "intangibles." Based on your reading of the Couger paper would you agree that these benefit categories are not capable of being quantified? Record your evaluation of each of the factors under:
    (a) Accounting Reporting System
    (b) Manufacturing Control System
13. Appendix E of the paper entitled "Business System Planning" (BSP) shows a column for "intangibles." Based on your reading of the Couger paper would you agree that these benefit categories are not capable of being quantified? Record your evaluation of each of the factors under:
    (a) Product Design and Release System
    (b) Personnel System

INDEX

Abstract system semantic model (ASSM), 369
Accounting theory, 10, 429
Accuracy, 477
Activity decomposition, 140
Ad hoc inquiries, 471
ADS (accurately defined systems), 4, 10, 40-42, 60, 62, 432, 442
  automated, 4, 11, 56, 432
Aggregation of data, 465
Alford, M., 347, 385
Analytical feasibility, 356
Anthony, R. N., 224
Applied statistical analysis, 10, 429
ARDI, 46
A-SPEC, 42
Attributes, 372
Automated ADS, 4, 11, 56, 432
Automated problem statement languages, 9
Automated tools, 375
Automatic code generation, 418

Backus-Normal Form (BNF), 388
Bayes, T., 490
Bayesian analysis, 490-491
Bell, T. C., 365
Benefit/cost analysis, 324
Benefits and costs, 307
BEST, 51
BIAIT, 213, 226, 231
BIAIT/BICS, 4, 11
BIAIT process, 221
BICS, 223
BISAD, 4, 45-46, 432
Bixler, D. C., 365
Board wiring diagram, 10, 429
Boehm, B., 379
Bottom-up approach, 244
Bottom-up design, 101
Boundaries, 154
BSP, 4, 11, 223
BSP, BIAIT/BICS, 432
BSP definition phase, 250
BSP definitions, 304-307
B-SPEC, 42
BSP identification phase, 246-250, 252-258
BSP processes and subprocesses, 302-303
BSP study team, 247
Bubble charts, 78, 81
Budgeting techniques, 10, 429
BUNDLES, 60
Burnstine, D. C., 215, 222
Business data processing, 397
Business information characterization study, 223-233
Business processes, 258
Business systems planning, 236
Business-wide perspective, 240-242

Carlson, W., 213

Cartesian product, 61
CASE, 440
Chapin charts, 89
Chief executive officer (CEO), 214
Chief programmer team, 91, 113
Classical economics, 453
Clerical time standards techniques, 10, 432
Clerical work sampling techniques, 10, 432
CODASYL, 55
CODASYL systems committee, 441
Code generators, 11, 432, 445
Cohesion, 83
Cohesiveness, 169-172
Coincidental binding, 169-170
Combined activity, 20
Communicational binding, 170-171
Complexity, 137
Composite design, 164
COMPUTATION, 61
Computer-aided approach, 400
Computer-aided technique, 316
Computer-based techniques, 22-32
Computerized corporate planning model, 428
Computerized PERT/CPM, 11, 432
Computerized planning models, 432
Computerized system planning techniques, 11
Computerized systems/planning model, 428
Computer science, 2
Computer simulation, 10
Computer simulation techniques, 432
Concept definition (CD), 396
Condition table, 416
Console aids, 433
Constantine, L., 81, 357
Control coupling, 169
Control structure, 76
Control transfer, 416
COORDINATE DEFINITION, 61
COORDINATE SET, 60
Cost avoidance, 492-493
Cost benefit/analysis, 344, 449, 459
Cost/effectiveness analysis techniques, 449
Cost/effectiveness practice, 4
Cost/effectiveness theory, 4
Cost reduction, 492
Couger techniques, 4
Coupling, 83, 165-169
CPM network technique, 10, 432
Critical success factors (CSF), 219
Curriculum, 2

Data base, 412, 466, 495
Data base administration, 289
Data base optimizer, 11, 432
Data coupling, 168
Data decomposition, 140
Data dictionary, 126
Data elements, 126
Data flow design, 86
Data flow diagrams (DFD), 78, 81, 88, 123
Data flows, 76
Data Group (DG), 224
Data processing subsystem performance requirement (DPSPR), 352
Data re-organizer, 445
Data stores, 126-128
Data structure, 76, 83, 125, 126, 206
Data utilization, 415
DATUM POINT, 60
DB design tools, 11, 432
DB/DL systems, 246
DBMS, 335, 432
DB specification language, 11, 432
Debugging, 107
Debugging aids, 433
Decision model, 472
Decision support systems (DSS), 400
Decision table, 4, 10, 38-40
Decision table processor, 4, 11, 55-56, 432
Decision tables technique, 432
Decision tree, 129
Decomposition, 89
DEFINING RELATIONSHIP, 62
Definition of first subsystems, 251
Delay, 20
DETAB-X, 55
Detail diagram, 80
Dictionaries and directories, 412
Disaggregated data, 464
Display costs, 471
Documentation, 316
Dominant systems, 461
Driver modules, 90
Dyer, M. E., 365

Efficiency frontier, 452, 460
Elements, 371
Emery, J., 452, 459
Emery concepts, 4
Emulators, 432
ENTITY, 60, 63
Entity diagrams, 78
Estimating system benefits, 489
Evolution of system development techniques, 8, 10
Exception principle, 470
Execution efficiency, 74
Execution overhead, 172-173

Fault isolation, 107
Fault tolerance, 109

Flowcharters, 433
Flow diagram, 4, 10, 30, 31
Flow of information, 9
Forms flow chart, 4, 10, 22, 23
"Front-end" costs, 7
Functional binding, 171
Functional decomposition, 78-80, 85
Functional specification, 438
Function descriptions, 88

Gane/Sarson methodology, 4
Gantt scheduling charts, 10, 429
General structure, 76
Generations of hardware, 8
Generations of system development techniques, 4
Geometry, 10, 429
Glans, T. B., 49
GLUMPS, 60
GPSS, 376
Grad, B., 49
Graph model of computations, 386
Graph model of software requirements, 386
Graph theory, 10, 429
Gridcharting, 4, 42-45
Grindley, C. B. B., 57, 68
Grosch, H., 451
Grosch's law, 451

Hardware costs, 97
Hardware/software monitors, 432
Hershey, E. A., III, 315, 330
Hierarchical analysis, 190
Hierarchical input, processes, and output (HIPO), 79, 80, 85, 103, 344
Hierarchical methods, 80-83
HISTORY, 61
Holstein, D., 49
Homomorphic, 61
Hoskyns code generator system, 67
Hoskyns system, 4, 67-68

ICOM codes, 156
IHD, 224
Immediate access diagram, 128
Incremental delivery, 91
Incremental value of information, 465
Indifference sets, 454
Information algebra (IA), 4, 10, 57, 60, 432, 442
Information distribution, 416
Information-handling disciplines, 227
Information matrix and flow, 48
Information process chart (IPC), 4, 10, 27-29, 429
Information processing system (IPS), 437
INFORMATION SET, 61
Information system, 437
Information systems management review checklist, 301-302
Information systems management value, 310-313
Information systems network, 285
Information systems network benefits, 308-310
Information systems plan, 251
INPUT, 61, 332
Inspection, 20
Intangible benefits, 454, 481-482, 491
Integrated subsystems, 455
Integration, 496
Interactive graphics, 376
Interdependence, 51
INTERFACE, 332
Interface complexity, 166-167
Interfunction communication table, 80
Inventory model, waiting-line models, 10, 432
IPO chart, 88
ISDOS, 4, 327, 345, 369, 427, 433, 441
ISDOS-PSL, 99
Isomorphism, 61
ITEMS, 60

Jackson, M., 83
Jackson charts, 89
Jackson methodology, 80
Job accounting systems, 432

Kent, H., 57, 442
Kerner, D., 213, 222, 223
Knight, K., 451
Konsynski, B., 399

LA (Langefors methodology), 4, 10, 52, 60, 432
Langefors, B., 57, 68, 444
Language translators, 433
Large system optimization models, 10, 432
Leveling, 78
Levels of abstraction, 102
Levels of system sophistication, 497
Librarian, 91, 433
Life cycle, 84, 85, 98
Lifestream system, 491
Linear algebra, 10, 429
Linear programming, 10, 429
Linear responsibility charts, 10, 429
Logical and physical design, 412
Logical binding, 170
Logical data access path, 207
LOGOS, 101

McFadden, F., 457
Management complexity, 74
Management control, 224, 237
MAP, 4, 10, 29

MAP system charting technique, 429
Marginal cost, 450
Marschak, J., 451
Mathematical and statistical techniques, 9
Mathematical programming, 10, 432
Memory overhead, 172
Message specification sheet, 31-32
Meyers, D. H., 56
Meyers, W. E., 49
Minimal system, 189
MIS, 3
Modifiability, 75
Modified forms flow chart, 4, 22, 26
Modularity, 74
Modularization, 103
Module, 165
Module size, 179
Multi-column process flow chart, 4, 20-21
Myers, G., 79, 81, 357

Negative synergistic effect, 450
Network modeling techniques, 11, 432
Network structures, 409
Node index, 161
Nunamaker, J., 65, 68, 399

OBJECTS, 321
Operating systems, 432
Operation, 20
Operational control, 224, 238
Operational-level systems, 7
Optimizers, 432
Organization charts, 10, 429
Orr, K., 83
OUTPUT, 332
Output checking, 107
Overview diagram, 80

Parallelism, 416
PERT, 46
PERT network technique, 10, 432
PLEXSYS, 4, 11, 399, 432
PLEXSYS analyzer, 409
Point set theory, 10, 429
Precedence network technique, 10, 429
Precedence table, 415
Precomputer techniques, 10, 17-22
Probability theory, 10, 429
Problem definition, 348-349
Problem definition management, 442
Problem statement analyzer (PSA), 65, 320, 330, 442
Problem statement language (PSL), 65, 320, 331, 442
Problem structuring, 403
Procedural primitive language (PPL), 419
Procedure definition facility (PDF), 408
PROCESS, 332
PROCESSES, 321
Process flow, 194
Process flow chart, 4, 10, 17-20, 429
Process group matrix, 415
PRODUCING RELATIONSHIP, 62
Profit improvement, 493-494
Program design language (PDL), 104
Programming languages, compilers/assemblers, 433
Program optimizers, 433
Program proving, 108-109
Program verification, 108-109
PROPERTIES, 60, 321
PROPERTY SPACE, 60
PROPERTY VALUES, 321
PROPERTY VALUE SET, 60, 63
Pseudocode, 89, 418
PSL/PDL, 419
PSL/PSA, 4, 65, 315, 330, 369, 402, 432
PSL/PSA II, 4, 11, 432
Punched card machine flow chart, 22, 24, 429
Punched card machine process chart, 10, 429
Punched card machine process/operations flow chart and board wiring diagram, 4
Punched card operations documentation package, 25
Punched card operations flow chart, 10

Quality of information, 461
Quantifying benefits, 490-491
Quick retrieval time, 474

Real-time test control, 361
Recording and analyzing resources, 9
Relationships, 371-372
Reliability, 75
Report writers, 432
Requirements engineering and validation system (REVS), 352, 368, 385
Requirements nets, (R-NETs), 351, 366, 370, 386
Requirements statement language (RSL), 352, 368, 386
Response time, 473, 476
Return on investment (ROI), 456
Revised system life cycle, 432
REVS, 352, 368, 385
Right and left hand analysis charts, 10
R-NET, 351, 366, 370, 386
Rockart, J., 219
RSL, 352, 368, 386
Rule of omission, 151-152

SA diagrams, 137
SA model, 137
Sayani, J., 437

SCERT, 440
Schedulers, 433
Schmidt, R., 49
Scope of control, 178
Scope of effect, 178
SDL/MSL, 419
SDC software factory, 114
Semantic model, 374-375
Sequential binding, 171
SET, 332
Sharpe, W., 450
Short update time, 473
SIGPLAN, 55
SIMSCRIPT, 64
Simulation, 376-377
Simultaneous motion charts, 429
SODA, 4, 11, 65, 432
SODA/ALT, 66
SODA/OPT, 66
SODA/PSA, 66
SODA/PSL, 66
Softech, 114
Software:
- costs, 97
- design, 101-104
- engineering, 6, 73, 97, 98-99, 115
- management, 112-115
- reliability, 105-109
- testing, 105-109

Software requirements engineering, 99
Software requirements engineering program (SREP), 347
SOP, 4
- activity sheet, 49
- documentation package, 50
- file sheet, 49
- general section, 49
- message sheet, 49
- operational section, 49
- operation sheet, 49
- resource usage sheet, 49
- structural section, 49

SOP activity analysis technique, 10, 432
SOP operation analysis technique, 10, 432
SOP resource analysis technique, 10, 432
SREM (software requirements engineering methodology), 4, 11, 349, 365, 385, 432
SREM allocation, 361
SREM analytical feasibility, 361
SREM decomposition, 360-361
SRE methodology, 352
SREM translation, 357-360
SREP-RSL, 99
Stages of system implementation, 497
Static analysis, 377
Static code analysis, 106
Statistical sampling, 491
Stepwise refinement, 74, 89
Stevens, W. P., 357
Stevens methodology, 4
Stochastic process, 454
Stop watch study, 10
Storage, 20
Storage efficiency, 74
Strategic planning, 224, 237
Strategic systems, 7
Structure charts, 82, 83, 89, 165, 173
Structure clash, 83
Structured analysis, 74, 75, 78-79, 122-123, 136-137
Structured analysis and design technique (SADT), 4, 76, 78-79, 85, 136, 344, 432
Structured brainstorming system (SBS), 405
Structured design, 75, 79, 164
Structured design methodologies, 74
Structured English, 129
Structure design, 129-133
Structured methodology, 11
Structured programming, 74, 75
Structures, 372
Stubs, 90
Study organization plan (SOP), 37, 46-51
Subjective estimation of benefits, 486
Subsystem ranking, 276
Summary description table, 80
Surprise content, 466, 486
Synergistic effort, 3
Synergy, 9
Synthetic time standards, 10, 429
System analysis, 6
System Analysis Techniques, 3, 427
SYSTEMATICS, 4, 57, 442
System building, 438-440
System definer, 442
System design optimizer, 11
System development, 6
System flow chart, 4, 10, 29-30
- and flow diagram, 429

System life cycle, 6
System planning model, 4, 433
Systems architecture, 187
Systems director, 445
Systems requirement, 187
System simulators, 432
System structure, 76-77
System structuring, 403

Tactical systems, 7
TAG (Time automated grid), 4, 11, 56-57, 60, 62, 432, 441, 442

Tangible benefits, 480-481
TBD (to be determined), 354
Teichroew, D., 57, 65, 68, 315, 330, 437
Temporal binding, 170
TERMINAL SETS, 60
Test aids, 433
Test case preparation, 107
Test data generators, 432
Test monitoring, 107
Theoretical foundation of MIS, 3
Time study (stop watch), 429
Top-down analysis, 89, 243
Top-down construction, 89, 90
Top-down design, 89
Top-down testing, 89
TP monitors, 433
Track loop experiment, 362
Tradeoff information, 483
Transportation, 20
Tree chart, 4
Truth tables, 10, 429

Unique inventory (UI), 224
Utilities, 433

Value of information, 451
Virtual machines, 102
Visual table of contents (VTOC), 80, 88

Warnier, J., 83
Warnier diagram, 195
Warnier-Orr approach, 80
Warnier-Orr diagram, 89
Warnier/Orr technique, 4
"What if" questions, 498
Workbench, 400
Work simplification, 10, 429

Yamamoto, Y., 330
YK (Young-Kent methodology), 4, 10, 57, 60, 432
Young, J. W., 57, 62, 65, 442
Young/Kent methodology, 4, 10, 57, 60, 432
Yourdon, E., 79, 81